niversity of
nal Theatre
Discovering
nguage (1972),
of plays by
ster'. He has
anies, work-
a, and New

The Oxford
Illustrated History of
THEATRE

Edited by
JOHN RUSSELL BROWN

Oxford   New York
OXFORD UNIVERSITY PRESS

Oxford University Press, Great Clarendon Street, Oxford OX2 6DP

Oxford New York

Athens Auckland Bangkok Bogota Bombay Buenos Aires
Calcutta Cape Town Dar es Salaam Delhi Florence Hong Kong
Istanbul Karachi Kuala Lumpur Madras Madrid Melbourne
Mexico City Nairobi Paris Singapore Taipei Tokyo Toronto Warsaw

and associated companies in
Berlin Ibadan

Oxford is a trade mark of Oxford University Press

First published 1995 by Oxford University Press
First issued as an Oxford University Press paperback 1997

British Library Cataloguing in Publication Data
Data available

Library of Congress Cataloging in Publication Data
Data available
ISBN 0-19-288062-4

3 5 7 9 10 8 6 4 2

Printed in Spain by
Bookprint, S.L.

CONTENTS

PART FOUR: WORLD THEATRE

✒️ CONTRIBUTORS ✒️

Michael R. Booth is Chair of the Department of Theatre at the University of Victoria, British Columbia, and an Editor of *Theatre Notebook*. His books include *English Melodrama* (1965), *English Plays of the Nineteenth Century* (5 vols., 1969-76), *Victorian Spectacu- lar Theatre* (1981), and *Theatre in the Victorian Age* (1991). He has edited a selection of English nineteenth-century plays for the World's Classics series.

John Russell Brown (General Editor) is Professor of Theatre at the University of Michigan and was for fifteen years Associate of the National Theatre in London. His publications include *Discovering Shakespeare* (1981), *Free Shakespeare* (1974), *Theatre Language* (1972), and editions of plays by Shakespeare, Webster, Marlowe, and the 'Wakefield Master'.

Louise George Clubb is Professor of Italian and Comparative Literature at the University of Cali- fornia, Berkeley. She is General Editor of the series *Biblioteca Italiana* and her books include *Giambattista Della Porta: Dramatist* (1965), *Italian Drama in Shakespeare's Time* (1989), and (with Robert Black) *Romance and Aretine Humanism in Sienese Comedy* (1993).

Victor Dixon is Professor of Spanish at Trinity College Dublin. His publications include critical editions and translations of three plays by Lope de Vega, and many articles on Spanish drama. He has also directed or acted in some twenty productions of Spanish plays.

Martin Esslin was head of BBC radio drama department 1963-77 and Professor of Drama at Stan- ford University 1977-91. His publications include *Brecht—A Choice of Evils* (1959), *The Theatre of the Absurd* (1961), *Pinter—the playwright* (5th ed. 1993), *Artaud* (1976), and *The Field of Drama* (1987).

Peter Holland is Judith E. Wilson University Lecturer in Drama in the Faculty of English, University of Cambridge, and Fellow and Director of Studies in English at Trinity Hall, Cambridge. His publications include *The Ornament of Action: Text and Perform- ance in Restora- tion Comedy* (1979), editions of the plays of Wycherley and of *A Midsummer Night's Dream* as well as co-editing three collections of essays on drama and theatre: *The Play Out of Context, Reading Plays, and English Comedy.*

William D. Howarth is Emeritus Professor of the University of Bristol and Honorary Professor at the University of Warwick. His publications include *Sublime and Grotesque: A Study of French Romantic Drama* (1975), *Molière: A Playwright and his Audience* (1982), *Beaumarchais and the Theatre* (1994), editions of plays by dramatists from Molière to Anouilh, and (as editor) *Comic Drama: The European Heritage* (1978). He is a member of the editorial board of the Cambridge *Documentary History of the European Theatre*, and author, in that series, of the volume on *The Neo-Classical Era in France, 1550-1791.*

Christopher Innes is Professor of English at York University in Canada. A Fellow of the Royal Society of Canada and of the Royal Society of Arts, he has written widely on twentieth century theatre. His most recent books include *Modern British Drama: 1890-1990* (1992) and *Avant-Garde*

Theatre: 1892-1992 (1993). He is editor of the Cambridge Directors in Perspective series, and the Canadian Playwright series, and co-editor of *Modern Drama* and of Lives of the Theatre series.

Colin Mackerras is Foundation Professor in the School of Modern Asian Studies at Griffith University, Australia. He has written numerous books and articles on Chinese history, minority nationalities and culture, and in particular on theatre. His books on Chinese theatre include *The Rise of the Peking Opera, 1770-1870* (1972) and *Chinese Drama: A Historical Survey* (1990).

Michael Patterson taught at universities in Leeds, North Wales, and Northern Ireland before becoming Head of the Department of Visual and Performing Arts at De Montfort University, Leicester. He has published widely on German theatre, including *German Theatre Today* (1976), *The Revolution in German Theatre 1900-1933* (1981), and *The First German Theatre* (1990).

Leslie du S. Read is Head of Drama at the University of Exeter. His research interests centre on stagecraft and the origins of performance. He has contributed to the *Cambridge Guide to World Theatre* and has written on theatres in the ancient world and in the Elizabethan age. His publications include *Social Space in Ancient Theatres* (1993) and *Edward Alleyn and the Rose Playhouse: Biography and Repertoire* (1994).

Farley Richmond is Professor and Chair of the Department of Theatre Arts at the State University of New York, Stony Brook. He has published books and articles on Indian theatre and has developed a video tape series on Asian acting technique with international experts in the field. He is developing an interactive CD-ROM on kutiyattam and editing a series of CD-ROMs on Indian theatre.

Leon Rubin is Head of the School of Drama and Theatre Arts at Middlesex University, London. He is former Artistic Director of the Bristol Old Vic, Watford Palace Theatre, and The Lyric Theatre, Belfast. He has directed for theatre companies in the UK, Ireland, Greece, Japan, Thailand, USA, Canada, and elsewhere. He is the author of a book about theatre directing to be published in 1996 and *The Nicholas Nickleby Story* (1981).

Oliver Taplin is Professor of Greek Literature at Oxford University and a Tutorial Fellow of Magdalen College. His books on the theatre include *Greek Tragedy in Action* (1978), *The Stagecraft of Aeschylus* (1977), and *Comic Angels* (1993). He has also collaborated with productions in the theatre, including the *Oresteia* at the National Theatre and *The Thebans* by the Royal Shakespeare Company.

Peter Thomson is Professor of Drama at the University of Exeter. His books include *Shakespeare's Theatre* (1983; 2nd revised ed. 1992) and *Shakespeare's Professional Career* (1992), and he has edited the *Cambridge Companion to Brecht* (1994). He is an editor of the journal *Studies in Theatre Production*.

David Wiles is a Reader in Drama at London University, Royal Holloway College. His many books include *Early Plays of Robin Hood* (1981), *Shakespeare's Clown: Actor and Text in the Elizabethan Playhouse* (1987), and *Shakespeare's Almanac: 'Midsummer Night's Dream', Marriage and the Elizabethan Calendar* (1993). In the classical field he has published a book on performance of New Comedy, *Masks and Menander: Sign and Meaning in Greek and Roman Performance* (1991), and is preparing another on Greek tragedy.

The Oxford
Illustrated History of
Theatre

INTRODUCTION

JOHN RUSSELL BROWN

THIS book tells the story of theatre from its various beginnings to the present day, and provides a permanent celebration of theatre's greatest achievements. Neither task is easy, because theatre has its full life only in moments of performance before an audience. This means that all surviving evidence is fragmentary and much of it misleading: a dramatic text is only the bare bones of a play in performance; an empty building gives no sense of a crowded audience; a photograph may represent a posed group and can never reproduce presence, timing, breath—all the vivid impressions actually received in a theatre. But there is an abundance of clues, waiting to be pieced together so that the illusions, pleasures, and innovations of the past can be suggested for a reader and a history of theatre attempted.

The present time is especially right for a new history. Theatre is thriving today in many ways, with more new play-scripts than ever before and unprecedented forms of production which draw upon advanced technology and the changes of awareness and perception that have followed in its wake. Companies are being started or reorganized, and reach out to new audiences. Theatre is now a subject widely studied in schools and universities, and provides a model for thought in philosophy, psychology, sociology, and education. Audiences have access to productions drawn from countries across the world. Yet, at the same time, theatre is also under threat. In a time of recession it is accounted a labour-intensive industry, and therefore expensive and difficult to develop. Besides theatre has strong and well-financed rivals in television, film, popular music, and spectator sports, which provide more easily exploitable forms of entertainment. To some observers theatre is a fabulous invalid, relying on huge injections of public money to prolong its life past a natural death-date. While theatres are still being built and new companies formed, others are closing or hanging on to existence with shrinking audiences, smaller staffs, and fewer and less ambitious productions. Successful authors, actors, and directors move away from theatre to television and film, and frequently prove incapable or unwilling to move back again. Few people work exclusively in order to produce performances on stage for live audiences. At such a time, the study of theatre's history may bring a better understanding of its present state by showing how its problems and opportunities have been formed.

Attention paid to the past can be much more than a reminder of greatness no longer attainable. It can also spur forward new work by revealing the great range of theatre's possibilities and forgotten sources of its vitality. By showing theatre's distinctive qualities as they have emerged in successive centuries and in different cultures, it can suggest how companies might reform themselves and so enter the unprecedented future better able to serve their audiences. A history of theatre is a book with practical implications.

Because past achievements of theatre are often represented in printed texts of plays, and because those scripts invite new productions, theatre has always had a vested interest in history. By understanding the kind of theatre for which a play was written, present-day directors, designers, and actors will gain a clearer idea of how best to recreate it on stage in very new contexts and with new facilities and styles for performance. A theatre history, especially a generously illustrated one, can therefore be an active agent in the future life of an inherited repertoire of plays as each one is reimagined and transformed for new audiences.

As the varying fortunes of theatre at the present time demand that its story should be attempted anew, so do recent advances in knowledge. Within the last few years, new evidence has been uncovered about the Rose and Globe Theatres in Elizabethan London, and about the extensive provincial touring of theatre companies at that time; these new facts must alter our view of the performance conditions in which great dramas were created and repeatedly performed. Reassessment of documents and new directions in research have revised accepted opinions about the origins of drama in Athens and its continuance after that theatre's greatest epoch, about playwriting in medieval Europe, Spanish theatres, the staging of plays in Louis XIV's France, audiences in several countries during the industrial revolution, and much else; and new knowledge and perspectives have brought about revised critical judgements. Theatre practice has also contributed, as ancient tragedies and medieval dramas have been produced with a frequency unknown before; performances alive to the present and past have led to a revaluation of the plays' specifically theatrical power. In the last few decades scholarly study has developed the means to record and analyse the work of individual directors and designers, and to understand the part they have played in making productions and leading theatre enterprise. Most importantly, the timing of this history affects all its contributors by providing an opportunity to reassess

theatre when knowledge of its past and therefore of its possibilities has been immeasurably enhanced by computerized research facilities and by a greater ability to travel to see the varieties of theatre which are practised around the entire world.

Imitation, play, and performance have always been part of everyday life, instinctive sources of pleasure and the means of exploration, affirmation, and persuasion; they provide a shared experience in which personal and social existence can be defined and redefined. How all this has been developed into productions that can be enjoyed by large and small audiences is the vast subject of this inquiry, which reaches back in time and across the world. This book seeks to open up for its readers the various forms that theatre has taken and to reveal their distinctive powers.

Sixteen specialists cover the great distances of the story and respond individually to its very different episodes. United by a concern to represent the various achievements of theatre and show where and by what means these were created, they take up many themes: theatre's changing role within society, the grounds of its popularity or failure, varieties of form and means of production, interactions with other arts and with critical, political, moral, and religious thought. The stage, backstage, and front-of-house; personnel, organization, politics, audiences, and

A festival performance at Avignon. In the ancient courtyard of the Palace of the Popes, Bertolt Brecht's *Caucasian Chalk Circle* is in performance by the Théâtre de Louvain la Neuve (1978), drawing thousands of tourists and theatre-goers during a summer evening.

3

Facing: A twentieth-century repertoire. Programmes from the National Theatre of Great Britain show the international and historical range of the theatre's work, and the ways in which its publicists recommended plays to their audiences.

society at large—all come into focus as individual authors have judged most suitable for their parts of the story. Playwriting and theatre criticism are also involved, but attention goes beyond performances which are controlled by written texts and given at regular times in specially equipped buildings. Theatre is taken here to include any performance by actors for other persons, when what is performed is calculated to hold attention and give pleasure, as well as providing other benefits according to time, place, and opportunity.

To keep the book within practical bounds and yet deal with the heart of its subject as fully as possible, some theatre-events have been excluded. First of these are entertainments where music is a basic and controlling element in performance, such as opera and dance in all their variety: these will be considered from time to time as they affect other forms of theatre, but are not given attention for their own sakes—for that, separate histories would be required. Film, television, and radio, which have many elements in common with theatre, are also excluded because their final products are controlled by means of camera, sound-recording, and editing; these new and popular art-forms will be included only when they come within the main focus of this book, which is always on live performances by actors for audiences. Events staged primarily for political, religious, educational, or social reasons are also out of its range; these include processions, rallies, religious rituals, commercial shows, instructional and political demonstrations, all using theatrical means but serving primarily very specific ends other than the creation and enjoyment of performance. With all these restrictions the subject of this book is still vast and its ramifications numberless.

One recourse might have been to summarize as much as possible and list all names of theatres and artists thought worthy of remembrance, and all titles of long-running plays and all responsible and tested theories of drama. Another procedure would have been to survey the vast subject and make sense of it according to one person's view of history. But the first method would have provided a reference book, rather than a history; the second a partial account, sometimes lacking in specialist knowledge and conviction. Here a different course is taken. While telling the story of theatre from their very different viewpoints, the authors of this history have chosen what seems most vital in their various parts of the past and have tried to share their sense of its importance and pleasures. With each new chapter the presentation alters, so that a reader is invited to engage with theatre in ways which draw most effectively upon the present state of knowledge, and upon the first-hand experience of its author. Theatre is an art possessing a wide spectrum of possibilities, but all will not be equally pertinent to any one performance; to be effective, therefore, a history must change constantly in direction and mode of attention, paying careful regard to those occasions and places when different elements have been gathered together in strong and revealing performances. Each author has made choices of what to include and what to enlarge upon, and in

4

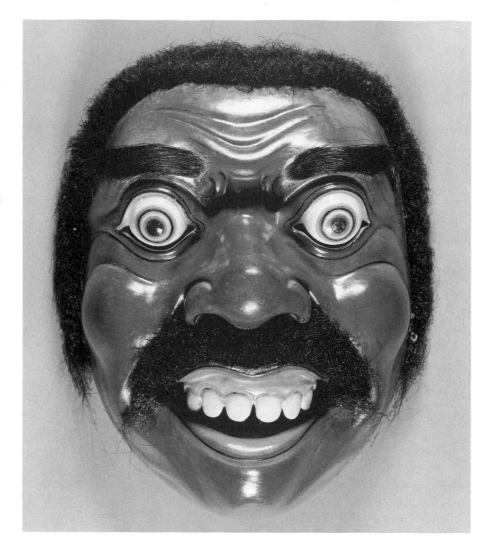

A modern mask in traditional manner. From a small workshop on the island of Bali, mask-maker I. B. Alit supplies both Western theatre companies and local dance-drama groups. He inherited his craft from his father, and is himself a performer. On this mask, curved slits under the eyeballs allow a wide view to its wearer.

doing so has been guided as much by how a production gained from consistency and continuity, as by the effects of conflict, crisis, and innovation. While taking measure of theatre's achievements in this way, they unfold the story of the art's varying fortunes and suggest the continuous dialogue of ideas that helped to shape them.

In some chapters the developing forms of writing for the stage provide the leading element in the narrative. In others, the visual images created on stage, the actors and acting styles, the art of directors, techniques of staging, or ideas implicit and explicit in play-texts will be given greatest attention. Yet another chapter will emphasize how theatre companies have reorganized themselves and managed their relationships to authority and audiences. Musicians, light and sound technicians, designers, star actors, supernumeraries, critics, kings, financiers—all at some time have leading roles in what is recounted.

So that a continuous focus could be maintained across centuries and across national and language barriers, some curtailment of perspective has also been necessary. This is a theatre history from the standpoint of British, European, and English-speaking cultures, a limitation that must falsify the account at some moments in time. But the alternative world-wide focus, if it were maintained throughout the book, would have proved unmanageable and made the narrative still more discontinuous; the history would have become a compendium of local reports and a bewildering collection of disparate narratives. A world-wide focus is attempted only for very recent times.

Any theatre history written in the West must start with the main strand of its narrative in Athens, but that will not account for all its origins. Theatre has started in many other places independently and, in a sense, it is still being created afresh today in hitherto unknown contexts. So while this book has a first chapter which takes time to establish what is known about ancient Greek theatre and then moves forward to Roman and early Christian theatres in the rest of Europe, a third chapter switches attention to the beginnings of theatre in Africa and the Americas, countries which have had comparatively long and close contact with Europe but whose theatre histories are distinctive and begin in significantly different circumstances and times. Theatres in Asia enter the narrative on their own accounts much later in the book, when its main narrative lines have reached the later part of the twentieth century; it was then that these vital and distinctive theatres could be more widely studied and enjoyed in the West and travellers could more easily value them for their own qualities in their own environments. By giving separate attention to these different sources of theatre tradition, some dislocation has been incurred. The story of theatre in Africa is found in widely separate chapters; events in India, Indonesia, China, and Japan are recorded without a simultaneous view of theatre in Europe from which some influence was being experienced. But, to offset this, the strategy gives scope for several independent accounts of the origins of theatre.

Boundaries between the various chapters cannot always be strict or easily defined by dates. Their authors are like relay runners who travel some short distance with those from whom they take over the baton. As medieval civilizations moved into the Renaissance, the Italian, Spanish, English, and French theatres, each at its own pace, took significantly different directions, according to social, economic, and artistic compulsions; and so, sometime in the fifteenth century, the story had to divide into four separate channels, and continue so for more than 200 years. Around the year 1700, although somewhat later in France than elsewhere, greater availability of the printed word, increase of travel and trade, and

Above: a nineteenth-century theatre in new surroundings. The Theatre Royal at Stratford in East London is now surrounded by later buildings and busy roads, but its unchanged interior still accommodates crowds and provides lively contact with its stage.

Right: a twentieth-century theatre in a park. Helsinki's City Theatre replaced two nineteenth-century theatres. The company's new building, overlooking the city and designed by Timo Penttilä, opened in 1967; its larger auditorium has an Elizabethan-type thrust-stage. A further wing, completed in 1989, houses a third auditorium and additional office-spaces.

developments in politics and society caused such fundamental changes that from here forward national distinctions are not so strong and the story demands at first a European and later an American context, extending gradually to other countries as they came under the influence of European languages and cultures. So the chapter on the eighteenth century considers theatre as it developed in a widening circle of European countries and in an increasing number of languages. By the beginning of the nineteenth century, the story is told more generally across Europe and here theatre makes a stronger entrance in North America, from where new forms begin to exert influence on the fortunes of other theatres.

As the story approaches the present day, a world-wide focus becomes increasingly necessary, and it is fully established in a final chapter that starts in the year 1970. An earlier date for this change would have been better in some respects, but by keeping the merger of all streams as late as possible, the theatres of Asia, Africa, and native America can be established in their own rights and with fuller regard paid to their distinctive natures. The point chosen for this last change of focus comes immediately following the widespread political unrest of the late 1960s which had brought a new generation and new concerns into political and cultural life. By this time too, the end of colonial dominance by the West had been clearly signalled; in rapid retreat since the end of the Second World War, it was by now broken or discredited. Not without connection to these other changes, 1970 also saw the beginning of the collapse of totalitarian communism in the Soviet Union and Soviet bloc, a beginning of the end for a political power which had seemed to hold the world in thrall and polarized much of its thinking. At this time when modern technology was bringing everyone into one instant view and reducing travel-time between nations, the barriers between people were being slowly dismantled too, and earlier certainties were giving way to new speculations, hopes, and fears. Now theatre was changing as decisively as the audiences which supported it.

In July 1969 the first man walked on the moon, and at about the same time computer-controlled micro-technology was changing the physical possibilities of theatre as never before. In the use of light and sound, in scenic presentation, and in organization and salesmanship, the post-1970 theatres are living in wholly different circumstances from any of their predecessors; this alone would argue for a separate chapter to consider their several entries into a new age.

But none of the chapters stands completely alone; each is meant to be read in sequence, moving both forwards and backwards. Of all the arts theatre is perhaps the one least able to escape the past, even as it is itself changing in terms of a changing present; never is its stage an entirely 'empty' space, free from echoes of previous performances. Even this history's concluding view of innovative theatre in the last few decades must refer back and draw upon earlier chapters. Old theatre buildings stand among new and not-so-new urban developments, with their own face on the world; and yet, inside, their old, accommodating auditoriums still draw spectators together and encourage active participation, as they did centuries before. Gleaming new theatres stand in parks on hillsides overlooking sprawling towns, having all the conveniences of modern planning and equipment, but their stages may well owe more in form to ancient Greek or Elizabethan theatres than to any model of the last 200 or 300 years. An experimental theatre company working in Europe, creating performances which reflect late twentieth-century attitudes and awareness, may none the less employ masks that have been purchased from a maker in a remote village in Bali who has always worked in an ancient tradition handed down from father to son; the company's acting style responds to the strength and eloquence of that newly discovered but almost timeless aid to performance.

If we walk into the foyer of any well-established theatre in West or East, the posters on its walls and the printed programmes we are handed will all assert the modernity of its repertoire, presenting us with images taken from the fashions and market-place of the present time. But beyond the theatre's contemporary sales-presentation, the play-texts spoken on stage may be centuries old and offer entrance to many theatres of the past, and to long-vanished ways of thought and behaviour. We may well find that the productions draw on ancient traditions of staging and acting, as well as on innovative procedures. Playgoing is an up-to-the-minute experience, created by people inhabiting our distinctive world, but it also involves us with a wide swathe from the harvest of the past. All the chapters of this book are written both to record the past and to enhance our present playgoing, reading, and exploration.

PART ONE

The First Theatres

I

GREEK THEATRE

OLIVER TAPLIN

THE very words *theatre*, *drama*, *tragedy*, and *comedy* are Greek in origin; so are *scene*, *episode*, *music*, *chorus*, *character*, *mime*, and *dialogue* (and so, come to that, is *history*). Given this primacy, it would be neat to be able to claim (as many have) that the origins of Greek theatre were derived from some primitive or primeval rite. But the sober truth is that, if they were, then we know nothing at all about the rite or the process. Tragedy and comedy, in any historically significant sense, were more or less the invention of one particular, far from primitive, community, in response simultaneously to new political developments and to the inter-Greek competition for cultural prestige.

Rivalry for influence and attention was fundamental to ancient Greek civiliza-tion—for good and for ill—both between city-states and within them. During the sixth century BC Athens, with its surrounding country and villages of Attica, became one of the largest and most powerful cities, with a total population of over 200,000. When the dynasty of 'tyrants' (or monarchs) was ousted in 511–508 BC, the re-establishment of the old landed families (oligarchy, aristocracy) was pre-empted by setting up an elaborate system which cut across old tribal and local loyalties and handed power over to the people as a whole (the *demos*), which meant the 40,000 or so free-born male citizens. This radical independent 'demo-cracy' lasted nearly 200 years and spread to many other cities. During this time Athens organized an 'empire' of cities which contributed a kind of 'protection fund'; and, despite set-backs, she became the most celebrated city in Greece. This standing was indivisible from spending vast quantities of energy and money on artistic and intellectual activities. This brief era produced the temples of the Acropolis and Sounion; the sculptor Phidias and the painter Polygnotus; the

The Theatre of
Dionysus and the City
of Athens.

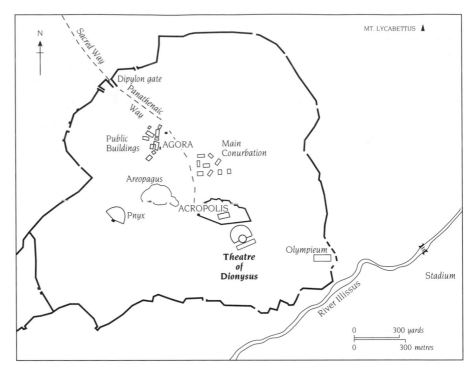

Sophists, Socrates, and Plato; Herodotus and Thucydides. This same era and society developed drama from its beginnings to a golden age that provided theatre to last for the next 700 years, and that was not rivalled, in Europe at least, until the London of Shakespeare.

First, the Athenians created a lavish new spring festival, the Great, or City, Dionysia, developed from older local festivals which celebrated the fermentation of new wine. And they dedicated for it a large area above the temple of Dionysus on the south-west slope below the Acropolis. At the foot they levelled a performance space (*orchestra* = dance-space); and the whole area was known as the *theatron*—the spectacle-place or viewing-place.

The typical processions and sacrifices of the Great Dionysia led up to competitions in honour of the god, held before a seated gathering of the people. Many such Greek festivals included competitions in music, song, and dance as well as the usual athletics (as at the Olympic Games). At the new Dionysia, however, the Athenians were determined to do something different.

In the early days there were just two 'contests', dithyramb and tragedy. Ten choruses of fifty men and ten of fifty boys (making 1,000 performers each year) competed in the dithyramb, an elaborate song with circular choreography, originally about Dionysus. Dithyramb long pre-existed this festival and was not special to Athens: *tragoidia*, on the other hand, seems to have been a purely Athenian artform, in effect invented when it was established at the Great Dionysia. There were only three dramatists competing each year, and each put on three tragedies. Before long a satyr play was added as a kind of appendix (more on this later).

The traditional date for the first competition was 534 BC and the first winner was said to have been Thespis; but this, even the existence of Thespis, may have been no more than popular myth. In any case, 508 is a far more significant date, and it may well be that the incorporation of tragedy and dithyramb in the Dionysia was all part of the setting-up of democracy. A magistrate of the city organized the festival and selected the year's playwrights; and the city treasury paid for the actors and the prizes. The choruses were trained and costumed at the expense of a rich citizen, the *choregos*; but this kind of taxation, which gave opportunity for civic display, looks like an institution of the democracy.

Tragoidia seems to mean 'goat-song', but no one knows what the relation between the song and the goat(s) was. Nor have we any idea what kind of quasi-ritual performances were put on at the rustic festivals of Dionysus, before *tragoidia* became part of this major festival in the city-calendar of the culturally self-promoting democratic Athens. But we do know something about other forms of poetic and musical competitions, which were already being held around Greece, and which fed into tragedy.

The Theatre of Dionysus at Athens. This is the very place where the great Athenian tragedies and comedies were first performed. The original performance-space (*orchestra*) was, however, considerably larger than this surviving semicircle, which dates from Roman times.

An early tragic chorus. Painted in Athens in about 490 BC, these six young soldiers sing and dance in unison (note the identical masks). They are presumably part of a tragic chorus, and are probably raising a ghost from the tomb on the left.

First and foremost there was the time-honoured narration of *epic* poetry, above all the *Iliad* and *Odyssey*. Tragedy drew on that same reservoir of stories about the distant heroic past, its great dynasties and wars, especially those set at Thebes and at Troy. The solo reciters of epic, the rhapsodes, specialized in enthralling their audience and moving them to strong emotions—apprehension, compassion, sorrow. Tragedy would go on to emphasize many of the same aspects of the stories: suffering, mortality, inevitability and responsibility, guilt, revenge, recognition, persuasion, anger, deceit, endurance, the difficulty of understanding the divine. In a sense, tragedy set itself up as a kind of rival to the Panhellenic epics, a rival which, unlike those fixed texts, produced fresh and varied versions every year. So it is a far from merely academic fact that the characters of tragedy speak in a poetic register of the local Attic dialect and in the iambic trimeter, a simple metre quite close to the rhythms of everyday speech (not unlike English blank verse, in fact). This is all in marked contrast to the elaborate dactylic metre and highly artificial and unlocalized dialect of epic.

Another crucial departure from epic narrative is that tragedy, in contrast to a single voice in a single metre, alternated sections of iambic speech and dialogue with dance-songs performed by a *chorus*. In fact, the alternation of scenes of action with choral dance-songs is the basic underlying structure of Greek tragedy. The standard tragic chorus had fifteen members—all Athenian citizens—and, though their leader would sometimes speak on their behalf, they normally performed in vocal and physical unison, as an indivisible group. Their songs were usually structured in pairs of stanzas with matching music and choreography, and the dance was based (unlike dithyramb) on linear and rectangular formations. The complex metres, which are always different, though known as 'lyric', were accompanied by the double-pipes (*aulos*). The music and choreography seem to have been expressive of mood, tone, and atmosphere, rather than representing any sort of 'narrative' features within the songs.

A flourishing tradition of choral dance-songs also pre-existed tragedy. They were performed for all sorts of occasions—hymns to the gods, weddings, funerals, displays by unmarried girls, dithyrambs, and, best known from Pindar's work, the celebration of athletic victories. The songs of Attic tragedy would combine and adopt features common to the tradition—snatches of myth, religious and moralizing material, elements specific to the particular occasion—to produce poetry that is usually discursive, allusive, and unpredictable in its sequence of ideas. It is of the essence that the choral songs do not go through sequences of narrative or argumentation, and that (with a few exceptions) they are not part of the plot. Their contribution is, rather, to respond, witness, compare, draw morals, make public, to try to make sense (in some ways they are comparable to the chorales in the course of a Bach *Passion*).

Some influential ancient sources claimed that tragedy was in origin purely choral. But that does not really makes sense: it was only when the iambic speeches of actors were combined with choral dance-songs that tragedy fully came into existence. When that vital cross-fertilization took place, whether in 508 or 534 (Thespis?) or earlier, we cannot know. It is worth noting, however, that at least once in every single surviving tragedy there is a scene set in song-metres (sometimes with an admixture of spoken iambics) which involves both the chorus and one or more actors (known as 'lyric dialogue' or as *kommos*). These song-scenes occur at crises of emotion and are part of the plot, quite distinct from the purely choral songs; at the same time they are far more free, more highly coloured, and less reasoned than the usual level of speech. This far from uncommon 'mixed' mode could only have been the product of the creative fusion of actor and chorus.

There is, of course, one more ingredient vital for the invention of drama in addition to the provision of a special time and place and to the combination of individual speech with communal dance-song: the word *drama* means a *doing*, an enactment. The virtuoso rhapsodes of epic no doubt varied their voice and body language to suit different speakers and situations, but there was always the narrator and the narrative framework. For drama to come into being 'The priest Chryses came to the ships and said to Agamemnon, "Please release my daughter"' had to become: 'I am Chryses the priest and I have just arrived here at the ships. You are Agamemnon, and I beg you to release my daughter.' The Greek chorus also, although a group, have a place within the world of the play—as citizens or servants or whatever—and they are not detached from the time and place of the action (as in Noh theatre, or indeed the Bach *Passion*). Once again we do not know in what ways or to what extent this kind of unframed impersonation pre-existed tragedy, only that it was an essential prerequisite for its birth.

It may well be that the *mask* enabled the performer to 'become' another person, even a god. Whether these masks were derived from primeval cults of Dionysus, or were invented for the theatre, the players, actors and chorus alike, in tragedy and satyr play and comedy, all wore whole-head masks, probably made with stiffened linen. Each role had a different mask, but not necessarily a different actor. At first there was, allegedly, only one actor, but by the time of the earliest

Facing: cast with
the musician
Pronomos. On this
elaborate vase, painted
in Athens about 400
BC, Dionysus is shown
in theatrical robes,
reclining above the
seated figure of the
famous instrumentalist
Pronomos. The chorus
and actors are in their
costumes for the satyr
play, the last of the set
after three tragedies.

surviving tragedy there are two, and before the death of Aeschylus, three. The number stuck at three (for comedy also), sharing all the parts between them. This was partly, no doubt, because exceptional qualities of vocal and physical technique were called for, but partly also because of the difficulty of making it clear in such a huge theatre-space who of several characters is speaking.

In fifth-century Athens the playwrights, often known as the 'teachers', were directors, composers, and trainers as well as writers. They were the chief celebrities of the spectacular success of the new art-form. But, as time went on, the actors also attracted more and more attention, and from 449 BC there was a separate actors' competition and prize. They were famed above all for the quality and virtuosity of their voices, but clearly the techniques of body language, especially of the arms, were also highly developed. Their costumes were splendid: while they would carry or wear role-markers, such as sceptres, garlands, oriental trousers, or mourning black, the standard outfit for the main characters was elaborately ornamented. With their tight sleeves, and for male roles fancy thin-soled boots, they were quite distinct from and grander than any everyday wear.

The musician was also finely dressed. He would enter with the chorus, and normally stand in full view to play his *aulos*, a kind of double-clarinet, good for the open air. The chorus were identically masked and costumed, sometimes at great expense for the *choregos*. Another expense for him was the provision of facilities and creature comforts during a long rehearsal period. Audibility, as well as skilled co-ordinated choreography, was at a premium. Delivery, acting, and general production must all have been developed to meet the huge size of the *theatron*. The *orchestra* alone, which may have been rectangular rather than circular in the fifth century, was probably some 20 metres across (larger than the entire auditorium of many modern studio-theatres). There was by 458 BC a substantial scene-building (*skene*), with a central door, on the far side of the *orchestra*, though it does not seem to have been there in earlier years. An art of painted scene-decoration developed, and became all part of the celebrated spectacle.

The audience-area was rocky and irregular, and we do not know how much seating was constructed (the surviving stone seating dates from not long before 300 BC). An audience number of 14,000 has become canonical; but it might have been possible to cram in something more like 20,000. Even if there were 'only' 10,000 it would still have been a huge gathering with some 25 per cent of all citizens, and bigger than any other gathering at Athens, including full gatherings of the democratic assembly. Indeed the Athenian Dionysia may well have been the largest annual gathering in the fifth century anywhere in the Greek world (the Olympic Games were bigger, but held only every four years). Advantage was taken of the occasion to confer honours, to display the tribute sent by the 'allies', and to conduct various other civic ceremonials.

The vast majority of the audience consisted of Athenian citizens; there were

A modern performance at Epidaurus. This gives some idea of the overwhelming size of an ancient Greek audience. This auditorium, in the north-east Peloponnese, was architecturally the finest of the many constructed, on the Athenian model, during the third and second centuries BC.

definitely boys there also, but probably not slaves. There was also an increasing number of allies, diplomats, and other visitors—this was a cultural show-piece. It is much disputed whether women were admitted to the *theatron* at Athens, but probably they were not. Their presence, if they were there, leaves no trace in the contemporary evidence, while there are plenty of references to, for example, foreigners, special seating for honorands, and to the judges of the competition, who were selected by lot from a panel. There were also officials who were responsible for orderly conduct. We do hear stories of unruly audiences, but most sound fictional and are, in any case, related to unsuccessful productions.

When a tragedy held that vast audience enthralled, the sense of collective concentration must have been very powerful. This would be increased by the communality of broad daylight (the tragedies began soon after dawn). The thousands heard the same words and focused on the same stage-picture. In this democratic theatre there were no limited-view seats; and a kind of 'funnel' of attention must have built up, all focusing down towards the *orchestra*. It was an enormous civic event, a purposeful political act of communality. Although so much has happened in the European theatre since then, there was only one Greek city, and

19968
792

only one particular stage in its political history, which could have got it off to such an influential start. While it is crucial to be aware that these plays were made for performance, and that they were performed as part of the 'political' life of the city, we unfortunately have very little direct corroborative evidence about the actual experience of theatre-going. Generally speaking it is the play-texts which have survived the intervening centuries that put us in touch with the Greek theatre—even though they are only a minute proportion of those that were put on year after year.

Athenian tragedies did not observe fixed 'rules' so much as build up a body of expectations and conventions that grew out of the precedents of what had succeeded. There does not seem to be any underlying pattern, whether ritual or of any other kind: on the contrary it is the unpredictable variety of shapings that seems remarkable. And the tragedies were not required to have any direct bearing on Dionysus. Some Dionysiac element may have been seminal originally, but the only surviving story of Dionysus happens to be one of the last (Euripides' *Bacchae*). Yet Dionysus may still be a particularly suitable patron-god, because he would induce in his worshippers mental states which freed them from their 'normal' everyday identities. He was, furthermore, a god who particularly combined the ecstatic and the liberation of the wild with the order and control of civilization—a combination characteristically contained within tragedy.

The story about satyr plays was that the complaint against tragedy that it was 'nothing to do with Dionysus' led to their addition. After their three tragedies each playwright and his company put on a fourth play which always had satyrs as its chorus, those half-human, half-animal, drunken, lustful followers of Dionysus. The plots were taken from the same heroic world as the tragedies, but through the satyrs they were displaced into a non-social, non-civic world and subjected to anarchic disorganization. It was a favourite plot-pattern for the satyrs to be confronted with a brand new invention—wine, music, or fire, for instance—and to subject it to their own misuse before it became an essential part of human social life. Unfortunately the only satyr play to survive complete is a rather late work by Euripides (*Cyclops*) which may well be untypical.

Returning to tragedy, it is almost impossible to make generalizations about the matter of Greek tragedy that are true without exceptions. Nevertheless, it remains true that almost all tragedies tell stories of suffering, of mental and physical anguish, of the waste of life and prosperity. Tragedies incarnate the cruelties and horrors of the human world, and somehow face up to them. Also they generally deal with conflict, and particularly family conflict, the extreme stresses of blood and marriage relationships. Yet most also contain issues of public or political conflict, within and between cities—and sometimes, most famously in *Antigone*, conflict between the claims of family and of city.

The main characters are, of course, of noble birth (in contrast to their

Prometheus and satyrs. This scene, painted in Athens in about 425 BC, may have been inspired by a satyr play of Aeschylus, in which the satyrs were confronted with the novelty of fire, contained inside the hollow stalks.

choruses). A remarkable proportion of them are women; and many of the most powerful Greek tragedies emphasize issues of gender conflict. That this happened in a society where women were strictly subordinated—and probably not even in the audience, let alone among the performers—might be suggestive about the function of tragedy within society. It would enact the terrors and forces that most threatened the security of civic and family life; and yet it could contain within its special time and place these issues, which were too horrific and dark to be aired or faced in ordinary discourse. Thus, within the male-only *theatron* the citizens of Athens could face gender issues with a starkness impossible elsewhere.

This may help explain the treatment of the gods as well. The religion of Athens was generally pious and conventional and aimed to keep the gods benevolent. In tragedy, on the other hand, the gods all too often bring about human suffering, and do so in ways that humans cannot control or even understand. The world is full of suffering—much of it undeserved—and the interference of the gods helps to account for this, but not necessarily in ways that make satisfying, let alone moral, sense. It is not true, however, that Greek tragedy is dominated by 'Fate', or that the characters are merely helpless victims of 'Destiny'. On the contrary, many plays tangle with issues of free will and responsibility and of human attempts to make the best of life despite ignorance. In retrospect we can see how the disasters 'had to happen', and from the divine perspective they can all be foreseen; but most of the tragedies are enacted by mortals who cannot see the shape of their sufferings until afterwards.

The tragedies dramatized deep, difficult, painful issues; they posed terrible questions, and far from conclusively resolved them. But the issues would not be successfully thought-provoking unless the play also captured the attention and the emotional engagement of the audience. The atmospheric and emotional range of the tragic experience was various and shifting. Everyone thinks of 'pity and fear' because these two emotions were canonized by Aristotle in his lectures on tragedy (his *Poetics*); but surely these are only a kind of shorthand. To judge from the plays that succeeded, the kind of register that Athenians expected from a tragedy was that it should be grave, sorrowful, poignant, horrific, dreadful, fascinating. They *wanted* to be distressed by things that would be utterly horrific outside the theatre, even possibilities so terrible that they could only be contemplated under these controlled circumstances.

Fifth-century Athens, not unlike Shakespeare's London, was a time and place of almost frenetic creativity. While so much about the practicalities can now be only glimpsed, we do at least have our texts of Athenian tragedy. Their very survival is some index of the variety and vitality of those productions which so obsessed so many thousands of citizens every year. To try to recover some shadow of their theatrical power I shall look at a couple of tragedies by each of the three greatest fifth-century playwrights. Three tragedians competed every year, and many who are now only names won prizes over Aeschylus and Sophocles, let alone Euripides. But since all the plays that have been preserved, except two, are by 'the great three', any history is bound to concentrate on them. (*Prometheus Bound* is probably by a mid-fifth-century follower of Aeschylus, and *Rhesus* probably dates from the early fourth century although it is attributed to Euripides.)

Aeschylus' epitaph singled out from his life that he fought at Marathon, the great Athenian defeat of the Persian invasion in 490 BC. The earliest tragedy to survive, *Persians* (of 472), 'celebrates' the defeat of the second invasion eight years before, especially the naval victory at Salamis, where Athens played a prominent part. The play enacts the return of the huge expedition to Persia, and this is treated in a remarkably unjingoistic way. There were other early dramatizations of recent events, but evidently the Athenians preferred to see the suffering of fellow Greeks. At the same time, they did not want something too close to the distresses of their own immediate world, and it became standard for the stories to be distanced by the temporality of the era of epic poetry.

In the years after 480 Athens's naval superiority built up her 'empire', and this was reinforced by the promotion of cultural achievements, including tragedy. Aeschylus became known outside Athens, and was even commissioned to compose a play for one or more of the 'tyrants' who ruled the prosperous Greek cities in Sicily. He must have been important in making the art-form known throughout the Greek world, and building it up into a serious alternative to epic. We could no doubt see this emulation better if only we had the trilogy of plays, probably early work, in which Aeschylus dramatized parts of the *Iliad*. The first, *Myrmidons*, was notorious for Achilles' stubborn silence when he refused to fight, and for his overtly homo-erotic laments over Patroclus. In the third, *Phrygians*, it is interesting that the deeply private *Iliad* scene when Priam comes to ransom the body of his son Hector was transformed by the spectacular presence of a whole chorus of Trojans.

There is good evidence, however, that his Orestes trilogy or *Oresteia* of 458, two years before his death, was regarded as the masterpiece out of his sixty or so tragedies (and twenty satyr plays). Aeschylus may quite often have composed trilogies or tetralogies of connected plays like this, though, as *Persae* shows, not invariably. If only the satyr play *Proteus* also survived, it would be very interesting to see how far it subverted the tragedies (it probably involved Menelaus' adventures in Egypt). The three tragedies form a work on a huge scale, calling for at least six hours of performance, plus two intervals. They also demand from any audience intense and sustained concentration, since so many thematic and theatrical features are recurrent across all three plays.

Whether or not most such trilogies followed the same pattern of construction, that of the *Oresteia* is that the first two plays form a kind of corresponding pair. In *Agamemnon* the great victor at Troy returns home only to be murdered, along with his captured 'concubine' Cassandra, by his wife Clytemnestra: in *Libation Bearers* (Greek *Choephoroi*) Orestes returns from exile and, in revenge for his father, murders his mother along with her new husband Aegisthus. As Orestes stands over the two bodies there is a clear 'replay' of the tableau where Clytemnestra stood over the two bodies in the first play. The pattern of vendetta is repeated from generation to generation—if blood calls for blood, there will be no end of it. There are, however, extra elements in *Agamemnon* which contribute to making it half as long again as its sequel. First, pieces of the past are patchworked together: the stories of Helen's desertion and of the sacrifice of Iphigeneia at Aulis are allusively contributed by the chorus, while Clytemnestra's powerful will, her ability to control people and words, and her desire for vengeance for her daughter are accumulated in a variety of subtle ways. Secondly, there is Cassandra, who in her prophetic visions ranges over the distant past, especially the murder by Atreus (Agamemnon's father) of his brother's children, the recent past, especially at Troy, the near future—Agamemnon's death and her own—and the more distant future which will bring vengeance. The barbarian female slave is the character who has most vision in this play of confusion and foreboding; and she has the most independence of Clytemnestra, even though she has to die at her hand.

Clytemnestra is the archetype of the powerful, dangerous woman who breaks out of conventional restraints and destroys the security of her menfolk. For most of *Libation Bearers* the audience is encouraged to see Orestes' revenge as a righteous act which will mend personal, sexual, and political wrongs. But, as soon as the deed is done, this false sense of security is exposed by the 'replay' of the murder scene. In contrast to the long scene after the murder in *Agamemnon*, Orestes is rapidly assailed by visions of his 'mother's angry dogs', fearsome creatures conjured up by her dying curse. The play ends abruptly on a question; and in the chorus's last lines Aeschylus allows through a certain self-consciousness about his trilogy construction, and about the 'ghost play' before *Agamemnon*:

> This is the third storm to gust against the royal house.
> First the child-swallowing agonies marked the beginning.
> Secondly the husband's royal fate,
> the slaughter of the Achaian war-lord, dead in his bath.
> And now third in his turn a saviour has come—or should I say doom?
> Where shall it all find an end, where be soothed and cease,
> this storm-force of disaster?

Apollo protects Orestes at Delphi. This is one of many representations, painted in southern Italy in the fourth century, which were ultimately inspired by the opening scenes of Aeschylus' *Eumenides*. In this one the ghost of Clytemnestra on the left tries to arouse some not very repulsive Furies.

An answer is unpredictably approached in the third play *Eumenides* (which means *The Kind Goddesses*—though not too much should be made of play titles, which may well not go back to the playwright). It is a highly untypical tragedy in many ways. One example is that early on the chorus is taken off the scene and brought back on, marking a change of place and a considerable lapse of time. The move is from Delphi, where Orestes has taken refuge with his protecting god Apollo, to Athens, where Athena is challenged to find a solution. The chorus consists of his mother's Erinyes (Furies or Curses). They have now become visible to everyone, including the audience, in the form of horrific snake-infested hags, and they pursue Orestes relentlessly.

Athena proposes to them and to Orestes a trial before a jury of human citizens of Athens, sworn to pass just judgement. This will be the archetype for the Athenian murder-court for all time. So the Erinyes become the prosecution, arguing that to murder the womb that bore you is an irreducible guilt; Apollo in defence argues that the father is the sole significant parent and the mother a mere incubator for his seed. With great suspense, the votes are cast and counted: they are equal. Athena gives her vote for Orestes, which, while it does not endorse Apollo's dubious case within the trial, breaks the deadlock, and provides an explanation (more strictly an aetiology) for the Athenian procedure that, whenever votes were tied, the defendant was acquitted by the 'vote of Athena'.

Orestes (and Apollo) leave the play, now that the family vendetta has been resolved and superseded by the civic judiciary and the establishment of homicide law. The primal Erinyes, female and fatherless, are left without any honours or functions, and now their anger is turned against the city of Athens. Though Apollo tried to paint them as mere blood-sucking monsters, they have emerged as in their own way champions of justice who have justly won half the votes. In the final scene Athena persuades them that, if they bless Athens, then her city will give them honour and cult; and the play ends with a torchlight procession to their new home beneath the Acropolis. The dangers and perversities and obscurities which have run all through the trilogy are in this final scene set straight, and human society at Athens is given a kind of 'fresh start'. Yet the Furies are not sentimentally transmogrified into good fairies. As well as their power to bless they retain their power to terrify: the just city still needs an element of deterrence. 'For some they shall provide poetry, but for others a life dimmed by tears,' as Athena puts it. On one level this is a kind of 'defence of Tragedy': the good citizens of the future may weep beneficial tears in the theatre, but the unjust shall shed real tears.

It is a strange but important fact in the history of Greek tragedy that the great model of the *Oresteia* was in many aspects not followed. The practice of the connected trilogy was, so far as we know, abandoned, and three separate plays became the norm. Tragedies were hardly ever again set in the city of Athens, although many make explicit Athenian connections with their settings elsewhere.

Triumphant processional endings became a feature of comedy rather than tragedy, which favours open, unhealed endings. And instead of following the lead of this active and participating chorus of Furies, later dramatists settled into a norm of weak women or old men, who sing as sympathetic witnesses, but cannot intervene. Last, but not least, later tragedians did not try to emulate Aeschylus' enormously rich, metaphorical, piled-up diction, but went for a style that is less grandiloquent and more flowing. For whatever reasons, the *Oresteia* was both the climax and the end of a chapter.

Two dramatists jointly dominate the new chapter. Sophocles was first accepted for the competition in 468, that is, his career overlapped with Aeschylus' by about twelve years, while Euripides first competed in 455, as if to fill the new gap. After long lives they died, Euripides first, within a few months of each other in 406–405—so neither lived to see the final defeat of Athens by Sparta at the end of the protracted 'Peloponnesian' War. We gather that Sophocles nearly always won first prize, while Euripides hardly ever did; also that Sophocles was a genial and influential citizen, while Euripides was more controversial. It would, however, be a crude over-simplification in both directions to characterize Sophocles' plays as always traditional and conventionally moral, or Euripides' plays as consistently subversive and anti-conventional. Sophocles could have a character complain in the closing lines of a play (*Women of Trachis*):

> Grant me plentiful fellow-understanding for all this;
> but see the plentiful lack of understanding of the gods,
> gods who beget children and are called on as father,
> yet look on sufferings like these with indifference.

while Euripides' chorus of Bacchants can sing:

> It may move slowly, but divine strength is still sure.
> It brings to book those humans who value the lack of understanding,
> and crazily fail to promote the divine.

It was not the place of tragedy in Athens to show edifying morality tales, or to ratify easy, pious platitudes. Both playwrights stir up uncomfortable questions, and neither suggests that there are comforting answers.

Seven tragedies still survive out of Sophocles' ninety or so (plus up to thirty satyr plays). Most cannot be dated, though *Philoctetes, Oedipus at Colonus*, and probably *Electra*, come from his last decade. We happen to know that he came second in the year that he put on *Oedipus the King*, though we do not know what year it was—I would be surprised to discover it was before 445 or after 420. This did not stop it from becoming, with *Antigone*, his most celebrated—and Aristotle's model—tragedy.

FROM SOPHOCLES' *OEDIPUS*

In this scene Oedipus finally recognizes that his life-story is quite different from the one that he has supposed, and is as disastrous as he had thought it prosperous. The chorus then responds in song.

OEDIPUS. Why then did you hand the baby to this old man here?

OLD THEBAN. Out of pity, master. I thought that he would take
him off to his own land. But he preserved him for the very worst—
because if you're the person this man says you are, then
you are, for sure, ill-fated from your birth.

OEDIPUS. Yes, O yes, then all would come out clear.
Light, light,
may this be the last that I look at you, the man
who has been exposed as born from those I should not,
living with those I should not, killing
those I never should have killed.   [*Exit into palace*]

CHORUS. O you human generations,
I reckon you and your lives as equal to nil, nothing.
What man, what man captures more of happiness
than an illusion,
an illusion that sinks and sets?
With your paradigm, your genius before me,
yours, unhappy Oedipus,
there is nothing human I could call blessed.

Scene from Sophocles' *Oedipus the King*. This vase (*above, right*), painted in Sicily *c.*340 BC, is most unusual because it represents a particular moment in a tragedy in performance. The old man from Corinth (left) reveals that he took the baby Oedipus from Mount Cithaeron near Thebes, thus making Jocasta realize that her husband must also be her son.

The play is not about how Oedipus killed his father and married his mother, but about finding this out. It charts the discovery, uncovering that his central assumptions about his life-story, which has made him the most fortunate of men, are quite wrong; and that the truth means that he is the least fortunate. The narrative line is effectively shaped to take Oedipus from secure confidence through a range of responses, as he attempts to accommodate each new fragment of information to his cherished life-story. The last piece falls into place, and he sees the other, the true, picture with dazzling clarity. He has returned to where he began; he left Thebes as a baby to return as a grown man for marriage; he has conceived children in the womb where he was himself conceived; he has killed his wife's husband, his own father.

Even though he blinds himself so that he cannot see his new world, the final scene of the play still compulsively explores its horrors with touch and with words.

Oedipus, like most of the great figures of Greek tragedy, never falls silent; there are always words in the attempt to confront the worst. The dynamic of the play excavates some of the root issues of human life: the will to knowledge, the pattern of a 'biograph' that cannot be seen even as it is being lived, the gulf between appearance and actuality, the unreliability of good fortune, the need to have fixed points, yet the impossibility of any final security. As the chorus sings of Oedipus as 'paradigm', the most terrible demonstration of the play may be that discovery and knowledge not only do not necessarily make life better, they might even bring unsuspected horrors. Yet it is still essentially human to need to know what can be known; and Oedipus remains the great 'hero' of that quest. The question whether there is any redeeming 'uplift' in the final scene of the play, or whether it presents unmitigated blackness, remains as disputed as it is for the ending of *King Lear*.

Oedipus the King has a kind of 'classical' quality to its construction. There are regular sections of dialogue and intervening choral songs; and the plot has a sense of almost ingenious inevitability. *Philoctetes*, dating from 409, is by contrast fluid, irregular, and full of surprise twists and turns. Aeschylus and Euripides (in 431) had earlier made plays about the fetching of the outcast Philoctetes from the island of Lemnos, because he and his bow were required to enable the Greeks to conquer Troy. Sophocles' greatest novelty was to reduce the cast in effect to three, and to make one of them the young son of Achilles, Neoptolemus. He is caught between the wily, unscrupulous pragmatism of Odysseus and the uncompromising, old-style integrity of Philoctetes. The techniques through which the play explores various angles, for example by testing out several alternative endings, seem extraordinarily agile for a playwright in his eighties. (The curious modernity of the play shows in its recent inspiration of Seamus Heaney, Timberlake Wertenbaker, and Derek Walcott.)

In Sophocles' later plays the dramatic texture tends to become more fluid and irregular: in Euripides, on the contrary, it becomes more formal, pushing conventions to extremes, and disjointing plays into episodes rather than breaking down the formal segmentations. We have eighteen (probably) of Euripides' tragedies, which is about a quarter of his output. Because the dates of several plays are recorded, and some features of his metre quantifiably change with time, we can more or less put our surviving plays in order. The two earliest (apart from *Alcestis*, which was unconventionally put on in place of a satyr play in 438) are the 'canonical' *Medea* of 431 and *Hippolytus* of 428. No fewer than eight of the surviving plays date from Euripides' last decade, and they include the other two plays most read and performed in modern times, *Trojan Women* (415) and *Bacchae* (first performed posthumously). It is more the pity that we do not have one of his most famous plays, *Telephus*, put on in 438 BC. Telephus was notorious for disguising himself as a beggar, and for taking refuge at the altar, while holding the baby Orestes hostage at sword-point.

Some of the features which make *Medea* so powerful are the very things which also make it untypical. The gods take little part, at least until the end and then with a twist; the central character is not Greek, but a highly articulate foreigner; and she is a woman who triumphs over men without being brought low. For the first half of the play Medea is presented favourably by comparison with Jason, who is exposed, not only as untrue to their vows, but as an opportunist unworthy of his associations of the heroic Argonautic myths. It is an interesting question how far Athenian men might have gone along with the 'feminist' re-evaluation of the early scenes. But Medea expresses sentiments as 'outrageous' as

> I would rather take my stand three times in the battle-line
> than go through child-birth once

in the course of a reasoned and persuasive speech. The chorus of respectable Corinthian women are carried along by her, and even sing that

> Story-telling shall turn my life to one of high glory.
> Honour approaches for womankind;
> disreputable stories shall no longer occupy the female sphere.

The play seems to be constructed to put Medea in a sympathetic light, at least as far as the scene when she secures a safe future harbour in her life-voyage from Aegeus, the childless King of Athens. It is only then that she springs her plan to kill her own children, as well as Jason's new bride, as the ultimate way to hurt him. Euripides wrings much pathos from bringing the children on stage in all their innocent defencelessness. After hearing the messenger-speech, memorable for its macabre description of the death of the princess and her father, eaten away by her gift of a magic robe, Medea goes, sword in hand, to find her children—a *crime passionnel* committed in cold blood.

The crime is so unthinkable that, when Jason comes hurrying to the protection of his sons, it is to rescue them from being victimized by the Corinthians. Medea's escape from them all is a *coup de théâtre*: instead of being trapped inside her ruined home, she appears above it all in the flying chariot of her grandfather the Sun. The play ends with the ex-hero broken and impotent, while Medea flies off triumphantly to her new life, more like a god

Scene based on Euripides' *Telephus*. This is an example, painted in south Italy not long after 400 BC, of the favourite scene of Euripides' play *Telephus* (now unfortunately lost): Telephus in beggarly rags takes the baby Orestes as a hostage. The 'stage-doors' on the left add a suggestion of theatrical associations.

than a human. In her human aspect Medea opens up a horrific threat for men: that their wives might so resent their maltreatment as to override the bond of motherhood and destroy their heirs. Human women do not have the means to escape exultant; but Euripides' combination of the human and the superhuman has produced a deeply disturbing sequence by luring the audience, even the male audience, into feeling for Medea, and then finding that she is unhuman.

It was perhaps above all Euripides' ability to produce plays that are both highly moving and unpleasantly disturbing that made him the dramatist that the Athenians 'loved to hate'. Not that his tragedies are by any means homogeneous—some are more political in their emphasis (e.g. *Suppliant Women*), some anti-heroic (e.g. *Electra*), some 'humanist' (e.g. *Heracles*), some melodramatic (e.g. *Hecuba*—his most admired play in earlier centuries). In the years 414 to 411 he evidently went through a period of producing relatively light tragedies, plays that look to the *Odyssey* rather than the *Iliad*, and which tend to grow out of the recognition of long-lost relatives and escape from long oppression. The most glossy of these is perhaps *Helen* (412). This exploits a kind of ironic countermyth to the most famous myth of all: Helen did not go to Troy, after all, only a phantom went with Paris, while the real faithful wife pined away in Egypt. In Euripides' play Menelaus is washed up in the Nile delta, and after a protracted recognition-scene (how can she be in two places at once?), an ingenious plot is staged, involving the pretence that Menelaus is dead, in order to escape from the clutches of the Egyptian King, who has designs on Helen. The tragedy is not trivial or banal; it airs challenging issues, for instance the relation between name and substance, foreknowledge and freedom, beauty and happiness. But it cannot be claimed that the suffering bites deep or that the dark ever seems impenetrable. Perhaps the most striking moment is when an old retainer exclaims incredulously: 'What? Do you mean that we went through all those troubles just for a cloud?'

But, after this group of escape plays (we have three and know of three others), Euripides reverted to grimmer tragedies, though with a newly unpredictable kind of narrative construction. Two of his best were 'on his desk' when he died: *Iphigeneia* (*at Aulis*), with its heart-rending portrayals of a mother and innocent daughter deceived to mobilize an army, and *Bacchae*. In this play, not unlike *Medea*, the audience is lured into thinking that the divine makes human sense, and is then confronted with the human suffering exacted by divine pride. Dionysus proves to be a mixed blessing for his mother's city of Thebes. Does the play also raise the question whether theatre is not a mixed blessing?

The deaths of the two great playwrights, followed closely by the defeat of Athens in the war, may make it look as though the fifth century saw the demise as well as the infancy of tragedy; and this myth has been fostered by Nietzsche's account of its *Birth*, which makes Socrates and Euripides into the murderers. As will be seen,

ARISTOPHANES AND THEATRE

An anthology of passages from Aristophanes' comedy
Frogs that bear on the theatre.

CHORUS.
  Demeter, grant that . . . we perform plenty that is funny,
  and at the same time plenty that is serious . . .

  Muse, . . . come and see this great mass of people,
  sitting in their sophisticated thousands . . .

  It's right and proper for the festival chorus
  to give good advice and sound lessons to our city.

—Right here—amazingly—poetics will be weighed out.
—What, they're going to test tragedy by the pound?
—Yes, they'll bring out rules and measures for poetry.

EURIPIDES. I brought on familiar matters,
  the things we handle and live with . . .
  I induced all these men to think about things like this,
  and by my art implanted in them the rationality
  and inquisitiveness to scrutinise and think about
  everything, and especially how to run their households
  better than before, and to question: 'why is this thus?',
  'where has this got to?' 'who's got that?'

EURIPIDES.—Answer me this: what reason is there to admire the poet?
AESCHYLUS.—For being clever and thought-provoking,
  and because he makes men better citizens in their cities.

AESCHYLUS. Little boys have schoolteachers to tell them things,
  while grown men have poets.

DIONYSUS. Now look, I came down here after a poet . . .
  so the city may be preserved for the future presentation
  of plays. So whichever of you two can give the city
  good advice, I reckon I'll take him.

this was far from true. At the same time, Aristophanes' comedy *Frogs* is good evidence that in 405 BC there was a sense that the best and most creative might be past. In *Frogs* Dionysus himself is—not unreasonably—a great tragedy buff; and he is desperately missing Euripides, compared with all the second-raters still left. He is even prepared to go to the kingdom of Hades to fetch Euripides back. His mission turns into a search for the best tragedian of all 'so that the city might be saved and still put on its dramatic choruses', and he ends up bringing good old Aeschylus back to Athens. But this is not the 'real Aeschylus' any more than it is the 'real Dionysus'—they are the conjurings of comedy. In a sense Aristophanes is making a bid for *Comedy* to rival or even supplant Tragedy as the purveyor of wisdom and poetry to the city.

Tragedy in no way presupposed Comedy: but Comedy—or at least Athenian *komoidia* (which seems to mean 'revel-song')—presupposed Tragedy; and it is highly self-conscious about its relationship with its handsome older sibling. Comedy is said to have been introduced into the programme of the Great City Dionysia in 486 BC. Recognizable forerunners may well have gone back further in time and have been more widespread through Greece than any recognizable fore-runners of tragedy. There were, for instance, processions accompanied by ribaldry which carried models of phalluses in honour of Dionysus; there were dances performed wearing grotesque padding; and we know from vase-paintings that at Athens men dressed up for bizarre choruses, accompanied by the *aulos*, including such curiosities as men riding on ostriches or dolphins, or walking on stilts wear-ing pointed hats. And at Syracuse in Sicily, about the same time as *komoidia* became official at Athens, Epicharmus put on performances in the local dialect (probably called *mimoi*), which told burlesque adventures of Odysseus or Heracles.

But Athenian *komoidia*, which was eventually to become mainstream comedy, did not properly come into existence until it had completed some of the same steps also taken on the way to creating tragedy. These included the combination and alternation of actors speaking iambics with the *aulos*-accompanied song-dance from a chorus, which for comedy numbered twenty-four. They also had to put on masks and costumes. The hallmark of the comic mask, male and female, was that it was ugly—all its features were distorted from the ideals, which were typical of the tragic mask. The body was uglified as well: actors wore gross padding in front and behind, and male characters wore a large dangling phallus made of leather. It is much less likely that this has something to do with ritual origins (it is not erect) than that it was a gender marker in keeping with comedy's vigorous exaggeration of the unpretty features of humanity.

It was also of the essence that comedy fully became comedy when it was incor-porated in the City Dionysia, and became part of the civic calendar. Five poets (at times reduced to three) competed each year. Prizes and the cost of the actors—not

Wierd choruses, forerunners of comedy. These choruses, with their musicians, are on the two sides of a cup painted in Athens about 500 BC. This is before any record of the official recognition of comedy, but seems to reflect some features which became characteristic.

the same men as for tragedy—were at public expense. Another *choregos* was found to pay for the chorus. Their performance was less complex than that of their tragic counterparts, and there are symmetrical sections of song-and-dance that may well have been largely routine. But there were more of them, and they often had weird and unpredictable roles that would call for special costumes, like Clouds or Centaurs. And sometimes there were twenty-four differentiated costumes, e.g. of wild birds or islands.

Comedy does not, however, seem to have made much impact in its first fifty years or so at Athens. Little is recorded of those times and few fragments survive. The genre began to make more of a mark towards 430 BC, and a share of the credit should go to the long-lived Cratinus, who died before 420. It is a particular shame that his *Pytine* (*Wine-Flask*) does not survive, as it seems it took to extremes the inventiveness of self-reference (metatheatre) that is characteristic of Athenian comedy. Cratinus had a reputation as a heavy drinker, and in this play he apparently made himself a character, married to Comedy herself. She, however, kicks him out of the house as long as he carries on an affair with a tart called the Greek equivalent of 'Tipsy'.

There was also a common kind of comedy that mounted incongruously unheroic burlesques of well-known myths. Not only would the story be ludicrously twisted (no real deaths in comedy, of course), there would also be political allegory detectable. Thus Pericles was evidently the target of Cratinus' *Dionysalexandros*, probably performed in 429 BC. In this Dionysus disguised himself as Paris (Alexandros) so that he could get his hands on Helen. When he got into trouble, he changed himself into a ram—the mysterious god turned into a pantomime-animal!

The only comedian of this period whose works survive is Aristophanes, and we have eleven, about a quarter of his output. He was under 20 when he composed his first comedy in 427 BC (someone else produced, or 'taught', for him in his early years); he was precocious, too clever by half according to old Cratinus, who coined the word 'euripidaristophanizer'. In fact he came from a completely different generation from the grand tragedians, and lived to see the rapid revival of Athens after 400, dying in the 380s.

Scene from a comedy (with words). Painted about 400 BC, well within the lifetime of Aristophanes, this evidently represents a particular scene, because the characters have words coming from their mouths (somewhat like cartoon bubbles). It may well be that the old woman on the stage is persecuting the old man (centre) for killing her goose.

While tragedy captures the attention of its audience through a wide variety of responses, comedy's main route is always the same—laughter. Yet it is hard to make generalizations which are true of all of Aristophanes' plays, other than they are all amusing, robust, fantastical, and more sophisticated than they may seem at first sight. *Acharnians* of 425, the earliest comedy to survive, already displays many of the leading features of the set of family resemblances which are spread among Aristophanes' plays. For a start, it grows out of the immediate political situation. The war with Sparta had been dominating life for six years, and had gone far from well for Athens—how would it ever end? The chorus consists of fiery old men from the town of Acharnae in the foothills not far from Athens; and we know independently from Thucydides that they were particularly hard hit by the war. There is also, as usually, a robust individualist who is not prepared to sit around doing nothing—in this case called Dicaeopolis ('Justicity'). And, as often, he has a fantastical idea, or rather two: to make a private peace treaty with Sparta, and to argue against objectors, like the Acharnians, that the Spartans were not wholly to blame for the war.

Another of Aristophanes' favourite ploys is to take some image or conceit and literalize it. Thus the Greek for 'peace treaty' (*sponde*) is the word for 'wine-offering' since the offering was poured to ratify the treaty: Dicaeopolis is offered flagons with peace-vintages of five, ten, or thirty years. Similarly, in Euripides' tragedy, Telephus had said that he was prepared to have his head chopped off if he did not speak justly: Dicaeopolis brings out a real chopping-block to make his big speech to the Acharnians. The whole comedy is full of various allusions to tragedy, especially Euripides' *Telephus* (see p.29 above). Dicaeopolis even calls on Euripides to borrow Telephus' rags and other convincing paraphernalia from his plays. The tragic material offers entertainment by parody and deflation, yet it also helps to supply a level of seriousness. When Dicaeopolis makes his big claim for the possible seriousness of comedy he coins the word trugedy—*trugoidia* (literally 'grape-press-dance'?)—for comedy: 'trugedy also', he claims, 'knows what is just'.

The chorus of comedy contributes in a variety of less orthodox and more assertive ways than the chorus of tragedy. In the middle of the play there is one (occasionally two) particularly substantial set piece, with elaborately symmetrical song-and-dance routines, known as the *parabasis* (also as 'the anapaests'). This is the chorus's display piece: they often stand up for their particular role; promote their playwright and indulge in literary polemic; attack current political targets; advise, and sometimes abuse, the audience. The *parabasis* of *Acharnians*—getting on for 10 per cent of the play—moves from commending the advantages of a playwright who tells the truth and gives just advice, to an invocation of the fiery Muse of Acharnae, to complaints about young career politicians abusing good old stalwarts of Athens.

The last third of an Aristophanic comedy is often more like a series of revue sketches. In *Acharnians* Dicaeopolis sets up a solo market and gets various desirable goods from traders who are still enemies of the rest of Athens; he also sees off a series of undesirables who try to denounce his private world of peace or to take advantage of it. In contrast to Lamachus the soldier, who represents the warmongers and who ends up wounded and humiliated, Dicaeopolis ends up with a girl on either arm as he celebrates victory in a drinking contest. In a sense it is a triumph of selfishness; but Dicaeopolis is also politically representative as well as outrageously individual.

It is precisely because Athenian comedy and tragedy had so much in common that they were also able to define themselves by mutual exclusion. The most basic antithesis is that tragedy tends towards the universal, while comedy tends towards the particular. Also comedy is set in contemporary Athens and names proper names, while tragedy is set in the past in such a way that its politics do not allude directly to topical debates or individuals. Tragedy exploits but does not explicitly acknowledge the fact that it is a performance in the theatre: comedy continually and blatantly plays with its own theatrical fiction—as when, to give just one example, Dicaeopolis explains his disguise to Euripides: 'The spectators must know who I am, while the chorus-men just stand there dumb-founded . . .'

It is all part of the same polarization that it is much easier for a theatre historian to make big statements about tragedy than about comedy. Comedy's constant deflection from the literal and sustained towards a clutter of detail and fantasy make it resistant to sweeping generalizations. So, rather than attempt further characterization of Aristophanic comedy as a whole, I shall pick out a couple of the theatrical turns in his career.

In 411 BC, with little or no precedent, it seems, Aristophanes suddenly put on two plays dominated by women (one at the Dionysia, one at a lesser festival called the Lenaea). In *Lysistrata* it is actually a woman who has the great idea which is taken through to triumphant, fantastical success. Lysistrata, like Dicaeopolis, is determined to make peace despite the warmongers—but she enlists the help of the women of all Greece. They conspire to refuse sex to their husbands; and they occupy the state treasury on the Acropolis. Absurd though all this is—and unrealistically monogamous—Lysistrata herself is presented relatively seriously; and the notion that women might have bright and constructive ideas is not finally negated.

Women at the Thesmophoria is very different and is more concerned with drama than politics. As with Socrates in *Clouds* (423), it centres round a real individual, in this case Euripides. He has the big idea of infiltrating the women-only festival of the Thesmophoria to defend himself against their attacks, but gets a buffoonish old relative to do the dirty work for him. His sex is soon exposed (in every sense), and he resorts to a series of ideas from Euripides' tragedies in an attempt to escape. For example, he takes a baby hostage, as in *Telephus*—but it turns out

Scene from Aristophanes' *Thesmophoriazousae.* This is clearly an iconographic parody of the famous hostage-scene from *Telephus* (see illustration p. 30); it is also clearly an illustration of a scene from Aristophanes. Euripides' relative, disguised as a woman, finds that the baby he has seized is in fact a wineskin!

to be a wineskin; and he tries out the role of Helen from the previous year's tragedy (see p.31 above) in the hope that Menelaus/ Euripides will rescue him. Euripides is actually treated quite affectionately in the play, though his happy-ending escape plays are given a rough ride. In the *parabasis* the male performers of the chorus, representing the women at the festival, stand up for women and challenge the men of the audience: 'why marry us and be jealous over us, if we're so terrible as you make out?'

Two of Aristophanes' plays date from the post-404 era. They are not so different as they are often made out to be, and, while there is less topical politics and less invective than in most early comedies, they still recognizably belong to the same family. In *Women going to the Ecclesia* (?392), the women of Athens, led by Praxagora, dress up as men to pack the democratic assembly (*ecclesia*), where they vote that all properties, including sexually desirable young men, should be equally available to everyone. *Wealth* (388) is both less public and more allegorical. Chremylus finds the god Wealth, who has been blinded by Zeus in anger so that prosperity will not be distributed according to deserts. Once he has cured Wealth's sight, he is pestered by a series of characters who want to share the benefits; and he ends up taking the god to live on the Acropolis. The only feature of these plays which is conspicuously unlike those from pre-404 is the treatment of the chorus. While they are quite active at the beginning and end of *Women going to the Ecclesia*, there is no *parabasis* and several of their songs are simply indicated as 'choral song here' (*chorou*), with no words supplied. And this is done throughout *Wealth*. So, for whatever reasons, the chorus was losing its place in comedy as a major, unpredictable, and assertive feature. There still always was a chorus in comedy, but the playwright would merely leave slots in the structure for others to fill with songs.

The same tendency affected tragedy, and this is just one of several ways in which tragedy and comedy grew together in the fourth century rather than hardening their polarity. Aristotle (writing in perhaps the 330s) complains of tragedians merely inserting interchangeable 'interludes'; and some fragments of text have been found with the instruction '*chorou*'. It was not a simple transition, however, since *Rhesus*, probably our one and only complete fourth-century tragedy, is an action-filled play, set on a battlefield at night where the chorus are military guards who actively keep watch and even pursue enemy intruders. So far as we know, performances of tragedy in Greek theatres *always* had choruses (possibly reduced in number); and at Athens the role of the *choregos* remained important through the fourth century.

It is a problem that our two main sources for this next period are both negative. Aristotle grumbles that artists and audiences have become vulgarized; and Plato thoroughly disapproves of the wide accessibility of theatre. Yet the Dionysia undoubtedly remained a big event at Athens, and the new tragedies remained the highlight of the festival. It would be intriguing to see a work by Chaeremon or Astydamas or one of the other celebrated later authors, but none survives.

Plato and Aristotle both complain even more about the cult of actors. They were already important in the fifth century, but in the next several rose to be superstars, and even powerful diplomats as well. They acted in new plays, but increasingly their fame was based on a repertoire of the great fifth-century classics. The Athenian Dionysia officially introduced the performance of 'old' tragedies in 386 BC. This probably gave the crowning sanction of the 'metropolis' to a practice that was already widespread, both in the small local theatres of Attica and elsewhere in Greece. For, although there is little detailed evidence, it is clear that the performance of Athenian theatre outside Athens began in the fifth century, and then spread widely throughout the Greek world. The companies of travelling players were presumably Athenian at first; but it is interesting that many of the famous actors of the fourth century were not Athenian.

There is some evidence of actors putting up temporary wooden stages. A few permanent theatres, outside Attica, go back to the fifth century, most notably Syracuse and Corinth, and during the next century more and more cities built one. With time these became increasingly grand; the most celebrated architecturally, at the remote sanctuary of Asclepius near Epidaurus, dates from about 340 BC (the *orchestra* was definitely round by then). And King Philip of Macedon set a precedent by being assassinated in the local theatre in 336 BC!

Tragedy also became an ever-increasing influence on the visual arts. The best evidence to survive from the fourth century is pottery painted in the Greek cities of south Italy, especially the huge funerary urns painted in Apulia in the mid-century. Many of these reflect tragedies, though they do not—unfortunately for us—directly illustrate the performances. It is interesting that Euripides is clearly the most popular tragedian, though by no means the only one known—there is, for instance, a painting inspired by a Medea-tragedy which is quite different from the Euripides version. We also have paintings of actors holding their masks, and terracotta figurines of actors and of masks. No doubt if more survived in metalwork, painting, and other art-forms, we would have much more evidence. Certainly the influence of tragedy is prominent in the wall-paintings of Pompeii and Herculaneum, buried by Vesuvius in AD 79.

Theatre was, then, already big business before the 280s, when, for some reason,

The spread of Greek theatres by the third century BC (only those with archaeological confirmation are included).

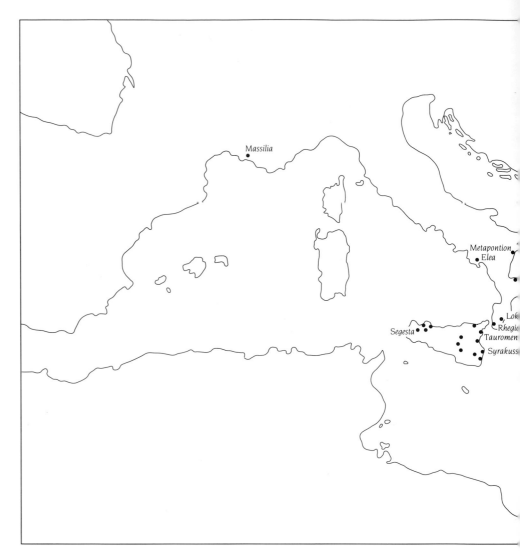

the performers organized themselves into professional guilds, known as the Artists of Dionysus. The most important were based at Athens, Isthmus (by Corinth), Thebes, Teos (in Asia Minor), and at Alexandria (in Egypt). These guilds employed large numbers of performers and other personnel, and worked with huge budgets. They travelled with special immunity to perform at a great variety of festivals, and at the commission of the various potentates who had become a feature of the greatly expanded Greek world after Alexander (died 323). By the time the Romans took all this over, theatre was, then, a highly prestigious and organized and expensive activity. It continued in this style for several more centuries, though there seems to have been little significant new writing, in Greek at least. Generally the Artists played a canon of great 'classics'. So the comfortable,

complacent old age of Greek tragedy lasted far longer than its vitally creative period in fifth-century Athens.

A final illustration, however, of the pervasiveness of tragic theatre is provided by a peculiar play of which sections happen to have survived. It is called *Exagoge* (which is the same as *Exodus*) and is by a Jew called Ezechiel. He probably lived in Alexandria, which had a huge Jewish community, probably in the second century BC; and he certainly made use of the Septuagint, the Greek translation of the Hebrew Old Testament. Here, for the edification of Jews and Gentiles in Egypt, we have a chorus of the daughters of Raguel, a messenger-speech telling of the inundation of Pharaoh's troops, and the voice of God from the burning bush addressing Moses in Attic iambics in the style of Euripides.

Actor with masks. This was painted in south Italy in about 375 BC. It looks as though this is an actor (or playwright?) choosing between a pair of masks of a young woman.

Comic scene with a 'tragic' Aegisthus. This vase, painted in south Italy in the early fourth century, was first published in 1992. In some ways it is typical of such comic scenes, except that one character, labelled 'Aegisthus', has the appearance of a figure from a tragedy-related painting, and that two characters have the label *choregos* (rich sponsor).

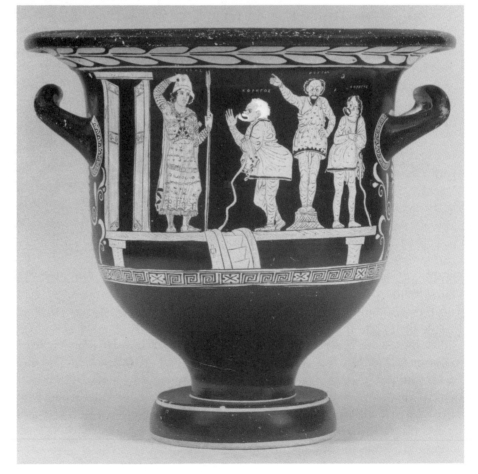

The story of comedy was left at the death in the 380s of Aristophanes, leading light of the so-called Old Comedy. It is three-quarters of a century before we have any more plays, and they are all by the leading light of the so-called New Comedy, Menander. All we have from in between is hundreds of quotations and titles. During that time comedy—unlike tragedy—went through a great change, so great that the name of the competition at the Dionysia is the most evident continuity. To judge from the fragments, the changes were not steady or continuous, and may not have seemed so obvious at the time.

This impression is borne out by the excavation of 200 or more comic vases painted in south Italy, especially Apulia and Paestum, in the period 400 to 320. These entertaining pictures, often known as '*phlyax* vases', have conventionally been associated with a local tradition of comic plays, but it has now become clear that at least some are closely related to Athenian comedy, including one direct illustration of Aristophanes' *Women at the Thesmophoria* (p.38) and another which looks like the beginning of *Frogs*. Unlike the vases associated with tragedy, these openly flaunt their theatricality, and often include a stage as well as comic masks and costumes. Their playful variety include quite a few scenes of mythological burlesque, scenes with metatheatrical allusions, parodies of tragedy, and, in one case, lines issuing from the characters' mouths. The most Aristophanic in spirit and content nearly all date from the first third of the century. As time goes on domestic scenes predominate and fantastical elements fade out. This reflects the mainstream shift from Old to New Comedy.

Aristophanes had already begun dislodging the chorus including the *parabasis*. It seems that after his death there were still sometimes novel and participating choruses, but by the time of Menander there was always simply the instruction '*chorou*'. This went hand-in-hand with a structural neatening to five fairly equal acts simply divided by four interlude songs. Gradually the fantastical, bizarre, and allegorical plots were dropped along with the burlesques of myths. All comedies came, instead, to be set in the 'real world', and they played out the earth-bound confusions among the households of the (usually) Athenian bourgeoisie. The topicalities of politics were phased out—and, indeed, towards the end of the fourth century freedom of expression was limited under Macedonian pressure. Instead, the foibles and intrigues and quandaries of New Comedy, though often involving legal technicalities, were relatively time-free. The plots almost always involve a love story which ends, after problems, with boy-marries-girl. But as often as not this plot is secondary to the mistakes and machinations of angry old men, besotted old men, ingenious slaves, stupid slaves, slaves in a hurry, blimpish soldiers, unscrupulous tarts, tarts with hearts of gold, pimps, uncouth rustics, flamboyant caterers, pseudo-scientific doctors—in fact the cast-list of comic types which feeds down to Ben Jonson, Molière, and beyond.

For better or for worse, the exuberant, gross, metatheatrical, and wildly

Statuettes of comic actors. Terracotta statuettes of this kind were mass-produced throughout the Greek world (these were made in Athens in the fourth century). They are often found in 'sets', and it may be that they were meant to be grouped to reconstruct particular celebrated scenes.

unpredictable comedy of Aristophanes became sober and tasteful and domesticated. All that sex and obscenity were excluded, the costume-padding was reduced, the phallus gradually made less conspicuous until it was concealed completely. A complicated and important 'cast-list' of masks was developed. While the masks of slaves remained ugly and grotesque, those of the young lovers, at the other extreme, were not very unlike the tragic masks (though the tragic mask became more distorted as time went on). This may point to one reason why tragedy did not develop more: comedy took over much of its territory, especially the exploration of family conflict—at least in some of its less horrendous, more sentimental aspects, such as recognitions, lost babies, suspected infidelity, issues of adoption, proving worth, forgiveness.

Several comic playwrights achieved great success in this period, and, though they gravitated to Athens, not all were Athenians like Menander—Alexis came, for instance, from Thurii in south Italy and Diphilus from Sinope on the Black Sea. Menander's work, like all the others', did not survive from antiquity, and was, until this century, known only through fragments and Latin translation-adaptations (see p.50 below). He was, however, very widely read in Greco-Roman Egypt, and excavations of papyri have brought to light one virtually complete play, *The Dyspeptic* (*Dyscolus*), substantial parts of five others, and bits of yet more.

Menander was immensely popular. As well as all the papyri, many mosaics, paintings, and terracotta figurines portraying scenes from his plays have been excavated from all over the Greek world—from Bulgaria to Syria to Libya—and spread across a span of 600 years. Modern taste has found it hard to see just what they found so great in him. It is evident that his plots are neatly constructed, and his dialogue is not only elegant but able to convey subtly more than is actually said. There are scenes of suspense and of pathos—but nothing to change the world, nothing to confront beyond the familiar. Perhaps the problem is that the territory of Menandrian comedy has been so thoroughly taken over since the eighteenth century by the domestic novel that it lacks the unfamiliar freshness of Old Comedy (let alone Tragedy).

Samia (*The Woman from Samos*) must have been one of Menander's best plays—we have had most of it since the publication of a papyrus in 1969. Its plot is a good illustration of Menander's skills. Demeas is a wealthy Athenian and 'good' man. His two most important relationships are with his adopted son Moschion, and with Chrysis, a beautiful young woman from Samos who is his 'live-in lover' (Athenians could not actually marry non-Athenians). The love interest of *Samia* is that Moschion has fallen for Plangon, the daughter of their poor neighbour Niceratus—and has got her pregnant (typically she appears only at the end, and then does not speak). The play will, of course, end with their wedding: the unusual twist is that everyone is in favour of it right from Act I. The two fathers have been several months away on business, and have meanwhile agreed to the match, provided the couple approve—it is typical of Demeas' magnanimity that her poverty is no barrier. But in their absence two babies have been born: Chrysis' baby has died, but Plangon's is fine. Moschion is too bashful to tell Demeas; and the womenfolk of Niceratus' family are terrified of his return to find an illegitimate baby—he is evidently a proud and hot-blooded character. Chrysis offers the solution: she will treat the baby as her own and persuade Demeas to keep it. By the end of Act II this has all worked well, and the wedding is to be this very day.

While bustling about the wedding arrangements indoors (like Old Capulet) Demeas sees Chrysis feeding the baby (which seems to prove it is hers), but he also overhears that Moschion is the father. He does not simply fly into a rage; he checks with a slave that Moschion really is the father, and then decides that the young man is not to blame, that he has been seduced by his beautiful Chrysis, his 'Helen'. Before the eyes of the astonished wedding-caterer he shoves her and the baby out into the street. He really loved her and is too pained to explain in so many words, so she has no idea what has happened. The failure of communication is beautifully conveyed.

45

Soon Niceratus finds Chrysis crying in the street and takes her into *his* house, where she is, of course, already well known and liked. Next it is Niceratus' turn to get angry, and he explodes in a much more violent and indiscriminate style. First, when Demeas tells his version, he is furious with Moschion and Chrysis; then he finds his own daughter breast-feeding the baby; as her mother protects her, and so does Chrysis, he will kill the baby to find the truth. Again Chrysis comes hurriedly out of doors with the baby, this time hotly pursued by Niceratus. But by now Moschion has confessed the truth to Demeas (if only he had done so earlier!), and so now Demeas rescues Chrysis by hurrying her back into *his* house, the very door that he had kicked her out from shortly before. Demeas calms Niceratus down, breaks the news about their shared pre-legitimate grandchild: the wedding is safely back on, and it is the end of Act IV.

It is a favourite device of Menander's to reach the happy ending early, and then

FROM MENANDER'S *WOMAN FROM SAMOS*

Demeas ejects Chrysis, the Samian mistress he loves, on a false suspicion of infidelity. She is still protecting the baby although it is not hers. The Caterer witnesses everything with wry astonishment.

DEMEAS. I'll put a stop to your . . .
CHRYSIS. . . . to my what?
DEMEAS. Nothing. You've got the baby and a nurse.
  Now . . . get the hell out.
CHRYSIS. Just because I kept this child?
DEMEAS. Because of that and . . .
CHRYSIS. . . . and what?
DEMEAS. Because of that.
CATERER. (Ah, it's that sort of trouble)
DEMEAS. You didn't know how to be well off.
CHRYSIS. Didn't know? What are you talking about?
DEMEAS. You came here in a cheap linen frock, Chrysis—
  see what I mean?—cheap.
CHRYSIS. Well?
DEMEAS. When you were poor, I was everything to you.
CHRYSIS. And who is now?
DEMEAS. Don't talk to me. I'll let you have all your own things
  and I'll give you a maid, Chrysis. Just get out of my house.
CATERER. (The whole thing's some sort of display of anger.)

46

bring in a new twist of trouble before the final closure. In a display of a third sort of anger Moschion now works himself up into a fit of indignation that he was ever even suspected of such a betrayal of his father: he will go off and be a mercenary soldier instead of getting married. Again Demeas is big enough to overcome the problem: 'Moschion, I love you for this anger . . .'

It is worth recalling that by Menander's day the schools of philosophy were becoming Athens' chief claim to fame (as they would be for the next 700 years). There is arguably a philosophical dimension to *Samia*, reflecting Aristotle's ideas on emotion. On one level the play is a dramatization of different ways that *anger* can be aroused, responded to, acted upon, soothed, aggravated, and extinguished. It is also a study in different kinds of love and of failures of communication. It also shows the possibility of true relationships across the barriers set up by the world: father and adopted son, man and mistress, woman and a friend's baby, Athenian and Samian, rich and poor. It may be this—the demonstration that heartfelt

Menander with masks. The poet has the standard 'portraiture' of Menander in this marble relief. He holds the mask of a young man, while those of a woman and an old man (slave?) rest on the table.

An Italian actor? This actor, painted in Campania in the late fourth century BC, has a relatively dignified costume. The label says SANTIA, which is probably the Oscan version of the Greek slave-name Xanthias, common in comedy. (Oscan was an Italian language similar to Latin.)

relationships between basically good folk are more important than worldly encumbrances—that gave Menander his lasting and wide appeal.

Like tragedy a 100 years earlier, comedy, after the golden age of Menander, settled into a long, complacent old age. Three short stories from southern Italy will, finally, point to some of the roads taken and not taken. Rhinthon worked in Tarentum (Taras) in about 300 BC and wrote a kind of drama known as 'tragic *phlyakes*' or as 'hilarity-tragedy'. He may have had no chorus, he may have included epic-type hexameters; his plays were definitely in the local Doric dialect instead of the proper Attic, and they defied the canonical categorization into either tragedy or comedy—they were neither and both. They probably also incorporated elements of local folk-theatre. Rhinthon's lack of significant successors in Tarentum or elsewhere accentuates the fact that local theatrical traditions did not ever grow enough to rival seriously the Athenian plays which were purveyed by the Artists of Dionysus.

On the other hand there is a late fourth-century jug in the British Museum from Nola in Campania with an actor wearing a Greek-style comic mask and costume, except that his outfit is rather smart and there is no visible phallus. His stage name, written in the local Italian alphabet, suggests that by 300 BC Greek comedies were already being adapted into Oscan-language versions.

In the course of the third century BC a Greek from Tarentum called Andronikos (Romanized as Livius Andronicus see pp. 49-50 below) produced the first-ever translations of Greek epic, tragedy, and comedy into an up-till-then rather minor Italian language spoken around Rome, viz. Latin. Generally speaking, Rome did not, or could not, develop its own culture of artistic and intellectual life, but adapted the achievements of Greeks, already so impressive and widespread before the time that Rome became a place of any special significance. So in most respects Roman theatre was an offshoot of the contemporary Greek theatre. In Europe it was to be the later vernacular theatres that were able to develop the old Greco-Roman tragedies and comedies into something which became both their descendants and independent of them.

2

THEATRE IN ROMAN AND CHRISTIAN EUROPE

DAVID WILES

FOR the Romans, 'theatre' was always a Greek activity. The first play written in Latin was probably a translation rather than in any sense an original work. Its author came from a Greek colony in southern Italy, and his name, Livius Andronicus (in Greek: Andronikos), implies that he enjoyed the patronage of the Roman Livius family. Andronicus was commissioned to write his play for the 'Roman Games', an annual event which required some new ingredient in 240 BC in order to celebrate the end of the first great war against Carthage. There had of course been dramatic activity before this date in the form of dancing, jesting, and improvisation, but not 'theatre' as we now, thanks to the Greeks, understand the term.

Theatre was thus an element of a Greek festival transferred to a Roman festival. While Rome had military strength and a will to expand, Greece retained its preeminence in the fields of art, literature, and philosophy. The Greek world had no dynamic political centre such as the city of Rome, but because Greek-speaking colonies were dotted all around the western Mediterranean the Greek language was an international lingua franca. Images created by Greek artists were exported to all corners of the Mediterranean world. For the Romans to have invented an alternative culture would have been ridiculous, not to say impossible. If they wanted to aggrandize their small republic, and assert its moral right to control the Greek world, then they had to appropriate what already existed, cultural activity as conceived and defined by Greeks. They needed moral and religious legitimacy if their armies were to succeed in controlling the sprawling Greek world, and political strategy therefore required a rigorous cultural strategy. The Romans imported hundreds of statues of Greek gods and equated those gods with their own; and, in a similar way, they imported, translated, and adapted hundreds of Greek texts in order to make the Roman theatre part of Greek achievement. Early playwrights like Andronicus, Naevius, and Ennius were not just employed to write plays: they also wrote epic poems to glorify the city of Rome.

Although a genre of plays with Roman settings did eventually emerge, most plays, including all that have survived, were notionally set in Greece. The actors wore Greek costume and Greek-style masks. A double-edged image of the Greek world was presented in the theatre: on the one hand, an ideal to which Romans should aspire, and on the other a negative image which defined the superiority of the Roman way of life. The aesthetic code of Greek dramatists such as Menander had been based on the idea that art should imitate reality, but the new Roman aesthetic was that art should imitate Greece.

There was an explosion of theatrical performance after Andronicus' production of 240 BC. Plays were performed as part of annual festivals, victory games, funeral games, and temple dedications, and these events were often given a second time when ritual details were incorrectly performed. Actors and writers had now to be accommodated as a new component of Roman society. For political reasons, they could not be allowed to live in the pockets of particular aristocratic patrons like the Livius family; so, in honour of the ageing Livius Andronicus, a guild of actors and writers was established in a temple of Minerva in 206 BC. The Roman name for Dionysus was 'Liber'—'freedom'—and his orgiastic cult was hard to reconcile with Roman traditions of order and hierarchy. The patron of the Roman theatre, at least officially, was the respectable goddess of crafts. It is interesting to note that Andronicus, like Plautus a generation later, was an actor as well as a writer.

The work survives of only two Roman dramatists from the period of the Republic. We have twenty and a half plays by Plautus (who wrote or co-wrote many more) dating from c.210–184 BC, and six plays by Terence (his entire output) dating from 166–160 BC. Ostensibly, all these plays were adaptations of Greek originals. Terence, the later of the two writers, kept much closer to his Greek models, adhering to the realist aesthetic of his sources, while eliminating choral interludes and soliloquies, and skilfully rolling two plays into a single narrative. Plautus took a more cavalier attitude to his sources, perhaps because he was not of Greek extraction, and drew on other popular performance traditions such as the Atellan farce performed in the region of Naples. Although he was happy to translate passages when it suited him, he radically restructured and subverted his Greek models. The world created on his stage was a utopia that could be Athens at one moment and Rome the next. The visual references, from costumes to silver coins, are all Greek, but the language is Roman, and so is the terminology used to evoke the gods, law, and the political system. This stage world made a mockery of Greece, but it also made a mockery of Rome.

There is no such thing as a typical Roman play. However, perhaps the most distinctive feature of Roman comedy is the figure of the trickster slave, and Plautus' *Pseudolus* is a fine treatment of that motif. The play was performed in April in 191 BC, a month which actually fell in midwinter thanks to the state of the Roman calendar. Rome was flush with success for Hannibal had been driven from Italy, victories had been won in Spain, and Philip of Macedon had recently been defeated. There is a striking contrast between the mood of *Pseudolus* and that of Plautus' only other dated play, the *Stichus* of nine years earlier. That earlier play was preoccupied with urban poverty and the return of long-absent relatives in the immediate aftermath of the war against Hannibal. *Pseudolus* is written as a celebration of victory, and in Plautus' dramatic fiction the victory belongs to a downtrodden slave. The 'spring' of 191 saw the Romans mount an unprecedented sequence of festivals. These were not pure acts of celebration, for the aristocracy wished to rouse a war-weary population and make them agree to another big military push. This time the aim was to drive Syrian Antiochus out of Greece and western Asia Minor, and so to establish absolute Roman dominance in the Greek-speaking world.

The precise context of the performance of *Pseudolus* was a festival called to celebrate the completion of a temple to Cybele, the 'Great Mother'. This Greek cult was associated with Mount Ida, the mountain next to Troy, and Troy was the home of Aeneas, mythical ancestor of the Roman people. Thirteen years earlier, the Romans had shipped to Rome a sacred stone that represented the Great Mother, together with attendant priests, eunuchs, and musicians. The importation of this cult was a typical instance of Roman cultural colonialism, as the Romans took for their own another slice of Greek culture. Having acquired the cult and its trappings, the Romans proved typically ambivalent, for, finding the Greek eunuchs and shamanistic dances hard to accept, they forbade Roman citizens to participate in the goddess's processions. *Pseudolus* was performed in front of the temple which housed the image of Cybele, a pseudo-Greek play in honour of a Graeco-Asiatic goddess.

Viewed in retrospect, Plautus seems to stand at the start of a secular comic tradition, yet all Roman plays were in fact given in honour of the gods. Livy tells us that stage entertainments of a quasi-dramatic kind, borrowed from the Etruscans, were first introduced at a Roman festival in order to bring divine protection from the plague. A play by Plautus was a Greek rite, likewise enacted to please the gods and give protection. It is hard to separate out reasons of *realpolitik* from reasons of piety that impelled the Romans to bring an ancient stone and a group of dancing eunuchs from Asia Minor to Rome. Whatever those reasons were, the same reasons impelled them to perform *Pseudolus* in honour of the Great Mother. The Romans saw morality as a matter for state regulation, for their gods were not moral beings but a source of danger and protection. Plautus' plays, in

Pompeian fresco evoking the grandiose stage facade of a temporary Roman theatre. The figures are not in stage costume. In the central doorway, Iphigenia emerges from the temple of Diana. The statue of Diana is visible behind her. To her right is her captor, and to her left her rescuers.

their joyous amorality, may be understood as instruments of protection. One who laughed at slaves, or senators, or foreign mercenaries, or male cowardice, might subsequently feel protected from those dangers.

Early Roman theatres were temporary wooden auditoriums, erected by wealthy politicians for the duration of a festival, and archaeology offers few clues about their shape. We do know that seating arrangements became a matter of intense controversy three years before *Pseudolus*, when the senators claimed an exclusive right to the seats at the front. In the dark days of the war against Carthage festivals were occasions for a display of solidarity and the social classes were free to mix, but now the nobility felt that they could assert their position. Another symbol of wartime solidarity had been the regulation of female clothing. In contrast to Greece, women in Plautus' Rome came to the theatre alongside their husbands. Four years before *Pseudolus*, aristocratic women mounted a demonstration to regain their right to wear what they chose in public. The regulation of seating would have helped both to protect and to draw the eyes of the poor towards women who put their wealth on display. Disputes about privilege within the audience do much to explain the obsessive concern of Plautus' plays with status.

The law of Plautus' stage world is that the underdog triumphs and status barriers crumble. The penniless youth finds money for a prostitute, the hungry man shares in the feast, the slave is rewarded rather than crucified, the rich nobleman

pays for all. In the pseudo-Greek world of the dramatic fiction, Athenian egalit-
arianism triumphs over Roman hierarchy. The plays of Menander were love-
stories, but in Plautus the theme is power. The hero seeks possession of a girl, but
not necessarily marriage. In many plays, wives economically empowered by their
dowries struggle against their husbands. At the centre of *Pseudolus* is a power-
struggle between father and son, a sensitive subject in a society which denied a
man financial and legal independence until his father's death. The dominant role
of the slave in the play must reflect a widespread fear of rebellion as slave gangs
were imported to work the soil, replacing Roman peasants conscripted into the
army. It is probable that most of the actors of *Pseudolus* were slaves, working

PLAUTUS' AUDIENCE

This prologue written for a performance of Plautus' *The Little Carthaginian* gives a
vivid picture of conditions in the auditorium. In contrast to the Greek theatre, admission was free, and women were not barred or segregated.

ACTOR AS PROLOGUE. Herald, on your feet and hold the ear of the people. It is
time to see if you are up to your duties. [*The Herald's voice proves to be inaudible.
Perhaps he is drunk, or choking on food.*] Exert the voice that is your daily bread.
Bellow! Or starve in silence. [*The Herald remains inaudible. Prologue applauds.*]
Back to your seat, I'm going to double your wages. [*The 'wages' prove to be lashes
of the whip. Prologue turns to the audience.*] You'll do well to obey my orders. No
seller of his body may sit on the stage, if older than the appropriate age. The lictor
will keep his words and his whips to himself. No usher is to block the view, or
show someone their seat while an actor acts. Leisured latecomers who lingered
too long, learn how to stand. Or how to get out of bed sooner. No slave may sidle
in: keep seats free for the free. Or pay up and purchase freedom. Failing that,
home boy! Avoid a twofold tragedy, a whipping here, and a flogging there when
master gets back and finds no work done. Nursing females, see to your tiddlers
at home, and do not bring them to the play, otherwise your milk will run dry and
the little lad will starve, or be so parched he'll bleat like a billy-goat. Married
women: watch in silence, laugh in silence. Here your mellifluous nattering is
prohibited. Take your prattle home again, don't harass your husbands at home
AND here too. As concerns the festival organizers: let no performer win the crown
through corruption. Lobbying should not prejudice the issue, so the worst man
wins the day. And by the way, I almost forgot: while the show is on, valets are
free to smash their way into the bakery. Now's your chance, the cakes are still hot,
run for it. These orders are given upon the authority of the actors. You will do
well, yes indeed, if all of you bear them in mind.

alongside one or two free lead actors, and securing their own place of honour in Roman society.

As we should expect in a heavily militarized state, the central metaphor of the play is a military one. Pseudolus presents himself as a Roman general laying siege to a brothel, and he dreams of being granted an official 'triumph' in recognition of his victories. His overt object, the unofficial object of all Roman generals and common legionaries, is loot. Pseudolus also pictures himself as the wily Ulysses, and the idea that he is recreating the siege of Troy acquires a special resonance because of the Trojan associations of Cybele's festival.

The Greek linkage between Dionysus and theatre assumed that the spectator underwent a change of mental state. The Roman actor by contrast kept drawing attention to the performer beneath the role, and to the overtly fictive nature of the stage world. In *Pseudolus*, the most striking physical characteristic of the slave Pseudolus is the large size of his feet. Since the name 'Plautus' means 'flat-foot', it is tempting to imagine that Plautus created the figure of Pseudolus as a symbol of himself. The plot tells how Pseudolus' young master wants money to buy the prostitute he loves, and how the slave provides him with his route to happiness. Pseudolus tells the audience that he will be no mere entertainer but a 'poet', and he promises to provide his own dramatic festival. He fulfils this promise by dressing up a fellow slave to play an improvised role for the benefit of the villainous pimp. Thus is Pseudolus gradually metamorphosed into Plautus, the purveyor of pleasure through deft improvisations. Plautus' text is laced with comments reminding the audience that a fiction is being constructed for its benefit. When the plot requires Pseudolus to impart information to another character that is already known to the audience, the slave breaks the fiction to say: 'These spectators are the reason this play is being acted—and they know, because they were here at the time. So I'll explain all later.'

Plautus' plays never hide the fact that they are actually part of a Roman festival. They are concerned with immediate sexual and digestive gratification, not with eternal love, and their climax is some kind of off-stage banquet or orgy. In *Pseudolus*, a sense of festival is built up in three layers. First we hear that the notional time setting is the Greek festival of Dionysus. Then we find that it is the pimp's birthday party: a bevy of his prostitutes flaunt themselves before the audience, and a cook proffers a menu guaranteed to restore youthful virility. The third layer is provided by Pseudolus, who constructs his own dramatic festival, and performs an erotic drunken dance to celebrate victory. The play moves from the fictive world of Greece to the real festive world of the spectators, who are invited to return for another show next day.

The cast of Plautus' plays comprises stock types adapted from the Greek repertory. We happen to know that the pimp in *Pseudolus* was played Greek-style with a bald-headed mask because the role was later taken by Roscius, the most famous of all Roman actors. The text refers to the pimp's goatee beard and crab-like movements. Besides big feet, Pseudolus has red hair and a pot belly in the manner of a Greek stage slave. All stage slave masks of the period were given grinning mouths which resembled a megaphone. The clever slave seems to have been the *sine qua non* of a Plautine play, and other mask-types were selected from the repertory as required. The most important of these are: *the handsome youth*: cowardly and concerned only with sex; *the father*: usually very rich, and often presented as a senator; *the mother*: wielding the power of her dowry; *the prostitute*: a positive figure in Plautus, just as the pimp is always negative; *the braggart soldier*: the antithesis of Roman military values, and thus helping to define those values; *the parasite*: in Plautus, a symbol both of the poor citizen bound in a client relationship to a rich man, and of the dramatist who earns his keep by entertaining. Roman dramatists like Plautus brought these mask-types together in varied combinations in order to build up the action of their plays.

The style of Roman acting is an elusive thing to reconstruct, for we can never tell from pictures whether scenes are being performed in Latin or in Greek. A consideration of it must begin with the fact that most plays were designed for musical accompaniment. Pseudolus typically flouts Greek illusionism by calling attention to the stage piper. He demands a pause to allow him to think out a plot,

Scene from a Graeco-Roman comedy. A rich man wants to punish his drunken son, but is restrained by a friend. A slave supports the son. The musician is a boy, miming. Painted scenery has been curtained off. Beneath the masks, the actors' mouths are visible.

abandons the stage, and orders the hapless piper to entertain the audience during the hiatus. Only four scenes in the play (slightly over one-third of the text) are designed to be spoken. Such scenes offer a pause for breath in the manic rush of the play, and are often concerned with filling in the plot. Shifts of mode help Plautus to play games with reality. The impostor who pretends to be the stooge speaks, but the real stooge who is mistaken for an impostor sings. The shift makes it easy for the audience to accept the pimp's confusion about where reality lies. The operatic nature of Plautine drama explains why the great Roscius, a century later, liked to take the part of the pimp, and not the longer part of Pseudolus. The pimp's opening scene is an aria of ever-increasing complexity. The male slaves are whipped and the female slaves humiliated in a display of physical and vocal extravagance. The virtuosity of the pimp's role is entirely lost in an English prose translation.

The transient nature of the building in which *Pseudolus* was performed helped to signal that the utopian world of the play was no mirror of normal life. Plays were never the only form of entertainment on offer in these temporary theatres. Half a century later, at the same festival of Cybele, a play by Terence was disrupted because boxers and a tightrope walker were due to appear, and the audience (or rather, the competing audiences) became over-excited. Festivals that began with plays usually ended with chariot-racing—a larger-scale event that more people could watch. Temporary theatres were economically inefficient, and the senators who insisted that the wooden theatre should be demolished at the end of each festival did so for political reasons. Theatres in the later Greek world doubled as places of political assembly, where the population in their thousands could sit on seats and listen to lengthy democratic debates. The Roman Senate preferred the electorate to gather in groups, say yes or no, and disperse quickly. The theatre in Rome did nevertheless become *de facto* a political arena. Plays, gladiator fights, and chariot races provided a unique opportunity for the masses to confront politicians and, from the safety of the crowd, to make their feelings known.

Cicero's description of a play performed at the festival of Apollo in 59 BC illustrates the political sensitivity of Roman theatre at the end of the Republican period. The actor Diphilus (the name is Greek) gave emphasis to certain lines in the tragedy such as 'our misery has made you *great*'. Prompted by a claque of knights and senators, the crowd took these lines to refer to the temporarily unpopular figure of Pompey the Great. The lines were applauded, and encores were given. When Pompey's political ally Caesar entered the theatre, a hostile silence ensued, but when a prominent opponent of Pompey arrived many rose to their feet and cheered. Two years later Cicero himself was the focus of applause when he made his return from exile. A tragedian pitched his lines and moulded his text in order to link an arson attack upon Cicero's house to the cruel burning of Troy.

Theatre in Rome had become a substitute for the democratic forms of assembly that were normal in the Greek world.

Cicero is not the only eye-witness to have left us an account of a Roman performance. The Greek historian Polybius describes a travesty of a Greek play in 167 BC, when games were given by Anicius to celebrate his victories in the Balkans. Anicius' victories were overshadowed by those of Aemilius Paullus in Macedonia. Paullus was a philhellenist, two plays by Terence were later performed at his funeral, and we may assume that Greek-style plays of similar sort formed part of his victory celebrations. He took his role as a sponsor of entertainments seriously, and Polybius records his motto: 'a man who is good at organizing games is good at running an army.' Anicius plainly decided to compete for popularity by setting up a ludicrous travesty of the kind of festival favoured by his rival. He hired the finest Greek pipers, choral dancers, and tragedians, and broke with convention by setting up a stage inside the Circus, the arena designed for hunting and chariot-racing. After the performance had begun, he subverted the event by making the dancers engage in an improvised fight. The pipers charged each other and created musical mayhem, egged on by an enthusiastic crowd. All degenerated into fisti-cuffs as some professional boxers joined the fray. Polybius declares that no one would believe him if he described the behaviour of the Greek actors. The tale illustrates the ambivalent nature of Roman attitudes to theatre.

A Sicilian chronicler describes for us a performance which took place in an 'allied' city on the western side of Italy on the eve of a major uprising. The 'allies' were indignant at their client status, and were demanding the privileges of Roman citizenship. Many Romans were in the audience, and when an actor in a comedy gave vent to anti-Roman sentiment, a group of these climbed on to the stage and killed him. The atmosphere became tense, and members of the local population plotted to kill the next performer, a great favourite of the Romans called Saunio (or 'Zany')—described as a 'satyr-like' figure. The chronicler pays tribute to the comic skills of this actor which allowed him to quench the anger of his would-be assassins. The actor evoked his servile life as an itinerant, likening himself humorously to a swallow that can build her nest anywhere. Again we see theatre functioning as a political arena. We are re-

Tragic actor in his dressing-room. The actor has finished his performance and his mistress (or Muse) is dedicating his mask to the gods. His hair has been tousled by the mask. In the background another actor warms up for the next performance.

minded of the diversity of dramatic modes, and of how little we know about non-literary dramatic traditions.

The first stone auditorium in the city of Rome was inaugurated in 55 BC. The Senate continued to oppose the building of a permanent theatre, but in this last phase of the Republic the Senate was less powerful than the 'triumvirate' of Pompey, Caesar, and Crassus. Having obtained his wealth, reputation, and power-base through campaigns in Spain and the east, Pompey's objective, in defiance of Republican tradition, was to install himself permanently in the capital. So, to boost his prestige and attract political support, he spent six years building a complex which included a stone theatre, a park with fountains and topiary, an art gallery, a modest house for himself, and a new Senate House dominated by a statue of himself. Statues represented the fourteen nations conquered by Pompey, and probably formed part of the stage façade. Religion was not neglected, for Pompey's theatre was constructed as a kind of temple. The seats were conceived as steps leading up to a temple of Venus the Conqueror at the head of the auditorium. Adjacent shrines above the seats were dedicated to Honour,

Roman theatre at Orange. Notice the beam-holes which once supported a roof, the few columns which remain of those which once covered the façade, and the statue of the Emperor above the central doorway. High-ranking officials could occupy the boxes above the entries on either side of the stage. Behind the stage façade was a colonnaded garden.

Virtue, and Success. Pompey's object, it seems, was to ensure that no one could condemn him for mere self-glorification, and that no one could in later years demolish his monument without committing an act of impiety against the goddess. Just as *Pseudolus* was performed under the eye of Cybele, so plays in the new theatre were performed before the gaze of Venus' statue. Vitruvius, town planner to Augustus, disapproved of Pompey's strategy, and urged that subsequent theatres be associated according to Greek tradition with the cults of Apollo (literature and song) and Dionysus/Liber (performance). A temple of Venus, he wrote, belongs outside the walls so that young men and married women may not be encouraged into thoughts of sex. The Christian Tertullian also lamented Pompey's dedication, seeing Dionysus and Venus as confederate devils of drunkenness and lust. Pompey's theatre, like those of Elizabethan London, was in fact built just outside the walls, adjacent to pleasure gardens, in an area free of traditional jurisdiction. The implication of the dedication was that, in the spirit of Plautus, theatre should be dedicated to pleasure.

The dedication of Venus' temple was marked by a series of grand events. The stage was huge (about 95 × 18 metres), and it was used for a festival of plays in three languages: Oscan (the dialect of the region around Naples), Greek, and Latin. The Latin plays included two dealing with the Trojan War. Cicero mentions 600 mules in *Clytaemnestra*, and their job was plainly to bear Agamemnon's victory spoils. He also tells of 3,000 bowls in *The Trojan Horse*, which again must have represented plunder from Troy. The significance of these tragedies is plain. Pompey's theatre, with its statues of the conquered and its dedication to Venus the Conqueror, celebrated Pompey's conquests. The objects displayed as stage properties must have been real spoils, carried back by Pompey from the east. There was a powerful emotional charge, as we have seen, to the idea that the ancestry of the Romans could be traced back to Troy. Although Cicero felt it unfortunate that the voice of the star actor gave way, and that drama was swamped by spectacle, Pompey's major miscalculation concerned the elephants who subsequently appeared in the Circus. The violence and cries of the elephants, unlike those of tragic heroes, provoked emotions of pity and fear in the audience, and Pompey ended up on the final day of the festival cast in the role of murderer.

Architecturally, Pompey combined an indigenous Italian practice of building auditoriums around the steps of temples with a Greek scheme allegedly based on the theatre of Mytilene, a city which he was proud to have liberated from Mithridates. Pompey's theatre, in common with later Roman theatres, differed fundamentally from the post-classical Greek theatre. It was a free-standing structure, not an excavation into a hillside, and accordingly we do not find in Roman plays any link to the soil such as we find in the *Oresteia*. It was an enclosed structure, for a roof covered the stage, and a high wall linked the massive stage façade to the auditorium, creating an architectural unity. The Roman aesthetic assumed an absolute divide between the festive world created inside the theatre and the every-day world of political order which prevailed outside. There was no separate space for dancing, and one vast stage accommodated both actors and chorus. The semi-circular orchestra was used to seat senators, who found political advantage in being a focus of attention. The stage façade was more decorative than in a Greek theatre. Early façades were faced with wood, and painted with *trompe-l'œil* deco-ration, while later façades were made of stone columns and niches. Probably from the beginning, there was no use of representational scenery. While plays of Menander commonly rely upon a suggestion of off-stage action, in Roman theatre the emphasis is upon visible spectacle.

Pompey's senatorial rival, who won all the applause in 59 BC, countered the stone theatre with a pair of semicircular wooden auditoriums that could rotate on their axes to form a single amphitheatre—but his efforts were in vain. The arrival of the stone theatre coincides with the demise of the Republic and the commence-ment of Empire. When Caesar achieved quasi-imperial power, he set about build-ing a theatre that would emulate Pompey's. Transient theatres had been bound up with the idea that power should rotate; permanent theatres were built by men who meant to hold power permanently. The old prejudice against stone theatres vanished under the Emperor Augustus, and enormous numbers of theatres were built across Europe and along the north African coast. Augustus attempted to reorganize the Roman audience in order to make the auditorium a microcosm of an ordered society, and to obliterate the crude divisions of rich and poor which had vitiated the Republic. Women, foreigners, and serving soldiers were all se-gregated, and male citizens were ordered to wear their togas.

From the period of the Empire, the play-texts of only one dramatist survive, and this dramatist cannot be taken to typify his age. Seneca was a philosopher, and served first as tutor then as minister to the Emperor Nero. Of the ten tragedies attributed to Seneca in the manuscripts, one and a half are probably by an imita-tor. The theatre historian faced with these texts is in something of a quandary. One theory has it that these are literary compositions, probably written during Seneca's years of exile in Corsica, and intended for dramatic readings rather than full performance. The second theory associates the plays with the reign of Nero, at a time when the latter became obsessed with his own skills as a theatrical performer. We have no information apart from the texts themselves, and a single crucial record of a debate between Seneca and a practising tragedian.

The plays do not conform to the conventions of surviving Greek tragedies. In the matter of violence, for example, we find that in Seneca's *Medea* Medea kills her sons on stage and spills her own blood on an altar at the climax of a scene of witchcraft. The plays are equally hard to reconcile with what is known about Nero's histrionic activities. Nero loved to put on a tragic mask, and play roles like Oedipus, the weeping Niobe, the castrated Atthis, and incestuous Canache. These

Tombstone commemorating funeral games. On the right is the dead aristocrat of the third century AD, with death-masks of his ancestors behind him. In the foreground from right to left are: a citharist; a poet; a tragic hero in raised buskins and mask, holding a club; a boy playing the water organ; a youth in the costume of comedy holding a marriage torch. In the background are an unmasked man (perhaps a mime) and an actor in a female mask.

performances seem to have been supported solo renderings of an operatic nature. Nero took them to the Olympic Games, and of course emerged victorious. Seneca's writing is more austere, and there is no sign that he is seeking to create a vehicle for a single star actor. In his *Medea*, music belongs only to the chorus, and to Medea herself in the incantation that culminates in her blood-letting. Otherwise the mode is speech.

It would seem that Seneca was pursuing his own aesthetic principles, in accordance with his Stoic philosophy. Stoicism sharpened his interest in the ideal of self-abnegation and in the dangers of emotion. His *Medea* is a study of uncontrolled anger, represented within a highly disciplined artistic framework. Seneca's precise references to Greek myth assume a far closer knowledge of literature than any surviving Greek tragedies, and suggest that his intended audience was an intellectual élite. We know from his letters that he had strong reservations about gladiator fights, pantomime, and other forms of mass entertainment. The Emperor and some of his nobles kept private companies of players, and dramatic performances were given in private houses. Seneca's plays should not be seen as typical Roman drama, but rather as a reaction against typical Roman drama.

Seneca's plays remain the product of a Roman imagination. The voyage of the Argonauts which brought Medea to Greece becomes a metaphor for the expansion of the Roman Empire. The plays belong to an imperial world where individuals could not choose their fate. Seneca's interest in Medea lies not in her choice to kill her children, and in what the Greeks would call her 'character', but rather in the elemental qualities of her anger. This lack of interest in character makes the plays hard for the modern reader or actor to approach. The Senecan chorus are likewise devoid of personality, and have become detached commentators. While Euripides' chorus generalize the plight of Medea so that hers is in some measure the plight of all women, Seneca's chorus isolate her so that she becomes the exceptional larger-than-life heroine. In a world where emperors like Nero called themselves gods, there could be no question of Olympian deities intervening in human affairs. Seneca declines to follow Euripides and introduce a spectacular dragon-drawn sun-chariot. Humans destroy each other through their emotions in Seneca's world, and there is no possibility of salvation. In the final shocking image of the play, Medea stands at a window and dashes the bodies of her sons down on to the stage beneath.

The disciplined structure of Seneca's plays reflects the discipline of his philosophy. He observes, like Menander, the rigorous unities of time, place, and action—which is to say that his plays are set in a single day, in a single location, and have a single plot. Such dramaturgy reflects élite artistic values rather than a Graeco-Roman norm. Seneca is a highly misleading guide to what late classical tragedy was like, and a better one is the Jewish/Alexandrian *Exodus* (see pp. 48, 49-50 above), a play which is epic rather than unified in its structure. The narrative here covered a long time-span, and the settings included the house of Moses' father-in-law, the burning bush, and the desert of Sinai. The Greek tradition was capable of developing in many different directions.

The mainstream dramatic tradition stemmed from the festival of Dionysus at Athens, and was mediated by texts. The play, according to this tradition, is an expression of the writer, and the actor is merely the vehicle of the writer. We are relatively well informed about author-centred classical drama because texts had the capacity to survive, to be read by later generations. There was a vast amount of actor-centred dramatic activity in the ancient world which we know little about because no one was concerned to keep records. We have the merest glimpse of performers such as Saunio, the man who had to clown to save his life. The Floralia, the spring flower festival, was a kind of fringe theatre festival, with erotic playlets performed by comediennes—but we have no means of reconstructing what the performances were like. In the period of the Empire, the emphasis shifted towards 'mime' and 'pantomime'. The term 'mime' did not necessarily imply silence, and was used to cover most forms of drama outside the traditional modes of comedy, tragedy, and satyr play. The term 'pantomime' implies that one actor played every part. Pantomime was perhaps a logical development from Greek tragedy, where three actors doubled all the parts between them.

Lucian, a Greek from Asia Minor, gives us our fullest account of pantomime. He mocks traditional tragedy for its grotesque appearance—body padding, the gaping mouth of the mask, and raised boots that impeded movement—and claims that the tragedian's skill is exclusively vocal. Far superior, he claims, is the visual art of the pantomime who dances to vocal accompaniment, and he supports his argument by recounting the tale of how a Cynic philosopher in the reign of Nero condemned pantomime for its reliance upon external pleasures such as costumes, masks, singing, the beat, and the pipes. A famous pantomime actor responded to the Cynic by giving a performance without music and sound. Adultery was the favourite theme of pantomime, and the story in this instance was taken from Homer. The smith-god Hephaestus, tipped off by the sun-god who sees everything, caught Aphrodite his wife and Ares her lover in the act of love, and entangled them in a net. The shame of Aphrodite and the fear of Ares were made so vivid that the Cynic philosopher responded: 'I hear your play, I do not merely see it; you seem to talk with your hands.'

From roughly the same period, we have part of the text of a Greek mime found in Egypt. The theme is again adultery, but the mode is domestic rather than mock-heroic, and the character-types are taken from comedy. This is the one dramatic text from antiquity which *may* have been written for a female performer, although it may equally well have been written for a man in drag. The text contains the words of the woman who plays the central role, supported by a number of non-speaking characters. The woman's speeches are punctuated by gaps which signify musical interludes when stage business is accomplished. The plot is as follows. The woman is in love with a male slave, but finding that this slave loves a slave girl she punishes both, and their supposed corpses are brought on stage. After embracing the body of her lover, the woman goes off to drown her sorrows in wine, and returns with another slave who is in a state of semi-undress—a detail which suggests the Aristophanic convention of the phallus. She then plots with the parasite (the mask-type of the glutton) to prepare a feast at which her husband will be poisoned. A third corpse is brought on, and, gloating to her accomplices, the woman embarks on a sarcastic funeral lament for her husband. Then, to her horror, one of the corpses starts coming to life—at which point our papyrus breaks off.

It is easy to see why plays of this kind aroused the antagonism of moralists. Mimes have enjoyed a bad historical press because we know about them through legislation aimed at crowd control, through biographers interested in homosexual affairs between mimes and emperors, and through the complaints of Christians. Christians found it hard to understand why mime actors made fun of their own pagan gods, and were outraged when Christianity came in for attack. We hear, for example, of a mime play which mimicked the

Pantomime actress. The actress holds the three masks with which she will perform. The mouths of the masks are closed because she will dance to vocal accompaniment. The props of sword, lyre, and sunhat suggest that the masks may belong to Ares, Apollo, and Aphrodite.

Entertainments in Byzantium. These plaques announce what the Consul will offer the people in 517 AD. Right: hunting in the amphitheatre— two lions and a leopard. Top left: women compete at horse-racing. Bottom left: four mimes and three tragedians. One of the mimes is a woman, one is a hunchback. The woman and robed man may be impersonating priests. The shaven heads of the male mimes are a kind of uniform. The three tragedians in masks, wigs, and raised boots may be representing a version of the Phaedra story, with Phaedra seated, Hippolytus approaching, and the nurse between as intermediary.

baptism of a dying man panicked about salvation. We need to read between the lines of our sources, and consider that a context of obscenity created a milieu in which some political freedom of speech remained possible. Actor-centred theatre could evade censorship in a way that text-centred theatre could not.

The centre of Graeco-Roman culture shifted in the later period of the Empire from Rome to Byzantium (now Istanbul), and it was here that the theatrical tradition retained its greatest strength. As late as the final decade of the seventh century, we find the Church declaring that it will ban:

dancing and mysteries performed by men and women according to an ancient custom alien to the Christian life. No man is to wear female dress, nor woman to wear what belongs to a male. No-one is to don the masks of comedy, the satyr play or tragedy. No-one is to shout the abominable name of 'Dionysus' while treading grapes in the press, nor celebrate the wine-pouring.

The ancient modes continued to thrive. From the point of view of the Church, theatre was not simply a source of obscenity and slander, it was part of an opposed religion. Since it could not be eliminated, the choice was either to restrain it or to appropriate it. The main means of restraint was the withdrawal of state funding for dramatic festivals. The effect of this was to drive drama from its traditional home in the theatre out into the Hippodrome, where hunts and races could be interspersed with performances by mimes. The change of venue helps to explain why open-air theatre in the Middle Ages was performed in amphitheatres rather than to audiences grouped in a semicircle. Appropriation proved more successful than restraint. Christianity overcame paganism by a process of absorption, turning pagan festivals into Christian festivals, pagan temples into churches, and pagan sanctuaries into cemeteries. The pleasures of theatre were introduced into church services, first through antiphonal music, then through formal dramatic enactments, and in a final phase through allowing certain holy days to become anarchic festivals of role reversal.

A remarkable example of appropriation is the undated Byzantine tragedy *Christos Paschon* (*The Passion of Christ*). The author has culled more than half his text from plays by Euripides, principally from the latter's play about Dionysus, *The Bacchae*. It is clear from spatial pointers not taken from Euripides like 'See, we have reached the house of Mary!', or, 'Look, the tomb is near!' or, 'This is the house on the right!' that the author is writing for performance. The Virgin Mary is the heroine of this Christian tragedy, and the action is presented from her point of view. She is, as theology requires, an intermediary between the spectator and divine truth. The first section of the play builds up suspense as Mary and the chorus of Galilean maidens hear from messengers about off-stage events: the betrayal of Christ and the nailing of his body to a cross. The action begins when Mary first glimpses, then approaches the cross. Although Christ dies in full view, the agony is hers rather than his. It is through her that the audience apprehends emotionally the idea of God becoming human. The audience sees the body removed from the cross, anointed, and placed in the sepulchre. It sees Mary and Mary Magdalene, after a long vigil, approach the tomb, where they encounter the resurrected saviour. Another messenger scene introduces an interlude with the soldiers and Pilate, confirming the fact of the resurrection. In the final scene, like a *deus ex machina*, Christ displays his wounds inside a locked house. Mary is now silent, but she remains the subject of the play because the locked house is a metaphor for her virgin body. The fact that Mary uses the words of Euripidean mothers like Agave, Medea, and Hecuba adds a dimension to her role, making her the universal sufferer. Where the playwright diverges from the stagecraft of Euripides is in his use of space. Although much of the action takes place in a neutral central area, specific locations are required for the cross, the sepulchre, and the houses of the Virgin, of Mary Cleophas, and of Pilate. The obvious setting is a church, where a permanent structure represented the sepulchre, and a cross stood above the altar.

When we pass from the classical world into the medieval, the writing of a continuous history becomes fraught with difficulty. In the classical world there were major cultural centres like Athens, Rome, and Byzantium, and major authors like Euripides and Menander, that influenced theatrical practice across Europe. In the Middle Ages the nation state was no more than a loose confederation of autonomous communities. Europe had no clear political and cultural centre, and no seminal playwrights or dramatic texts acquired classic status. Although one cannot in the medieval period identify dramatic traditions in relation to biographical continuities or national boundaries, one can at least identify and describe a homogeneous European tradition, tied together by Christian writings and by the shared language of Latin. In the classical world, architectural structures like the Theatre of Dionysus or the Theatre of Pompey determined the spatial conventions of dramatic performance, and a drama can be defined as a narrative enacted in a building called a 'theatre'. In the medieval world there were no purpose-built 'theatres' of a permanent or standardized kind, and it is therefore often difficult to determine what is and is not classifiable as medieval 'theatre'. Medieval records refer to *ludi*, the Latin term which embraces both plays and sports, and obliterates what for the modern world is a crucial distinction.

But perhaps the most serious problem for the historian of medieval theatre is the lack of scope for defining patterns of historical change. Theatre historians used to hold that medieval drama evolved progressively from the churches, via street performance and via secularization, towards the sophistication of the Renaissance—but this Darwinist model has now been discredited. The profusion of records now available makes it clear that the picture is far more complex. Quasi-theatrical performances were a constant part of church services from the ninth century through to the sixteenth. It is only in the twelfth century that the records allow us to form some kind of overview of the main dramatic modes, and there seems no case for maintaining that drama at that time was at a lower evolutionary stage than it was in the fifteenth. There is scope for writing local historical narratives in the medieval period, but not for tracing a single line through the drama of one nation, or of Europe as a whole. A safer course will be to identify the major genres of medieval theatre which we find flourishing in the twelfth century, and by this means give some idea of the sustained and continuing vigour of medieval drama.

The drama of the period was by no means exclusively Christian. There seems to have been an unbroken tradition of performing Roman comedy, as we glean, for instance, from a ninth-century dialogue (complete with stage directions) written for an aged Terence and a young player who thinks him out of date. This text was plainly written as the prologue to a performance of a play by Terence, defending the play against critics. From the very beginning of the twelfth century, a good example of the genre of pagan comedy inspired by the Roman tradition is *Pamphilus*. The playwright shows with acute psychological observation how a young man called Pamphilus succeeds in seducing a girl of higher social class. The young man takes advice from Venus, and bribes a bawd to act (like Juliet's nurse) as go-between, but in the end abandons self-control and rapes the girl. The piece was probably written for a solo performer working in the tradition of the ancient 'pantomime', for the scenes are all two-handers with infrequent changes of speaker and there is much scope for the speaker to use bodily expression. The rape is more easily performable if the victim alone is visible to the audience. A twelfth-century Frenchman described the play as sexually provocative, and performed through dance and gestures. By the end of the century it was considered suitable for the repertory of a solo entertainer in Provence and Aragon, alongside visually

Medieval reconstruction of Roman comic performance. From a raised booth, the poet (or his substitute) chants the text accompanied by pipers, while masked mummers dance the action. The amphitheatre is circular, and there is no boundary between actor and spectator. These conventions are all medieval. In the lower section the artist portrays the road which leads from text to performance.

challenging material such as the story of the Minotaur, the flying of Icarus, Solomon imprisoning the Devil, and Virgil hiding in a tub. An example of pagan comedy from England is *Babio*, a farce requiring several performers and written for the court of Henry II. The climax is a castration scene, described as an 'orgy of Bacchus', and it seems clear that the performers retained the Aristophanic convention of the phallus.

At one end of the pagan spectrum are scripted plays with courtly auspices, at the other improvised folk-drama. A typical account taken from the thirteenth century is that of a Scottish priest who participated in a local spring fertility rite. He celebrated the Easter resurrection by carrying a phallic emblem through his village on the end of a pole. Using obscene language and actions, he encouraged the maidens of the village to dance around the phallus in the churchyard. It is worth recalling in this context that in Britain medieval church altars were often built around ancient phallic stones that were thought too sacred to be destroyed. The medieval world was bi-cultural rather than monocultural, and Christianity had to keep renegotiating its position in relation to a suppressed pagan religion.

Most surviving medieval plays amplify stories drawn from the Bible. A representative twelfth-century example is *The Holy Resurrection*, almost certainly performed at Canterbury in around 1175. The play was written for open-air performance, and its audience was 'the people'. A political context is vividly evoked, with Joseph of Arimathea able to secure the body because of feudal ties with Pilate, while the penniless converted soldier finds himself in goal. It is relevant to this political context to note that the play was written in French, which in Canterbury was the language of the ruling class. The manuscript refers to a series of carefully placed structures. Some are visual representations of places like Hell, Heaven, and Christ's tomb, while others are merely platforms where groups of actors stand in view of the audience to await their moment. Structures representing 'Galilee' (perhaps a lake) and the inn at Emmaus (described as a 'castle') are placed in the middle of the acting area. Medieval staging allows no scope for reported off-stage action of the classical type used in *Christos Paschon*. God is not an unseen force as in the Byzantine play, but is visible on one side of the acting area. There was nothing that medieval dramatists felt themselves incapable of representing, when a story had to be told. The emphasis was not upon individual emotion, but upon narrative, and the narrative centre in this instance is not Mary but Joseph of Arimathea, probably because of the legend that brought Joseph to England with part of the cross. Plays of this kind were normally mounted by guilds of laymen working in close association with the Church, and the subject-matter suggests that the play would have been performed at Easter.

Theatrically simple but musically complex dramatizations of biblical stories were performed as part of church services at the appropriate moment on the calendar. Priests dressed as the three Marys, for example, would approach the empty tomb on Easter Sunday, and announce the resurrection. A German nun in the twelfth century accepted the value of church plays about Bethlehem, Herod, and the Magi, but complained of drunken orgies when priests dressed as soldiers. She refers to Innocents' Day, when the biblical story of the murder of the Innocents became an annual excuse for the anarchy of choirboys and the parodying of church rituals. Not even church services were free from pagan invasion.

Allegorical or 'morality' drama has attracted much critical attention because of its survival during the Reformation at a time when other modes were condemned as idolatrous. Over the period as a whole, allegorical techniques seem more significant than the 'morality play' as a genre. We shall see later, for example, how allegorical techniques shape the telling of the Noah story at Mons. The first surviving allegorical play is the *Order of Virtues*, dating from the late 1140s. Its author was Hildegard, a distinguished German abbess, visionary, and intellectual. The cast were predominantly nuns, and the place of performance was a convent church. God was represented by some kind of icon placed above the altar and suggesting both sun and bridegroom. On the steps below God's icon are assembled sixteen Virtues, together with their leader Humility, and these Virtues are conceived as clouds in front of the sun. Each is identified by an emblem: a lily for Shamefastness, a pillar for Patience, an ever-flowing stream for Faith, and so forth. The protagonist, called the Soul, begins the play sewing her white garment of immortality, but casts the garment off when she encounters the Devil. The Devil is the one character in the drama who speaks rather than sings, and it is likely that the part was written for Hildegard's secretary, the one necessary male retained in

Allegorical figures as conceived by Hildegard. This illumination illustrates a work by Hildegard closely linked to her play. Holiness has three heads like the Trinity. Strength tramples a devil. Righteousness stands on high. Wisdom holds a text. Twelve men (including four evangelists at the base) build a tower which symbolizes the Church.

a convent for purposes of administering communion. The title of the play refers to an elaborate dance performed by the Virtues as each presents herself and interprets her emblem. The climax comes when the Virtues descend into the body of the church and enchain the Devil so that Chastity can trample his head beneath her heel. Now physically grotesque and covered in painted wounds, the Soul mounts the steps, puts on a new costume called 'the armour of light', and ascends to Heaven. Although a performance by women, this was not exclusively a performance for women, for a few lines are given at the beginning to a group of astounded Patriarchs and we may assume that these marginal roles were played by men. The body of the church is shared by the audience and by the Devil, and at the end of the play the audience are asked to kneel so that the Father can reach them. Such techniques physically involve the audience in the spiritual drama enacted by the nuns.

A very different allegorical play from the twelfth century, which might equally be described as a historical play or a political play, is the *Play of Antichrist*, performed in Bavaria in the reign of the 'Holy Roman Emperor' Frederic Barbarossa. Stages for this play were set up around an open arena, creating a visual opposition of Church and State. A large stage and throne on the west side represented the Roman Empire, and a similar stage and throne on the east side represented the 'temple'. The layout was as shown below. The play dealt with relations between sacred and secular powers at a time when the Church was in schism. One Pope was supporting Barbarossa, while a rival Pope enjoyed the support of the French and English. The first part of the play is historical, and shows how Caesar conquers the French, the Greeks (Byzantium), the Jews (Jerusalem), and the Gentiles (Babylon). Having conquered his empire, Caesar places an actor who personifies the true Church inside the temple, relinquishes his crown, and retires to the stage representing Germany. Symbolically this marked the claim of Barbarossa and his predecessors to be the new 'Holy Roman Emperors'. The villain of the play, Antichrist, now appears, accompanied by a chorus of Hypocrites, and through a series of tricks, arguments, and bribes wins the support of Byzantium, France, and finally Germany. The heroine of the play, called 'Synagogue' (i.e. the Jews), is joined by two Old Testament prophets who strip off her blindfold, allowing her to see Antichrist for an impostor. After she has been martyred, a thunderclap from Heaven sends Antichrist scurrying away. The political message of the play is that the rival Pope opposed by Barbarossa is a version of Antichrist, and is responsible for the anti-

```
                    N.
               [? ANTICHRIST]


       FRANCE                    JEWRY

  W.   ROMAN EMPIRE         TEMPLE    E.

       GERMANY                   JERUSALEM


          BYZANTIUM    BABYLON
                   S.
```

Emblems of faith. On the tomb of the resurrection Brueghel places an allegorical character who personifies Faith, with the tablets of the law on her head and the gospel story in her hand. In biblical plays, stage objects had enormous emotional impact. Included here are the cross with the crown of thorns, Veronica's hand-kerchief, the lance and sponge, the pots of ointment brought by the three Marys to the tomb, and the scourge on top of the column to which Christ was tied.

Semitism of the crusading movement which caused huge disruption to urban economic infrastructures. The Jews are championed in the play as monotheists who first pointed the way to Christianity.

The Play of Antichrist uses a variety of theatrical techniques. Allegorical drama provides a repertory of visual symbols: the Church, for example, is accompanied by Mercy identified by her oil, and Justice identified by her scales and sword. But the play is also an inverted 'saint play' (Antichrist performs miracles like a saint, and Synagogue is martyred), and it has in addition a metatheatrical dimension, for Antichrist is an actor, and his miracles are false, theatrical miracles. The difficulty of telling truth from fiction and Christ from Antichrist is the crux of the drama. We see a characteristic medieval use of stage space, with the 'temple' being located in the east, the position of a church altar, while the transient political world is located in the direction of the sunset. France is set in a polar relationship to the throne of Jerusalem, established by a Frenchman after the First Crusade; and the Jews of the Diaspora are in a polar relationship to the throne of Barbarossa. Despite the break with classical dramaturgy, the action of the play shows how closely the medieval world still felt itself linked to the Roman world. Roman popes replaced Roman emperors and aspired to control the same territory, using the same Latin tongue. The play reminds us that the medieval Church was no monolithic entity, but a locus of conflict in which different groups competed for power. Medieval drama never offered its audience *the* Christian message; it offered an interpretation that was open to challenge and argument.

The final genre that remains to be considered is the saint play. Although plays dealing with the miracles and/or martyrdom of saints were quite as popular as biblical plays during the medieval period, all but two English saint plays have been lost because Protestant reformers found the genre more offensive than plays based upon Scripture. In Catholic France, by contrast, over a hundred have survived. The attraction of such a play was the local and specific nature of any given saint's cult. Members of a parish would celebrate the feast-day of their patron saint with worship, plays, and markets; cathedrals housing the relics of saints were keen to become regional centres of pilgrimage; each craft guild had its patron saint, and its chapel dedicated to that saint; kings sometimes tried to create national saints, like George of England. Saints, like Graeco-Roman gods, were essential tools in the building of group identities.

The earliest saint plays which have survived are those of Hroswitha, a German nun of the tenth century, and her six plays give us a unique glimpse into the theatre of that period. It would be wrong to think that as a 'nun' Hroswitha was on the margins of cultural activity. She had close links with the court of the Holy Roman Emperor, and could thus write with the confidence that her voice would be heard. As a 'canoness' she retained her personal wealth and freedom of mobility; marriage remained a possible option. The convent allowed aristocratic women independence in a way that marriage did not. Nor should we imagine that Hroswitha was isolated in a female environment, for her prefaces refer to male teachers and male patrons. Her plays are concerned with the role of women in the Church. One local battle which concerned her was lost soon after her death when the convent to which she belonged was forced against its will to become a satellite of a male-ruled monastery.

The Roman world and the medieval world meet in the plays of Hroswitha, and she must be a crucial link in any attempt at an overview of the history of the theatre. And yet, a series of historiographic biases have led historians of theatre consistently to marginalize her work. It may be useful to pause for a moment and consider what these biases within the academic community have been. The first is a simple preference for literature. The fact that Hroswitha's plays have no stage directions is held to demonstrate that her plays are literary compositions—despite the fact that no classical manuscripts and far from all medieval manuscripts contain stage directions. The second is a patriarchal bias which holds that women are not likely to write for performance, particularly if their plays seem to be full of humour and violence. A sign of male appropriation has been the renaming of five of the plays after male rather than female protagonists. Thirdly, a Darwinist view of history, which conceived of drama emerging slowly out of the ritual practices of the Dark Ages, had no space for a sophisticated playwright working in the tenth century. And finally, because she wrote in Latin, Hroswitha has been of more interest to scholars with a classical training than to students of medieval theatre practice. For rhetorical purposes, Hroswitha in her prefaces suggested that she was writing a Christian version of Terence, and critics have taken her remarks at face value, assuming that, if she were a competent dramatist, then she would have organized her plays like Terence around the three unities of time, place, and action. Yet, as we have seen, even the classical purist who wrote *Christos Paschon* wrote episodically for a series of locations. Unity had long vanished as an aesthetic ideal by the time Hroswitha wrote her plays. They are bold in their theatricality, for she is happy to stage a bonfire, a mountain, a battle, a dragon, a vision in heaven, and torture. Her most vivid images, even an on-stage mastectomy, are paralleled in later medieval drama. Her theatrical talent emerges as soon as we recognize that she is using medieval rather than classical stage techniques.

The Conversion of Thaïs the Prostitute, like all Hroswitha's plays, is written as a series of episodes. Between each there is a significant movement as characters travel from one acting area to another. The locations required for this play are two hermitages, an open area to represent a market-place with some means of presenting a bonfire, the house of the prostitute with an inner room, a convent, and a cell. The play begins with Paphnutius the hermit giving his ignorant disciples an academic lecture about music. The learned Paphnutius demonstrates in the course of the play that he understands none of the truths set out in his lecture: that body and soul should be in a relation of musical concord; that the music of the heavens is inaudible to humans; and that ignorance is sin only in those who have been accorded knowledge. He proclaims his moral outrage at the sin of Thaïs the prostitute, complaining that men impoverish themselves for her sake and break the peace, never considering that these men too may be at fault. In a plot motif that has echoes of Terence and lends itself to comic visual effect, Paphnutius resolves to dress as a young lover and go to seek her. When he finds Thaïs, he accompanies her to her chamber, and asks with obvious innuendo if she does not have a place that is yet more secret. She responds that her most secret place is known only to herself and God, implying that her soul is untouched by the prostitution of her body. She is plainly waiting to be rescued from her way of life, and her conversion is instantaneous. Paphnutius, however, later gives a highly tendentious account of

Scenes from Christ's Passion. As in a processional performance, the viewer of this crowded Dutch painting experiences a sequence of events happening simultaneously. There is a special emphasis upon female figures in the story because the painting was commissioned by a woman.

the conversion to his brother hermit. Amidst much talk of mercy and hell-fire, Paphnutius takes Thaïs to a convent, where she is locked up in a tiny cell that has a window but no door. After three years have passed, Paphnutius is disturbed when a vision from God reveals that a couch in heaven is destined not for a hermit but for Thaïs. Covered in excrement after three years without facilities, Thaïs dies while Paphnutius speaks of the resurrection of her body so it may join the snow-white sheep in heaven.

Hroswitha was an ironist, and her plays are full of savage humour. The critical tradition which renamed her play as *Paphnutius* took it at face value as a moralistic tract in which the Christ-figure Paphnutius converts a sinful woman. Hroswitha's feminism cannot be dissociated from her dramaturgical skill. The crucial scenic image of the play is that of the locked cell, a visual emblem not only of the chaste female body emitting the stench of corruption when entered, but also of Christ's tomb. While Christ is enclosed for three days, Thaïs is enclosed for three years before being resurrected from living death. The play leaves space for the patriarchal spectator to see Paphnutius as the authentic Christ-figure: his donning the clothes of a lover *could be* an analogue for the incarnation, and his entry into the brothel *could be* an analogue for Christ's descent into Hell. Hroswitha skilfully protects herself from attack, leaving the spectator free to interpret the play according to his or her preconceptions. There is a complex relationship between the pretentious philosophizing of Paphnutius and the buried philosophy of Hroswitha. The latent argument, supported by a discussion of music and presumably by the playing of music, is provocatively Neoplatonist. The beauty of body and voice which draws men to Thaïs betokens the beauty of her soul, not its corruption. The play leaves its audience with a philosophical enigma: if musical harmony is based upon difference, how are we to interpret the relationship of body and soul? As discord or as harmony?

In the tenth century, records are scarce. We simply have no idea how many other dramatists were writing plays of the quality of those of Hroswitha. When we move ahead to the fifteenth and sixteenth centuries, records finally become abundant. Across Britain, France, Catalonia, Switzerland, northern Italy, Germany, and the Low Countries, there seems to be a remarkably consistent picture. Drama is performed in every great household, and in every town. The medieval drama of England has been better documented than that of other European countries, thanks to the interest stimulated by Shakespeare, and to the industry of anglophone American and Canadian scholars; but it would be a mistake to conclude from the weight of scholarship that Britain has the richest tradition.

Late medieval drama can conveniently be divided into summer activity, performed out of doors and involving the participation of an entire community, and winter drama, performed indoors and based upon the household. This does not imply a sharp divide between public and private drama, for the head of a household, whether monarch, abbot, lord mayor, lord of the manor, or innkeeper, always remained in some degree a public figure. Plays continued to be performed as part of church services, and the church was potentially available as an indoor venue for community events of all kinds.

If we take the city of Coventry as a typical example, records of summer activity include the following: copious references to a cycle of biblical plays performed at Corpus Christi by members of the craft guilds; regular references to processions with giants at midsummer (a paratheatrical event organized by the craft guilds); a reference to maypoles erected on 1 May, presumably the focus of some kind of dramatic activity; occasional performances of saint plays; presentations of St George and the dragon before royal visitors, allowing us to infer that a folk-play of George and the dragon was linked to the annual St George's Day procession; a detailed account of a play (or mock battle) about Coventry people defeating the Danes, with women arresting the runaways, a play linked to the rites of Hock Tuesday (the second Tuesday after Easter), when women gathered money for the parish by seizing men and releasing them for a ransom. We build up our picture of medieval dramatic activity from records of this kind, and records can be highly selective. We know about the Corpus Christi plays because they were organized by the guilds, who kept account books and inventories. The Hock Tuesday play

COSTS OF MOUNTING A PLAY, 1490

Every year the Guild of Smiths at Coventry presented a play on a pageant wagon. The play dealt with the arrest and torture of Christ and with the dream of Pilate's wife foretelling that her son would die. The fees paid to the actors reflect the length of their parts, and there was a star role for the beadle, Pilate's comic servant. Joint payments reflect the way actors rehearsed in partnership.

This is the expense of the first rehearsal of our players in Easter week . . . 2s.2d.
Paid at the second rehearsal in Whitsun week in bread, ale and cooking . . . 2s.4d.
For another quart of [*wine at the*] hiring of Procula's [*Pilate's wife's*] gown . . . 2½d.
Spent at the repairing of the pageant and the expenses of having it in and forth . . . 14d.
Paid to the players for Corpus Christi day:

to God *Jesus*	2s.	to Peter and Malcus	16d.
to Caiaphas	3s. 4d.	to Annas	2s. 2d.
to Herod	3s. 4d.	to Pilate	4s.
to Pilate's wife	2s.	to Pilate's son	4d.
to the beadle	4s.	to another knight	2s.
to one of the knights	2s.	the minstrel	14d.
to the devil and to Judas	18d.		

Pilate's wife (played by an apprentice) hired a costume in order to wear the latest fashions. Hats are the basic symbols of rank. The torturers have different costumes for the buffeting, the scourging, and the procession with the cross, and the hammers and nails on their costumes are signs of the Smiths' craft. It is not clear why the torturers are unpaid: perhaps they belonged to an affiliated guild.

Memorandum that these be the garments that were new repaired against Corpus Christi day:

4 jackets of black buckram for the tormentors with nails and dice upon them
other 4 for the tormentors of another suit with damask flowers [*emblems of blood?*]
two of buckram with hammers crowned [*the torturers again: hammers for the nails, and a crown of thorns*]
two parti [*-coloured*] jackets of red and black
a cloak for Pilate
a gown for Pilate's son
a gown for the beadle
a hood for the beadle
two burlets [*hoods*]
a crest for Herod
a falchion [*curved Mohammedan sword*] for Herod

a hat for Pilate
a hat for Pilate's son
two mitres for the bishops [*i.e. Caiaphas and Annas*]
two hats for two princes
four hats for the tormentors
other two hats for the tormentors
a pole-axe for Pilate's son
a sceptre for Herod
a mace [*for the beadle*]
a sceptre for Pilate's son
four scourges and a pillar [*for the torture scene*]
two chevelures [*wigs*], gilt, for Jesus and Peter
the devil's head [*i.e. mask*]

The sum of all costs and workmanship draweth to 15s.

was not organized by any official body, and in 1575 it was a common stonemason with a taste for popular culture who led the players to the Queen, asking for the right to continue performing it in defiance of the wishes of the city authorities, the keepers of records. We know about the play only because of the Queen's momentary involvement. When we study the history of medieval drama, we have to consider whose history we are studying.

The official drama of the medieval city was often based around a procession in which the city put its social order on public display. Processions played a crucial part in the life of medieval cities, as they had done in the life of classical cities. We possess a vivid account of the Festival of Dionysus in Alexandria where actors processed through the city accompanied by wagons on which were displayed tableaux from the life of Dionysus. Christ, as so often, stepped into the shoes of Dionysus. Tableaux of biblical scenes displayed in a procession gave way in many medieval cities to short biblical plays. A graphic account of this mode of staging comes from Florence, where a visitor in 1454 described the midsummer procession. Groups with pageant wagons each performed a play before the city authorities when their section of the procession reached the central piazza. A separate processional group often preceded the entry of a wagon in order to introduce the play that was to follow. The arrivals listed by the visitor give an idea of the scale, the formal overall structure, and the actual heterogeneity and unpredictability of the components. The order was as follows: (1) *The cross* with choir boys. (2) Two groups of craftsmen process with *angels*. (3) A play: *Lucifer and his angels cast from Heaven*. (4) Two more groups with *angels*. (5) A play of *Adam and Eve*. (6) *Moses and Israelites* process on horseback. (7) A play of *Moses receiving the law*. (8) *Prophets* process. (9) A play of the *Annunciation*. (10) A play about the *Roman Emperor*. At this point proceedings were interrupted. A religious fanatic took exception to this play in which the pagan Emperor saw a vision of the Virgin and Child, and hurled the pagan idol into the square, pushing the player of the Emperor into the crowd. He then started to climb up to the roof of the wagon to reach the children who represented angels in the Emperor's vision. The crowd finally

Pageant wagon. An image of St Sebastian is drawn through the streets of Cuzco, Peru. The parish of St Sebastian was responsible for the wagon. An Inca *cacique* leads the way, and a priest follows with the host, the emblem of Corpus Christi Day, while spectators watch from the street and from windows. It was a small step to substitute human actors for such lifelike statues.

Scenes of torment and death. In the upper panels we see the torture and death of the son of God. In the lower panels, King Herod oversees the massacre of the innocents, and finds that his own son is accidentally killed. The flames of Hell await him. This non-biblical legend was popular with dramatists.

realized that these events were not part of the drama, and seized the intruder. It is clear that the Florentine crowd was prepared for innovation and the realistic simulation of violence, and fully expected that an actor might emerge from the audience if dramatically appropriate. (11) *Nativity play*. (12) *Herod* together with the *Purification of the Virgin*. The visitor supplies us with numerous details about this play. The pageant wagon represented an octagonal temple, of which seven sides were occupied by seven personified virtues and the eighth was occupied by the Virgin Mary. Mary was to the east, and Herod performed his part all around the temple. As in *The Play of Antichrist*, the audience are gathered on all sides, and the topographical orientation is of the first importance. (13) The *Three Kings*. We get an idea of the scale of the event from the fact that 200 horses accompanied the Kings. At this point the visitor expected, on the basis of a previous visit, a representation of the crucifixion, but in this particular year it was omitted. Next in sequence, (14) *soldiers* process towards the tomb. (15) A play of the *resurrection*. (16) The pageant wagon represents *limbo*, and Old Testament figures are rescued by Christ. (17) The next wagon represents *Heaven*, and the Old Testament figures take their place aloft. (18) Procession of *Apostles and Marys* in preparation for the next wagon used for (19) a play of the *Ascension* of Christ. (20) Procession of *the living*. (21) Wagon displaying *the living and the dead*. (22) A play of the *Last Judgement*, with representations of both Heaven and Hell.

The most ambitious medieval plays follow this model: a sequence of episodes focused upon the crucifixion and resurrection of Christ, commencing with the Creation and ending with a spectacular representation of the Last Judgement. The system of performing such plays on pageant wagons was especially popular in the north of England, and it will be helpful to look closely at one episode or 'pageant' from a play of this kind. At Newcastle upon Tyne stages were carried rather than propelled on wheels because of the steepness of the streets. The performance was given on

Corpus Christi Day, an appropriate day for celebrating the idea that the urban community was a single body, analogous to the body of Christ. The procession started at the city gates and moved towards the centre, performances being given *en route*, each by a different craft guild. The pageant of *Noah* was performed by the guild of shipwrights, and celebrated the place of that craft in the divine scheme.

For this *Noah* pageant we must visualize an upper level representing Heaven, a platform stage where Noah begins the play asleep, and an acting area at street level where Noah's wife and the Devil merge with the crowd of spectators. The crowd are not separated from the dramatic action, but are made to feel that they are the sinners whom the flood is destined to destroy. The central action of the play is the transformation of the stage into an ark. Deliberate comedy is made of the fact that Noah, unlike the actor playing his part, is no professional shipwright. An angel appears with the tools of the trade—pitch for sealing the seams, tar, nails, and rivets—and while Noah builds his ark, silently chopping and hammering, a parallel action is played at street level. The Devil supplies Noah's wife with a drug so that she may elicit from her thirsty husband what he is plotting. The experience of the audience is necessarily conditioned by the fact that other parts of the biblical narrative are being played simultaneously (and perhaps audibly) in other parts of the town. Noah's wife is a second Eve, supplying her husband not with the apple of knowledge but with a magic truth drug. The erecting of the mast is likened to the crucifixion. When Noah breaks secrecy, God makes his work audible, and the first sound of the shipbuilding craft which the audience hears is the evocative noise of two nails hammered into wood, a noise which anticipates the nailing of Christ's limbs. By such techniques, what appears to be an isolated Old Testament story becomes an integral part of a single unified theatrical event.

As an alternative to processional performance, circular theatres could be set up on open ground—a spatial conception taken over from Roman and Celtic amphitheatres, and reflecting the ancient idea of the cosmos whereby planets orbited about a central earth. The system of erecting miscellaneous booths around a central open acting area (or 'place'), known technically as 'place-and-scaffold' staging, allows the action to oscillate between tiny booth stages representing specific locations and a neutral central area. A French miniaturist has left us a vivid representation of a saint play performed in the round using this principle, and his painting is a crucial historical document. For technical reasons the artist has only depicted half of what should be understood as a closed arena. The viewer is looking from the north. On the left (east) is Heaven, on the right (west) is Hell, and in the centre (south) is the throne of the Roman emperor, who has descended from his booth via a ladder in order to direct the torture. The saint is bound like Christ to a piece of wood, and two of the four torturers pull at the rope just as, in plays of the crucifixion, they stretched the limbs of Christ to reach the nail holes

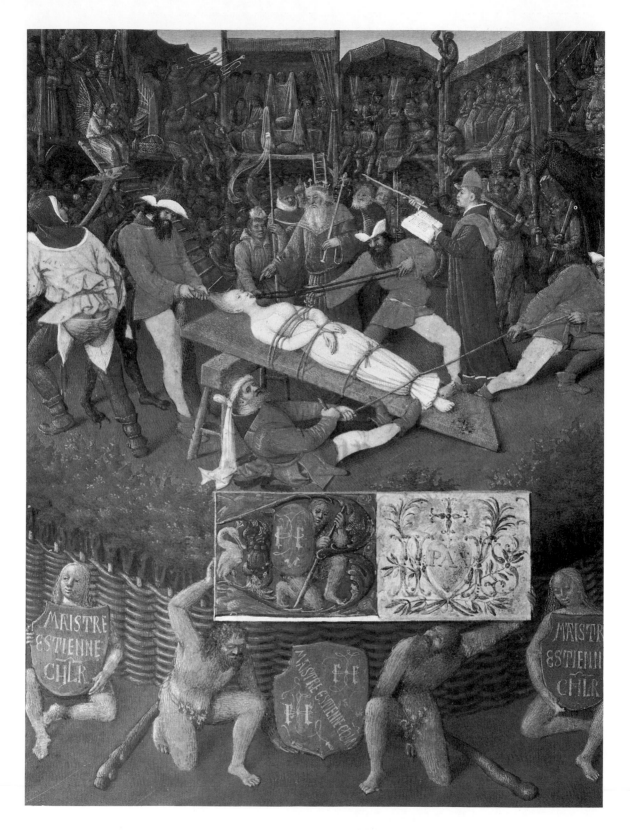

bored into the cross. The other two torturers pull the saint, in a literal sense, between Heaven and Hell. Her spiritual part is oriented eastwards and upwards towards Heaven, her feet and body are oriented westwards and downwards towards Hell-mouth. In the finale, we can be sure that her soul will ascend the broad gangplank up to Heaven. Next to Heaven are angelic musicians, competing with the beauty of their music against the threats and antics of the devils, and next to Hell are five female figures who perhaps represent the five senses assaulted by the torturers. At the centre of the booth between the 'Five Senses' and the emperor's throne are a lord and lady who seem to be privileged spectators. The deference shown by the lord towards his lady contrasts with the brutality of the emperor. From the point of view of common spectators, these nobles are also actors. The Roman emperor is dressed as a contemporary French monarch, so there is no historical distance between Rome and the world of the audience. The actors are not professionals but members of the community gathered to watch the play. The author or stage manager stands with book in hand to direct the large group of untrained actors, no more intrusive than a conductor in an orchestra or a referee in a sporting contest.

A three-day play from Penrhyn in southern Cornwall was performed in this manner, as we know from diagrams in the manuscript. The plan for the first day is shown opposite. The diagram omits crucial information about structures erected inside the central acting area, which included property trees, a river, two mountains, and a temple that had to be built in the course of the play. Like medieval painters, sculptors, and architects, medieval stage designers abhorred an empty space.

The Old Testament story on day one of the Penrhyn play tells of the tree that grew from seeds of Eve's apple to become the cross of the crucifixion, and again we can perceive through the Noah episode the artistry of the whole. The ark is a metaphor for the Church, and the building of the ark prefigures the climax of day one, the building of the temple in view of the audience. The ark is also an emblem of Christ's tomb, for the door is nailed up, and access is from above through the cloth roof. Noah thus reappears from the ark in the manner of Christ resurrected. When Noah fells trees to build the ark, the audience are reminded that it like the cross is made of wood. In the next

Facing: a saint play. Actors portray the martyrdom of St Apollonia, who became the patron of dentists. The artist, Jean Fouquet, has not shown the face of God, and has obscured the face of the Devil beneath the mask of folly. The author (or stage manager) gazes at the Devil and points his baton at the hidden Godhead. The fool's inanimate marotte gazes at God, while his anus suggests that through Hell-mouth the world excretes its waste.

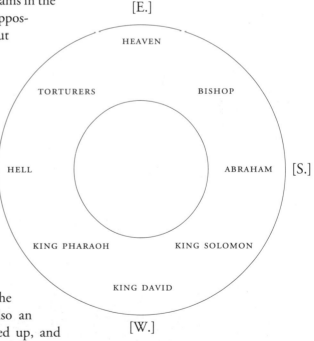

[E.]

HEAVEN

TORTURERS   BISHOP

[N.]   HELL   ABRAHAM   [S.]

KING PHARAOH   KING SOLOMON

KING DAVID

[W.]

episode Isaac probably uses the walls of the demolished ark as the firewood which he carries to the mountain in a simulation of Christ carrying the cross to Calvary. Such visual patterning gave the Cornish treatment of Noah its power. After the flood, Noah offers sacrifice on a raised part of the acting space identified as Calvary, scene of the crucifixion, and by this means links himself to the next saviour of humanity. Noah's wife sacrifices a cow, and his sons and their wives sacrifice birds—a dove first and a capon last. In the following scene Abraham sacrifices on the same altar a lamb in lieu of his son. The linkage of cow and lamb of God, of capon and dove of peace, made for comic effect, purposeful comedy that might make an audience think afresh about traditional Christian symbols. Noah, like Cain and Abel, interprets sacrifice as the paying of tithes. The people of Penrhyn paid tithes to a monastic institution nearby, and the author of the play was probably a canon from there, concerned to emphasize the sanctity of his income. The author is not, however, a spokesman for the Church at large, for he is savage in his portrayal of the 'bishop' (a crown appointment) who controls the temple of Jerusalem. The real local bishop was doubtless not a Cornishman. The bishop and king in the play reward their supporters with Cornish land around Penrhyn, a dramatic device which related the world of the Bible to political corruption in the present. The play was performed in the Cornish language, and became a cultural bastion against the encroachment of the English.

In certain areas, as in Cornwall, particular places were set aside by tradition as playing places. More often, the spaces of everyday life were theatricalized for the duration of a festival. Classical theatres, broadly speaking, were conceived as places where art inside the theatre could imitate reality outside the theatre. In the medieval world, true reality was other-worldly. God and the Devil were omnipresent spectators of human lives, and the actor/audience boundary was fluid because all humans were conceived as ultimately players. To portray reality was to portray the divine scheme, and one's own place within that scheme. By turning a market-place, the focus of communal everyday life, into a theatre, medieval makers of plays showed how everyday life concealed a deeper meaning. In Lucerne, for example, the administrative and commercial centre of the city became the centre of the cosmos for the two days of the Easter play, performed approximately every five years. As a sign of humility, the mayor and councillors sat on the right hand of the Devil.

The sketches drawn by the designer of the Lucerne play in 1583 are a useful guide to the way plays were staged in other market-places such as that of Mons in northern France. Here an *ad hoc* group in 1501 persuaded the council to set up a stage in part of the market-place. Seats were placed above and behind the actors, and many of the rich got an enhanced view by looking through windows. One of the builders of this stage was 'John the Turner', the carpenter cast in the role of Noah. The man who built the ark on stage was thus known to the audience as a

Above: the Valenciennes stage of 1547. An artist recreated these settings thirty years after the event. From left to right: Paradise (God surrounded by Peace, Mercy, Virtue, and Justice), multi-purpose room, Mary's home in Nazareth, garden, temple, gate of Jerusalem, palace (with prison beneath), bishops' palace, Golden Gate of Jerusalem (where Mary's parents met), Sea of Galilee, limbo with Old Testament figures, Hell-mouth and Wheel of Fortune. The settings are a miscellany from different parts of the twenty-five-day event, and the play was probably performed with an audience on four sides.

Below: the Valenciennes play: Nativity sequence. The fourth day of the play, evoked by the same artist. Joseph's bedchamber, the temple, and Herod's throne-room are theatrical settings, while the inn and stable are based on neo-classical artistic convention. Notice the moving star, suspended angels, and stage mountain required in performance. As in performance, the viewer sees constant movement between locations, some of which are near and some distant.

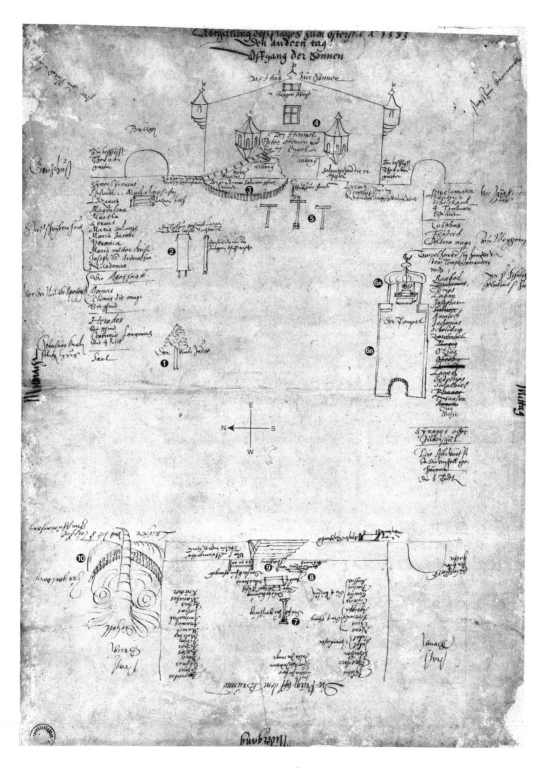

man who did similar work in Mons. Again the divine order and the order of every-day life coalesce. Theatre-in-the-round using multiple locations was able to present several actions simultaneously, and the Noah episode at Mons shows us how this technique worked. In Heaven God gave his orders to Noah, while in another space the latter's family waited with birds and animals. Meanwhile in Hell devils gloated, and on or in front of a stage belonging to 'The World' a character called Human Race dined on a shoulder of lamb alongside the seven deadly Sins in a parody of the Last Supper. As Noah entered the ark, Human Race danced in the 'place' with Lechery. The ark was positioned directly beneath Heaven on the east side of the square. A cut-out raincloud obscured God from view, and allowed rainwater to be dropped from four large wine vats to cool the ardour of Human Race and Lechery. Spectacle was not allowed to obscure symbolic meaning, for the water was interpreted as the water of baptism, blessing Noah's family as they entered the emblem of God's Church. Proximity to Heaven was used to further effect with the birds. A live dove flew from the ark tethered to a piece of string and was pulled back. It was then properly released, and an artificial dove with olive in its beak was slid down a string from Heaven. This action imitated part of the mass in French churches, where the host was raised inside a model dove and lowered on to the altar. Noah's dove thus became an emblem of Christ. A second live dove was released to complete the illusion.

Facing: the Easter play at Lucerne. 1: tree for Judas' suicide. 2: Last Supper. 3: Mount of Olives in garden. 4: Heaven. 5: crucifixion (scale misleading: actually more central). 6A: altar with pinnacle above. 6B: low wall enclosing nave of temple. 7: the scourging. 8: tomb for Lazarus' resurrection. 9: tomb of Christ; also his reappearance at Emmaus. 10: Hell-mouth shown with jaws closed. The lists give starting positions for actors. On the east the features of a building are used, and on the west a stage is built to cover the fountain.

The most famous version of *Noah* derives from Wakefield in Yorkshire, and makes an interesting contrast with the three that we have examined. Wakefield was a small town in the fifteenth century, under the jurisdiction of a lord of the manor, and it had no infrastructure or resources for mounting a large-scale event. The play of *Noah* seems to have been written in the first instance as a self-contained work, making few references to a wider dramatic and theological context. The building of the ark seems to rely upon mime and the audience's imagination. The focus is the farcical human relationship of old Noah and his harridan wife, and the mood is carnivalesque, not to say subversive. God has clearly made a mistake in saving as the best of humanity such a risible pair. He does not reappear at the end to promise a better future, and there is no final scene of sacrifice to represent the submissive paying of tithes.

Noah. On this twelfth-century capital Noah fells a tree, and his son (or an angel) carries on a structure of simple basketwork which represents the ark.

From summer plays in the streets we must turn to winter plays in the household, events that were necessarily small in their scale. For the scope and variety of such activity, we cannot do better than turn to the memoranda given by the Earl of

Northumberland to his steward, who was of course principally interested in the question of fees. The annual events on the Earl's schedule at the start of the sixteenth century were as follows: *1 November: All Saints' Day.* After the eighth lesson at Matins (Matthew 25: 1–13), choirboys wear veils and carry lamps in order to impersonate the five wise virgins journeying to Heaven. *6 December: St Nicholas' Day.* One of the choirboys is appointed 'Bairn Bishop' and takes responsibility for services in the chapel, which are conducted in a more or less parodic manner. Special robes and ornaments are kept for him, and, at some point over the Christmas holiday, a boy bishop from York is an honoured guest. *25 December: Christmas Morning.* The chaplain and choir mount a nativity play in the chapel. *25 December–6 January: the main festive season.* A Master of Revels is appointed to oversee plays, interludes, and disguises played before the Earl. The almoner (part of the chapel establishment) is expected to be a 'maker of interludes', writing plays for the four men who constitute his lordship's players, and a secretary is paid to copy out parts. The Earl's bearward provides entertainment with his lordship's beasts. *1 January: New Year's Morning.* The Earl's musicians play at the chamber doors of the Earl, his lady, and his sons as they awake to a new year. Later in the day the 'Abbot of Misrule', another mock dignitary, is rewarded for his services. *6 January–2 February:* The season for visiting players: players who belong to kinsmen or to an earl receive 20 shillings, players of mere lords 10

Salome dances in front of Herod. Short entertainments such as dances and plays were often presented between the courses of banquets. The execution of John the Baptist is depicted by the artist in a manner which allows us to see how the illusion of an execution could be simulated in a dramatic performance. Castles of wood and canvas were a common feature of indoor plays.

shillings. *2 February: Candlemas.* A groom of the vestry carries a candle into the chapel to represent the entry of the infant Christ into the temple. *Shrove Tuesday.* A play is performed at night by members of the chapel and other servants: a celebration, probably non-religious, to mark the entry of Lent. *Easter Morning.* A play of the resurrection is performed in the chapel. Actors impersonate the three Marys arriving with ointments at the empty tomb, and an angel asks them whom they seek. We see from the Earl's memoranda that he was accustomed to seeing both newly written plays and traditional plays that had taken an unchanged form for centuries. There was space in his household for piety, but also for mockery and farce.

The East Anglian *Mankind* of *c.*1470 is a vigorous example of the 'interlude'— the genre for which the Earl's almoner bore authorial responsibility. Seasonal details like an obscene Christmas carol, Shrovetide football, and winter sowing confirm that the play was designed for the winter festive season. The cast are as follows: a Christ-figure called Mercy, a peasant farmer called Mankind, a comic Devil-figure called Mischief, and three layabout retainers who represent the World. It is apparently Mischief who doubles as Titivillus, a masked and horned devil conjured up for the audience as protagonist of a play-within-a-play, and forever invisible to 'Mankind'. Mercy and Mischief compete for the loyalty of Mankind, who fights a losing battle against obdurate soil, illness, and poverty. He abandons farming and family life, and attempts suicide, before finally being rescued when he calls on God for 'Mercy'. The play seems to be written for performance in the dining-hall of a manor before an audience of 'sovereigns' who are seated and 'brothers' who stand, though the hall of an inn could equally have been used. We should imagine the lord, his family, and guests seated on high table at one end of the hall, while lower-class spectators sit at tables and stand at the sides of the hall, and the players use the floor in the centre. A dining-hall allowed the actors to merge in with the audience and then separate themselves as they pleased. The playwright assumed that the audience would know who the actors were, and plays on double identity when the actors/characters name prominent gentry of the neighbourhood and jest that they are going to rob their houses. When the actors pass round the collecting bag, they proclaim that this is payment to see the Devil. *Qua* entertainers, the actors are almost by definition servants of the Devil. Their play begins when a pious sermon is subverted, and when evil is banished at the end of the play theatricality vanishes too.

It is an important medieval paradox that goodness grows out of evil. Spectators were never taken to be objective and detached beings, viewing from a safe position of virtue. In *Mankind* the audience were positioned as confederates of the villains, and were seduced along with the protagonist into evil deeds and thoughts. We find the same paradox in the way the Earl of Northumberland celebrated his Christmas and Easter, watching pious plays in chapel on Christmas morning and

Man in a devil costume. A knight is required by his lady to put on the guise of evil in order to rescue his charger from the island upon which it has been imprisoned. The charger must in some way be associated with sexuality. While the knight with his sword is in stage costume, the devils who guard the island are not.

Easter morning, but also employing a Bairn Bishop and an Abbot of Misrule to conduct blasphemous revels. It was assumed that human beings could never be perfect, or hope to be, and therefore without an admixture of impiety there could be no acceptable enactment of piety. The annual round of dramatic and paradramatic activity allowed medieval people not merely to view the cycle of sin, repentance, and redemption but actually to live that cycle. Though they may have differed in their emphases, both upper and lower classes participated in the two polarities: sacred drama integrated with a church service, and carnivalesque revelry.

The problem with studying carnivalesque activity is the lack of texts. An annual battle between Carnival and Lent was a feature of Shrove Tuesday celebrations in France, Italy, Germany, Spain, and Sweden, and is familiar through Brueghel's famous painting. A fat man dressed as 'Carnival', after voracious eating and drinking, would lead his army into battle with 'Lent', played by a very thin man or woman. In Brueghel's painting Carnival rides on a hogshead of wine, and wears a meat pie as his crown, while his followers make 'rough music' by beating on pots. Whoever wins the battle, Carnival will necessarily die at the end of the day, and Lent will impose her austere rule. Two texts of such plays survive, but, being texts, are at a remove from an activity that celebrated freedom from constraints of every kind including textual. In the Italian text (published in Florence in 1554), Carnival and his cook die on a bonfire like martyrs in a saint play, uttering mock pieties like 'O fat Saint Capon, to thee I entrust myself . . . O marinated hares, from you I take last unction.' The play is thinly moralized at the end when the Devil takes Carnival's soul, and an Angel admonishes the audience to virtue. The fifteenth-century French text was written for a group of educated young men in a town on the Loire who formed themselves into a 'society of fools', and dubbed Carnival 'the glorious Saint Pensard' (from *panse* = 'paunch'). They used food for missiles in the fight, and the play ended not with a bonfire but with a peace treaty, as Carnival's army humbly and ceremoniously munched the dish of turnips offered them by Lent. The play was staged in the place-and-scaffold style, with God's place being usurped by Bacchus, who naturally views Carnival as his saint and martyr, bidding his three angels—Noah, Lot, and Architriclin ('ruler-of-the-dining-couch')—to ply Carnival's army with wine. The classical god of theatre and wine, Bacchus/Dionysus, here found his place within a Christian framework as an anti-God, an embodiment of all the joyous values that Christian piety opposed.

In England and Scotland, for reasons of climate, Shrovetide was not a focus for outdoor celebrations. A more popular figure than 'Carnival' in the late fifteenth and early sixteenth centuries was Robin Hood, whose reign began on May Day, and whose green costume symbolized spring vegetation. The example of the Robin Hood play allows us to see how folk-plays or folk-games flourished on the various levels of parish, manor, court, and city. Within parishes, Robin Hood plays were often organized as a means of raising funds. The church-wardens' accounts of Kingston upon Thames show us how the play changed from year to year according to financial necessity and spurts of creativity. On May Day four young unmarried men were appointed as Robin Hood, Little John, Friar Tuck, and Maid Marian. The play extended over the eight days of the Whitsun fair when the town was full of visitors. Spectators were invited to spend a penny to buy a badge, which would symbolically enroll them as members of Robin Hood's gang, and upwards of 2,000 badges could be sold. Some kind of chal-lenge must have been issued to men who refused to pay. The players

The Battle Between Carnival and Lent. Brueghel depicts Lent as a woman played by a man. The bee-hive on the man's head suggests that he has been cast in this role as some kind of ritual punishment.

Feast of misrule. The participants are probably French students of the sixteenth century. Bottom left, a group of mummers in bird costume arrive with an entertainment. Bottom right, unpalatable foodstuffs are doled out for the feast.

visited neighbouring parishes along the river, and on one occasion the King himself at his nearby palace of Richmond. There was a display of archery incorporated in the events, but more important were displays of morris dancing by Robin's followers. Lords and gentry were often treated to Robin Hood plays presented by their servants, and a unique skeletal text survives from Norfolk from the year 1475, with a story-line adapted from a popular ballad. In the first part of the play Robin and a Knight compete at archery, stone throwing, and wrestling in sequences that must have been improvised. After a fight to the death, Robin cuts off the head of his opponent, wraps it in his green hood, and puts on his opponent's 'clothes'. The symbolism was both seasonal—the Knight representing winter—and social—Robin becoming a mock lord. The second scene is again one of role reversal. The outlaws are captured and placed in the Sheriff's prison, but Robin turns the tables and imprisons the Sheriff instead. Plays of this kind performed by retainers before their lord must have embodied complex messages about power. One of Henry VIII's courtiers complained at the time of the Reformation about the popularity of Robin Hood plays 'wherein, besides the lewdness and ribaldry that there is opened to the people, disobedience also to your officers is taught whilst these good bloods go about to take from the Sheriff of Nottingham one that for offending the laws should have suffered execution'. At a time of political tension, the ritual overturning of authority was seen as unacceptable, and the plays were gradually suppressed. Earlier in his reign, however, Henry VIII had cultivated his image as a man of the people by participating in Robin Hood plays. In 1515 four gentlemen impersonated Robin Hood, Little John, Friar Tuck, and

Maid Marian, and feasted the King in a bower. The King was dressed in green, and was accompanied by the Earl of Suffolk, who had made himself unpopular by marrying the King's sister, and whose playful outlaw disguise must have helped in his rehabilitation. The King was also accompanied by his young Spanish bride, for May Day was always associated with rites of courtship.

In Edinburgh before the Reformation men of consequence were chosen annually to 'make sports and jocosities in the town', and were given such names as Abbot of No-rent, King of May, Abbot of Unreason, Lord of Inobedience, and Robin Hood. In the crisis of 1561, while Mary Queen of Scots was being urged to return from France to bring political stability, a decree was issued by the Puritan city fathers to all apprentices and craftsmen that there should be no gathering 'efter the auld wikit manere of Robene Hude'. In defiance of the edict, a tailor was chosen as Robin Hood, and his followers marched through the town on the afternoon of Sunday, 12 May. A cordwainer's apprentice was seized by the authorities, and for the sake of example was condemned to be hanged for 'playing with Robin Hood'. On 21 July fiction became reality and the Provost of Edinburgh became Sheriff of Nottingham, as Robin Hood's outlaws seized the apprentice from the foot of the gibbet and smashed the gates of the gaol. The authorities locked the

Butchers greet the spring. The guild of butchers of Nuremberg performed a ceremonial dance every Shrovetide. Trees were often the focus of ritual. Here, mirrors hung on bare branches reflect the sun. Like the English Robin Hood play and morris, the dance required a single symbolic virgin. The hobby horse is similar to the one used in the English morris. The peasant top right rides a ram, because this was the symbol of the guild. A ram is also carried by the leader of the dance.

Plays on a parish feast day. An image of St George, patron saint of this Dutch village, hangs above the door of the Crown Inn. Youths perform a sword dance. Outside the walls of the churchyard, we see a folk-play of St George and the dragon. Top right, outside another tavern, a booth stage has been erected on barrels for the performance of a farce.

city gates and fired at the gang. Unable to escape from the city, the 'outlaws' turned on the Provost and Bailiffs, who took refuge inside the gaol. After a siege of three hours, the masters of the apprentices promised the 'outlaws' continued employment and freedom from prosecution, the siege ended, and of course the promise was broken. It was forbidden by Act of Parliament that anyone in Scotland should thereafter be chosen as Robin Hood or Little John. In this instance the subversive possibilities of the Robin Hood play were fully realized.

In a divided society, carnivalesque drama become a focus for the expression of discontent, and events at Edinburgh demonstrate why such drama was suppressed. The reasons for the demise of religious drama are complex, and later chapters of this book will explore what happened in different parts of Europe. Broadly, the story is more one of suppression than lack of participatory interest. In Catholic countries we find increasing repression and regulation thanks to the Counter-Reformation, the great attempt to counter Protestantism and restore discipline within the Roman Church. The context for the demise of medieval drama, both Christian and pagan, is a transfer of power from local communities to the monarchy and to the political centre. The context is also in some areas one of economic individualism, as people became reluctant to pour their personal wealth into community ventures. Within divided communities, there was always the fear of religious dissent becoming political dissent. In Spain, in Spanish colonies, and in pockets of stability across Europe, drama in the medieval style continued to be performed well into the seventeenth century.

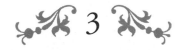

3

BEGINNINGS OF THEATRE IN AFRICA AND THE AMERICAS

LESLIE DU S. READ

THAT theatre existed in ancient Egypt is often denied, while any relationship between this 'high' civilization and Black Africa is contested and pre-colonial performance traditions throughout the continent are redefined in terms of cult, ceremonial, dance, spectacle, the plastic arts, music, or rhetoric. These are issues of interpretation and perspective, made more complex both by the relative paucity of the historical record and by the privileging of literary reception and values. As, during the twentieth century, *theatre as performance* has been emphasized increasingly in the West and *drama as a process of social development* has flourished in the Third World, so the lineaments of a distinct African theatre and its history have been affirmed. Mimesis and ritual have been found to be misleading and inappropriate concepts with which to define this theatre. The resort to such terms as ritual drama, ceremonial drama, festival drama, and dance-drama reflect the inadequacy of Western categories in the face of different performance skills and expectations, while notions of the 'primitive' and pre-theatre defer to the dominant Western mode of theatre and distort the particularities of indigenous traditions.

In ancient Egypt the context for theatre practice was the festival. Public celebrations are listed in inscriptions from the beginning of the Fourth Dynasty (*c.*2600 BC) onward. These were occasions when the gods were carried in procession out of the temples, through the urban quarters, into the country, and back. Music and dance were integral elements, as was dramatic play. The journey itself, as well as events staged on the way, enacted the experiences and doings of the gods and took place over a number of days. We get a glimpse of one such performance at Abydos in the autobiography of Ikhernofret (*c.*1820 BC):

I acted as 'his beloved son' for Osiris . . . I conducted the Procession of Wep-waut, when he goes forth to champion his father. I repulsed the attackers of the *neshmet*-bark, I felled the foes of Osiris. I conducted the Great Procession, following the god in his steps. I made the god's boat sail, Thoth steering the sailing . . . (trans. M. Lichtheim)

Performers were members of the temple élite, in the service of the Pharaoh. They wore animal masks alluding to divine roles in the governing myths. As steersman, the figure of Thoth would have been ibis-headed. In the combat-scene, the follow-ers of Wep-waut ('Opener-of-the-ways') would have worn jackal-shaped helmet masks, probably assisted by the falcon-headed followers of Horus. Such masks operated on a number of referential levels simultan-eously: the jackal mask, for example, spoke to the polit-ical union of Upper and Lower Egypt in representing the gods of the southern town of Nekhen and hinted at mortuary mysteries in representing the embalming god Anubis, at the same time as it celebrated the safe passage of the *neshmet*-bark in the figure of Wep-waut.

Anubis, the embalming god. This is the only example of an ancient Egyptian helmet mask to have survived. Note the eyeholes under the chin which enable the wearer to see out.

This allusiveness, dependent on a lore that per-formers and spectators shared, is a distinctive feature of the festival scenarios that have survived, which some Egyptologists have viewed as examples of dramatic literature. It is central to the forty-six scenes, recorded in 138 vertical columns of text and thirty-one designs on a papyrus excavated at the Ramesseum in 1896, which were performed for the Jubilee (Heb-sed) of Senwosret I (*c.*1918–1875 BC). Each scene in this 'production notebook' consists of an account of an action or actions, a mythological explanation, a short dialogue involving two or more characters, together with directions concerning subsidiary roles, the inclu-sion of song or dance, the properties needed to stage the scene, and—when neces-sary—the virtual location for the enactment.

The reaffirmation of the accession of the Pharaoh, of the maintenance of the status quo, is asserted and celebrated in this processional drama with reference to a core myth which underpins every manifestation of ancient Egyptian theatre of which we have a trace. Osiris—dismembered by his brother Seth—is *re-membered* by his son Horus. The Pharaoh, as Horus overcoming Seth and recovering 'the

Eye of power', ensures the continuity of order (*ma'at*) through the burial and transformation of Osiris, his embalmed ancestor. Each scene is discrete and, although there is much repetition and embellishment from one scene to another, there is little sense of narrative development. The dialogues are pithy, full of puns which reinforce the associations to be evoked within and among the spectators at specific moments and places on the processional route. Every symbol, gesture, sound, and embodiment seems shaped to trigger a chain of significations in the onlooker, so that each episode compacts and resonates with shared assumptions. The trampling of grain evokes Seth's dismembering of Osiris, dissension between Upper and Lower Egypt, fratricidal strife, and the ever-present infertility of the desert. The royal barge on the life-supporting Nile evokes the solar boat in the sky as well as the ship of death steered by the moon-god Thoth. Continuity and coherence depend on the network of social, religious, and political relations which are articulated and sustained, even taught, by these festival enactments.

The most complete text of an Egyptian festival play can be found, with accompanying reliefs, on the walls of the Ptolemaic temple of Edfu. Although engraved as late as about 110 BC, the record relates to performances which had taken place annually for over 1,000 years. Allusion and multiplicity of reference again predominate. Narration, dialogue, and mime, however, are more consistently developed than in any other extant text, structured by duplication and repetition, as the Egyptologist H. W. Fairman has observed, 'to cater for a strong element of audience participation'. For instance, a response—'Hold fast, Horus, hold fast!'—is encouraged from spectators at regular intervals in a repetitive sequence of scenes depicting Horus triumphantly defeating Seth by harpooning a hippopotamus. In each of these scenes Horus appears on two boats simultaneously, as representative of Upper or Lower Egypt, wielding two harpoons in a struggle to sustain order, a struggle in which the audience are co-actors.

The ideological function of ancient Egyptian theatre was paramount. Rooted in the practice and lore of a priestly élite, the persistence of its small core of themes, symbols, and personages over the millennia is remarkable. As fragmentary as our knowledge of this theatre practice is, it is clear that it was part of a network of institutional dynamics which contributed to the control and integration of feelings and values, to the generation of communal beliefs. In this, it is closer to traditional theatre practice in the rest of the continent than it is to the more product-oriented modes of the West. Formally, as well as functionally, these theandric performances foreshadow many aspects of ritual theatre attested in the tribal, cultic, or court festivals of Black Africa. Their commingling of religious and theatrical celebration in a fluid rearticulation of social space parallels, for example, the vibrant patterning of processions and masked 'operas' found among the Dogon. This spatial representation is fundamental to African ritual theatre, which, according to the Nigerian playwright Wole Soyinka,

Dama festival. Something of the dynamics of festival processions is captured in this moment in which a Dogon mask is led by a 'teaser' through a mobile audience.

establishes the spatial medium not merely as a physical area for simulated events but as a manageable contraction of the cosmic envelope within which man fearfully exists. And this attempt to manage the immensity of his spatial awareness makes every manifestation in ritual theatre a paradigm for the cosmic human condition.

Knowledge of sub-Saharan traditional theatre is based on oral history, early European descriptions, and colonial or post-colonial practice. Forms of theatre, it is claimed, existed in Ethiopia during Axumite times as well as in many later kingdoms, like the Mandingo Empire of Mali (1200–1500 AD). The diversity of indigenous traditions recorded throughout the sub-continent reflects more than 800 distinct language groups.

Naturalistic mimicry is usually said to be the source for what is most theatrical in these traditions. The mimetic plays and dances of the pygmies of equatorial Africa or of the Khoisan of Botswana, performed without masks and with an uncanny lifelike accuracy, seem to give support to the theory of a hunting/gathering origin of theatre. Whether the detailed and festive enactment of a successful elephant hunt is performed before an actual hunt, as with the Beku pygmies of Gabon, or after one, as with the Mbuti of Zaïre (where the struggles of the elephant are staged so realistically that by the end of the play the group's temporary dwellings have been utterly destroyed), the links between mimetic and productive action are strong. In such 'mimes for survival' there is a unity of subjective expression and objective imitation which seems to adumbrate some ur-drama out of which more differentiated forms might have grown.

Story-telling is also cited as an important source of theatre. From the Ivory Coast to the Kalahari Desert, African story-tellers impersonate characters and concretize detail to a remarkable extent. Using call-and-response patterns, they involve an audience in the histrionic process through its verbal participation as a group. In some instances, the narrator's song or speech is also enacted in dance or mime by assistants (as the *Ozidi* saga is among the Ijo of the Niger delta, where the story-teller becomes the protagonist in a seven-day masked dance-drama) or may be interrupted and embellished by the physical participation of individual members of the audience, as with *Anansesem* among the Akan. Character interaction and dialogue can occur when the story-teller, remaining on the margins of the action like some master of ceremonies, elicits information from characters, guiding both spectators and actors through a developing improvisation, as in the *koteba* theatre of Mali.

However, although mimicry and narrative can be important elements, neither are formative nor unifying in the synthetic mode of theatre most characteristic of Black Africa. The mask, together with the high display essential to festival, is a more potent source and resource. Masks appear in the context of periodic celebrations connected with the ancestors, with the mythical or historical origins of the people, with the critical moments in communal work, or with the events of group-life such as initiation and death. In place of, or alongside, mimicry the mask can not only invoke the presence of the 'other' but also distance it and contextualize it through abstract or composite shapes, signs, and colours. A mask used for funeral festivals in Burkina Faso, for example, symbolizes the fusion of the Nakomse-Mossi (the 'first people' with those who founded the state of Mossi in the late fifteenth century) at the same time as it recalls a myth which tells how the knowledge of cotton and weaving came to humanity, and also suggests the shape of a weaving sword. When masked dancers simulate the motion of the shuttle passing back and across the warp, mimicry and story cohere, but only as a moment in a larger non-narrative theatrical event which embraces social, religious, and therapeutic—as well as aesthetic—concerns.

The referential and transformational powers practised by the masquerader synthesize disparate theatrical skills—which, besides mimicry and story-telling, can include singing, dancing, acrobatics, stilt-walking, tightrope-walking, puppetry, and magic—in a form which is distinctly African. Masquerades take place in the open air, often with a fluid sense of stagecraft. 'The action may wander down the streets and even out onto boats on the river.' Fictional and non-fictional events mingle in an extended game. There is rarely any rigid separation of spectator from event, even when the power of a mask or the intensity of possession invokes a spirit to join in the festivities of the community.

Dancing and drumming are regular features. As African languages are tonal, the drums often supply a verbal context or commentary in which the rhythmic gestures, forms, and steps of the dance can operate with precision and subtlety. Together with the iconography—often complex and stylized—of the masks and disguises, this drum-language creates a performative 'text' which the interaction of spectator and actor can articulate, embellish, and revitalize during the course of the festival. It is a text which hints at the stories, experiences, and relationships which identify the community as a people, shaping the enactment through evocation and allusion rather than through any unifying narrative.

Egungun Apidan theatre exemplifies many aspects of this fluid form. Presented by troupes of up to fifteen actors, each performance consists of eighteen or more scenes in which a single mask or a group of masks improvises on a synopsis hinting at a story, a situation, a character, a state of being, or a magical power. Each

Apidan theatre. Baba Agba, Old Man, challenges the audience who have just ordered him to revive a boy he has killed. He shows them that *he* is the one to decide whether the boy can come back to life or not.

Facing: Nuhlimkilaka. Audience participation was central to this Kwakiutl mask of a forest spirit who was thought to cause one to become confused and lose one's way. In performance, anyone who pointed a finger at her could make the mask fall asleep immediately (whatever situation or position the performer might be in).

scene is independent. Continuity depends upon a chorus of female singers and the drums of the *bata* orchestra. The drums give directions to the actors, information to the audience, and lines of dialogue to the masks. Apidan (a performer of tricks), or Alarinjo (a vagrant dancer), as this itinerant theatre is sometimes called, developed in the Oyo Yoruba Empire, probably towards the end of the sixteenth century, alongside the lineage masquerades associated with the Egungun cult of ancestor worship. The earliest Western accounts of an Apidan masquerade (witnessed on 22 February 1826) describe scenes which are still played today, except that nowadays the time taken to play them is much shorter.

Throughout the Americas, masking also seems to have been a primary source of theatre—in hunting economies, among agricultural tribes, as well as in archaic urban civilizations. From the north-west coast to the Great Island of Tierra del Fuego, masks are associated with a variety of indigenous theatricals connected with religion and social organization. The abundance of natural resources around northern Vancouver Island and the adjacent mainland enabled the Kwakiutl, for example, to create a repertoire of aesthetically elaborate masks and costumes for their indoor sacred spectacles and entertainments which often involved extensive use of theatrical props and devices—speaking tubes, underground tunnels, marionettes, trick knives, etc.—as well as lively audience participation. Some of these embodiments (of animals, qualities, mythical beings, or humans) were intended simply to entertain the audience, while others sought to impress them with the seriousness of their relation to natural and supernatural forces. Many of the names given to these masks refer explicitly to the play and simulation involved—'embodiment of the personation of otter', 'acting the embodiment of mercilessness', 'pretending "just about to eat"'.

Even where hunter-gatherers have had far fewer natural materials than the Kwakiutl at their disposal, complex masking aesthetics have been developed. Among the Selk'nam, one of the southernmost hunting groups in the world, considerable time and energy was lavished on the staging of masked performances during the Hain, a periodic initiation festival which in pre-contact times might have lasted well over a year. Though some enactments were confined to rites conducted in the ceremonial hut, witnessed only by men, most were performed on a flat treeless area between the

Ulen. This Selk'nam mask created an illusion of moving with superhuman speed and then remained immobile for several minutes staring steadfastly at the audience. The contrast between speed and immobility was heightened by the abstract design of white stripes on dark red.

back of the hut and the encampment, set up specially for the festival, where women and children formed an audience. Many of the scenes of the Hain were designed purely to amuse, provoke, and involve this audience, as with the homosexual pantomimes of the Hayilans and the clownish dexterity of the Ulen mask. The Hain, last performed in 1933, was central to the communal life and identity of the Selk'nam, who were first encountered by Europeans in 1579.

In the *Kachina* cult, among Pueblo agriculturalists in the south-west of the United States, there is a similar division between 'private' indoor sacred dramas—like the Hopi *Palulukonti* (in which mechanical serpents are skilfully manipulated through holes in a screen to interact with *Kachina* impersonators, while a masked clown mimes a heroic struggle with a horned water serpent whose lifelike movements he also simulates, disguising his control of the serpent by wearing a false arm)—and outdoor 'public' dances and improvisations. The former take place in semi-subterranean chambers called *kivas*, the latter in plazas enclosed by buildings with flat roofs on which some of the audience can sit or stand.

Kivas are entered through a hole in the ceiling, an entrance used to great effect at the beginning of the Tewa *Nu-hi* or *Rain-God Drama*, when two sacred clowns, having made the sound of thunder with their feet on the top of the *kiva*, tumble through the hole. They look around at the assembled community:

FIRST CLOWN. Hey, Little Twin Brother, what kind of place is this?

SECOND CLOWN. Looks like hell to me—just take a look at all those devil-headed people all around us!

FIRST CLOWN. You are right, Brother, just get a load of that woman over there. . . . she's staring at me.

SECOND CLOWN. Maybe she's got the hots for you.

(trans. V. Laski)

There is no drumming or singing in this particular ritual drama, which relies throughout on pantomime, the spoken word, and an extraordinary range of sounds and noises. For instance, each Rain-God (*Kachina*) has his own particular vocal repertoire that identifies him.

In plaza performances *Kachinas* dance in a long line with a side-dancer encouraging the group, emphasizing the rhythm of the song, or highlighting its words through gesture and action. All the performers may

Above: Hopi mudheads. These sacred clowns entertain onlookers in the intervals between plaza dances. The improvised plays that they perform often reverse the norms of everyday communal behaviour.

Left: Casa Rinconada, Chaco Canyon. Ruins of a great *kiva* isolated on a swell of ground across from Pueblo Bonito. Its roof would have afforded a sweeping view up and down the canyon. A double stairway allowed for separate entrance-ways on to the floor of the *kiva* and into the tunnel under the floor.

Pueblo Bonito, Chaco Canyon. This reconstruction shows how the tiers of flat roofs, on which a mass audience could gather, embrace the great plazas where early pan-village enactments played a vital role in maintaining social cohesion.

represent a single *Kachina* or the line may be an ensemble of mixed masks and costumes. Between dances, clowns improvise satires on village life and the strange habits of outsiders, or mime sexual, scatological, and anti-social activities (often engaging the audience physically in their farcical play). The number of known *Kachina* masks is considerable; the Hopi recognize at least 500 distinct types. These cult performances, which were in existence by 1280, although influenced by literate civilizations to the south, have their roots in local pan-village rituals for which the great kivas like Chetro Ketl and Rinconada or amphitheatral plaza-oriented communities like Pueblo Bonito were built during the eleventh century. They developed as part of the co-operative institutions and exchange systems essential to the equilibrium of aggregated farming villages.

Ritual dance-dramas and farces were a conspicuous feature of pre-Columbian Meso-American civilization. Early Spanish accounts speak of comic and dramatic *bailes* with actors dancing in a variety of human and animal masks, of *cantares*, *entremeses*, *farsas*, and *representaciones*, of training schools, professional buffoons, humpbacked clowns, and contortionists. Post-conquest suppression of religious ceremonies, the burning of indigenous texts, and the substitution of Christian *bailes* ensured that most of this rich panorama of native theatre was lost. Traces survive—in the Tigre masks and costumes, for example, of contemporary Mexican fiestas, and in ceremonies such as the *Play of the Voladores*. Depicted in early Aztec codices, this spectacle involves four, five, or six performers, including a musician and a man dressed as a woman, who—after dancing on a frame at the top of an 80-foot pole—swing out on ropes and whirl through the air, head first, in ever-widening circles to the sound of their gourd-rattles and the musician's pipe-and-tabor.

A few Mayan and Nahuatl dance-dramas have survived relatively uncontaminated by later influences. Farces like the Nicaraguan *Gueguence* and the Guatemalan *Charamiyex*, involving an old man's feigned deafness and stupidity, continue to exploit the stock situations and characters, as well as the scatological humour and satire, attributed to native *entremeses* by early witnesses. *Rabinal-Achi* (one of a cycle of tragedies about a captured warrior, which once included the *Quiche-Vinac*, the *Tum-Teleche*, the *Oxtum*, and the *Loj-Tum*) still maintains, in the ceremonial vaunting, repetitive dialogue, and stately motions of its masked actors, gesturing to the beat of the *tun* and the music of horns, the form and intensity of pre-Columbian spectacles. The *Patzca* or *Dance of the Men with Goitre*

Below, left: Juego de los Voladores. The weight of the 'flyers' sets the cap slowly revolving in a spectacle which has changed little over the years. 'We are the sacred birds that fly with the four winds to the four cardinal points.'

Below, right: Códice de Azcatitlan. An early image of this pre-conquest performance. In more recent times, the number of 'flyers' has often been increased to five or six 'to make a finer show'.

Patza. A scene from a *baile* or dance-drama performed at Rabinal, Guatemala, in which goitre masks beg for relief from their illness.

preserves the symbolism of ancient rain festivals in the humour and pathos of the maskers with their grotesquely enlarged buttocks and snake-shaped canes. The persistence of such masking traditions among Native Americans, in spite of a history of suppression and destruction, is yet further testimony to the power of theatre as a tool for collective identity and self-definition.

PART TWO

Theatre in Europe
from the Renaissance
to 1700

4

ITALIAN RENAISSANCE THEATRE

LOUISE GEORGE CLUBB

EVEN as the word 'Renaissance' is being dropped from the vocabularies of many historians, most of Burckhardt's myth lives on, to the benefit of historical synthesis. One corollary that is disappearing to no one's regret, however, is the denial of creative vitality to the Italian theatre because it produced no Shakespeare.

All European countries where Renaissance drama appeared had strong traditions of medieval theatrical forms, religious and festive plays, mumming, municipal pageants, court spectacles, and, eventually, humanistic neo-Latin school drama. In millennially disunited Italy the regional varieties of these forms were many. When a new generation of humanistically educated writers undertook to surpass antiquity by constructing vernacular genres out of Roman ruins, they made avant-garde models of classical comedy and tragedy which gradually established a standard for 'national' scripted theatre, while coexisting with older local or popular styles.

Italy was unique in producing the technology of modern theatre. From the Cinquecento (the 1500s) into the Seicento (the 1600s) it developed a new system of play-making, comprising generic structures, methods of acting, and innovations in scene design, as well as theoretical principles and vocabulary. For sheer mass of archival data—printed and manuscript plays, scenarios, repertories of poems, speeches and dialogues, *intermezzi* texts, and descriptions of court festivities—Italian theatre is singular. Published plays alone amount to nearly 6,000 in Lione Allacci's *Drammaturgia* to 1755. Long before plays in other countries were signed or printed, Italian presses were publishing theatrical texts by Ariosto,

Machiavelli, and Trissino. Before theatrical companies existed elsewhere, travelling Italian troupes were creating a foreign market for the *commedia dell'arte*.

A plenitude of forms was the result of this theatrical energy and a range of potential functions which would be increasingly conjoined: the play as holiday, as lesson, as display of and bid for power by competing signories and city-states, and as commodity. A sophisticated art of staging contributed to the functions of the play as mirror of reality and as simulacrum of the cosmos. Finally, when long cohabitation of drama and music produced opera, the generating power of the Italian Renaissance theatre had performed its last act.

'Today you'll see a new comedy called *Calandria* (*The Follies of Calandro*): not in verse, not ancient, not in Latin. If anyone says it's stolen from Plautus, he can search Plautus but he won't find anything missing.' Thus was introduced to a glittering audience of prelates, noblewomen, and courtiers surrounding the new Pope Leo X in 1514 a play already reputed a paragon of fashionable humanistic wit. It had been performed a year earlier at the court of Urbino under Baldassar Castiglione's direction and was destined to be a model for generations of playwrights bent on appropriating classical drama for the triumph of modern Italian culture.

The work of the pontiff's most powerful counsellor, Bernardo Dovizi, Cardinal Bibbiena, *Calandria* was sumptuously produced. *Intermezzi* were inserted as diversions to be danced and sung between the five acts. Baldassare Peruzzi designed an innovative perspective set to give a view of contemporary Rome. Acted by beautifully costumed male amateurs speaking a Tuscan prose modelled on Boccaccio's *Decameron*, the play was a concentration within a unified action, place, and time of 'theatregrams', units of structure (characters, situations, actions/words, thematic patterns) that would eventually grow into a universal theatrical repertory. The intrigue is formed by combining a version of Plautus' comedy of the twin Menaechmi brothers with a fusion of several Boccaccian stories of ill-served wives and silly old cuckolds. The struggle toward reunion of the boy-and-girl Greek twins separated in childhood produces transvestite disguises for both and erroneous identifications, in counterpoint with the deceits and dodges to which Calandro's love-starved wife is driven in her passionate affair with the male twin.

Bibbiena and his immediate contemporaries united such elements in a verisimilar but stylized model of urban middle-class domestic conflict between youth and age, masters and servants, love and money, wit and fortune. Comedy so constructed was called *erudita*, *grave*, or *osservata* because it was written in observance of rules derived from the Latin plays that these authors had read in school and acted at court. To the nucleus of characters populating their ancient exemplars—fathers and mothers, sons, scheming servants and parasites, nurses, brag-

gart soldiers, prostitutes, procurers, and cooks—they added more recent figures from the novella tradition or from contemporary society—the frisky young wife, the humanistic pedant, the charlatan magician-alchemist.

This avant-garde comic form was a shapely and capacious container for elements from the vernacular tradition of narrative and of theatrical representation, a wide cultural variety of masking customs and carnival shows, sacred plays, song, dance, mime, farce, dramatized game, eclogue. The organizers and actors of the new comedy were men of courtly or academic ambience but the various kinds of entertainers for hire who were perennially on hand for public and private shows were quick to add these play-making structures to their wares.

The brief and brilliant papacy of Leo X nurtured the modern theatre even as it did the Protestant Reformation. In the act of providing Luther with new fuel for accusing the Church of Rome of licentiousness, paganism, worldliness, and aggrandizement of its wealth and temporal state, the court of the first Medici pope offered conditions enormously favourable to the growth and proliferation of drama. Among the first events of his reign was a spectacular bestowal of Roman

A stage-set by Baldassare Peruzzi. His perspective set for the 1514 *Calandria*, later described by Giorgio Vasari, would have structurally resembled this one, which shows the Roman Colosseum and Castel Sant'Angelo in the background. The predominance of palatial architecture in this sketch, however, could have accommodated a tragedy.

citizenship on his brother Giuliano and nephew Lorenzo, for which a special wooden theatre was constructed on the Campidoglio. In addition to a lavish performance by aristocratic youths of Plautus' *Poenulus* in Latin 'with the proper pronunciation that can be acquired only in Rome' there were plays, rituals, and processions employing the city itself as a stage for glorification of the Medici and publicizing their return to rule in Florence and assumption of world power in Rome, equating the Medici reacquisition and expansion of power with the revival of the imperial and cultural glory of the Roman Empire.

In the fifteenth century the Latin dramatists had been played in the humanists' schools and imitated in neo-Latin comedies by schoolmasters and their alumni. Duke Ercole d'Este had supported humanistic education in Ferrara and performance of Plautine plays, in Latin and in translation, interlaced with mythological tableaux, eclogues, and dances, as features of the famous Ferrarese classical festivals. Pomponio Leto's academy in Rome had developed an élite audience for performances of Plautus, Terence, and Seneca.

Not segregated from other sources of theatre, academicians in the Rome of Leo X frequented the company of fashionable courtesans such as the celebrated Imperia, for whom music and recitation were part of an elegant erotic persona, and Niccolò Campani called Strascino, a one-man show in his role of Sienese clodhopper. This social mingling swelled the tide of drama.

The Pope's legendary appetite for theatrical entertainment was seconded by awareness of its political uses. His reign illustrates a continuity in the tradition and variety of dramatic forms, and the vitality of the humanists' enthusiasm for recovering and competing with every aspect of the classical culture whose material remains lay about them thick on the ground. Moreover, Leo's exercise of patronage encouraged theorizing on structures of secular vernacular theatre and contributed to the new system of play-making that would be one of Italy's gifts to early modern Europe.

In his pre-papal years Giovanni de' Medici had been accustomed to theatre as part of municipal and religious life in Florence, as of his humanistic schooling and his sojourns at courts outside Tuscany. The carnival season was the principal time for entertainments of every sort, but religious feast-days, aristocratic weddings, and visits of dignitaries were also taken as occasions for theatre.

His father Lorenzo il Magnifico had written a *Sacra rappresentazione di SS. Giovanni e Paolo* (*Sacred Play of Saints John and Paul*) (1491) in the medieval rhymed octave format, for performance by confraternity boys. Such plays, chanted in public spaces by the sons of citizens of various social class, contained scenes from biblical and hagiographical tradition, history, romance, and local custom. Elaborate stage machinery made possible a multiplicity of settings from Heaven to Hell and all places and times in between. The entire community was the audience and its lore the subject; the genre was a vehicle for reassuring messages, political and ideological, and its popularity continued in the sixteenth century even as new kinds of theatrical styles were being generated by fashion, politics, and humanistic scholarship. The *sacre rappresentazioni* had some counterparts elsewhere in Italy. Among flourishing secular contemporary forms were the Ovidian myth play, *favola mitologica*, and other hybrid works in vernacular rhyme written and performed by courtly amateurs, with accompanying spectacle in which professional musicians and dancers took part.

The future pope had been tutored by the poet and humanist Angelo Poliziano, editor of Terence's *Andria* and author of *Orfeo*, a *favola mitologica* in mixed verse-forms with song that has been hailed variously as the restoration of the Greek satyr play, as Poliziano's version of the Venetian festival mumming called *momaria*, and as the precursor of Italian Renaissance tragedy and pastoral drama. *Orfeo* was performed about 1480 on a hill-shaped float carried or wheeled into Cardinal Francesco Gonzaga's banquet hall in Mantua. By pre-empting structures from the *sacra rappresentazione* to re-enact a symbolically charged Greek myth, with a finale of Bacchantes tearing Orpheus apart to the rhythm of popular carnival songs, Poliziano probably aimed at pleasing an audience of humanistically educated ecclesiastical connoisseurs. But the shifting elements in this prophetic work illustrate the dangers and difficulties of distinguishing between high culture and low, for a two-way motion is visible here at the outset of Renaissance drama.

Before his accession Leo X was friendly with Ludovico Ariosto, who was engaged from his youth in Ferrarese court productions of Plautus and Terence in both Latin and translation. Ariosto's comedies *La cassaria* (*The Coffer Comedy*) (1508) and *I suppositi* (*The Pretenders*) (1509) established the primacy of the genre in which Bibbiena would follow, the *commedia erudita* constructed on the principles of Horace and Donatus but in contemporary Italian settings. In *Suppositi* these features are matched with traces of the schoolroom and of the undergraduate Goliardic spirit that in the preceding century had been cultivated in humanistic neo-Latin comedy; Ariosto modernizes the Latin comedy stock type of the *adulescens* as a Sicilian student at the University of Ferrara who exchanges identities with his servant in order to become 'a student of love'.

The innovation, identified admiringly in the comedies of Ariosto and his generation by imitators and commentators, inhered in the design, the structural coherence of beginning, middle, and end, the mechanism of presenting fiction on-stage so as to capture and embody in the vernacular the causality and illusion of quotidian reality summed up in the supposedly Ciceronian definition of comedy as 'imitation of life, mirror of custom, image of truth'. A technical advance in dramaturgy, this feat of theatrical engineering depended on analysing and translating Latin comedy.

More immediate and less intellectual pleasure was also to be had in these comedies. When *Suppositi* was produced for the Roman carnival of 1519 with singing and mythological morris dances in the *intermezzi*, French emissaries to the Vatican were scandalized by the Pope's delight in the prologue's puns about sodomy.

Niccolò Machiavelli also turned his pen to avant-garde comedy in hopes of Medici patronage. His *La mandragola* (*The Mandrake*) was considered a masterpiece from the time of its first appearance about 1518. Machiavelli had honed his stagecraft by translating Terence's *Andria* and now he welded narrative elements from the *Decameron* into a dramatic *tour de force* of adultery triumphant. The supposed generative power of a mandrake potion is the basis of a wickedly hilarious and profoundly Florentine trick perpetrated by a brilliant parasite, aided by a corrupt friar, to the satisfaction of the lover, the initially reluctant wife, and the oblivious cuckold himself. More than his Plautine *Clizia* (1525), Machiavelli's *Mandragola* reflects his interest in the Attic Old Comedy model of political satire, much discussed in theoretical circles but discarded by the majority of dramatists as potentially libellous and dangerous to civic discourse. Machiavelli's reported

Scenes from Terence's *Andria*, Venice, 1524. The order of the woodcuts follows the progress of the plot; the back-curtains represent entrances to the habitations of the various characters, who are identified by the first letters of their names.

imitation of Aristophanes, *Le maschere* (*The Masks*), has been lost but the capacity for satirical political allegory visible in *Mandragola* hints at his Aristophanic tendencies.

While classical models and precepts were searched for the basis of a new vernacular comedy, tragedy was under discussion by the same literary intelligentsia, the restored text of Aristotle's *Poetics* providing an impetus. Giangiorgio Trissino, a noble Vicentine frequenter of Leo's court, with his *Sofonisba* (1515) in unrhymed hendecasyllable verse made a vernacular claim on the territory of Greek tragedy, fashioning a version of Aristotelian 'tragedy of pathos' around the figure of the Carthaginian queen who was a casualty of imperial Roman policy administered by Scipio Africanus. Dramatizing a subject of Livy's history and of Petrarch's Latin epic *Africa* in a theatrical form evoking Euripides and Seneca, Trissino's tragedy was an act of cultural self-assertion typical of early Cinquecento literary playwrights.

Also in 1515 the Pope's kinsman Giovanni Rucellai, host to the critical discussions in the Florentine Orti Oricellari, put a Gothic legend into Senecan shape with *Rosmunda*, in which appear the political conflict and consequent violence contemplated in the theory of statecraft that emerged from the same intellectual context in Machiavelli's writings. Although the custom of performing tragedy would come only later, especially in the ambience of Ferrara, Padua, and Venice in the 1540s, these experiments in regular tragedy were soon imitated and published.

Just out of the incunabular stage, the printing industry was revolutionizing culture high and low. The place of theatrical entertainment in the culture changed accordingly. The practice begun in the preceding century of publishing illustrated editions of Terence's comedies continued in the Cinquecento, but now vernacular plays also were printed as texts for reading and acting or as commemorations of specific performances.

The first editions of Rucellai's tragedy, of Bibbiena's *Calandria*, and, very probably, of *La commedia di Callimaco e Lucrezia*, as Machiavelli's *Mandragola* was entitled in its original undated appearance in print, were all enterprises of Giovanni di Alessandro Landi, a beadle at the University of Siena, whose involvement with the printing of plays illustrates the multiplicity of theatrical activity tangent to the papal court.

Among Leo's favourite entertainments were performances by Sienese actor-authors like Strascino the bell-maker, Mescolino the paint-mixer, and Mariano the farrier, whose dialect plays on the antics of the peasantry, sometimes joined with fantasy and legend, belonged to the municipal ludic tradition of Siena. An example of 1516 is the eclogue of *Mezucchio* by Pierantonio the cloth merchant (P. A. Legacci, called Stricca). Its basis is the genre of the *mogliazzo* or 'wifing play', of which this one is typical. A dramatized mating contest in one act for five peasants, it is a dispute between two yokels for the love of Vica, conducted by dialogue in tercets or octaves and songs accompanied by a rustic cittern. Vica decides at last to take both suitors and add Menichella, another of her lovers. The finale is a round dance sealing the four-way accord with an invitation to the audience to join Vica in bed.

Though it would be an exaggeration to say that in the first half of the sixteenth century theatre throve in every city of the politically fragmented Italian peninsula, there were pageants and plays of many sorts and lines of communication open by which news of them travelled to the principal centres and variously allied courts:

A rustic eclogue, Siena 1518. On the title-page to the second edition of Pierantonio dello Stricca Legacci's *Egloga rusticale di Mezucchio*, a woodcut shows Vica with her quarrelling suitors and a view of the countryside.

Venice, Ferrara, Milan, Mantua, Padua, Bologna, Florence, Rome, Naples. Among them Siena figured more prominently in the theatrical life of the peninsula than its size and political power warranted, owing in part to geography and to its connection with the papacy. Almost any event in Siena was an occasion for theatre. Local rejoicing at the election of a Piccolomini pope, Pius III, in 1503 called forth not only processions, orations, and adornment of the city but also a spectacular enactment in the Campo of the rites of papal coronation, with music, fireworks, masked gentlemen, and priests playing the roles of bishops and cardinals for the edification of the populace.

The custom of the *veglia*, the evening pastime of story-telling, dialogue, and verbal games, long cherished in Siena and its countryside, helped develop rapport between performer and audience as a natural social bond. The local tradition of theatre was institutionalized by the founding of the Congrega dei Rozzi and the Accademia degli Intronati. This double event occurred in 1531–2, early in the city-state's long and finally losing struggle against Spanish-Florentine power.

That year, in a move toward specialization that seems an inevitable trope of cultural change, the Rozzi or 'Rough-cut' artisan play-makers officially established the Congrega, and thus limited their range but protected their territory by laying claim to the rustic *commedia villanesca*. In this genre the lower middle-class urban Rozzi represented the characters of semi-literate peasants, their dialect, desires, quarrels, and trials, portraying them in rhymed octaves and tercets with gusto, sympathy, or derision, but always at a distance that made it possible to use the peasant as an immutably Sienese voice or mouthpiece for local political sentiment, anti-Florentine or anti-Spanish. Despite bans imposed in the later Cinquecento, the Congrega would survive into the early Seicento but its members then would be drawn from the genteel classes and the forms of its production would change accordingly.

Simultaneously with the institution of the Rozzi, a group of aristocratic university wits declared a different theatrical mission in the Intronati Academy. They chose the name in reference to their pose of thunderstruck silliness and took as emblem a gourd, signifying sexual jesting and empty-headedness hiding wit. On the first day of carnival, or Twelfth Night, 6 January 1532, for an audience of ladies, the scions of such families as the Piccolomini, Sozzini, Tolomei, and Landucci staged a 'Sacrificio' in which they renounced love, and then repentantly followed it up later in the carnival season with the famous comedy of atonement *Gl'ingannati* (*The Deceived*). With an up-to-the-minute setting and a plot based on social upheavals following the sack of Rome in 1527 by the army of the emperor Charles V, *Ingannati* offered a mood of restorative conviviality, laced with badinage and bawdy, reminiscent of the traditional *veglia* blended with the *Decameron*.

Years later Girolamo Bargagli's *Dialoghi*, idealizing the great early days of the Academy, described the Intronati's taste for romances and novellas containing 'beautiful examples of constancy, of greatness of spirit and of loyalty', like those in some favourite *Decameron* tales, especially stories of 'great virtue and endurance in women who after persecution and calumny are found to be chaste and innocent'.

Earlier loosely built rhymed octave plays written for Siena by the Aretines Bernardo Accolti and Giovanni Pollastra had set precedents for dramatizing such romance narratives. *Ingannati* was the first of the modern line of Sienese comedies that bore witness to this taste. Incorporating the avant-garde methods and stage devices launched by the generation of Ariosto and Bibbiena, the academic playwrights, probably working in committee, gave the new idiomatic prose language and the unified verisimilar five-act intrigue form to the story of Lelia, whose spirit is scarred and family scattered by the Sack but whose love and audacity are strong. In male disguise she serves the man she loves, wooing for him another woman, who falls in love with the wooer; the return of Lelia's lost brother provides the peripety that concludes in double weddings and reunion of families.

With their carnival entertainment of 1532 the Intronati produced a romantic strain of avant-garde *commedia grave*, now as grave in content as in structure, that would become internationally successful after accumulating several Sienese exemplars that disseminated the model throughout Italy. Contemporaries of Shakespeare, the playwright's playwright, recognized the kinship of *Twelfth Night* to the *Ingannati* family.

Women are central to this prophetic innovation. Ladies functioned in the theatre of the time primarily as spectators, whether as guests of honour or onlookers at male display. Female singers and dancers took part in spectacles and *intermezzi*, accomplished courtesans like Imperia and Tullia d'Aragona entertained the literati with music and recitations, nuns and convent girls engaged in cloistered performances, but the day of the professional actress was some thirty years in the future. It is the more significant, therefore, that the Intronati comedies not only concentrated on the adventures of romantic heroines but were fashioned for the approval of feminine audiences. The annals of law enforcement show, moreover, that women sometimes joined in the private performances. Among those attending a comedy at a *veglia* in 1542 and convicted of infringing prohibitions against secular gatherings and wearing disguise, three ladies were cited for performing, one in servant's costume.

To the north, during the same carnival season of 1542, the sudden death of Angelo Beolco, the celebrated actor and writer famed as creator of Ruzante, leading character in many comedies, wrecked the grand programme planned by the Accademia degli Infiammati, founded the year before by gentlemen of the University of Padua (among whom was Alessandro Piccolomini, a key member of the Sienese Intronati who was continuing his studies in the Veneto). The crown of the occasion was to have been a performance of the tragedy *Canace* by Sperone Speroni, professor of philosophy and elected 'Principe' of the new academy. Rehearsals started under Beolco's direction, but the production was cancelled when he died before what would have been his first appearance in a tragic role.

It is interesting for theatre history that high culture by this decade regarded tragedy as so important. Intense discussion of the restored text of Aristotle's *Poetics* in such company produced experiments with a mythic incest plot and irregular verse lengths in *Canace*, and the attacks on it from the Ferrarese circle of Giovanbattista Giraldi Cinthio were also couched in Aristotelian terms. Giraldi's own tragedy *Orbecche*, blending Thyestean horrors from Seneca with ruthless Machiavellian political motivation, was performed in 1541 for Duke Ercole d'Este and later on several great occasions, but in his *Discorso* on playwriting Giraldi championed the happy ending as more pleasing to audiences and he went on to write several tragedies in which virtue triumphs and villainy is punished.

The debates about Aristotle and the kinds, construction, and features of tragedy are revealing of the intellectual passion for theory in this age of academies. The invention of a new science of dramatic criticism was another achievement of sixteenth-century Italian culture. The various types of tragedy that followed, whether on historical, mythical, or chivalric subjects, would share principles of regularity and unity, and features in which Aristotle's analysis of structures and Seneca's practice were visible.

Tragedy was less often produced, less popular than comedy, and eventually both were less attractive to audiences than pastoral plays and anything served up by the professional *comici*. It is nevertheless instructive about the way theatre was produced, and about the flourishing of drama in the Veneto, to observe that the academic environment in which *Canace* evolved was inhabited by Ruzante, the supreme impersonation of the comic peasant.

Beolco was a comet from Padua whose fame was to remain for centuries limited to the Veneto because of his linguistic localism. Baseborn but well-connected, he had rare histrionic, literary, and musical gifts with which he created the character of Ruzante, who sings and comments on his world in a Paduan peasant dialect fortified and complicated with a parodic erudite lexicon. Beolco had been first observed playing the role of Ruzante in a 'comedia alla vilanesca' during the Venetian carnival of 1520. His performance was part of a grandiose entertainment with banquet, allegorical floats, and bull-running sponsored by the Immortali, one of the Compagnie della Calza, or clubs designated by coloured hose, whose patrician young members organized various kinds of spectacle in Venice.

Beolco followed the tradition of satirizing the grossness of peasants, a source also of the artisan plays in Siena and the Cava farces in Naples, but he simultaneously adopted the fashion of burlesquing academic and elegant literary forms such as the courtly pastoral eclogue. Even his earliest works, *La pastoral* (1517?) and the *Prima oratione*, a dramatic monologue performed in 1521 for a new bishop of Padua, reveal Beolco's command of rhetoric and of dramatic technique, as well as a bond with the earthy world of the bumpkin he played. His sympathy would ferment into compassion and protest when the Wars of Cognac and bad harvests devastated the countryside around Padua, driving many peasants to sell out to large landholders and flee to town slums.

Of his seven full-length plays, the desperate comedy *La moscheta* (*The Fly-Specked Phony Lingo*) (*c.*1532) best exemplifies Beolco's power to catch grim reality and comic character in complex linguistic nets of monologues and dialogues. *Moscheta* shares some assumptions of *commedia erudita* in its five-act division, unity of time and place, and 'theatregrams' of disguise and eavesdropping, but its setting is a down-and-out underside of the regular urban scene and its pitiful scrap of a plot the threadbare events that occur there: with the help of Menato, another refugee peasant, Ruzante tries to pass himself off as a Tuscan-speaking foreigner to test his wife Betia's fidelity, but ends beaten by the soldier Tonin and sharing Betia with him and with Menato. The characters define themselves in vital speech and the compelling Ruzante is alternately parodic, craven, bitter, coarse, touching, and, above all, funny in his linguistic creativity.

In *L'Anconitana* (*The Woman from Ancona*) (1534–5?), a lighter comedy with a Decameronian plot and multiple love interest reminiscent of Bibbiena's *Calandria*, Beolco moves closer to *commedia erudita* and presents a different Ruzante, well fed, with a steady job in town, filling the role of clever mocking servant to a dodderer who lusts for a courtesan and is cuckolded by his wife. Ruzante is all fun here, dancing, making love, and singing Paduan songs; his Venetian master Tomao uses his own dialect and the elegant lovers speak Tuscan.

Beolco's fellow actors in these comedies were gentlemen amateurs in the Paduan circle of Alvise Cornaro, the patron in whose palace garden the Loggia and Odeon built by Falconetto provided a place for Ruzante's plays and music. The milieu was upper class, close to the university, moved by intellectual currents and practical concerns, given to discussions of Lutheranism and projects of land management. The role of Ruzante was a genre figure with special resonance for this audience, but Beolco was also in demand elsewhere and was invited to act as *choregos* in more than one theatrical centre.

He took his group to Ferrara in 1529 to perform rustic *intermezzi* at a ducal banquet preceded by a performance of *Cassaria* and in 1532 again collaborated on a production with Ariosto, who was still functioning as the Este court *choregos* near the end of his life. In the role of Ruzante folk and high culture touched, and its creator belonged to the private gentlemanly ambience in which theatre was sponsored, as the failed project to produce Speroni's *Canace* attests. Beolco's reputation is so exclusively theatrical, however, as to explain why he should sometimes have been incorrectly presented in theatre history as a *comico*, or player of the *commedia dell'arte*. Capering Ruzante and fatuous, rich Venetian Tomao prefigure Arlecchino and Pantalone, but Ruzante's status as the director, writer, and leading actor of a group of amateurs should not be mistaken for the later one of the *capocomico*, or actor-manager, of a professional acting troupe. That Beolco's example led the way, however, is more than likely. The first extant professional contract binding a group of men to travel about playing comedies for money was signed in 1545 in Padua. The documented *commedia dell'arte* begins here.

The long-remembered festivities for the wedding of Ferdinando de' Medici, Grand Duke of Florence, with Christine of Lorraine in May 1589 included a marine spectacle, the *Naumachia*, in the courtyard of the Pitti Palace, and in the Uffizi a Sienese *commedia grave* with *intermezzi* that were a milestone in theatre music and stage design, as well as two *commedia dell'arte* performances with well-known actresses in competition. This occasion offers a convenient chronological vantage-point for viewing the immediate past and future of the theatre in Italy, for seeing what was in vogue, how the technology had developed, how the immemorial variety of entertainment from all parts of the peninsula had contributed to a self-consciously mature institution of the theatre. It was an institution that still provided élite and corporate patrons with instruments of self-celebration and political competition, but now was proliferating and organizing forms of entertainment to sell to many markets, and in the process expressing the cultural paradigm of the spirit of Catholic Reform, the charge to revitalize and internalize the substance of doctrine by means of theatrical images.

The dramatic productions on the varied programme took place in a hall of the Uffizi that in 1586 had been decorated by Bernardo Buontalenti for another

Illustration to a printed *commedia grave*. A nurse remonstrates with a cross-dressed *innamorata* in Oddi's *Prigione d'amore*, 1591. One of thirty-one woodcuts in a series used by Venetian printers for different regular comedies in 1591 and 1592.

Medici wedding and was redisposed when great occasions demanded. Here gentlemen of the Intronati Academy played one of their romantic regular comedies, Bargagli's *La pellegrina* (*The Pilgrim*) (*c.*1568), revised to compliment the French bride. It bore the features of standard literary comedy: five acts of intrigue plot, disguises, and deceits, prose ranging from impassioned love speeches by the high-minded pilgrim Drusilla, seeking her lost husband, to satire and scurrility in the mouths of gluttons and bawds—all contracted into a one-day, single-set dramatization of the emotional narrative of feminine heroism typical of the Intronati and in the late Cinquecento cultivated by many other playwrights such as Giambattista Della Porta and Sforza Oddi.

Intermezzi, compact of verse, music, and dance, had now achieved the status of genre in themselves, although the cultural prestige of the dramatic text was unchallenged and a noble entertainment would have been thought insubstantial without a regular five-act play. Thematic connection between comedy and *intermezzi* was desirable in theory, as Bernardino Pino asserted in a treatise on comedy in 1572. At a wedding feast, the occasion itself offered the theme—union, harmony, divinely providential plan. Suitably, the plot of *Pellegrina* is a domestic intrigue combining several levels of love and contemporary manners, with a denouement in which wit and Christian forgiveness reconcile families, uniting Tuscan and French lovers.

Seated on movable tiers around three sides of the Medici theatre, the spectators watched the comedy unfold on a painted perspective set offering a single composite view of Pisa rising beyond a proscenium arch. But between each act Pisa was eclipsed by cosmic visions, as lighting and machinery illustrated the Platonic theme of music as a magic influence on gods and humankind, symbol of nuptial harmony and binding force of the universe in all its elements, air, fire, earth, and water. Conceived by Giovanni de' Bardi and staged by Buontalenti, the *intermezzi* displayed celestial cloud-riders, terrestrial gardens, infernal fire-demons and a Dantesque Lucifer, a flying Orphic sorceress, and sea waves with deities, barges, and ships afloat. Danced and sung to texts by Ottavio Rinuccini, Giovanbattista Strozzi, and Laura Lucchesini Guidiccioni and music by Bardi, Luca Marenzio, Cristofano Malvezzi, and others, these *intermezzi* are theatrical agglomerations representing long cumulative experience of the various resources of municipal and courtly diversion.

By this time professional acting troupes had multiplied, and for at least two decades had included women, a few of whom quickly achieved fame, becoming the first leading ladies of the modern theatre. The presence in Florence of the celebrated Gelosi Company resulted in an invitation to perform their speciality, an

'Hades', the fourth *intermezzo* to Bargagli's *Pellegrina*. Engraving by Epifanio d'Alfiano, 1592, from the original scene designed by Bernardo Buontalenti for the 1589 performance

DESCRIPTION OF A *COMMEDIA DELL'ARTE* MAD SCENE

A spectator of Isabella Andreini's command performance at the Medici wedding of 1589 wrote a rapturous account in his diary.

Finding herself the victim of Flavio's deceit and with no remedy for her sad case, Isabella abandoned herself to grief. Defeated by passion and yielding to rage and fury, like a mad creature she roamed the city scene, stopping one passer-by, then another, speaking now in Spanish, now in Greek, now in Italian, and in many other languages, but always irrationally; and among other things she began to speak French and to sing French songs, which gave the most inexpressible pleasure to the bride, Her Most Serene Highness. Then Isabella fell to imitating the manner of speech of all her fellow actors, Pantalone, Gratiano, Zanni, Pedrolino, Franca-trippa, Burattino, Capitan Cardone, and Franceschina, all so naturally and with such hilarious absurdities that it is impossible for tongue to tell the matchless worth and powers of this Woman. Finally, by the Action of magic art and certain waters she was given to drink, Isabella was brought to her senses, and here, with elegant and learned style explicating the passions and the ordeals suffered by those who fall into love's snares, she brought the comedy to its close, demonstrating by her acting of this madness the sound health and cultivation of her own intellect; leaving her audience in such murmuring of admiration and wonder that while the world lasts, the eloquence and inestimable worth of Isabella will ever be praised.

improvised or non-scripted comedy in three acts, for the wedding guests. Two actresses vied for the limelight; the contention was resolved by performing two improvised comedies, using the *Pellegrina* set and repeating the hugely successful *intermezzi*. The wedding guests thus saw both Vittoria Piissimi's gypsy role in *La çingana*, and Isabella Andreini's multilingual mad-scene in *La pazzia d'Isabella*.

Theatrical spaces also were increasing in number and function in the last quarter of the Cinquecento. As early as 1576 there was a 'stanzone', or large room, named for the nearby Baldracca tavern behind the Uffizi, where, under the administration of the customs office, travelling troupes performed for payment to bourgeois spectators and to the ducal audience, which had access through a palace corridor and a point of vantage behind a grate. In Spanish-ruled Milan also a small theatre for commercial comedy was built in 1598, attached to the Palazzo Ducale.

Venetian carnival maskers *c.*1600. Depicted in tempera in the Paduan Codicetto Bottacin, the revellers are costumed as if for comedy. The lady walks on high wooden clogs to keep her feet dry and is supported by companions disguised as a lover and a *zanni*.

The modern centralized stage space, as opposed to the scattered or consecutively ordered 'luoghi', or multiple places, of medieval staging, had been defined before 1508, when *Cassaria* was performed at the Este court on a perspective set painted by Pellegrino da Udine. But experiments in scenic illusion and research on Vitruvius and Roman theatre had not all moved in the same direction, and, even in cities where theatre was most abundant, permanent stages were slow to appear. A wooden theatre built in Ferrara for Ariosto had burned down and was not replaced. Within the usual theatrical venues—courts, municipal buildings, private houses—different spaces were adapted for individual performances.

When Leone De' Sommi, author and producer of Hebrew and Italian plays for the Jewish community's actors and the Gonzagas' Invaghiti Academy in Mantua, wrote the earliest Italian treatise on practical staging, *Quattro dialoghi* (*Four Dialogues Concerning Stage Representations*) (1556), he was thinking of temporary stages at court. In 1567 De' Sommi applied, unsuccessfully, for exclusive right for ten years to provide space for commercial players. The impresarial approach would eventually make theatre regularly available at prices to attract a broad public. The Spanish system of linking theatre revenues with rentable space owned by charitable institutions was followed in Milan and Naples. In Venice the mercantile aristocracy, families like the Giustinian and the Tron, bypassed the hospitals and entered directly into commerce, fitting up their properties as theatres and pocketing a profit from the sale of tickets and refreshments. As early as the 1570s there were two theatres in the vicinity of San Cassiano frequented by the public for comedies at carnival, although in subsequent periods playhouses were closed and commercial players banned.

A permanent theatre was established and an epoch of research on classical theatre architecture was ended in 1585 with the completion by Vincenzo Scamozzi of Palladio's plan of the Olympic Academy's theatre in Vicenza. The way of the future would lie with the box stage and movable flats used by Buontalenti and earlier Cinquecento scene-painters, rather than with the fixed architectural perspectives receding from the arched openings of the *frons scenae* built by Scamozzi, but the Teatro Olimpico was both a monument to the past and a herald of the permanent playhouse as an urban necessity. The academicians used their theatre for many kinds of events, such as the reception of the first Japanese visitors to Europe. For the inauguration of such a building, however, drama was required. A pastoral play was first intended, but decades of intense theorizing about Aristotle's *Poetics* favoured his prime exemplar of tragic structure and the Teatro Olimpico opened with *Edipo re*, Orsatto Giustiniani's verse translation with choral music by Andrea Gabrieli.

The first Japanese visitors to the Teatro Olimpico. A monochrome fresco records the welcome of the Olympian Academicians in 1585 to new Christian converts who were being escorted around Italy by Jesuit missionaries.

The general conviction of the superiority of *Oedipus rex*, manifest also in Torquato Tasso's challenging it with an elaboration of its structure in *Il Re Torrismondo: tragedia* (1587), collided with the Church's condemnation of Protestant tenets of predestination—the ironic reversal in Sophocles' 'perfect' plot being a figure of the inescapability of fate. This conundrum challenged Catholic imagination and called forth theatrical counter-demonstrations of providence, a benign pre-vision that leaves intact the concept of free will. Writers of *commedia grave* in the period often adduced Sophoclean principles for their labyrinthine intrigue structure illustrating the workings of a heavenly plan for human happiness. The genre that best embodied the concept, however, was the pastoral play, which, after Ferrarese experiments culminating in Tasso's *Aminta* (1573), became the dominant form of dramatic literature. Italian stages and presses were inundated with varieties of *favole pastorali* in occasionally rhymed verse, set in a rustic landscape of the mind distant from the urban realism required of comedy and the courtly solemnity of tragedy.

The most influential example was Guarini's *Il pastor fido: tragicommedia* (*The Faithful Shepherd*), published at the end of the 1580s amid polemics about the mixing of the two regular genres. Its happily ironic Sophoclean plot demonstrates the inescapability of a divine providence that uses the power of faithful love to lift the ancient curse on Arcadia. With song and emblematic dancing, philosophical choruses, and thematized metamorphic *intermezzi*, *Pastor fido* continued the humanistic programme to surpass the ancients by creating a mixture of tragedy and comedy that could pass muster with the 'new science' of dramatic criticism and with Catholic doctrine. After a long gestation from the distant time of Poliziano's *Orfeo*, the idea of a pastoral world gradually permeated every branch of theatrical enterprise, offering freedom to enact psychological change and opening new symbolic and festive spaces. A practical advantage of the new genre, as Angelo

Harlequin. Zany Corneto. Il Segnor Pantalon.

O la belle chanson, Pantalon chantons bien, | Accordons nous tous trois, si bien & proprement | Courage (mes amis) ie chante le dessus,
Si voulez esgayer voftre maiftreffe belle, | Que puiffions l'endormir au doux fon de ma lire, | De ce plaifant trio, compofé pour madame,
C'eft le moyen certain pour en fin iouïr d'elle, | Encor que comme vous ie n'aye apris à lire, | La douceur de ma voix luy penetrera l'ame:
Qu'eftre mufeau de chien, dy-ie muficien. | Ie ne laifferay pas de ioüer brauement. | Mes paffages ne font ni tortus ni boffus. j.

Music in the commedia dell'arte. Pantalone serenades a lady, helped by Arlecchino and Zanni the cuckold. A scene from the late sixteenth-century *Recueil Fossard* containing images of the Italian players in France in the time of Henri III.

Ingegneri, playwright, theoretician, and director, noted, was that pastoral plays could be produced inexpensively.

As the Seicento approached, music drama was on the horizon. Contributing to its development was the interest in the lost music of Greek tragedy that stirred discussion in Bardi's Florentine circle, the Camerata. But the works of its members, the pastoral *Favola di Dafne* (1597) with Rinuccini's verse sung to music by Iacopo Peri and their *L'Euridice* (1600), partly scored by Giulio Caccini, were not the only announcements of the coming *melodramma*. It had been prepared in the union of text and melody in *intermezzi,* and in the long cultivation of Neoplatonic theory concerning Orphic music. In their 'madrigal comedies' Orazio Vecchi and Adriano Banchieri commandeered characters of the *commedia dell'arte* for extended polyphonic compositions. Music had had a place in or around the earliest and humblest plays of the Cinquecento, and became essential to the art of the professional *comici.*

Descriptions of professional acting companies and their comedy improvised on a plot sketch began to appear in the 1560s, witnessing the presence of women on-

Facing: Guarini's *Pastor fido*, II, Venice, 1602. In the augmented 20th edition G. B. Aleotti's engravings represent each act's scenes simultaneously, disposed narratively in order of occurrence from foreground to background, with characters labelled at every appearance. Spatial relations signify time's passage, so that the primary subject illustrated is the structure of the plot, rather than the spectacle of action on stage.

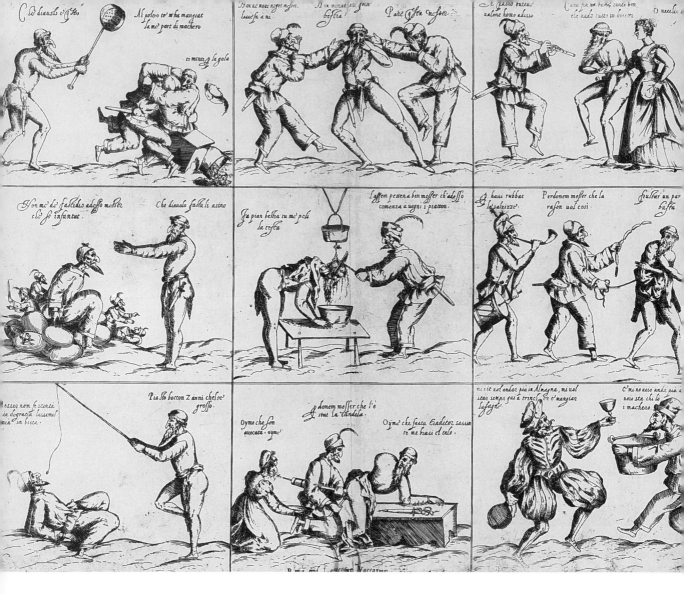

Early images of *zanni* and Pantalone played by acrobatic street-players. The composite of figures etched by Ambrogio Brambilla about 1580 was printed in Rome by Lorenzo Vaccari with snippets of characteristic patter. The figures were later cut apart and used in the *Recueil Fossard*.

stage. Isabella Andreini's *tour de force*, admired at the Medici wedding, as the crazed *innamorata*, mimicking the Venetian, Bolognese, and Neapolitan dialects of other characters and singing in French to please the bride, represents contemporary *commedia dell'arte* performance in its most brilliant vein and circumstances. But there were other sides to a profession often described as a lower-class alternative to the private theatre of courts and academies.

Like the medieval minstrels, mountebanks, and hawkers of cure-alls who were their predecessors, the *comici* made their living by selling entertainment wherever they could, in public or private places, adding whatever they found there to the store out of which they made theatre. There was a class system within the world of the paid players and a social abyss between all of them (including the Andreinis, though they maintained familiar correspondence with royalty and eventually set up a family crest, or Pier Maria Cecchini, who held a patent of nobility from the Emperor) and the philodramatic amateurs. But the encounter between humanistic writers and courtiers who wrote and performed plays as pastime or by command as courtly function and the vendors of entertainment was a generative event for the *commedia dell'arte* and for the modern theatre.

When travelling troupes were organized from the mid-1540s on, they ranged in quality from those who set up trestle-stages in piazzas, selling medicines, pulling teeth, and passing the hat, to the companies who were invited to Medici weddings. Professional players had their own specialities and eventually would be identified with one of them—the improvised three-act comedy of masked Pantalone and Doctor Graziano, various *zanni*, and the non-masked *innamorati*, maidservants, and swaggering *capitano*—but they participated in as much of the literary private theatre as they could and offered for hire the widest possible range of genres. In the 1570s they were carrying their wares to France, Spain, and England.

Among the early troupes known by name, the Gelosi, under the sometimes oppressive patronage of the Gonzaga dukes of Mantua, travelled in 1571 to Paris, where the tradition of the *comédie italienne* would become a formative resource of the French theatre. Often called to Ferrara by the Este duke, the Gelosi company performed such exquisite pieces as the lyric and literary five-act pastoral *Aminta*, which Tasso wrote for the court in 1573. At the other extreme of their repertory, in 1579, they gave a command performance to Guglielmo Gonzaga of a comedy of *gobbi*, with Zanni, Pantalone, Graziano, and the whole cast playing hunchbacks.

The variety of spectacle in Italian show business at this period is illustrated by a unique and recently discovered watercolour picture book containing 117 images of entertainments. Some of the figures depicted were staples of the scenarios for improvised comedy: a masked Zanni (or one of the *zanni*, perhaps Francatrippa) pursues the lean and slippered Pantalone; a captain with cape and sword, bearded like the pard, accosts an elegantly dressed lady; a doctor adjusts his spectacles to inspect the bottom of an untrussed bumpkin.

Other images in the codex are far removed from comedy: grotesque nether-world regents ride a float drawn by dragons, the skeleton Death stalks his domain, damned male and female souls are ferried in flames by Charon, herded toward Hell-mouth by bat-winged devils, and toasted on spits. Such scenes might have figured in banquets and carnival pageants, infernal *intermezzi* like Buontal-enti's, traditional *sacre rappresentazioni* of the Last Judgement, or even, with appropriate modifications, of the martyrdom of St Lawrence on a grill. Fit for tournaments and some kinds of tragedy were the suits of armour, Moorish costumes, and tents depicted in images of chivalric battle, and the parades of festooned and caparisoned horses, some of them trained, seemingly, to fall down and die.

Also represented are scenes of *gobbo* farce, in which humpbacked friars preach to humpbacked congregations, humpbacked barber-surgeons attend humpback

TEATRO FATTO IN FIRENZE NELLA FESTA A CAVALLO PER LA VENVTA DEL SER.ᵐᵒ PRINCIPE D'VRBINO
Qui fecero 42 Caualieri diuersi abbatimenti e dipoi un ballletto ci si uido ancora una battaglia a piedi di 300 persone, oltre i Carri e l'altra gente per diuersi seruitj
Iullius Parigi Inu: Callot delineauit et

Festival tournament for the Duke of Urbino's visit to the Grand Duke of Tuscany, 1616. *Guerra di bellezza, festa a cavallo*, with poetic text by Angelo Salvadori, machines and costumes by Giulio Parigi. Printed the same year in Florence by Zanobi Pignoni with engravings by Jacques Callot.

clients, acrobatic humpbacks and dwarves dance, walk on stilts, and grab at a goose hung high. Descendants of court buffoons and sideshow clowns, tuned to the grotesqueries of Arlecchino's early stage postures, the comic hunchbacks prefigure Shakespeare's Launcelot Gobbo and the Neapolitan mask of Pulcinella. In the scripted *commedia ridiculosa* of later years, the longevity of the *gobbo* was demonstrated by the cast of Margherita Costa's *Buffoni* (1641).

Among the images of entertainment are many of dancing and making music, from a pastoral scene of Orpheus charming the animal kingdom with his viol on through an astonishing series of solo performances on instruments—strings, winds, brasses, keyboards, harps—to which ladies dance, one with a tambourine. A busty peasant woman jigs to bagpipe music.

The publication in 1611 of fifty scenarios in Flaminio Scala's *Teatro delle favole rappresentative* (*The Theatre of Stage Plots*) was an event of the first importance for theatre history. Although it is the repertory of a company that never existed, Scala's compilation gives a fuller idea of the dynamics, tropes, and variety of the improvised comedy than any other single text has done, and was the only such collection to be printed in the period.

There were numbers of troupes in the public eye, their names no more stable than their constitution, and there was rivalry among them and for control of them. Courtly patrons like the Duke of Mantua and Don Giovanni de' Medici made demands, Venetian impresarios made offers. Cecchini, the great Frittellino, director and principal *zanni* of the Accesi, competed bitterly with the leader of the Fedeli, G. B. Andreini, whose acting of Lelio *innamorato* matched the fame of his mother as Isabella and his father as Capitan Spavento. Scala had retired from his role of Flavio *innamorato* when he published the *Teatro*, but soon returned to manage the Confidenti for Don Giovanni, opening the 1613 season of the Giustinian family's Teatro San Moisè in Venice.

Scala furnishes his 'theatre of the theatre' with an imaginary company, including the late Isabella and the retired Francesco Andreini, disposed in scene-by-scene plot summaries with stage directions and prop lists useful for improvising amateurs. The

LI.BVFFONI
COMEDIA RIDICOLA DI MARGHERITA COSTA ROMANA

Grotesque finale of a 'comedia ridicola'. Stefano Della Bella's engraved frontispiece to Margherita Costa's *Li buffoni*, printed by Amador Massi and L. Landi, Florence, 1641.

scenarios represent the most characteristic genre of the *commedia dell'arte*, the three-act improvisation, a technique requiring reading, memory, timing, and regular practice in dialogue. Training and exercise were as necessary to this kind of acting as to the dancing and singing which were intrinsic to the performance, although only occasionally are they indicated.

Whereas the written regular drama in which the *comici* also participated shares with the genres of classical music an aspiration to form and immutable structure, the improvised comedy is akin to jazz: the scenario provides the guiding modulations for the ensemble, mood sets a tempo, solo flights are sustained and anchored by individual resources and by the habitual give-and-take of collaboration.

A standard cast required about ten or a dozen players for the basic comic relationships: two pairs of Tuscan-speaking young lovers, two senior men to oppose them as parents or husbands, the Venetian merchant Pantalone and the Bolognese

A *COMMEDIA DELL'ARTE* SCENARIO

Flaminio Scala's collection of scenarios offered to readers and actors the plots, required cast and properties lists, and scene-by-scene indications of exposition, action, and dialogue for improvised performances.

The Two Fake Gypsies
Argument
There was in Rome a Venetian merchant named Pantalone de' Bisognosi, who had two legitimate children, one called Orazio, the other Isabella. She was loved by a young man named Flavio, son of a Bolognese doctor named Graziano, and she as ardently loved him. It happened that Flavio was sent by his father on important business to Lyons in France, and was captured at sea and enslaved by Turkish corsairs. When this news reached the loving Isabella, she determined to go with a servant to seek him throughout the world. To buy her lover's freedom she took jewels and money from her father, and when the money ran out, to save life and honour, she and the servant disguised themselves as gypsies. After long wandering, they returned at last to Rome, where her father recognized them. After many occurrences she found her lover and cured her brother, who had gone mad because of her flight, and she married her lover as she had desired.

Characters of the comedy
PANTALONE, a Venetian merchant
ORAZIO, his son
FRANCESCHINA, his servant
DOCTOR GRAZIANO
FLAMINIA and
FLAVIO, his children
ISABELLA, in gypsy garb, then recognized as Pantalone's daughter
PEDROLINO, her servant, in gypsy garb
CAPTAIN SPAVENTO
ARLECCHINO, his servant

Properties for the comedy
Three women's dresses
Gypsy costume for Pedrolino
Many lighted lanterns
Clothes for a madman

Scene: Rome

First Act
FLAVIO tells Captain Spavento how ten years ago he was sent
   to France by his father Graziano and on the way was enslaved by Turks, and how two years ago he was freed from a
   Maltese galley, and on returning home couldn't find his
   beloved or his most confidential servant, and that since
   then he has had no joy. The captain consoles him with
   friendly words and they leave together.

FLAMINIA, Graziano's daughter, tells Franceschina that she loves
   Pantalone's son Orazio even though he's mad. Franceschina
   weeps, remembering her husband Pedrolino and her mistress
   Isabella, lost many years ago. Flaminia asks her to help
   Orazio, and goes inside. Franceschina says she's in love with
   a captain, believing her husband to be dead. Enters now the

CAPTAIN, who makes his usual boasts. Franceschina tells him her
   love. He mocks her. Franceschina taunts him, saying that
   she'll make him love her willy nilly. They dispute. Enter now

PANTALONE, who asks what the quarrel's about. Franceschina says
   that the Captain tried to force her honour. They berate the Captain
   and chase him away. Franceschina goes into her master Pantalone's
   house. Graziano tells Pantalone that his son Flavio has been
   melancholy since returning home. . . .

doctor Graziano, two *zanni* for male servants, one or two female servants, and the captain, who might also be a lover. One of the maidservants, the older Franceschina, could be played by a man, and the troupe among them covered the occasional roles by doubling as Moors and Turks, Ragusan seafarers and whore-mongers, gypsies, innkeepers, constables, fake magicians, French, Greek, and German travellers. In Scala's ten pastoral or tragic-heroic scenarios the nuclear cast was redisposed as nymphs and shepherds, kings, real magicians, lions, bears, and aerial spirits.

The identities of the stock roles survived the ceaseless disguisings of the plots, bringing their enduring functions and cumulative histories to each ephemeral scenario. Together with their immediate social and linguistic features, the lovers, the elders, and the captain manifested their descent from the classical Roman *adulescens, senex*, and the *miles* respectively, by way of *commedia erudita*, while some of the *zanni*—notably Arlecchino—had folk-festival roots on to which qualities of the *servus* and *parasitus* had been grafted.

Within the stylized and specialized functions shared by all *commedia dell'arte* casts, each player invested original features in his roles and many became permanently associated with their interpretations of one or the other of them. Thus the *zanni* were subdivided into Arlecchino, Francatrippa, Pedrolino, Frittellino, Scapino, and others, and thus arose the differences among the military blusterers, Captains Spavento, Coccodrillo, Rinoccronte, Matamoros. Plurilinguism made for another variable: Bergamasque for Zanni, Mantuan for Arlecchino, Neapolitan for Pulcinella, Spanish for some captains, pidgin Slavonic for Ragusans, and so on.

The plots in which these characters meet were built of condensed and recombined structures from the repertory developed in regular comedy, stories of crossed love, mistaken identity, disguises of sex and status, runaway wives and children, adventurous rescues, madness, apparent death, reunion of separated families, clowning, ingenious tricks, ridicule of jealous husbands and lustful old men, mocking of masters by servants, witty extortion, gulling, and unlimited opportunities for mayhem, erotic play, and coarseness. The actors fleshed out the scenes by drawing on their stores of stage business, slapstick, double-takes, and quick changes, their tirades and dialogues, their verbal and gestural *lazzi*, or comic quips and turns. The situations are entangled but the action on-stage is usually unified and limited to a short time and a single place, ordinarily a contemporary Italian city.

Some *comici* had published plays in the late Cinquecento, using the five-act form of regular drama to expand and fix in literary form material used in improvisation on-stage. Isabella Andreini's pastoral *Mirtilla* (1588), Adriano Valerini's tragedy *Afrodite* (1578), and a few comedies by other actors set a precedent for the printing in the Seicento of comedies by Scala and Cecchini, of the *Bravure* or bragging dialogues from his former stage practice by Francesco Andreini, and of a very mixed series of eighteen plays by his son Giovanni Battista.

A few of them wrote defences of play-acting. *La supplica* by Nicolò Barbieri, acclaimed for his improvisations of the Lombard *zanni* Beltrame, shows where the profession had arrived and how its members were seen by the society at large in 1634. The demand for their services was widespread but the obstacles to providing them were many. Following the Council of Trent, campaigns against the theatre caused old prohibitions to be taken more seriously. In Milan between 1565 and 1584 Archbishop Carlo Borromeo had inveighed against the theatre, urged that *sacre rappresentazioni* be replaced with oratorios, and warned against traffic with dissolute actors. Promoting the same end by different means, some religious orders, especially the Jesuits, offered a substitute drama in their schools and invited the community to attend. Sometimes travelling troupes were denied local permission to perform, customs officers confiscated their costumes and props, townspeople feared them as criminals, magicians, and prostitutes. The actors often supplicated the intervention of higher authorities and noble protectors.

Barbieri's defence takes the form of a supplication to the world in general, presenting the instructive and recreational value of theatre in the most positive way, appealing to the educated and worldly-wise against the superstitions of those who suspect all outsiders, particularly those who create magical illusions. He relies on anecdotes of how the actual performances of his company have repeatedly won over such ignorant provincials, especially among the clergy. He appeals to higher minds who recognize the wholesomeness of theatrical entertainment and can distinguish the grain of his *métier* from the chaff. The Jesuit theologian G. D. Ottonelli, author of the massive *Della Christiana moderatione del theatro* (1646–52), agreed in principle but deplored the immorality to be found even on Barbieri's stage.

Barbieri masked as Beltrame. Title-page of *La supplica*, or *The Supplication: Familiar Discourse of Nicolò Barbieri called Beltrame, addressed to those who write or speak of Actors without recognizing the merits of their virtuous actions. To be read by gentlemen of parts who are not compleat critics nor in any way dullards* (1634).

Stardom did not guarantee status. Although upper-class dilettantes sometimes joined their ranks and exceptional players like Cecchini obtained honours, the profession itself would remain less than respectable even after it was somewhat bureaucratized. But theatre was increasingly desired at all levels; moreover, it was a commodity and a profitable one. The *comici* would not achieve the economic independence of a guild, as Cecchini recommended in his *Brevi discorsi* (*Brief Discourses on Comedies, Comedians and Spectators*) (1621), but the success of ticket-selling theatres in Seicento Naples, Milan, and Venice multiplied audiences and put money into the pockets of impresarios. Actors depended increasingly on the kind of patronage from which the security of pensions might be expected, and freedom both of movement and of improvisation was restricted by their progressive need for organization under directors able to deal with finances, logistics, and politics as well as programmes. Great, though not always enviable, success in this line was achieved by G. B. Andreini, whose ability to manage a company and maintain the favour of patrons at the courts of Mantua and France, even more than his celebrated acting and writing of plays and treatises, made him a model *capocomico*, the theatrical administrator around whom the organization of professional troupes would depend long into the future.

The *comici* continued to play all genres, but the defining image of the *commedia dell'arte* established by their travels around Europe was of the improvising style. With its invention nearly played out at the end of the Seicento, though with decades of popularity still ahead, the style was codified for scholars and amateurs by Andrea Perrucci, a Sicilian lawyer in Naples, city of rich theatrical tradition.

Perrucci was also a playwright and the artistic director of the Teatro San Bartolomeo, but he insisted on his status as a dilettante rather than a professional. Half of his analytical treatise *Dell'arte rappresentativa* (*The Art of Staging Plays, Premeditated and Improvised*) (1699) is devoted to improvisation and its superiority to other acting techniques. The work is a mine of examples of types, topoi, dialogues, and specific language and dialectal locutions used for playing the various stock roles in solo and ensemble scenes. There is also a whole scenario called *La Trapolaria* (*The Play of Trapola the Trapper*) taken in part from Della Porta's eponymous comedy published more than a century earlier, a vestige of the creative exchange that had once linked the professional improvisers with the best Neapolitan literary comedy. By Perrucci's time, the *commedia dell'arte* style was a subject for history.

It is significant that so quintessential a *capocomico* as G. B. Andreini should in 1613 have written a closet drama, a religious one on the ambitious subject of the Creation, *L'Adamo*, dedicated to Maria de' Medici, Queen of France. Although subtitled 'sacra rapresentatione', the play is laid out as regular tragedy, with five acts, the unities observed, dialogue in unrhymed verse, a chorus, and as much decorum and verisimilitude as its subject permits. Andreini's preface defines his purpose as 'representing internal conflict by means of images and words in the Theatre of the Soul, with the heart as spectator'. To assist the inward spectacle the printer provided illustrations of every scene, many of them depicting the characters treading the boards in front of a backdrop in perspective.

L'ADAMO
SACRA RAPRESENTATIONE.
DI GIO. BATTISTA ANDREINO
FIORENTINO.
ALLA M.CHRIST. DI MARIA DE MEDICI
REINA DI FRANCIA.
Dedicata.

Con Priuilegio
Ad' instanza di Geronimo Bordoni libraro.
in Milano. 1618.

Andreini's sacred tragedy of Adam and Eve. Title-page of *L'Adamo*, first published by Geronimo Bordoni, Milan, 1613, engravings by C. Bassano from drawings by C. A. Procaccino.

This is no penitent repudiation of the stage; Andreini was launched on a career that would be thoroughly theatrical from beginning to end—in the 1650s his *Maddalena* was a multimedia event. Rather, *Adamo* shows the importance assumed by sacred drama in the Catholic sphere and gives another example of the cultural paradigm of the world-as theatre which increasingly dominated the imagination of seventeenth-century Europeans, regardless of sect.

Beginning in the mid-Cinquecento a redesigning and overhauling of the old-fashioned *sacre rappresentazioni* had become evident. In a process which repeated the fifteenth-century humanistic progression from recitation to oration to dialogue to playacting, moving from Latin to the vernacular, the newly founded Jesuit schools especially encouraged drama. Their reasons were pedagogical, promotional, and social, aiming to kindle devotion and to appropriate theatre as a weapon in the battle for Catholic reunification. Wherever they established ministries in Europe and Asia they continued and elaborated on spectacular Corpus Christi processions and introduced theatrical performances by student actors. In the Seicento sacred drama became a major genre, proliferating into subgenres in several venues, most brilliantly in the Rome of the Barberini Pope Urban VIII.

The critical treatise printed with *Ermenegildo martire: tragedia* (*Hermenegild the Martyr*) (1644) by the Reverend (later Cardinal) Pietro Sforza Pallavicino after the first of several performances at the Jesuit Seminario Romano measures the cultural distance travelled since another cardinal produced *Calandria* for another theatre-loving papacy. Pallavicino wrote in an age of debate over the theatre, Ottonelli and 'Christian moderation' on one side, Barbieri and the defenders of professional acting on the other. No ideological battle was waged, there was no argument for liberty against censorship or over desirable content, but merely over who was fit to produce socially responsible drama. No one defended the lascivious productions of the lowest strolling players. Champions of sacred theatre, like modern exponents of superior television programming, aimed at displacing vain theatre, maintaining the high genres and appropriating them as instruments of education. Like Tasso's and Milton's undertakings in the epic form, *Ermenegildo* with its

CLASSICAL TRAGEDY CHRISTIANIZED

Invoking the authority of Horace and Aristotle and the model of Sophocles, Pallavicino drew on Catholic history for his plot and theorized on the ideal modern tragedy.

From the *Breve discorso*:

Disposing the plot to observe unity of time and verisimilitude:
'The precept of Horace is appropriate here: *Nec quodcumque volet poscat sibi fabula credi*. Thus we see that Sophocles, in the tragedy that Aristotle indicated as the ideal of perfection, did not introduce Oedipus in happy state but in ordeal, as king of a city devastated by plague. The dramatist preferred to weave a plot that would be verisimilar on careful examination rather than excite wonder at first careless glance. Wonder without verisimilitude is easily achieved and gives no pleasure except perhaps that of laughter to those who hear the plot, nor does it merit the name of poetry, for it is not imitation of the truth, whereas verisimilitude, though not exciting wonder, has these virtues and is used with great art and mixed with wonder by the wisest makers of fictional plots. . . . '

Defending the occasional mixing of rhyme with the blank verse appropriate to 'Tragedia regolata':
'For a long time our language knew not blank verse, which was rediscovered in the last century by Trissino. Lodovico Castelvetro held that no Italian verse was without rhyme. Without mentioning the writers of Italian epic who wrote their poems in rhymed stanzas, which would have seemed strange to the Latins and Greeks, we see that our finest dramatic authors have not wished to deprive their plots of this beauty. The example was first given by the learned Sperone in his *Canace*, and followed by Guarino, Rinuccino, and Chiabrera . . . . likewise Andrea Salvadori in *Santa Orsola* and the lovely Muse of Monsignor Giulio Rospigliosi, of whom my pen must also record how he has deserved applause for grafting the most fragrant roses of Parnassus on the thorns of Calvary, consecrating to holiness the theatres of Rome usually given more to license.'

accompanying treatise incorporates the formal achievements of neo-Aristotelian theory of tragedy into a reading and writing of history that was for its times a representative declaration of the spiritually imperial mission of the Church and the civilizing power of Christianity.

Pallavicino's subject was timely. The Church was encouraging the cult of martyrs in the cause of orthodoxy. Uniting contemporary innovations in regular genre with traditional biblical and hagiographical subjects from *sacre rappresentazioni*, the new sacred tragedy also invited legends of national conversion and of modern sufferers for the faith. Tragedies were written about Thomas More and Mary Stuart. Pallavicino's Ermenegildo, son of the sixth-century Visigothic King Levogildo, is converted from the Arian heresy by his wife Ingonda. She overhears a wrong message and therefore fails to save her husband from execution, but by divinely providential irony (turning inside-out the Sophoclean pattern admired by Aristotle and held to be the highest aim of serious tragedy) her error begets triumph, for it brings Spain to Roman Catholic orthodoxy, as Ingonda's ancestress Clothilde had done for France.

Naturally sacred tragedy also found its way into music drama, the theatrical genre that finally surpassed all others, as Italian opera assumed the shape in which it would conquer Europe. The famous private theatre in Palazzo Barberini to

The stage as funerary chapel. The saint lies dead amid splendour in the niche where he has lived in anonymous humility. Rospigliosi's *Sant'Alessio*, III, printed by Paolo Massotti, Rome, 1634, with engravings by F. Collignon.

which a large though select audience of Romans and visitors had frequent access opened its first season in 1631–2 with *Il S. Alessio: dramma musicale* by Giulio Rospigliosi and the composer Stefano Landi. As a cardinal, Rospigliosi found time to write various dramatic texts for the Barberini, many of them on secular subjects, employing styles and characters from Spanish drama and from the *commedia dell'arte*, and even after he became Pope Clement IX he had his *La comica del cielo* (*The Actress of Heaven*) performed at Palazzo Rospigliosi.

Reputed to be the most dazzling theatrical event of its time, *Sant'Alessio* represents an intensely inward experience by means of striking outward display and variety of verse and musical forms, dances by devils, peasants, and others, disguises and metamorphoses, *intermezzi* with transformation machinery, splendid architectural sets for each of the three acts (the city of Rome, Hell, the saint's tomb), and a final view of Paradise with angelic hosts. A story from an old *sacra rappresentazione* frequently retold is here made an affirmation of the religious life, a rejection of the world, of Rome, wealth, family, bride, and honours. Alessio's inner struggle with the devil and spiritual redefinition of honour are made visible and universal. Roma personified appears in the prologue, as mother of heroes and owner of slaves, and proposes a new concept of heroism, based not on arms but on imitation of Christ, humility, self-knowledge, with anonymity in the world; Roma herself is transfigured and frees her slaves in order to rule within their hearts. Alessio, torn by the grief of his parents and bride at his flight, admits to a 'fierce battle in the theatre of his heart' but resists the temptation to return to the world's joys.

The stage transformations and spectacles most admired at the time were those of the versatile Gian Lorenzo Bernini, whose sculpture, architecture, and dramatic texts also reveal him as the emblematic exponent of the baroque age. An iconic *summa* of its governing metaphor is provided by the *mise-en-scène* of his Roman Cornaro chapel, where St Theresa's ecstasy observed by Cornaro prelates in theatre boxes is represented in marble with lighting by heaven.

While Bernini's doubling of illusions to make the spectator a part of the spectacle—by comedies within comedies or simulated floods that threaten to wash away the audience—has been seen as reducing humankind to the level of other merely natural phenomena and the spectacle of life to a theatrical illusion, the resulting emptiness of earthly existence need not be received as a tragic vision. As the dramatic genre of tragedy was then understood, life as a vale of tears and the human condition as limited by its terrestrial end were fit subjects for non-Christian tragedy, but the stage effects of Bernini and the later baroque era, like Calderón's dramatizations of the insubstantiality of life, ultimately express joyful transcendence.

Meanwhile the extreme development of *intermezzi* effects now extended into or sometimes simply displaced dramatic action. The material products of Italian

stage designers' technical ingenuity in this and in the succeeding generations of Ferdinando Tacca, Giacomo Torelli, Ludovico Burnacini, and the Bibiena-Galli family, and the lavish expenditures possible at great courts, produced marvels of landscape, action scenes, and sumptuously complex architecture, which did much to weaken the hold of the concept of unity of place, already loosened by the influence of Spanish drama. Increasingly Italian artists who worked wonders at the courts of Parma, Modena, Mantua, Rome, and Florence were called to Paris and Vienna and wherever there was a demand for splendour.

The spectacles and the rise of music-drama caused many to deplore the diminishment of the literary text, even in the act of welcoming the hybrid forms. Although overshadowed by developments in scenery, acting, and music, however, the writing of plays did not cease in Seicento Italy. Hispanoid three-act prose cloak-and-sword plays by Giacinto Andrea Cicognini in abundance were performed and printed. Regular five-act verse tragedies romanticizing foreign history in the tradition of Prospero Bonarelli were projected with all the advantages of the latest scene designs, as Girolamo Graziani's tragedy on Cromwell reveals. For audiences of literary connoisseurs the severe Aristotelian tragedy remained the great tradition and highest aim of the serious dramatist; Carlo de' Dottori's uncompromisingly intellectual and classical *Aristodemo: tragedia*, first performed in Padua in 1654, went into several editions and aroused great critical interest.

The liveliest phenomena, with the greatest following, were the commercial theatres of impresarios and the *melodramma* or *dramma per musica*, as early opera was commonly called by mid-Seicento. Here the emphasis was shifted from special-effect machines and costumes of court spectacle to more remunerative and economical programmes of singing and scene-changes. After debates about music in Greek tragedy, Aristotle on *melos*, and the propriety of sung dialogue, and following the success of the Florentine collaborations of Rinuccini, Peri, and Caccini that had produced *Dafne* and *Euridice*, the demand for more and more use of music in drama prevailed. Even the opponents of commercial theatre were more lenient toward music. In Rome singers like Adriana Basile and her daughter Leonora Baroni prospered, but, because of a papal ban against women in plays, *castrati* usually took the feminine roles on-stage.

In Venice, with *L'incoronazione di Poppea* (*The Coronation of Poppaea*), the three-act *opera musicale* with Giovanni Francesco Busenello's text and some of Claudio Monteverdi's last music, performed in the 1642–3 season of the Grimani family's Teatro SS. Giovanni e Paolo, with the noted singer Anna Renzi as Ottavia, we see fully established the commercial music drama and the institution of the operatic diva. Compared with Monteverdi's earlier *Orfeo* at the Gonzaga court of Mantua, the text of *Poppea* is kaleidoscopic, inclusive of favourite elements from different sides of the theatrical tradition: from comedy the amatory intrigue, the

Facing, above: pastoral landscape and Ovidian transformation. Aurelio Aureli's *Il favore degli dèi: drama fantastico musicale* (*The Favour of the Gods*), III. ii. Printed by the Farnese ducal press in Parma, 1690, with engravings by D. Bonavera from scene designs by Domenico Mauro and Ferdinando Galli Bibiena.

Facing, below: staged siege with gods in cloud machines. Giovanni Andrea Moniglia's *Ercole in Tebe: festa teatrale* (*Hercules in Thebes*), IV. xix. Printed at the Insegna della Stella press, Florence, 1661, with engravings from scene designs by Ferdinando Tacca.

transvestite disguise, the comic nurse Arnalta with a specific *commedia dell'arte* flavour borrowed from the Franceschina role, from tragedy the stichomythic debate, the narrating messenger, the *sententiae*, and from pastoral drama the triumphant figure of Amor.

The score matches this variety with its range of song-forms, contrasting moods, brilliant artifice, diversity of musical structuring, and lyrical expansion. Monteverdi's superlative gift for melodic expression of emotion was no impediment to a brilliant musical externalization of Busenello's Marinistic and intellectual text. With ingenious musical structures and interweaving dialogue to display conflict, Monteverdi even manipulated the text to make more theatrical effects. His stylistic diversity recalls the linguistic gamut for which the best of the *comici dell'arte* were admired.

Seen not as words set to music but as a vital organ of a theatrical body, each part inseparable from the other in performance, the text, written for the music but later published as if to be read, is very revealing for theatre history. With its mixed genres and multiple effects, its intellectual bent informed by the philosophical scepticism of the Venetian Academy of the Incogniti to which Busenello and other aristocratic dramatists belonged, the libretto of *Poppea* is also typical of Seicento

Cromwell in his bedroom having bad dreams. Girolamo Graziani's *Il Cromuele: tragedia*, v. xi. Printed by the Manolessi press, Bologna, 1671, with engravings of scene-changes for each act, designed for a performance which may not have taken place.

ideology. Moods and emotions are evoked, but primarily in relation to a formal inclusive plan that represented logical relationships seen from the distance of the audience, though not by characters confined within the play.

The triumph of love over fortune and virtue in *Poppea* echoes the Incogniti's free-thinking rhetorical debates. Their motto 'Ignoto Deo' was a declaration of philosophical scepticism and also of a more mainstream concept: the unknowability of the divine mind. At a still further distance the triumph of love and ambition over goodness and wisdom illustrates human beings' ignorance of what lies beyond them in time. With tacit reference to the pseudo-Senecan *Octavia*, Busenello selects a few events and excludes the rest of Nero and Poppea's bloody history from the plot. Addressing an audience aware of the whole story, he adds a definitive layer of irony to the ostensible victory of love. The godlike superiority of view thus established invites the spectators by extension to recognize themselves as actors in a play of which God alone knows the ending. The theatrical variety, the stylized series of scenes, characters, styles, and emotions from which Busenello builds his *libretto*, functions as a cultural whole with Monteverdi's theatrical and stylized structural music.

The only direction that the theatre could take in the next age would be away from the hybrid globalism of baroque symbol and spectacle, toward refinement and reason, toward Metastasio's lyric restraint, Goldoni's realistic and nuanced bourgeois comedy, and Alfieri's austere tragedy. For the time being, total theatricality had gone as far as it could go in Italy.

5

SPANISH RENAISSANCE THEATRE

VICTOR DIXON

SPAIN'S most important drama belongs to her Golden Age, when she gave her own proudest performance in 'the great theatre of the world'. Its history truly began in the era of Ferdinand and Isabella, who, by uniting the kingdoms of Aragon and Castile and imposing a stable, centralized regime, by completing the Reconquest of the Peninsula from the Moors, and by initiating the discovery and conquest of America, can be said to have laid the foundations of the greatest empire since Rome. The evolution of both its characteristic playhouse, the public *corral*, and its major dramatic form, the three-act *comedia*, took place in the reign of those monarchs' great-grandson Philip II. Most of its finest plays were performed when Spain's over-strained empire was already in ill-concealed decline, under Philip III and Philip IV. By the time the decline was all too obvious, in the reign of Charles 'the Bewitched', that drama was dying too. Only one major playwright, Calderón, could prolong its period of glory until 1681.

Before the end of the fifteenth century, most of the Peninsula saw only the most sporadic manifestations of drama. Things were different in Catalonia, but that story can merely be mentioned here. Quickly reconquered from the Muslims in the reign of Charlemagne, it developed a tradition of religious theatre, in the vernacular as well as in Latin, not unlike that of northern Europe. Indeed, its drama was both innovative and precocious, but would also prove conservative and long-lived. The performance of plays at Easter, for instance, survived till the nineteenth century, and the *Mystery of Elche*, though medieval in its staging, can still be witnessed today.

Medieval Castile, by contrast, produced no native liturgical drama in Latin, and has left only limited evidence of vernacular performances, though some may have been engendered by the major feasts of the Church. From the mid-fifteenth century, however, a variety of factors—the persistence of balladeers and other such strolling performers, the appearance of pastoral conventions and debates in courtly verse, an increasing fondness among the nobility for significant ceremonial, and a new respect for learning—set the scene for the appearance of a generation of accomplished court entertainers.

The first, 'a quintessential Renaissance man', was the poet and musician Juan del Encina. In 1496 he published four pairs of 'eclogues' he had devised and himself performed in at the court of the Duke of Alba, near his native Salamanca. Six or seven later plays, written mostly after various trips to Rome, portray more genuine conflict between characters and ideas, insisting in particular on the power of erotic love. The longest and most experimental is *Plácida and Victoriano*, performed at Rome in 1513. Replete with scabrous interludes, liturgical parodies, and pagan invocations, it could hardly be more irreverent; not surprisingly, it was to appear, like numerous early dramas, on the Counter-Reformation *Index* of 1559. By contrast, the *Auto de la Pasión* (*Passion Play*) written for Salamanca cathedral by his rival there, Lucas Fernández, is movingly devout.

Another all-rounder, the goldsmith Gil Vicente, was meanwhile surpassing the Salamancans both as a poet and as a playwright at the bilingual court of Portugal. Between 1502 and 1536 he produced over forty works: shepherd plays like most of theirs, but also moralities, allegories, romantic comedies, and farces. His masterpiece in Spanish, 'the most inspired of all nativity-plays', is *The Sybil Casandra*. Its spectators realize only slowly that the man-hating heroine and her aunts are not simply peasants but also pre-Christian prophetesses. Her rejected suitor is Solomon; his uncles are Moses, Abraham, and Isaiah. When at last she reveals her tragic mistake—she thinks, knowing God will be born of a virgin, that she is the chosen vessel—the others all rebuke her with truthful prophecies. The last of these hints that Judgement Day may be near, but at its climax the infant Redeemer and his true Mother are revealed. Vicente has interwoven, in exquisite verse laced with lyrics set by himself, a host of diverse traditions, and produced a play which proves in turn amusing, intriguing, and awesome, a theatrical jewel analogous with the Sistine Chapel frescoes.

More dramatic and more prophetic of later theatre, however, are the plays written mostly in Italy by Bartolomé de Torres Naharro. In 1517 he prefaced a collection of these, *Propalladia*, with the first important manifesto to be printed in the Renaissance. A self-assured, no-nonsense redefinition of humanistic comedy, it proposes for example just two kinds of drama, one based on factual experience and the other on plausible fiction. His own best-known play of the second type, *Himenea*, portrays a romantic passion that threatens to end in the death of its heroine.

Only her lover's nick-of-time arrival and promise of marriage can save her from her honour-conscious brother. Indebted both to Plautus and Terence, and to the *Celestina* (a turn-of-the-century masterpiece in dialogue, too long to be played entire, but probably meant to be read aloud), it anticipates both the cloak-and-sword comedies and the tragedies of honour so common a century later.

This first generation of playwrights, unfortunately, had few able successors. Many writers aped Torres Naharro, but their works, though widely read, were very rarely put on. The nobility remained addicted to pomp and spectacular display, but such plays as were performed for them were only occasional pieces. The Church, by contrast, was increasingly active in promoting religious plays for significant days in its year. The best are the forty-seven of extremely varied type by Diego Sánchez de Badajoz, published in 1554. But amid this predominantly amateur activity, professionals were beginning to appear. By the early 1540s, for instance, the most famous, Lope de Rueda, was leading—and writing plays for— a company of players. He and similar actor-managers were essentially itinerant,

The Prince of Wales in Madrid. In 1623 the future Charles I of England was welcomed with the pomp and pageantry long associated with royal entries into cities. But the performance taking place on the simple stage to the left of this engraving gives a good idea of how companies like Lope de Rueda's acted in the mid-sixteenth century.

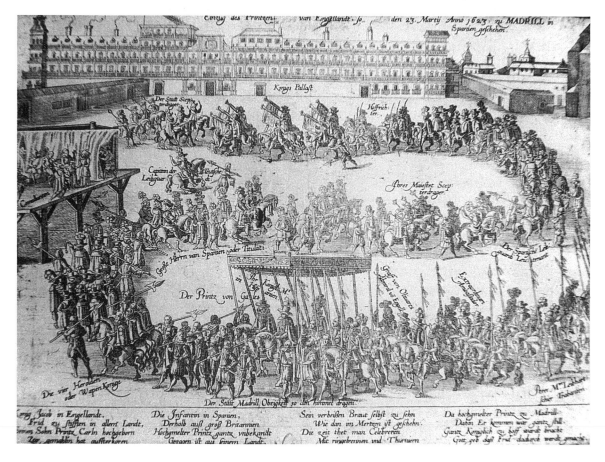

performing with minimal costumes and props wherever they found an audience. But in 1552, significantly, he was offered a long-term contract to direct and act in Corpus Christi plays for the city of Valladolid, and as early as six years later he seems to have planned to build a theatre there. At least from the 1540s, moreover, and throughout the rest of the century, Italian troupes, like that led by 'Ganassa', were constantly seen in Spain. They of course brought *commedia dell'arte*, and the occasional erudite comedy, but they too performed for the nobility and at religious celebrations. Indeed, Italy played an important role—how important, scholars dispute—in the rise of drama in Spain.

On the other hand, a native humanistic drama had also begun to emerge. The universities of Alcalá and Salamanca saw frequent performances of neo-classical comedy; initially Plautus and Terence, but later imitations in Spanish. At Jesuit schools, moreover, from their establishment in the 1540s, many teachers systematically wrote improving works—saint and martyr plays especially—for their pupils to perform to a variety of audiences. This trend towards learned drama was destined not to prevail, but it did have a recrudescence. For some fifteen years before 1590, a dozen serious playwrights attempted to stem the tide of popular drama by composing numerous tragedies of a rather more classical kind; in Valencia, for example, these included Andrés Rey de Artieda and Cristóbal de Virués. Even they, however, in different ways, were willing or forced to compromise and experiment. The same may be said, with even more truth, of two like-minded coevals, Juan de la Cueva and Cervantes. The author of *Don Quixote* has rightly been said to have 'missed the bus' as a dramatist, except in his farcical interludes.

This story so far has been complex. In sixteenth-century Spain court entertainers performed in the palaces of the great; the amateur actors of religious plays, in churches and public squares; students, in their colleges and schools; professional players, anywhere they could. But their different kinds of drama overlapped and influenced each other. Theatre flourished in every kingdom, as it would for another 100 years. Toledo, Salamanca, Valladolid, and Seville were especially lively centres, and Valencia was a somewhat aristocratic hotbed, towards the end of the century especially. Three critical factors, however, led to the crystallization of what was to be Spain's most characteristic and dominant form of drama: the evolution of permanent professional theatres; the establishment of the seat of government, in 1561, at the oversized village of Madrid, and its mushrooming growth thereafter; and the birth there the following year of a writer who (like Shakespeare) was a genius both as a poet and as a playwright.

Playhouses evolved in the way they did for two reasons. The professional actors needed places to which they could charge admission, and the charitable brotherhoods who ran hospitals for the poor found from the 1560s onwards that they could hire out their own yards or others for the purpose. These were mostly simply adapted, though some were custom-built. Entrepreneurs cashed in too, but each brotherhood was usually granted a monopoly in its own town. As attendances grew, they came to depend very heavily on this increasing source of funding, with very important results. Moralists were to demand incessantly that such dens of iniquity be closed, not allowed to open again after periods of public mourning, or at least more strictly controlled; but the economic argument, as well as public demand, always won out in the end. By the close of the century such theatres could be found in every sizeable town, and some in Spanish America. In recent years over a dozen have been researched, or even reconstructed, like those at Almagro or Alcalá de Henares.

A late sixteenth-century *corral.* This theatre, built when Almagro in central Spain was an important provincial town, is now the centre of an annual drama festival. Its spruce appearance is misleading; the restoration has been essentially accurate.

We shall have to concentrate here, however, on how drama developed in Madrid, about which we know most and where plays came normally to be written and given their first performance. After various disputes and false starts, by 1584 (and for a century and a half thereafter) Madrid had two *corrales*, the Príncipe and the Cruz.

If we imagine ourselves at the former in its heyday, the 1630s, and go in by its principal doorway, we see a roughly rectangular yard, about 65 feet by 60, hemmed in by adjacent buildings and ringed by projecting roofs. Its centre remains exposed, except that an awning (which will also improve the acoustics) is often used to shield spectators from the heat and glare of the sun, though this is its only source of light. (Some *corrales* elsewhere, at Valencia or Seville for instance, are different in shape, fully roofed, and rather more customer-friendly, but have similar staging resources; smaller towns of course have more modest theatres.)

At the far end stands a platform-stage, with traps, some 6 feet above the ground, about 28 wide by 16. Behind a series of pillars, it goes back another 8, as a tiring-house for the women (the men's being under the stage). This area is concealed by curtains with openings at either side to serve as entrances and exits, and those at the centre are often parted to reveal a discovery-space. Above, there run two galleries, the full width of the stage, providing more acting areas that the players can reach unseen; in some plays, though, they will have to ascend or descend in full view, using movable ramps or steps. On both of these galleries, in the discovery-space, and on platforms beside the main stage, tableaux or pieces of scenery may sometimes be displayed. For more spectacular plays, moreover, a wide range of stage machinery, concealed in an attic above, can afford many special effects.

Spectators are seated, some on benches, along both sides of the yard, on long banks of steps. These are topped, successively, by grilled windows, rooms we might call boxes, and attics, all of which have been opened out from the adjacent buildings. (Part of the attic on the right is the *tertulia*, mostly occupied by clerics.) At the back beside us are two boxes; in front of these various refreshments are sold, like *aloja*, a mixture of honey, spices, and water. Above us is a railed-off *cazuela*, or 'stew-pot', to segregate female spectators, with more boxes at either corner. On the floor above, there are still more boxes, with a large one in the middle reserved for city officials; higher yet, another *cazuela*.

In the yard, at the edge of the stage, stands a raised row of stools. Between these and ourselves stand the groundlings, called 'musketeers', since they often loose off missiles. To command their attention and approval, the actors and their material will have to be dynamic and compelling. They pay least, perhaps a fifth of their daily wage; the gentry, for a box, may be charged nearly twenty times as much, though they may lease one for the season. In fact the spectators in each *corral*, some 2,000 at capacity, are a microcosm of Madrid society. But like that society they are as far as possible divided; by rank and role, financial status, and (among the common folk) gender. Many are first- or second-generation immigrants from a countryside ever less able to sustain them; hence, no doubt, the fact that many plays offer a false, idealized image of the peasant and country life.

By now the Madrid *corrales* were run by lessees under contract to the town hall, but their operation was controlled by the Protector of the Hospitals, a member of the Council of Castile. The Council recognized only a limited number of troupes, and issued licences to perform to their actor-managers, called *autores*. In deference to public demand, and despite some early bans, women were allowed to act, and even to dress as men (though in theory above the waist only) whenever the plays required it (as remarkably often they did). Many actresses were indeed as able and famous as any man, and a very few ran their own troupes. A high proportion of major roles were written therefore with leading ladies in mind. Censorship was strict; every new play-text had to be vetted by the Protector's appointee. Moralists often complained that of course the way they were actually performed could not be controlled, but police were always on duty to enforce the regulations and try to keep public order. Demand having also led to performances all year round, starting times were fixed to ensure that spectators could be home before dark; at 4 p.m. in summer, but in winter two hours earlier. During Lent no performances were permitted (except by puppets and acrobats). This was

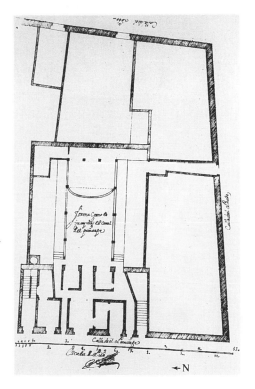

The Corral del Príncipe. This ground-plan of one of Madrid's two public playhouses was drawn up in 1735, but the theatre's basic structure had not changed throughout the century and a half of its previous existence.

A MORALIST DEFENDS ACTORS

A response to the grumbles of groundlings, from Juan de Zabaleta's *El día de fiesta por la tarde* (*The Holyday, in the Afternoon*), 1660.

Why do these men insult the actors? Because they don't appear as soon as they themselves arrive? Because this makes them waste the time they need for other vices? Because waiting is a bore? . . . If the actors were sleeping in their lodgings, there might be some cause for complaint; but they are always in costume long before it is time to begin. If they delay, that is because there are not yet enough spectators to make up for the losses on weekdays, or because they are waiting for some personage to whom they owe such respect that in order not to upset him they upset instead those they need so much to please, the common folk . . .

They prepare a play so thoroughly that rehearsals are many days of torture. At its first performance, any one of them would very gladly give a whole year's victuals to shine that day. When they come on stage, what fatigue, what harm do they shrink from to perform their task to perfection? If they have to fall from a height, they fling themselves down those mountain-sides they feign as vehemently as if they were in despair, though their bodies are human like others, and feel blows like the rest of us. If the play contains a scene in which an actor dies in agony, the one concerned rolls on those boards, full of spittle turned to mud, of ill-sunk nails and spiky splinters, with as little concern for his costume as if it were leather, though most often it is worth a fortune. . . . I remember seeing an actress, one of the most renowned (she died not long ago), who when acting a scene of high passion and happening to have a handkerchief in her hand tore it into tiny pieces to portray more precisely the emotions she was feigning, though its value was twice as much as she was earning. She did even more; since on that occasion it went down well, she tore such a hand-kerchief every day for as long as the drama was played.

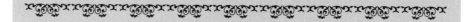

therefore when the *autores* had to form or reform their troupes, acquire and rehearse new plays, and negotiate with the town hall. Those contracted to perform at Corpus Christi the extremely well-financed sacramental plays we shall be considering later would usually get exclusive use of the *corrales* from the start of the season at Easter. Thereafter they might have to give place to rivals who had been playing in other towns, or (like a range of smaller outfits) simply touring. In practice, however, they often contrived to remain. Thus the Corpus festivities have rightly been described as the keystone of the drama of Golden Age Spain.

Increasing stability naturally led to higher performance standards, and the founding in 1631 of a craft guild, the Cofradía de la Novena, indicates the emergence of a corporate sense of acting as a profession. Actors in general might still be classed as vagabonds, as public sinners unfit for Christian burial, but the best were widely admired. The leading troupes had between a dozen and twenty-odd members, and so could cope with lengthy cast-lists, though all resorted to doubling.

Their company structure was hierarchical, with implications for their material. The *autor* and his leading lady, invariably his wife, expected meaty parts in every play. Other actors were hired each year for supporting roles, with at least one *barba* for older parts and above all specialists, male and female, as *graciosos* (clowns) or servants. Versatility was at a premium, for every show was a whole afternoon's entertainment. The main play, preceded by music and perhaps a festive prologue, often had its own songs and dances, and both its intervals were filled with farces, jigs, or other interludes. Thus musicians and dancers were needed, as well as stage-hands (called *metemuertos* or 'corpse-shifters'). Their financial rewards varied greatly, from a labourer's daily wage upwards. Successful leading actors could make a very respectable living, but the splendid costumes their public demanded (though invariably 'modern dress') were a very large expense.

The *autor* not only directed and acted, but ran the whole operation. Only rarely now a writer, he commissioned and bought plays outright, adapting them as he alone saw fit. Initially at least, their shelf-life was very short. First-time runs might be only a day or two; in 1631 three weeks was claimed as an all-time record. As in television or cinema today, the public expected a constant stream of new works. The *autor* might try to revive them, take them on tour, or sell them (though piracy was rife) to be played by others or printed. But the published versions were often so poor (or ascribed to a different author) that several writers

A Lope de Vega manuscript. Over forty plays have survived in Lope's own hand. This page from Act II of *El primero Benavides* (1600) is typical. It shows corrections during composition, brief stage directions, and also changes of verse-form. Between two scenes in octosyllabic stanzas, Mendo soliloquizes in a Petrarchan sonnet.

retrieved their texts and saw to their printing themselves. The original sale, however, was the playwrights' only serious source of profit, and one alone was prolific enough to make the major part of his income from plays. He was surely the first popular dramatist of modern times who ever managed to do so.

Lope de Vega, throughout his long and amazingly active life, was exceptionally fertile as a playwright, though also as a poet and writer of prose. Over half a century he must have written a full-length play every three or four weeks, on average—more than 700, we conjecture, about half of which have survived. These are so very varied that none can be said to be characteristic, let alone singled out as his masterpiece. Very many are comedies, some of the best having spirited leading ladies, like *La moza de cántaro* (*The Girl with the Pitcher*), *El perro del hortelano* (*The Dog in the Manger*), and *La dama boba* (*The Dumb Belle*). But one of his favourite subjects was the history and legends of Spain, and his most famous play is undoubtedly *Fuente Ovejuna*. Critics often bracket it with two superficially similar dramas, *Peribáñez* and *El mejor alcalde el rey* (*The King the Best Judge*), as if he meant them to form a trilogy, though he undoubtedly did not. As he wrote it, in about 1612, he had open before him five pages of a chronicle. These described two separate incidents of the 1470s, when Ferdinand and Isabella (whom Lope worshipped as the founders of the Spain he knew) were seeking to control the unruly noblemen at the head of its three great Military Orders. The young Master of Calatrava, persuaded to support the King of Portugal rather than Isabella as the rightful ruler of Castile, seized but was expelled by her forces from the strategic Ciudad Real; his Order's Grand Commander, having tyrannized a village, was murdered by its inhabitants, but when questioned under torture they replied as one that 'Fuente Ovejuna did it'. Lope's play has therefore been said to have not a single action but two. Some modern directors, like García Lorca, have even suppressed the first. In fact they are so indissolubly conjoined that, as so often in Golden Age plays, complexity enriches a unity that it in no way undermines.

Lope tells his story in sixteen scenes, packed mostly with vivid action. We can readily imagine how they must have been performed in a *corral*, and to what intense emotion. For example, at the end of the first act the Commander assaults the peasant heroine Laurencia, but is foiled by her lover Frondoso; at the end of the second, he disrupts their wedding, and has them both abducted. In the scenes that follow, Laurencia, who has apparently been raped, bursts in on the peasants in council, and we see the whole village attack and murder the Commander. The absence of a permanent setting allows a continuous flow of action, since shifts of time and place are effected by exits and entrances only; but this by no means implies a lack of visual impact. The peasants' costumes and props distinguish them sharply not only from the knights, whose crosses and swords are symbolic of both their rank and their true obligations, but also from the Monarchs, who at all their three appearances are accompanied by standard-bearers. When the Calatravans, with the banners of their Order, are expelled from Ciudad Real, those of Ferdinand and Isabella appear upon its battlements (the gallery above the stage); and the villagers, as they celebrate their murder in an arresting parody of the earlier wedding-scene, replace the head of the tyrant and traitor with the royal coat-of-arms. The climactic torture-scene is a supreme example of rapid changes of mood, through philosophical reflection, unselfish passion, suspense, fear, triumph, farce, and sentiment. Could one find anywhere in drama a more complex fusion of the serious and the comic?

No European theatre of Lope's time could have portrayed a popular rising in so positive a light. But the play is far from subversive; the peasants learn solidarity, not only as a community, but *pari passu* with the Monarchs who symbolize the state, and bestow at the conclusion a cautious pardon. Like his audience, Lope has no doubt that such monarchs are the Lord's anointed on earth. He may well be hinting that Philip III shows less direct concern than his forebears for the welfare of all his subjects, but his message—as in another fine play, *El villano en su rincón* (*The Countryman in his Corner*)—is not so much political as moral and philosophical. As all but the villain and his henchmen either know or learn, no man is an island; self-destructive self-regard must give place to its opposite, love.

A possible source of *Fuente Ovejuna*. One of Sebastián de Covarrubias' *Emblemas morales* (1610) may have influenced Lope's play. As well as soldiers and instruments of torture, it shows the villagers (unhistorically, but as in Lope's last scene) appearing before the King.

Lope's universal appeal, his phenomenal output, and not least his evident genius, were crucial to the development of Golden Age drama. Antecedents can be found for most of the elements of his art, but those he adopted became the norm, and little in later plays had not been foreshadowed in his. Above all, he gave the 'New Comedy' a fixed but flexible formula that would last for over a century. That formula he codified to some extent in his *Arte nuevo de hacer comedias* (*New Art of Writing Plays*), published in 1609. On the surface his 389-line poem is a shamefaced apology to his neo-Aristotelian critics. Not far beneath, it proclaims the self-assurance and know-how of the popular practitioner. Deeper still, it reveals a familiarity with and respect for the ancient writers characteristic of all his work. While Lope rejects, in the face of previous Spanish practice and contemporary taste, the rigidity of the neo-classical rules, he proposes modifications still full of Horatian decorum, 'striking a mean between these two extremes'. Those modifications, nevertheless, were destined to open the flood-gates.

He declines, in the name of nature, the variety of experience, to separate 'the comic and the tragic'. He and almost all his successors in Spain break the artificial mould of classification by genre. Most of their plays are called *comedias*, and a large proportion are in fact romantic comedies. The commonest are those 'of cloak and sword', set mostly in cities, in which middle- or upper-class courting couples are beset by a series of problems, almost always to be joined in marriage in the final scene. The majority are simply entertaining, though many are mildly subversive, and some decidedly dark. More essentially serious dramas too tend to have a happy ending, in that virtue is rewarded and vice condignly punished in accord with poetic justice. This, however, is hardly peculiar to the drama of Golden Age Spain; mass audiences always favour romance, and prefer to be reassured that right will eventually triumph. It is absurd to suggest, however, that Golden Age playwrights never project a tragic sense of life, or indeed that their Christian faith in divine justice hereafter precludes an agonized awareness of human suffering. Lope himself wrote few tragedies, but some are deeply moving, like *El caballero de Olmedo* (*The Knight of Olmedo*), or *El castigo sin venganza* (*Punishment without Revenge*), which almost conforms to neo-classical canons. Other dramatists chose more often to write in sombre vein, but in general they too saw life through a window, not a distorting prism. Above all they were free, from play to play and scene to scene, to vary the mix, to experiment. Thus their works defy academic attempts to pigeon-hole their plays into kinds they rarely sat down to write.

Paraphrasing Aristotle, Lope insists on the unity of action, though as *Fuente Ovejuna* illustrates he may interpret it broadly (and indeed he had often breached it). Those of time and place can however (with some caution) be subverted, for instead of concentrating only on a crisis a play must tell a whole story, 'from Genesis to the Last Judgement'. It does not occur to him to mention the possible objection that some kinds of story are not suitable for the stage. He and his contemporaries, faced by the constant demand for new plays, could contemplate no such restriction. Another of their contributions to theatre, more important than the plots and characters other nations were to borrow, was to demonstrate that drama could be made from what no one had dramatized before. Any source might provide an engaging story, and their first concern was to tell one. Rather than to embody in an action (as has been influentially argued) a preconceived theme or message, they usually sought to excite, move, and edify their public by staging extraordinary series of events. Their titles are full of superlatives and enigmatic phrases, and many of their plots have improbable complications. Very often these involve disguises (women dressing as men especially), or errors based on the premiss that scenes are occurring at night which of course are played in broad daylight. Their spectators must have loved such conventions, but were rarely themselves in the dark; in both comic and tragic situations, they revelled (Lope tells us) in irony, enhanced as it invariably was by the ambiguous use of language. Dramas-within-the-drama, interjections by the clown, constant changes of verse-forms, and probably of acting style, must frequently have reminded them that the play was simply play; but they must have known too that the image in the 'mirror of life' was a poetic distortion, offering both an escape from mundane reality and an awareness, typical of their time, of the illusoriness of appearances and the transience of earthly existence.

The story is often also more important than its persons. But another idea that has gained too much credence is that Golden Age dramatists were not concerned with character portrayal. That misconception was advanced to defend them in the face of another still prevalent: that any writer's primary task is to create 'real human beings'. In fact a high proportion of *comedias*, as again their titles show, are biographies of heroes, villains, or saints, or have at their centre a person who is in some way exceptional. That person may be a type, more exemplary than quirkish, or indeed an archetype; but some dramatists display a deep understanding of

SOME OUTSTANDING SPANISH ACTORS

An extract from José Alcázar's *Ortografía castellana*, *c.*1690. Damián Arias and María Riquelme played the young lovers in Lope's *Punishment without Revenge* (1631). The end of the passage suggests how one of its soliloquies may have been performed by the leader of their company, Manuel Vallejo.

Arias was a great actor. His voice was clear and pure, his memory reliable and his performance dynamic. Whatever he might be saying, he seemed to have the Graces in every movement of his tongue and the Muse in every movement of his hand. The most excellent preachers flocked to hear him to learn perfection in pronunciation and performance.

Amarilis was an actress of prodigious artistry. She acted, sang, played instruments, and danced; all she did won praise and acclaim. Another actress very famous a few years later in the theatre was La Riquelme, a good-looking girl and one so powerfully responsive that as she spoke, to the wonder of all, she could change the complexion of her face. If happy events were being described on stage, her features as she listened were suffused with pink, but if some ill-omened circumstance were related, she turned pale at once. In this she was unique and inimitable . . . There are few who can well express the turbulent passions of the spirit. An actor once came on stage in Madrid reading a letter to himself. For a long time he kept the audience in suspense. At every line he registered shock. Finally, in an access of fury, he tore the paper to shreds, and began to declaim the most vehement verse. And though he was praised by all, he provoked more wonder that day by his action than by his speech.

LA VILLA DE MADRID CORTE DELOS RE
RIO DE MAN
MARES
El Parque y mouro de los Venados y otras Caças
El Palacio Real del Su Mag Catholica
La Puente Segoviana
Puerta del Sol
Plaça Mayor
Calle mayor
El Rastro
1
2
3

A map of Madrid, *c.*1630. Plays were performed in (1) the royal Alcázar; (2) and (3) the Plazuela de la Villa and Plaza Mayor; (4) and (5) the Cruz and the Príncipe *corrales*; (6) noblemen's gardens (as on Midsummer's Eve 1631); and in the Retiro Palace, soon to be built at (7).

motives and behaviour. The whole-story formula permits their creations, like Laurencia in *Fuente Ovejuna*, to show both complexity and development. The finest playwrights strike a perfect balance between characters, action, and theme. Typically indeed they deal with the interaction of an individual with society, and beyond that society with a cosmos conceived as divinely ordered, and through imagery, allusions, and aphorisms they seek to clarify the universal import of each particular story.

As subjects for drama, says Lope, 'cases of honour are best, for they powerfully move all people'. Certainly Golden Age Spaniards are concerned with the need to defend their reputation, especially against attacks on it through their daughters, sisters, and wives. But preoccupation with one's image is not exclusive to Mediterranean peoples, and questions of honour are often a pretext for exploring other issues, like the nature of justice or prudence. Unless understood more broadly (as embracing for instance integrity, valour, loyalty, and faith), honour has surely been overrated as a theme of the *comedia*. Lope's other suggested subject, 'virtuous actions', showing positive regard for one's loved ones, one's community, and one's God, is certainly no less central.

Lope by 1609 takes it for granted also that a play should have three acts, and be written entirely in verse of several different types. Polymetry, thanks no doubt to his own poetic virtuosity, becomes another of Golden Age drama's most distinctive traits. Lope and his successors have always to hand a very wide range of forms, any one of which they may choose to create a particular mood, style, or pace, as a composer changes key or tempo from one passage to the next. (*Punishment without Revenge* exploits no less than thirteen.) Mainly they use, in diverse rhyming patterns, the fast-flowing octosyllabic line of Spanish ballad tradition, but for special effects they often turn to forms of Italian origin with lines of eleven syllables or seven. Thus their rapid dialogue or narration is often interrupted by passages of rhetoric or lyrical contemplation, and music, song, and dance are frequently introduced also. The *comedia* is always a lyric drama, indebted for many of its subjects and its intricate use of language to literature and especially to poetry ancient and modern. Its audience, usually called 'listeners', said they were going to 'hear' plays written for them by 'poets'. But those plays, like all great theatre, were also conceived for both the eye and the 'mind's eye'. Spanish dramatists, like their contemporaries in England, were skilled at exploiting not only descriptive language and costume but also all the other apparently limited but emblematically potent resources of minimalist theatre to work by suggestion rather than illusion on their spectators' imagination.

Castile alone had dozens of dramatists (Montalbán in 1632 listed seventy-seven), and some were scarcely less prolific than Lope; both Tirso de Molina and Calderón wrote well over 100 *comedias*. One critic has conjectured, not too wildly, that seventeenth-century Spain produced around 10,000 plays. Golden Age drama is characterized therefore by its unparalleled bulk, and to some extent was a collective enterprise. As early as 1622 we find nine dramatists combining to write a single play, and some of the finest works were in fact reworkings. One highly talented and active craftsman of the middle of the century, Agustín de Moreto, may never have written a single wholly original play, and its second half saw even more collaboration and rewriting. This massive response by the playwrights to the appetites of their public, and the actors' demands for a predictable range of roles, meant of course some standardization. Both expected for instance that every play would include at least one *gracioso*, who was usually a street-wise schemer or an ignorant country bumpkin. But the range of purposes, apart from humour and satire, for which this Protean figure was used (as a contrast to other characters, as adviser, messenger, Chorus . . .) is only one illustration of how flexible the formula could be. Since the writers' sources were so varied, since many had distinctive styles or treated similar subjects differently from one play to the next, the monolithic *comedia* crumbles whenever we examine individual works.

Lope's reputation, regrettably, has consigned to near-oblivion his numerous able coevals. One of these was Guillén de Castro, remembered now in the main for *Las mocedades del Cid* (*The Youth of the Cid*), the source of Corneille's *Le Cid*. Another was Mira de Amescua, the author of a powerful and influential drama, *El esclavo del demonio* (*The Devil's Slave*); yet another, Luis Vélez de Guevara, whose tragedy *Reinar después de morir* (*Queen after Death*) is the finest play ever written on the legend of Inés de Castro. By the early 1620s Lope's pre-eminence was more clearly challenged, however, by Tirso de Molina; many critics indeed would rank the disciple with his master and Calderón. His plays, more uneven than Lope's, are often ill-constructed, but his range is almost as wide, and he has a distinctive voice. His characters are often unusual, especially in their disconformity with gender stereotypes. Some of his heroines are endowed with great intelligence and resolution, as for instance in *La prudencia en la mujer* (*Prudence in Woman*), which, like *Privar contra su gusto* (*A Favourite against his Will*)—and many other plays of the time about ministers and monarchs—has a clearly political slant. His romantic comedies, like *Marta la piadosa* (*Martha the Pious*) and *Don Gil de las calzas verdes* (*Don Gil of the Green Breeches*), are distinguished by witty satire and complexity of plot, and his biblical plays, like *La venganza de Tamar* (*Thamar's Revenge*), have a wealth of arresting scenes. But his chief claim to fame undoubtedly rests (though his authorship of both has been questioned) on

A famous clown, Cosme Pérez, 'the funniest actor Spain has seen', was popular both in the *corrales* and at court, and numerous comic interludes were written for him. The frog he is dangling here alludes to the pseudonym by which he was better known, 'Juan Rana'.

a pair of plays whose protagonists deserve the unusual fate of damnation. The magnificently visual *El condenado por desconfianza* (*Damned for Despair*) tells the tale of a double-dyed villain who is ultimately saved, in counterpoint with that of a doubt-obsessed hermit consigned to the fires of hell. The latter, however, stays in the mind as a tragic, tormented figure. *El burlador de Sevilla* (*The Trickster of Seville*), which gave the world Don Juan (and the Stone Guest into the bargain), is in essence a similar sermon; its hero is damned less for lechery than for spiritual procrastination and self-deceiving deceit.

Juan Ruiz de Alarcón is in many ways much less typical of his time. A lawyer born in Mexico, he wrote only some twenty *comedias*, and the best known are thesis plays. Their construction is taut, their language direct—rarely lyrical or rhetorical—and their themes are crystal clear. Rather than romantic love, they celebrate honesty, integrity, loyalty, and friendship, and castigate those who lack them. Mistakenly, therefore, he is often considered unusually moralistic; in fact his preaching is simply more specific, and much more worldly-wise than transcendental. His most famous play by far is *La verdad sospechosa* (*The Truth becomes Suspect*). Its protagonist loses the girl he fancies through a series of misunderstandings, but these would all have been dispelled were he not a compulsive liar. Unlike his successor in Corneille's *Le Menteur*, he is therefore properly punished. Yet the exuberance of his inventions, and deft hints at his motivation, sorely tempt us to forgive him; most of his society, moreover (as in *The Trickster of Seville*), is shown to be no less mendacious.

Alarcón's precision and polish reflect one aspect of a transition, from the era of Lope de Vega to the era of Calderón, though for a dozen years they coincided and influenced each other. Calderón, born in 1600, was writing plays now famous in his twenties, though his greatness would not be fully acknowledged for at least another decade. The apparently spontaneous Lope is often and increasingly in full control of his inspiration; but calculating

A court jester. Pablos de Valladolid was one of the many buffoons who were kept to entertain the court of Philip IV. Velázquez's portrait shows him in histrionic pose, no doubt in imitation of contemporary actors.

craftsmanship is the hallmark of the more intellectual and philosophical Calderón. He perfects the established formula and intensifies its impact by imposing a more rigorous though also more stylized discipline. His technique reveals more method; fewer shifts of scene, and a less experimental use of verse-forms. Dialogues become more patterned, and long, rhetorical speeches more frequent; on the other hand, more use of soliloquy reflects an interiorization of conflict. He focuses more single-mindedly on the exploration of ideas, more clearly though sometimes more laboriously spelt out. Novelty and extravagance are replaced by incessant variations on similar devices, symbols, and motifs, giving an impression of predictability that is nevertheless misleading. His elemental imagery too becomes an almost mechanical system of cosmic correspondences. But his universe is characterized less by harmony and order than by chaos and disruption. His Christian orthodoxy offers no facile or complacent solutions to the problems his characters face, dilemmas no less intense because they are of their own and their fellow mortals' making.

His most approachable plays today are his comedies of cloak and sword—like *La dama duende* (*The Phantom Lady*) or *Cada uno para sí* (*Each Man for Himself*) —or his rural drama *El alcalde de Zalamea* (*The Mayor of Zalamea*), with its admirably prudent protagonist. His religious dramas—*La devoción de la cruz* (*Devotion to the Cross*), or *El mágico prodigioso* (*The Wonder-Working Magician*)— and his much-debated wife-murder tragedies—*El médico de su honra* (*The Surgeon of his Honour*), or *A secreto agravio secreta venganza* (*Secret Vengeance for Secret Offence*)—may be less to modern taste, but each is more profound and subtle than may at first appear. *La vida es sueño* (*Life is a Dream*), usually regarded as his masterpiece, was first published in 1636, though an earlier version may have been written six or seven years before. Like many *corral* plays, it makes more scenic demands than Lope's *Fuente Ovejuna*, though unlike many others it does not call for mechanical devices. Recent studies have made it easy to visualize a performance.

The discovery-space has been covered by a door-flat and made to suggest a stone tower. The gallery above has been decorated with simulated rocks. At one end of it a set of steps has been placed, disguised as the side of a mountain, and affording access to one of the platforms beside the stage. A woman appears on the gallery, but she is dressed as a man. To the end of the first act no character, apart from the servant who follows her, Clarín, will be aware of her gender; we ourselves will not learn her name, Rosaura, until the second act. Miming having been thrown by a runaway horse, she stumbles down the mountain, lamenting her plight as a stranger newly arrived in the land. In the gloom and desolation of the wilderness evoked by their lines, she and Clarín approach the open door of the tower. She makes out a 'living corpse', and a curtain is drawn to reveal him: a man dressed in skins and chained, though with a light beside him. Bewailing his fate, the prisoner asks heaven how he deserves it, only to reflect that 'to have been born is man's greatest crime'. Why then, he protests, is freedom denied to him and not to 'the rest'? By 'the rest', though, he means all creation other than mankind. The play's first few minutes have confronted us, it seems, not with real beings, but with archetypes and images of our common human condition.

A play in performance. This engraving in J. Solórzano Pereyra's *Emblematum centum* of 1653, showing a king and courtiers on stage (as in the palace scenes of *Life is a Dream*), illustrates how simply many *comedias* could be staged.

In the second half of the act, the cave is concealed by a curtain. The mountain remains in place, but courtiers emerge from the gaps in the tiring-house curtain and we imagine a royal palace. The rest of the action—there are seven scenes in all—will move between these very different locations. As the protagonists' identities and situations are gradually revealed, we engage with them as persons with individual problems. Rosaura has been seduced and abandoned by a claimant to the throne; the prisoner, Segismundo, is the legitimate heir, but his astrologer father foresaw at his birth that he would lead a rebellion. Brought nevertheless to the palace by way of an experiment, he indeed seems an arrogant monster; he is returned to the tower and told that what happened was only a dream. In the last act, freed by rebels and meeting Rosaura again, he learns that this was a pious lie, but the experience has taught him the lesson of the play: that all life is a transient state that man must look beyond. Clarín thinks to hide on the mountain from the civil war that follows, but is, because of this, its only victim. His death persuades the King, in defeat, not to flee from but face his own fate, and submit to Segismundo. But the Prince in turn surrenders, and so proves his fitness to rule. He vindicates Rosaura's honour, but prudently orders the leading rebel to be imprisoned in the tower.

On reflection we see that the drama has been masterfully crafted. Rosaura's story and Segismundo's are deftly intertwined. Her character, her situation, and her reinstatement offer parallels to his; at the same time her complex impacts on his emotions, his mind, and (as a symbol of other-worldly values) his soul, all play a role in his reformation. But *Life is a Dream* is also, in more than one sense, a mystery. Its opening scene, by contrast with those of almost all other *comedias*, tells us little about its persons, and they themselves very partially understand their situations. As the mists are slowly dispelled, we see them beset by what seem insuperable difficulties. Our own experience of the play is of life as a nightmarish labyrinth, through which only some, 'with valour and with prudence', are able to thread a way.

Corral plays like *Life is a Dream* or *Fuente Ovejuna* were not, as has been argued, propaganda fomented by authority in support of the socio-political status quo. That their ideology is by and large conformist (though certainly not uncritical or unreflective) is attributable less to censorship or other official pressures than to their authors' and their audience's essentially conservative outlook. On the other hand, much drama was clearly promoted by patrons with particular axes to grind. As theatre became the main mass medium, the powerful saw that the entertainment they themselves enjoyed could also be used as a means of indoctrination and self-aggrandizement. The religious promoted piety by having plays performed not only at Corpus Christi and on local saints' days but to celebrate special events like beatifications. Noblemen advertised their famous forebears and themselves by sponsoring writers like Lope to dramatize their achievements. Above all the drama performed at court was invariably not only entertainment but also, in one way or another, propaganda.

The reigns of Charles V and Philip II had seen relatively little, except on special royal occasions, but the accession of Philip III marked the dawn of a different era. Far more than before, professional troupes were hired for command performances, called *particulares*. But concurrently royal and noble amateurs often took part in plays and elaborate masques. Philip's favourite, the Duke of Lerma, was particularly active; his pages performed on his own estate in a series of lavish productions: in 1614, for example, *El premio de la hermosura* (*The Prize of Beauty*), by Lope, and in 1617 *El caballero del sol* (*The Knight of the Sun*), by Luis Vélez de Guevara. Significantly, the former was later adapted for performance in the *corrales*, and Vélez was making increasing use in them of machinery and special effects. Henceforth, palace and popular theatre would flourish together, alongside religious drama, and each would exert great influence on the others.

Court patronage greatly increased in the reign of Philip IV. Incomplete royal accounts from 1622 to 1638 record payments for over 500 *particulares*. Most were given at the Alcázar; one hall there was dubbed the *salón de comedias*. The young King himself (though in 1613 the rocking of a chariot in which he was playing Cupid had made him violently sick) became a keen theatre-goer, sometimes attending performances in the *corrales*. In another sense, though, he

The Retiro Palace. This plan of the main floor of the Retiro Palace in 1712 shows (1) the Church of San Jerónimo adjacent to which it was built; (2) the Hall of Realms; and (3) the Coliseo, in which most plays at the palace were performed. Note how the back of the theatre was extended to provide deeper perspectives.

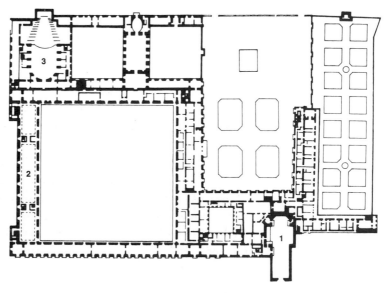

The Surrender of Breda. Velázquez's painting may have been influenced by Calderón's *The Siege of Breda*. In the last scene of the play Ambrosio de Spínola shows similar magnanimity in victory as he receives the keys of the city from Justin of Nassau.

was always an actor, the undisputed protagonist of a show stage-managed until 1643 by his minister and mentor the Count-Duke of Olivares. Olivares was to create for him in the 1630s the Palace of the Buen Retiro, not only (as its name suggests) as a place of retreat and recreation, but also as a magnificent showcase and performance space. The most ritualistic court in Europe, with the Planet King at its centre, was meant to impress on all comers (in the face of ever more apparent economic and military decline) the opulence and might of the Spanish crown. Such stress on display was a major motive for performances-within-the-performance; but these were often commissioned as vehicles for information or celebration. Information on current affairs was indeed one function of plays in general. With journalism still unknown, they complemented (or replaced, for the illiterate) the sporadically printed pamphlets that were often their source-material. Like these, they would only report the news if it was good. Recent Spanish victories especially were frequently either mentioned, described in irrelevant speeches, or used as the principal matter of documentary dramas.

In 1625, for example, an unusual series of triumphs (short-lived though they all were to prove) gave good reason for celebration. On 2 June the city of Breda in the Netherlands was taken by Ambrosio de Spínola after a nine-month siege. When the news reached Madrid two weeks later, Olivares supplied Calderón with the source-materials he needed for a play, *El sitio de Bredá* (*The Siege of Breda*); it was

162

put on at court no later than 5 November. By early July Bahia in Brazil, seized by the Dutch in May of the previous year, was known to have been recaptured by Don Fadrique de Toledo. Lope finished a play on the subject, *El Brasil restituido* (*Brazil Restored*), on 26 October; it too was produced at court, on 6 November. The two greatest living playwrights, simultaneously and almost as quickly as television writers today, had clearly been called upon to dramatize triumphs that other media would take longer to record. For one of the principal rooms of his new Retiro Palace, Olivares was to commission in 1634 twelve large paintings of Spanish victories, including no less than five achieved in 1625. One of these was Velázquez's *The Surrender of Breda*; another, *The Recapture of Bahia*, by Juan Bautista Maino. Both were almost certainly indebted to the plays that had preceded them. At the end of Act II of Lope's, for instance, a laurel-crowned Apollo lists the Spaniards who have been killed or wounded in a battle, and Heresy rises through a trapdoor to defy the Catholic Religion. At the end of Act III, the victorious Fadrique consults a portrait of the King, which seems to bestow a gracious pardon on the Dutch. In the unconventional foreground of Maino's painting, a Portuguese woman is tending a wounded Castilian soldier. This is subtle propaganda, like that in many of Lope's lines, for one of Olivares's long-sought aims: to persuade realms like Portugal to share the burden of Castile's imperial commitments. To the right stands a portrait of Philip, with Heresy under his feet, being crowned not only by Minerva but (at the Count-Duke's suggestion, no doubt) by that minister himself.

The court was beginning now to promote collaboration and competition, and some writers not primarily playwrights, like Francisco de Quevedo and Antonio de Mendoza, enjoyed its especial favour. Even more than before, though, it was looking to professionals to provide its entertainment and display. One of three spectacular plays performed by courtiers at Aranjuez in 1622 and 1623 had been written by the Count of Villamediana, but its scenery had been designed by an Italian engineer. In 1626 Olivares engaged another, Cosimo Lotti, who was to work a transformation effect in the staging of palace drama. In December the following year he astonished the court in his production at the Alcázar of *La selva sin amor* (*The Loveless Forest*), the first Spanish attempt at opera, by his stunning scene-changes, hidden lighting, and detailed perspective settings. Lope, as the librettist, was forced to confess, prophetically, that such spectacular effects made 'the ears defer to the eyes'. Lotti was apparently not involved in another lavish entertainment laid on by Olivares for Midsummer's Eve in 1631. Lasting the whole night, it moved from one nobleman's garden to another, and included two full-length *comedias*, one by Lope and one by Quevedo and Mendoza. But Lotti's innovations had come to stay. Before his death in 1643 he was to design a whole series of other productions, in the Alcázar and the Retiro, in the gardens of the latter and on its lake. Especially elaborate were his stagings, in 1635 and 1636, of

Facing: two scenes
from *Andromeda and
Perseus* (1653). The first
shows the proscenium,
with the curtain rising
to reveal a colossal
automaton, represent-
ing Atlas, which then
rose and sang. The
second depicts a scene
from Act III. Perseus
has just slain Medusa;
Pegasus is flying above,
and the *gracioso*, played
by 'Juan Rana', stands
stage left.

Calderón's *El mayor encanto amor* (*Love the Greatest Enchantment*) and *Los tres mayores prodigios* (*The Three Greatest Wonders*).

Lotti it was, moreover, who designed a new theatre at the Retiro, the Coliseo. It opened in 1640 with *Los bandos de Verona* (*The Factions of Verona*), a Romeo and Juliet play hardly worthy of its able author, Francisco de Rojas Zorrilla. The text we have suggests that the new theatre could accommodate the kind of staging characteristic of the *corrales*. In fact at the first performance, to amuse the Queen especially, their no less characteristic rowdiness was deliberately reproduced. Groundlings jeered and whistled, women pretended to fight in their *cazuela*, and boxes of live mice were released among them. Thereafter many similarly simple plays were put on at the Coliseo, and paying spectators were regularly admitted. On the other hand, some of the productions mounted there exploited to the full the spectacular scenic devices that Lotti had also provided, devices Italy had given by now to the whole of Western Europe: a proscenium with a front curtain; complex artificial lighting, painted wings offering deep perspectives, repeatedly changed to produce astonishing transformations; and a wide range of machinery, used especially for aerial effects.

The 1640s were nevertheless a bleak period for Spanish theatre, and a watershed in its development. The revolt of the Catalans and the war that gave Portugal independence were followed by periods of public mourning after the deaths of Queen Isabella and Prince Baltasar Carlos. Prohibition of the public performance of *comedias* (in 1644–5 and 1646–51) was a blow from which popular drama would never completely recover, not least because the court required those companies that survived to be at its own beck and call. At the end of the decade, by contrast, palace drama came truly into its own. Philip's new Queen, his niece Mariana, was welcomed from Austria with elaborate festivities, including several plays, and another outstanding designer was imported from Italy, Baccio del Bianco. In 1652 his production of Calderón's mythological drama *La fiera, el rayo y la piedra* (*The Beast, the Thunderbolt, and the Stone*) took seven hours to perform and was flocked to from all over Spain. Some idea of the spectacle he could achieve may be gained from eleven drawings, sent later to Vienna, of his staging the following year of a similar play by Calderón, *Andromeda and Perseus*.

Calderón, after entering the priesthood in 1651, had given up writing for the *corrales*, and works like these would constitute for the rest of his life a considerable part of his output. Most were based on Ovid's *Metamorphoses*, and most if not all were written for similar total-theatre productions, in which music, song, and ballet played an integral part. *La púrpura de la rosa* (*The Purple of the Rose*), in 1660, was indeed sung throughout, and *Celos aun del aire matan* (*Jealousy, even of the Air, can Kill*), of about the same date, came almost as close to pure opera. Others, combining arias, Italianate recitative, and spoken dialogue, came to be known, after one of the palaces they were performed in, as *zarzuelas*, very differ-

ent as they were from the Spanish operettas so called two centuries later. Scholars have long assumed that Calderón's texts were largely subservient to all the other elements, and his interpretations of myth either religious or philosophical. More recently it has been persuasively argued that the productions were conceived as coherent entities largely political in their encoded content; not merely, like such works elsewhere in Europe, supportive of royal authority, but critical or at least polyvalent in their reference to current issues. Golden Age playwrights were often adept at what an end-of-the-century dramatist, Bances Candamo, was to call 'saying without saying'.

Spectacles like these were sometimes revived, and plays successful in the *corrales* continued to be performed at court. The Coliseo had become the principal royal theatre, but productions were also mounted in other parts of the Retiro and at other palaces, like the Pardo, the Escorial, and especially the Alcázar. The *salón de comedias* there, refurbished between 1636 and 1640 and given in particular an elaborate gilded ceiling, was known now therefore as the *salón dorado*. A wooden stage, gilded to match, could be dismantled and stored, or even set up elsewhere. It was certainly in place, however, on 22 December 1672, for a performance in honour of the Queen Mother's birthday of a new two-act *zarzuela*: *Los celos hacen estrellas* (*Jealousy Makes Stars*), by Juan Vélez de Guevara, with music by Juan Hidalgo and settings by Francisco de Herrera the Younger. Scholars have unearthed not only the text (plus a prologue, interlude, and finale) but most of the music and six fine drawings by Herrera himself of how it looked in performance. This might seem enough to enable us to reconstruct the production. But further research has told us, more interestingly perhaps, how a not intrinsically distinguished work was experienced by those present. The *salón*, lined with portraits of former kings of Castile, was two storeys high but less than 30 feet wide. Its length at this time was about 120 feet, but the stage and a dressing-room beyond took up a quarter of one end. Court protocol, notoriously stricter than anywhere else in Europe, prescribed precisely the composition and disposition of the audience. The King was seated on a throne, on a carpet, between 10 and 12 feet from the back. Around him, on chairs and cushions, sat close members of his family. Covered benches lined each side wall. The ladies of the court sat leaning against these; the gentlemen stood behind, with pages kneeling before them. The stage space, though unraked, was restricted by three sets of wings, and two low steps at its forward edge afforded easy access. The numerous dances the texts provide for must therefore have been performed in the space between the spectators. Since this surely stayed lit, there was no clear distinction between stage and auditorium. All four texts moreover allude, directly or indirectly, to the occasion of the performance and the presence of the royal family. The latter, perfectly placed not only to see but also to be seen, were in fact the stars of the show, which truly began and ended with their ritual entry and exit.

Facing: Two scenes from *Jealousy Makes Stars*. In the first of Herrera's watercolours, Jupiter is attempting to seduce Isis. In the second Mercury, disguised as a shepherd, is singing the watchful Argos to sleep.

Scene representing Comedie.
Une Comedie representée devant le Roy et la Reine.

Above: Charles II watching a play. This engraving, *c.*1680, depicting a woodland scene in performance, illustrates how ideally the King was placed to appreciate the perspective effect created by a series of painted wings.

Facing, below: the *autos* in Madrid in 1644. That year, unusually, a special stage was built in the Plaza Mayor. These plans show how a two-tiered stand for official spectators was to be constructed in front, and how the two carts were to be drawn up behind when each of the four *autos* was performed.

At other palaces the audiences might be differently disposed, but the monarch was no less prominent. The Coliseo's auditorium had three rows of boxes each side, and a *cazuela* at the back; above this, in a crescent shape, rose 'His Majesty's balcony', with access from the King's apartment. But an engraving of about 1680 shows Charles II and his second wife observing a play in progress, flanked by their guard and privileged courtiers and enthroned on a canopied dais in the centre of the floor of the theatre. A description of the gala performance at the Coliseo that year of Calderón's last play suggests moreover that they often adopted just such an arrangement, 'to enjoy the ideal point of vantage for the perspective'. The stage-picture there must indeed have been impressive, and was to be even more so. The back of the theatre was extended, and a window constructed to allow a distant prospect of the Retiro gardens to form part of the setting; by early in the eighteenth century the stage could accommodate eleven wings either side.

For the last thirty years of his life, however, Calderón's main commitment was to a type of drama unique to Golden Age Spain: the *autos sacramentales* performed every summer throughout the country to celebrate the feast of Corpus Christi. The Corpus festivities have been described as 'the supreme example of that complete fusion of religious and secular life that characterizes late sixteenth- and particularly seventeenth-century Spain', and Calderón was by now the undisputed master-provider of the one-act plays at their centre. From 1648 to 1681 all the *autos* performed in Madrid were devised by him. But we must consider briefly how they had developed.

Early in the fourteenth century the papacy had decreed that the feast be honoured with rites and processions in every Christian parish. In Catalonia, by the end of the century, those processions had featured a series of tableaux on carts; in Castile, at least from early in the sixteenth, they often involved the performance of *autos* (then a general name for plays), based mainly on episodes from the Bible or the lives of the saints. The annual festival had become an occasion of public rejoicing and civic pride; but its profaner elements, in the view of pious critics, had obscured its central message: the redemption of mankind by the body and blood of Christ. By mid-century some *autos* were not only religious but truly sacramental, related more or less clearly to the mystery of the Eucharist. This may indeed have been a condition of such plays' survival in the Counter-Reformation.

Their performance varied, but usually a pair of carts would be drawn up at points on the procession route on either side of a third, or of a platform erected *ad hoc* to afford a raised acting area. Their scenic fittings might include not only curtains or doors (to permit discoveries) but often an upper level (to represent Heaven, for instance), and before long even machinery (to provide for special effects). Thus the outdoor productions of this type of play had features in common with those of *comedias* in the *corrales*, and later of spectacular shows at court. Alongside obvious differences, there are equally clear analogies between the three types of staging, especially in the symbolic use of space, both vertical and horizontal.

Financed by now, at huge expense, by rival municipalities, the Corpus *autos* inevitably came to be performed by the professionals, and to play a crucial role, as we have seen, in their economy. In early seventeenth-century Madrid, two com-

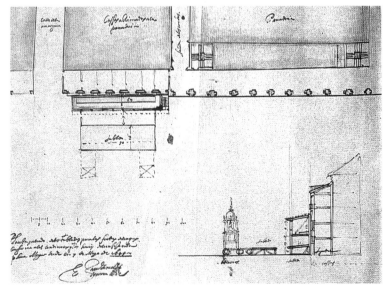

Above: design for a *tarasca* (1663). The Corpus Christi processions involved much street theatre, including a huge dragon; mechanical figures, on it's back worked by men inside, performed simple movements.

panies were each responsible for two, and were required to give several performances, in strictly determined order: first, normally, at the Alcázar before the King, then in the Plazuela de la Villa for the Council of Castile, and thereafter in other squares, like the Plaza Mayor. But they also put on adapted versions in the *corrales*, both before and after taking their productions to towns and villages round about. Thus the *autos* might keep them busy for a month or even longer.

No less inevitably, *autos* were written by all the leading play-

The Defence of Cadiz.
The seated figure in
Zurbarán's painting is
the veteran Don
Fernando Girón, who
though crippled by
gout insisted daily on
directing operations.
In Montalbán's *auto
The Relief of Cadiz*,
Faith is supported in
battle by
Understanding, 'an old
man with a very white
beard', who may be an
allegorized version of
Don Fernando.

wrights. Since the sacrament that these plays, as 'versified sermons', were designed to celebrate is central to all Catholic doctrine, they could keep to their brief by illuminating that doctrine in many different ways, but their favourite and characteristic mode was the allegorical. New Testament parables were ready-made, and Old Testament tales had long been regarded as primarily prefigurative, but with a little ingenuity (though sometimes a very great deal) they were able to take stories to allegorize from every kind of source: mythology, literature, history, everyday life . . . and even their own *comedias*. Thus one of Lope's cleverest and most lyrical is *La siega* (*The Harvest*), based on Matthew 13: 24–43; it refers, he says, 'through allegory, either to His church, or to the kingdom of heaven, or to the soul'. One of Calderón's earliest but most accomplished is *La cena de Baltasar* (*Belshazzar's Feast*), a re-enactment of Daniel 5 that is also a dramatization of the conflict within Man's psyche between his thought, his self-worship and worship of false gods, his conscience, and his awareness of mortality. But the range of subjects Calderón was able to allegorize is evident from some of his titles: *El divino Orfeo* (*The Divine Orpheus*), *La nave del mercader* (*The Merchant's Ship*), and *El pintor de su deshonra* (*The Painter of his Dishonour*).

Some *autos*, like many *comedias*, were written for special occasions or based on current events. For instance, one of the victories of that *annus mirabilis* 1625 displayed in the Hall of Realms was painted by Zurbarán in 1634 as *The Defence of Cadiz;* like others, however, it had been dramatized before. In 1623 the future King Charles I of England had been welcomed in Madrid (with lavish entertainments, including plays) as a suitor for the hand of Philip IV's sister Maria, but the project had been abandoned, in part because the Prince proved unwilling to convert to the Catholic faith. On his accession two years later he dispatched an expedition, which in November made an unsuccessful attack on the port of Cadiz. Its failure was due to poor planning and incompetent leadership, but for Catholic Spain was a God-given triumph over a Protestant foe. One of Lope de Vega's disciples, Juan Pérez de Montalbán, was moved (one hesitates to say inspired) to write an *auto sacramental. El socorro de Cádiz* (*The Relief of Cadiz*), based clearly on printed pamphlets, was completed in Seville on 6 April 1626, but was probably first performed in Madrid on 12 June, while Cardinal Barberini, as a papal envoy, was visiting the court. Its first half portrays the frustrated courtship, its second the defeat of the English. But the suitor and enemy is here called Error, the Prince not of Wales but of Darkness, and the Infanta has become Faith. Her defence of the City is sustained by the Church, who relieves it with bread and wine, and these are then transformed into the resurrected Christ. Spain's victory has not only been staged, but given an allegorical message both political and doctrinal.

The *autos*, like palace drama, became increasingly elaborate in conception, and above all in their use of music, dance, and scenic effects. In Madrid, from 1647, only two were performed each year, but each employed four carts. This clearly enhanced their magnificence and complexity, but made them also a spectacle rather less visible to the masses, since a platform rather than another cart was always needed now as a stage. Happily the municipal archives have preserved a wealth of documents, including from 1659 Calderón's memoranda for his designers. We can therefore reconstruct, for example, the performance of an *auto* he based in 1673 on his own play *Life is a Dream.* The four carts, drawn up at the rear of the stage, represent Earth, Fire, Water, and Air. In accord with the systematized imagery characteristic of Calderón's poetry, they are decorated with creatures and objects associated with each. The lower halves of huge globes mounted on the upper storeys fall forward on pillars to reveal four actresses, dressed to embody the Elements and seated, respectively, on a lion, salamander, dolphin, and eagle. Descending to the stage, they begin a vociferous struggle, symbolizing primeval chaos, for possession

Model of an *auto* cart (1646). The Corpus Christi carts carried massive two-storeyed structures of painted canvas over a wooden frame. Their decoration and the ingenious machinery they contained varied greatly. This drawing seems carefully executed, but the wheels and the height of the base above the ground are surely not to scale.

of a crown, but desist at the appearance, from the lower storeys, of Power, Wisdom, and Love. These attributes of the Trinity reduce them, as at Creation, to a conflictive accord, symbolized by songs and dances of praise.

It would be impossible here to summarize the whole of the action that follows, let alone to demonstrate how accurately but dramatically it mirrors the theology of the creation, fall, and redemption of man. Able to assume (like the writers of burlesque versions of other famous plays) that many spectators will know the story of *Life is a Dream*, Calderón follows it in outline, but by radically revising and reinterpreting it produces an entirely different work. When for instance Man, poisoned by Darkness (Guilt) and the Devil, has been returned to his prison on the lower storey of the cart of Earth, unable even with the help of Understanding and Free Will to break his chains, angelic voices herald the appearance, dressed as a pilgrim, of Wisdom. This Saviour, taking Man's place, is attacked by his enemies with the branch of a tree, but they fall at his feet. Grace, bearing as ever a light, returns to Man, and the alienated Elements serve him as before. Water provides baptism; Earth, bread and wine; Air and Fire, to transmute them, the Word and Love of God.

The allegory is one of a kind Calderón perfected late in life and (though anticipated to a degree by an *auto* specialist, José de Valdivielso) made especially his own: plays in which he contrived to make theatre out of theology itself, to present abstract and complex dogmas, as concretely as possible, through the analogue of a dramatic narrative. Such plays undoubtedly promoted deeper understanding and so more ardent fervour in the believers they were written for. In the view of their most lucid critic, they should convince even an unbeliever that the dogmas they body forth are tenable. They will not; but their fusion of intellect and art is itself, to all lovers of theatre, a revelation. The finest mystery plays of seventeenth-century Spain were conceived for complex and richly endowed productions by experienced professionals, and written by a dramatist of genius deeply versed in theology. They are the supreme examples of the genre.

We have seen, in broad outline, the rise in Golden Age Spain of a genuinely popular, national theatre, served by highly competent professional actors and at least two playwrights of world stature. That theatre, however, was increasingly privatized by the establishment of its day. This take-over, furthermore, coincided with and consolidated another. A drama centred on poetic language, speaking directly and intensely through its actors' bodies and voices to its public's imagination, was swamped in large measure by a tide of illusionistic display. Calderón alone truly rose to the challenge, extended his range, swam with and rode the tide. It is a measure of his greatness that he could. But even in his case, as in that of every other leading dramatist of his age, the *comedias* he wrote for the public *corrales* are the plays that have most to say to us today.

6

ENGLISH RENAISSANCE AND RESTORATION THEATRE

PETER THOMSON

WE can follow the development, during the sixteenth and seventeenth centuries, of something identifiable as an English theatre, centred on London but reaching out to the provinces. From one perspective, the narrative describes the trapping of the free spirit of the drama within the confining walls of a playhouse. From another, it celebrates the marriage of performance and literature through the enterprise of matchmaking projectors from John Brayne and James Burbage to William Davenant and Thomas Betterton. It is primarily a story of actors and their patrons or audiences, only secondarily of architects and playwrights, and it is conditioned by conflict. On the one hand is a delight in innocent pastime, on the other a suspicion that no pastime is innocent. The perils are clearly indicated in the chronicler Raphael Holinshed's account of Edward II, who 'furnished his court with companies of jesters, ruffians, flattering parasites, musicians, and other vile and naughty ribalds, that the king might spend both days and nights in jesting, playing, banqueting, and in such other filthy and dishonourable exercises'. Yet the Tudor monarchs, under whose patronage Holinshed wrote, kept jesters, musicians and 'other vile and naughty ribalds'. Henry VII's household included twelve trumpeters and a small company of actors. The King's players (*Lusores Regis*), four men and a boy augmented when occasion demanded, were versatile entertainers: actors, singers, dancers, acrobats. Only after a cultural shift would opera and ballet be separated from drama and a play forfeit its association with a game.

A Tudor monarch's progresses round the realm were marked by processions, street pageants, feats of arms. Even a mundane river-trip to Greenwich was serenaded by seamen who 'rowed upe and down syngying'. But the royal court had also to provide in-house diversions. Prestige and competition were involved. When the mighty Earl of Northumberland paid twenty troupes of players during his Christmas revels in 1511, the good host was masking the politician. Four years earlier, his brother-in-law the Duke of Buckingham made payments to forty-one minstrels for the entertainment of Yuletide guests in his newly built palatial home at Thornbury. Both these families, the Percys and the Staffords, had eased Henry's passage to the throne by turning against Richard III. Both had large households that were only a royal snub away from becoming private armies. Such lavish hospitality was a show of strength. The grandiloquent kitsch of Henry VIII's famous meeting with Francis I of France on the Field of the Cloth of Gold (1520) was a folly of nationalistic exhibitionism, certainly, but also a reminder to Henry's own regional warlords that anything they could do the King could do better. It was a theme to be taken up a century later in the court masques of the Stuarts.

King Henry VIII and his jester Will Somers. Henry VIII encouraged by example an aristocratic interest in the playing of musical instruments. He is here shown playing the harp behind his long-serving jester Will Somers. In common with Richard Tarlton, Somers was physically deformed and aggressively ugly.

The tense relationship between rulers and powerful families and the dangerous rivalries of the families themselves are recurring themes in Tudor and Stuart drama as they are in Tudor and Stuart history. The quantity of servants was a measure of strength, and liveried household players were a prized commodity. At home, they helped maintain the loyalty of a retinue by plying it with privileged entertainment; on tour, they enhanced prestige. Furthermore, their trade allowed uncommonly easy access to other houses or even to court. From there they could report back, enacting in real life the part of the ubiquitous intelligencer that many

Facing: the Brussels *Ommegang* of 1615. The annual Brussels *Ommegang* dates from the fourteenth century and is typical of religious processions throughout Europe. Denis Van Alsloot completed six canvases (two of them lost) depicting the 1615 *Ommegang*, which was dedicated to the Archduchess Isabella, governor of the Spanish Netherlands. This section of the fifth canvas shows the third, fourth, and tenth carriages in the procession. In the bottom one are Diana and her nymphs, not unlike the nine orders of angels Coventry displayed to Henry VIII; in the middle one are Apollo and the Nine Muses, 'divers beautifull Damsells'; and the top one, a ship drawn by sea horses and containing a Madonna and child, had so impressed on its first appearance at the funeral of Charles V in 1558 that it was brought out again in each subsequent year.

TRAVELLING PLAYERS

This extract from the play *Sir Thomas More* (*c.*1595) looks back to the composition and repertoire of the household players of the earlier sixteenth century.

MORE. My Lord Cardinal's players? now trust me, welcome.
You happen hither in a lucky time
To pleasure me, and benefit yourselves.
The Mayor of London, and some Aldermen,
His Lady, and their wives, are my kind guests
This night at supper. Now, to have a play
Before the banquet will be excellent.
How think you, son Roper?
ROPER. 'Twill do well, my Lord,
And be right pleasing pastime to your guests.
MORE. I prithee tell me, what plays have ye?
PLAYER. Divers, my Lord: *The Cradle of Security,*
Hit Nail o'th' Head, Impatient Poverty,
The Play of Four P's, Dives and Lazarus,
Lusty Juventus, and *The Marriage of Wit and Wisdom.*
MORE. *The Marriage of Wit and Wisdom?* That, my lads;
I'll none but that; the theme is very good,
And may maintain a liberal argument.
To marry Wit and Wisdom asks some cunning:
Many have wit, that may come short of wisdom.
We'll see how Mr Poet plays his part,
And whether wit or wisdom grace his art.
Go, make him drink, and all his fellows too.
How many are ye?
PLAYER. Four men and a boy, sir.
MORE. But one boy? then I see
There's but few women in the play.
PLAYER. Three, my Lord: Dame Science, Lady Vanity,
And Wisdom, she herself.
MORE. And one boy play them all? By'r Lady, he's loden.

of them would have played on stage. As successive Tudor Acts decreed harsher punishment for unlicensed playing, the livery was both passport and protection. It is clear, in hindsight, that household players were the immediate antecedents of the Elizabethan professional companies. Since the household troupes were small and doubling necessary, the actors had to be or become versatile. Increasingly, after Henry VIII's court set the fashion for audience involvement, they needed to develop the art of inoffensive interaction with social superiors, to improvise around interruptions, and to adapt to changing physical circumstances in unpredictable contexts. These demands are met by the professional troupe that visits the Danish court in *Hamlet*, but beyond the amateurs who take their play of Pyramus and Thisbe to Theseus' court in *A Midsummer Night's Dream*. Flexibility comes with practice. An established company of household players probably had a repertoire of six or seven plays at any given time, but they would take pride in their ability to prepare new work to order or to adapt old work to new circumstances, as Hamlet's players do. Nearly twenty years before the composition of *Hamlet*, during their patron's ill-fated campaign in the United Provinces (1585–7), the Earl of Leicester's Men, too, had performed in Elsinore. Some of them may have been retained by the Danish King Frederick II as part of his company of English players. There is no clearer evidence of the flexibility of English actors than their popularity on the Continent. Tokens of princely competition, Frederick II's company soon passed from his keeping to that of the Elector of Saxony. Other groups performed at fairs, where they caught the eye, even if they bemused the ear, by their agility, by the unfamiliar bawdiness of their clowning, and by the splendour of their costumes. The German John Sommer, deploring his countrymen's self-display in costly finery, complained that 'they strut along like the English comedians in the theatre'. Some actors took up permanent residence on the Continent. Others toured, as they might the English provinces, before returning to base.

Increasingly, during the last quarter of the sixteenth century, that base was London. The gradual suppression of the Corpus Christi plays in provincial centres was one outcome of the post-Reformation attack on the Catholic calendar. Plebeian culture was not, of course, eradicated, but, by 1600, its grip had been significantly loosened. The rapid growth in population led to economic instability. Under the Tudors, land-rents tended to rise and real wages to decline, so that rich and poor became increasingly polarized. It is unlikely that the urban poor were beneficiaries of the shifting pattern of dramatic entertainment that saw locally produced plays and pageants replaced by occasional visits from household companies. Some of the poorest took to the roads to swell the number of vagrants; some sought a change of fortune in the great city. Between 1520 and 1600 London's population quadrupled, from about 50,000 to 200,000 (by 1700 it would approach 600,000). The transformation of England from a feudal to a capitalist society was most rapid here. One typical Elizabethan Londoner is a speculator in search of a market for his commodities; and it was to marketeers that London owed the construction of the first purpose-built playhouse. In 1566–7 Thomas Gresham was planning and erecting the Royal Exchange in Lombard Street and John Brayne was planning and erecting a stage and scaffolds for the Red Lion playhouse. It is a historical coincidence that makes its own meanings. Four decades later, Thomas Dekker summarized the story so far in *The Gull's Horn-Book* (1609):

The theatre is your poets' Royal Exchange, upon which their Muses—that are now turned to merchants—meeting, barter away that light commodity of words . . .

We can only guess what gave Brayne, at the age of 26, the confidence to undertake the Red Lion venture. He had recently completed his apprenticeship as a grocer, married, and set up shop in Bucklersbury. He may, with business prospering, have succumbed to persuasion from his brother-in-law James Burbage. Unlike Brayne, Burbage, who had served apprenticeship as a joiner, is known to have been active as a player. There is more than a whiff of collaboration in the extraordinarily bold Red Lion project, and we cannot exclude the possibility that Burbage was already in Leicester's Men, of which he was certainly a leading member by 1572. Was it for that company that Brayne financed the erection of a playhouse on the Red Lion farm, just east of London, in the hamlet of Mile End? The immediate intention was to present a play about Samson, but there is no record then or subsequently of performances at the Red Lion. Since nine years passed before the building of London's next playhouse, we may conclude that it was not a manifest moneyspinner. Since Brayne and Burbage were again involved in the construction, we cannot conclude that it was a fiasco.

The legal documents (Brayne sued his carpenter) on which our knowledge of the Red Lion is based reveal the prescience of its design. The stage was about 40 feet wide, 30 deep, and 5 high, dimensions close to those of the Globe and Fortune. It was backed by a turret 30 feet high. In later playhouses, the turret was probably made continuous with the galleries to form the tiring-house, but at the Red Lion it was free-standing, though battened to the stage for security. There was standing room in the yard and seating on the scaffolds that enclosed it and the stage. Enclosure was of the essence. Collecting money from casual spectators in the open air is difficult. Whether or not the Red Lion served as a model for later playhouses, it both anticipated the shape of things to come and created the theatrical box office. Given the English climate, an indoor stage would have been better, but a comparably capacious roofed structure would have been prohibitively expensive to build and light. The size and design of theatres has almost always been determined by the needs and greeds of managers. Brayne was a businessman, not Maecenas.

We may reasonably read the nine-year hiatus between the Red Lion project and the construction of the Theatre as evidence of caution in the face of hostility from the city authorities. The constant theme of anti-theatrical rhetoric in the sixteenth century, echoed in modern England by press campaigns against football hooligans, is a fear of unruly assembly. That Tudor legislation in the matter is contradictory is unsurprising, since the inhibition of pleasurable pastimes might itself provoke a riotous response. In provincial towns, touring players were not infrequently paid *not* to perform. London's problem was that it had too many players. As two-faced as in their treatment of usurers, the legislators proceeded by a policy of trial and error, obstructing sometimes what at others they found useful. Thomas Cromwell's employment of the fiercely Protestant John Bale in his propaganda campaign is powerful evidence of his belief in the efficacy of drama. Had Cromwell not lost favour with the King in 1540, the English theatre's passage to legitimacy might have been speedier and its condition on arrival less wholesome. But in the religious turmoil of the mid-century, there was a perceptible hardening of attitudes, with the corporation of London especially vociferous in its opposition to plays and players. During the reign of Elizabeth, it became almost a routine for each successive lord mayor to appeal to the Privy Council for the suppression of theatrical performance within the city. To be sure, mayors barked more often than they bit, but actors who looked to court or to noble patrons for protection did not always find it. The rigour of the 1572 Act for the Punishment of Vagabonds brought to a climax years of hostility towards masterless men, most of them victims of economic depression, with whom common players were identified. It was an unintentional by-product of the Act to accelerate the transformation of household players into professional actors. If the London Corporation

PLAYHOUSES IN LONDON

This extract from John Stockwood's *Sermon Preached at Paules Cross* (1578) provides early evidence of their popularity.

Will not a filthy play, with the blast of a trumpet, sooner call thither a thousand, than an hour's tolling of a bell bring to the sermon a hundred? Nay even here in the city, without it be at this place and some other certain ordinary audience, where shall you find a reasonable company? Whereas if you resort to the Theater, the Curtain, and other places of plays in the city, you shall on the Lord's day have those places, with many other that I cannot reckon, so full as possible they can throng, besides a great number of other lets to pull from the hearing of the word of which I will speak hereafter . . . What should I speak of beastly plays, against which out of this place every man crieth out? Have we not houses of purpose built with great charges for the maintenance of them; and that without the liberties, as who would say: 'There, let them say what they will, we will play'. I know not how I might with the godly learned especially more discommend the gorgeous playing-place erected in the fields than to term it, as they please to have it called, a Theater, that is even after the manner of the old heathenish theatre at Rome, a show-place of all beastly and filthy matters, to the which it cannot be chosen that men should resort without learning thence much corruption . . . For reckoning with the least, the gain that is reaped of eight ordinary places in the city, which I know, by playing but once a week (whereas many times they play twice or sometimes thrice) it amounteth to two thousand pounds by the year.

recognized an ominous portent in the Queen's patent to Leicester's Men two years later, they concealed anxiety behind conciliation. An Act of the Court of Common Council, dated December 1574, reaffirms the city's command over performances within the city and suburbs, but welcomes 'Lawfull honest and commely use of plays'. Such vagueness of terminology is as typical of Tudor theatrical legislation as of its evident ineffectiveness. Legal prohibition was, perhaps, most useful as a prelude to income from the sale of legal exemptions.

It seems likely that the flurry of theatre-building in London in 1575–7 was a reaction to the increased status of household players. The Bel Savage inn was readied for performance in 1575—possibly the Bell, the Bull, and the Cross Keys, too. Enclosed inn-yards like theirs had provided a model for John Brayne at the Red Lion, and the legitimate business of hospitality had some chance of incorporating dramatic interludes, as in the great households, without offending the authorities. We have records of occasions on which they fell foul of the law, but can only guess at the number of performances that passed without comment. For Burbage and Brayne, embarking on a substantial building project in spring 1576, it was prudent to site the Theatre outside the city walls. The plot of land they leased above Finsbury Fields in Shoreditch was within the liberty of the dissolved Holywell Priory. Liberties were thorns in the flesh of the city council, since they fell outside their jurisdiction, or, if within it, so ambiguously as to render the sheriff and his officers innocuous. As atavistic relics of feudal independence, the liberties (or manors), inside or outside London's walls, were both marginal and peculiarly intrusive in the life of the city. In both respects they mirror the cultural impact of the playhouses they contained, not finally exempt from the law but elusive within it. The Theatre was more substantial than the Red Lion, with timbered galleries in place of scaffolding. Burbage surely intended it for Leicester's Men. It has been recently argued that the neglected playhouse in Newington Butts was built around this time for Jerome Savage, a leading member of Warwick's Men, and the equally obscure Curtain, situated close to the Theatre, may have been initially occupied by Sussex's Men. These were Elizabethan public theatres at their most basic. The actor made his entrance on to the platform through one of two doors in the tiring-house façade. Once there, he was exposed to an encircling audience and protected only by his own skill and an eye-catching costume. The stage was bare, but never drab. Actors were regularly exempted from sumptuary legislation on apparel, legislation which was designed to enforce a correspondence between style of dress and social status. His costume was the outward sign of an actor's place in the pecking order of a play, a visual aid to the spectators' immediate involvement in its narrative of power. Costume identified his role and liberated him to perform its function before auditors who pressed in on him from below and above. No one in the audience was far from the platform. For anyone who has stood inside the ring of the excavated Rose foundations, the startling impact is of smallness, of

proximity. In a modern theatre such proximity would invite intimate playing, but the Elizabethan crowd was quiet only if the actors silenced it. If they were ever to draw the audience in, they must first reach out. Expressiveness was their mode.

It is extraordinary, but possible, that Elizabethan speculators built playhouses in anticipation of a repertoire of plays to fill them. Certainly, few texts suitable to the public stage between the Red Lion project of 1567 and the opening of the Rose in 1587 have survived. But it is hard to reconcile the shrewdness of Burbage with the naïve optimism of a Caxton building a bookshop. Elizabethan London gathered from the universities and Inns of Court many young men with literary aspirations and empty pockets, eager to provide the new market with its commodity of plays. Actors were the entrepreneurs, and we can gain an impression of the demand for material from a contract dating from 1571–2. The signatories were on the one hand three actors and on the other a certain Rowland Broughton, who undertook to provide them with eighteen plays over thirty months. Broughton

Foundations of the Rose Playhouse. The bystanders at the bottom of the photograph are looking over the back wall of the playhouse, across the yard with its two modern concrete pylon-bases, towards the outline of the two stages, the lower one dating from 1587, the upper one showing the extension to the playhouse in 1592. The stage is not as deep as scholars had expected, and it evidently tapered.

failed to fulfil the contract and was sued by the notoriously quarrelsome Dutton brothers and their fellow actor Thomas Gough, but the rapidity with which a national repertoire was created is one of the Elizabethan theatre's most astonishing achievements. According to the hostile John Stockwood, London had 'eighte ordinarie places' occupied by acting companies in 1578, and this at a time when new plays filled theatres and six performances of any play marked it as a singular success. To meet the demand, writers clubbed together to produce texts by a simple division of labour—nothing so grandiose as a unifying artistic concept.

There were, of course, exceptions, and John Lyly, the first distinctly Elizabethan playwright, was one. Lyly wrote for and had a financial interest in boys' companies, the best established of which performed in an indoor, 'private' theatre in the precinct of St Paul's Cathedral. The 'privacy' was as much a law-dodging fiction as the conventional defence that public performances were nothing more than rehearsals for presentations to the Queen. Curiosity tinged with prurience fed a fashionable delight in watching pubescent boys tackle adult roles, and Lyly, whose aspirations were courtly, wrote for a sympathetic audience. His plays, from *Campaspe* (1583) to *Endimion* (1588), did nothing to advance the cause of a popular, public drama. Their staging harked back to a former age, with mansions mounted on temporary platforms to represent locations which are sites for debate rather than action. Lyly brought to plays the prose style that he had pioneered in *Euphues* (1579), in which a slender plot provides a pretext for elegant discourses on education, wit, friendship, religion, all embellished with alliteration, rhythmical antitheses, allusions to mythology, and elaborate similes drawn from natural history. An overriding intention is to please the court through lightly encoded flattery. The eponymous hero of *Endimion*, for example, might coincide but need not equate with the Earl of Leicester; but Cynthia *is* Elizabeth I. She is also the moon, since the monarch has a celestial as well as a human body; and, as the moon, she is remarkably like Elizabeth I:

Is she not always Cynthia, yet seldom in the same bigness; always wavering in her waxing or waning, that our bodies might the better be governed, our seasons the daylier give their increase . . . (III. iv. 177–80)

It was by 'always wavering in her waxing or waning' that Elizabeth kept her ambitious courtiers guessing. She would not have been offended by Lyly's characterization, but, just in case, he adds a final clause: 'yet never to be removed from her course as long as the heavens continue theirs.' *Endimion* transports us into the dangerous decorum of the Queen's theatrical court, but it is Marlowe who embodies the greater rashness of the public theatres.

The son of a Canterbury shoemaker, Marlowe, like many of the characters he created, was already competing for personal advancement when he entered the University of Cambridge. It was perhaps through espionage, as a minor cog in the revolving wheel of Walsingham's intelligencing, that he acquired his synoptic view along the corridors of power. Much of Marlowe's notoriety was posthumous, but he had certainly a talent to provoke. His plays are implicit dialogues with those in social, political, or religious authority on themes such as homosexuality, regicide (even to imagine the death of the Queen was treasonable), and damnation. They opened the public theatres to a complex discourse on matters of state. Only in the sixteenth and seventeenth centuries has the English theatre been allowed to be adult, and some of the credit for that belongs to Marlowe. *Tamburlaine the Great* was bought by the Admiral's Men and staged in 1587, when England was at war with Spain and Philip II was preparing his invincible Armada. Its hero is a barbarian warrior who overwhelms sophisticated kingdoms and who ends in triumph. London audiences flocked to it, and Marlowe agreed to write a sequel for the company. Here at least Tamburlaine 'the scourge of God' dwindles into death, but shamelessly. Had Marlowe been suspected of writing allegory,

after the style of Lyly, he would have been imprisoned, tortured, perhaps even executed. The nation was, after all, in crisis, and the Lord Admiral, patron of the players who presented *Tamburlaine,* had greater things on his mind than the defence of a repertoire. Because he went unpunished, and because he perceived the power-struggles in and between the courts of Europe in terms of choreography and rhetoric, Marlowe set the standard for fifty years of political drama.

It might also be argued that Edward Alleyn's performance of the title-role advanced the art of acting. To play Marlowe's heroes calls for a counterpointing of rhetoric with playfulness, something more subtle than the straightforward 'personation' of a fictional creation. It was in Alleyn's response to the challenge of Marlowe that the journey from personation to characterization began. In the years preceding *Tamburlaine,* the new professional companies were establishing their personnel and practices. Their status was enhanced in 1583, when the Queen appointed her own company, but their livelihood was under frequent threat from opponents in office and from outbreaks of plague which led to the closure of the city's playhouses. Partly to protect themselves, they formed joint-stock companies whose members underwrote loss in return for a share in profit. The actor-shareholders purchased plays, hired additional players, and negotiated with the authorities. A successful company might, by the end of the sixteenth century, have eight sharers, a wardrobe-master (tireman), a book-keeper in charge of prompt copies of the purchased plays, as many as six boy-apprentices to take the women's parts, and a pool of names from which to draw extra cast, musicians, stage-hands, box-office staff (gatherers), scribes, and so on. The likelihood is that the number of adult actors normally required for the performance of an Eliza-bethan play rose from twelve to sixteen between 1580 and 1600. Doubling was routine, and plays were broken into scenes to facilitate the alternation of charac-ters. The occupation of a playhouse eased access to financial pickings in the burgeoning consumer society of post-feudal London. Their residence at the Rose assisted the rise to prominence of the Admiral's Men. Their chief rivals, after 1594, were the Chamberlain's Men, initially at the Theatre and then, from 1599, at the Globe. Competition between these two prime companies is a feature of the last decade of Elizabeth's reign.

A stable membership was of importance to a theatre company. Each actor liter-ally possessed his part, written out by a playhouse scribe. The loss of an actor meant the loss of his parts with all the associated repertoire problems. Most of the men who formed the Chamberlain's company had worked together before, but Shakespeare, at the age of 30, was new to the rank of senior actors. Over the next decade the membership changed very little. Shakespeare's proven skill as a play-wright must have advanced his claims to a share. He was one of a new breed, resented by the university men who had settled in London with the intention of living by the sale of plays to their educational inferiors. Robert Greene's notori-ous abuse of Shakespeare as an 'upstart crow', which dates from 1592, is built around a charge of plagiarism. Actors had written plays before, but none with such a range of reference and evident learning. This rustic *arriviste*, in Greene's view, must be a cheat. In effect, though, working with and writing for a company gave Shakespeare a vital incentive. The talent of Richard Burbage, in particular, stimulated his creativity. Like Molière, Shakespeare responded intuitively to the histrionic temperament of his colleagues, and his plays made a major contribution to their triumph over the Admiral's Men. The competition had been heightened by the building of the Globe, a stone's throw from the Rose in the liberty of the Clink, and it is hard not to read the Admiral's Men's decision, taken in 1600, to move north of the river to the Fortune as an admission of defeat.

By the time the Chamberlain's Men took up residence at the Globe, Shake-speare was at the height of his powers, and his confidence in his colleagues was, by any standard, unusual. Whereas Marlowe had tended to compose his plays around the solo instrument of Alleyn's voice, Shakespeare wrote for an ensemble. Who,

The wedding celebrations of Sir Henry Unton. This section of the Unton Memorial Picture shows a circle of fantastically costumed masquers around a characteristically Tudor broken consort of instrumentalists. It was for such a group that Thomas Morley prepared his *Consort Lessons* (1599).

after all, is the 'star' of *Twelfth Night*? The overt plot centres on Olivia (played, presumably, by a boy-apprentice) and her quintet of 'suitors'—Orsino, Sir Andrew Aguecheek, Malvolio, 'Cesario', and Sebastian. The emotional centre, though, is the constant Viola (another boy-apprentice); and there are opportunities for virtuoso performances from Malvolio, Sir Toby Belch, and Feste, as well as scene-stealing invitations to Maria and Antonio. Nor should we neglect the consort of musicians for which the play cries out. *Twelfth Night* is animated by music, from the overture to Orsino's opening speech to the concluding dance and song. We might learn much about the different expectations from audiences of tragedy and comedy if any musical scores had survived. In their absence, we are in danger of underrating the role of musicians in the Elizabethan theatrical event. Shakespeare's responsiveness to the demands of the ensemble was matched by an equal responsiveness to the remarkable individual. The maturing talent of Richard Burbage, four years his junior, contributed to the creation as well as to the successful performance of the great roles in the Globe tragedies. We cannot say whether Shakespeare exploited or formed in Burbage the ability to make the vocal leaps from the colloquial ('Pray you, undo this button') to the operatic ('Blow winds and crack your cheeks'), but we can say that the whole course of Elizabethan drama was affected by it. The complex plots, marked off by a succession of contrasting scenes, had to be made clear to a busy audience. Clarity can be achieved only if the characters are clearly differentiated from each other. Costume is certainly an aid, but accurate delineation of attitude is also required. This, to the Elizabethans, was 'personation'—the making concrete of something so intangible as an invented personality. It is something a good professional actor can achieve even if the text is inadequate. But when a great actor, like Burbage, is invited to personate Hamlet, something remarkable happens. Mere personation cannot encompass a prince whose actual behaviour contradicts his invented personal impulse. In that contradiction, as in the later inventions of Macbeth, Antony,

The death of Sir John Oldcastle. An illustration from Foxe's *Actes and Monuments*: 'Then was he hanged up there by the middle in chains of iron, and so consumed alive in the fire, praising the name of God so long as his life lasted.'

Coriolanus, the anachronistic idea of 'character' is contained. It is not mere bardolatry, though bardolatry has tainted it, that has made Shakespeare's 'characters' part of our cultural heritage. In speaking Hamlet's lines, Burbage could represent simultaneously the duplicity and the human honesty of language under pressure from events. Shakespeare would not, otherwise, have written them. At the centre of *Hamlet* is *The Mousetrap*, performed by the capital's finest company of players. The first audience at the Globe *knows* that this is the Chamberlain's Men. It is not fanciful to suppose that the First Player was bearded and made up to look like Burbage. In a vivid, metatheatrical trope, the First Player as Burbage discusses with Burbage as Hamlet the art of acting. Quite soon, perhaps the next afternoon, Burbage may be playing Jonson's vicious Volpone, a brilliant improviser who speaks most of his lines as if they had sprung to his lips without the intervention of thought. Chicken and egg: which comes first, great acting or great writing?

The old image of Shakespeare as a man elevated by genius above the grave and petty quarrels of his age is, by now, tarnished. As an actor, he was occupationally associated with two Renaissance obsessions, the one with change and changeability, the other with the world as a stage. The licence to play a king, a lord, or a lady (there were many who read cross-dressing, scripturally, as transgression) was necessarily contentious at a time when it was easy to confuse a hunger for action among the nation's leaders with their propensity for acting. No one played the queen better than Elizabeth I or the hero with more flamboyance than the Earl of Essex. Anxiety about their capacity to be deceived by appearances was common to courtiers, diplomats, soldiers, and tradesmen. The actor was both an image of deceit and, because his deceit was avowed, the exposer of deceit in the world at large. The point can, of course, be over-stressed, but a sensitivity about the very idea of impersonation added to the quantity of abuse heaped on Elizabethan actors. Always under scrutiny, they were bound to offend on occasions. 'Let one but name breade', Nashe wrote in 1594, 'they will interpret it to be the town of Bredan in the lowe countreyes.' The famous dispute over the naming of Falstaff exhibits the neurotic factionalism of the Elizabethan court, the effect that that neurosis might have on acting companies, and, in detail, the alacrity with which one company might make capital out of the discomfort of another.

The officiousness of the dramatic censors impelled playwrights towards metaphor, and the metaphoric mode fed the Renaissance impulse towards explanation. The proper conduct of the modern state was rarely out of view in the writing, reading, or hearing/watching of 'histories', however ancient. It was under the disguise of history that the Protestant exile John Foxe compiled his *Actes and Monuments* (1563), a sensationalist compendium of Catholic brutality. One of Foxe's Protestant martyrs was the Lollard Sir John Oldcastle, Lord Cobham. Oldcastle was executed in 1417 during the absence of his sometime friend

Henry V in France. He had made a modest appearance as one of Prince Hal's merry company in the anonymous *The Famous Victories of Henry V*, which was in the repertoire of the Queen's Men in the late 1580s. It was allowed to pass without known comment from William Brooke, the Elizabethan Lord Cobham and one of the Queen's Privy Council. There is no reason to doubt that Shakespeare's redeployment of the character in the first part of *Henry IV* (1596–7) was innocent; but, in the context of court faction, it was read as inflammatory. William Brooke and his son Henry were aligned against the Earl of Essex and sensitive to his mockery of the Oldcastle connection. We do not know whether it was before or after the opening of *Henry IV* that pressure was put on the Chamberlain's Men to rename the fat knight, but the scandal, as well as the comic splendour of Shakespeare's creation, was the talk of the town. Brooke had succeeded to the office of Lord Chamberlain in 1596 and may have intervened ex officio, but it was too late. Whenever Falstaff appeared on the London stage, the shadowy figure of Oldcastle of the house of Cobham accompanied him. And Falstaff was a popular hit, not only in the *Henry IV* plays, but also in *The Merry Wives of Windsor*, his vehicle. The writing of *The Merry Wives* is a prime example of Shakespeare's professional pragmatism, but there is more than a hint of mischief in his having the jealous Ford claim the name 'Brook' when appearing in disguise. If, as has been argued, the play was commissioned by the second Lord Hunsdon to mark his investiture with the Order of the Garter, its first staging would have occurred a month after the death of William Brooke in March 1597. It was Hunsdon, not the ambitious Henry Brooke, who succeeded to the office of Lord Chamberlain and the patronage of Shakespeare's company. Perhaps Shakespeare, on behalf of Hunsdon's Men, was cocking a snook at an outfaced rival. It may have been Edmund Tilney, Master of the Revels, who enforced the emendation of 'Brook' to 'Broom', a forfeiture of the Ford/Brook pun which Shakespeare probably resented. It may also have been Tilney who pressed the Chamberlain's Men into a clumsy disclaimer in the epilogue added to *2 Henry IV* on its revival:

One word more, I beseech you. If you be not too much cloyed with fat meat, our humble author will continue the story, with Sir John in it, and make you merry with fair Katharine of France: where, for anything I know, Falstaff shall die of a sweat, unless already he be killed with your hard opinions; for Oldcastle died a martyr, and this is not the man.

Henry V, the play promised here, contains, in the choric prologue to Act v, an elaborate homage to the Earl of Essex, part of the persuasive evidence that the Chamberlain's Men were his adherents. Few people hated Essex more than Charles Howard, Lord Admiral and patron of the rival company. Howard's widowed daughter was eager to marry Henry Brooke, and it may have been at her instigation that, in late 1599, the Admiral's Men commissioned the designedly corrective *Sir John Oldcastle*, whose prologue promises the audience:

It is no pampered glutton we present,
Nor aged counsellor to youthful sins;
But one whose virtues shone above the rest,
A valiant martyr and a virtuous peer.

Sir John Oldcastle is the routine outcome of multiple authorship. It serves here to show how the Admiral's and the Chamberlain's Men played off and against each other for high stakes. Shakespeare's company's engagement in factional politics would reach a dangerous climax in 1601, when they were paid to perform *Richard II* on the eve of Essex's abortive rebellion (not, he would protest, to overthrow the Queen but to save her from such evil councillors, 'King Richard's Men', as the Lord Admiral). They escaped unpunished, but the theatre was under stricter surveillance during the last years of Elizabeth's reign. Plays about English history would be particularly vulnerable. It was probably with *As You Like It* or *Julius Caesar* that the Chamberlain's Men began their new era at the Globe in 1599. There is a story behind every addition to a theatre's repertoire.

It is unlikely that Shakespeare had free choice in what he wrote. He may have proposed, but the company's corporate management disposed. The preference was for a balanced programme. Shakespeare, Jonson, Heywood, Chapman, Fletcher, Marston, Middleton, and lesser associates wrote both comedies and tragedies. More shocking to the purer taste of the late seventeenth and eighteenth centuries, they embedded foolery in the midst of tragedy. It made economic sense to do so; Elizabethan audiences loved their clowns. When the Swiss Thomas Platter attended *Julius Caesar* at the Globe in 1599, he was impressed equally by the performers in the tragedy and in the post-play jig, the clowns' speciality piece. He may even have witnessed one of the last appearances with the Chamberlain's Men of Will Kemp. Kemp's audacity had a strain of defiance, and his crowd-grabbing antics, together with a readiness to substitute his own lines for the author's, could annoy his fellow actors. Whatever the reason, he left, carrying with him his skills as a self-publicist. His famous Morris Dance from London to Norwich, undertaken in February 1600 and

Will Kemp. This illustration from the title-page of his own narrative of his progress from London to Norwich shows Kemp with his taborer. Kemp is wearing the ankle-bells and ribbons of the morris dancer. The shirt may also record his actual costume. Its leaf-pattern is appropriate to the dance. But this is not intended as an accurate portrait of Kemp.

Kemps nine daies vvonder.

Performed in a daunce from
London to Norwich.

Containing the pleasure, paines and kinde entertainment of *William Kemp* betweene *London* and that Citty in his late Morrice.

Wherein is somewhat set downe worth note; to reprooue the slaunders spred of him: many things merry, nothing hurtfull.

Written by himselfe to satisfie his friends.

LONDON
Printed by *E. A.* for *Nicholas Ling*, and are to be solde at his shop at the west doore of Saint Paules Church. 1600.

Richard Tarlton. The title-page of *Tarlton's Jests* (edition of 1638) contains a woodcut copied from an earlier drawing. It shows Tarlton, as a contemporary recalled him, 'in russet, with a buttond cap on his head, a great bag by his side'. There are contemporary references also to the rustic 'startups' on his feet and to his squint and flat nose. He may have shared with Will Somers a deformity of the shoulders approximating to a hunched back.

written up for publication in April, is a startling parody of a royal progress. Such *lèse-majesté* was in the tradition of court jesters from Henry VIII's Will Somers through Lear's fool and on to Charles I's Archie Armstrong. The greatest Elizabethan exemplar was Richard Tarlton, chief prop of the Queen's Men during their brief supremacy, which ended with his death in 1588. It was Tarlton's genius that established the jig as something more than song-and-dance—a crafted, usually ribald, afterpiece featuring cross-dressing. Tarlton's clown was an oafish rustic with a wicked ability to score off urban sophisticates. In his misshapen body, the medieval Vice, servant to the Devil, merged with the Elizabethan stage clown, a popular troublemaker whose rapport with the audience allowed him to subvert not only the good order of the play, but good order. *Tarlton's Jests* (1611), posthumously published, preserve apocryphally the overwhelming personality of an entertainer who was temperamentally adversarial, even at times demonic. If Shakespeare pays homage to him as Hamlet's Yorick, the livelier recollection is in the grotesque and dangerous clowning of the deformed Richard III. Such entertainers operate at the liminal point between humour and effrontery. When the Chamberlain's Men replaced Kemp with Robert Armin, it was an emblem of the taming of the clown. Armin enacted the worldly-wise fool, a cultural commentator with the gift of song. Shakespeare's intuitive response was the creation of Feste in *Twelfth Night*, an unplaced social leveller, admitted everywhere, belonging nowhere. There was a measurable gain in dramatic coherence, but the sheer exhilaration of Tarlton and Kemp, their plebeian physicality, was lost to the stage. Jigs remained a popular feature of rougher playhouses like the Red Bull and dominant in Elizabethan fairs, but the historical process that led drama into the enclosure of high culture had advanced a little.

In our eagerness to make useful distinctions, we have tended to exaggerate the difference between Elizabethan theatre and Jacobean and between Jacobean and Caroline. Theatres, like people, change their style of dress as they grow older, but for London's professional actors the accession of James I promised a welcome end to the uncertainties of Elizabeth's fraught final months. In 1603, with plague raging and the city's playhouses closed, the Chamberlain's Men learned of their elevation to King's Men and the Lord Admiral gave way to Prince Henry as patron of their chief rivals. Changeable weather rather than the change of ruler governed the significant move towards indoor performance. The boys' companies, having lost political favour by their ill-judged intervention in the Marprelate controversy of 1589–90, were resurgent in the new century. They drew fashionable audiences to their indoor playhouses in sufficient numbers to alarm the adult companies. There was, in these audiences, an element of the perverse and of coterie complacence. It was easier to be noticed in these small (the capacity at St Paul's was probably less than 200) indoor

Robert Armin. The title-page of Armin's own play *The Two Maids of Moreclack* (written 1597–9, published 1609) carries a woodcut of the character Armin played in it, John of the Hospital. What is notable is that Armin allowed the character to dictate the costume. Clown has been subsumed by character.

THE
History of the two Maids of More-clacke

VVith the life and simple maner of IOHN
in the Hospitall.

Played by the Children of the Kings
Maiesties Reuels.

VVritten by ROBERT ARMIN, seruant to the Kings
most excellent Maiestie.

LONDON,
Printed by *N. O.* for *Thomas Archer*, and is to be fold at his
fhop in Popes-head Pallace, 1 6 0 9.

houses, where a stool on the stage could make a spectator as prominent as an actor. Nor was the difference between boy and adult performances as extreme as has often been supposed. Just as boys were employed by adults to play the women's parts, adults were mixed in with the 'little eyases'. Sir Tophas in Lyly's *Endimion*, for example, is an adult role, and Francis Beaumont's *The Knight of the Burning Pestle* (1607), written for the Children of the Revels at the Blackfriars, relies on the adult control of grocer George and his buxom wife Nell, supported by a senior chorister as the out-Micawbering Old Merrythought. Beaumont's brilliant burlesque failed with its first audience, but it tells us much about theatrical customs. Pipe-smoking and beer-drinking accompanied performances; there was entr'acte dancing, music, and tumbling; the actors' footwear was remarked on; and the play was not thought of as the day's last entertainment but as the prelude to further pastime. Beaumont's play sets out to exaggerate the effect of audience demand on the progress of the play in performance, but its emphasis on the liveliness of discourse between actor and audience in the roofed playhouses is instructive.

Once James I's honeymoon period was over and actors, like the country at large, found themselves negotiating between the divine right of kings and the liberty of the people, it was the boys' companies that most self-destructively trans-

Amateur musicians. This Gilling Castle frieze, dating from *c*.1585, shows fashionably dressed Elizabethans engaged in the favourite pastime of making music. This group forms a loose broken consort, somewhat after the style of the professionals who played at Sir Henry Unton's wedding.

gressed. Their managers and writers presumed too much on the mask of puerile innocence. After a succession of productions had given offence at court, their purchase on patronage slackened until, in 1608, the King's Men took over the lease of the indoor Blackfriars. Although they maintained the Globe over the ensuing decades, and although we cannot say that the indoor audiences were exclusively gentrified, the decision marks off another stage in the decline of a popular theatre. The sheer capacity of the Globe, the Fortune, and the Red Bull was an aid to survival, but, with the singular exception of the speculatively built Hope (1613), London's subsequent theatrical development was all indoor. The best evidence of popular resistance belongs to 1616–17. Having managed the Queen's Men at the Red Bull since 1612, Christopher Beeston chose, in 1616, to install the company in the refurbished Cockpit in Drury Lane. The following Shrove Tuesday, a mob of apprentices wrecked the Cockpit and burned all the playbooks they could find. Ruffled he may have been, but Beeston remained at the Cockpit, and the Red Bull's raggle-taggle reputation was sustained by a succession of raggle-taggle companies.

It is over-simple to ascribe to the Jacobean court the whole responsibility for the evident malaise of playwrights in the early seventeenth century. The majority of the theatre's wounds were self-inflicted. After the high times of the early profes-

sional years, when endeavour was likely to be rewarded, the nation's continuing economic decline had contributed to leaner recompense. There was growing uncertainty about the true purpose of playing, and a tendency to blame audiences for a decay of taste. However good-humouredly, *The Knight of the Burning Pestle* charges London's citizenry with a corrupting preference for violence and far-fetched fantasy, the very things for which Beaumont, in association with John Fletcher, would soon be renowned. It was the lack of a stable theory to control the chaos of dramatic adventurism that antagonized Ben Jonson, to whom the Jacobean hybrid of tragicomedy was anathema. But Jonson was strong enough to establish his own conventions, and stubborn enough to abide by them. The discernible edginess of much Jacobean drama was a product of both national and theatrical politics. The publication of plays was one of many concerns. The accession of the emphatically literary James I encouraged a print explosion which threatened to engulf the precious commodity of playbooks. The independent Jonson took the opportunity to declare the case for plays as literature. The 1616 publication of his *Works* in the proud folio format set a precedent for the Shakespeare and Beaumont and Fletcher folios of 1623 and 1647, but the authors were not responsible for these compilations. No company risked the permanent engagement of Jonson, whose self-belief might earlier have challenged the primacy of actors over playwrights. They plumped instead for more compliant dramatists. After Shakespeare's retirement, for instance, the King's Men turned first to Fletcher and then to the self-deprecating Philip Massinger. They had also to negotiate anew their relationship with the Revels Office. Tilney's successor, Sir George Buc, kept a dutiful eye on the London theatres from 1603 to 1622. Henry Herbert, whose long tenure began in 1623, was always mercenary and intermittently rigorous. But the King was unpredictable. His interventions were almost always on personal grounds and, though decisive, he was unafraid to change his mind. A vehement (and profoundly unpopular) fondness for his Scottish favourites explains his fury over a few passages of dialogue in *Eastward Ho!* (1605) and the brief imprisonment of two of its authors, Jonson and Chapman. The third, Marston, who probably wrote the offending lines, escaped punishment, but would soon retire from the theatrical fray. It may have been James's obsessive fear of assassination that led to the silencing of the King's Men's lost play about the Gowry conspiracy (1604) and impelled Buc, in the aftermath of the assassination of Henri IV of France, to censor portions of *The Second Maiden's Tragedy* (1611). These are examples that reveal a theatre centrally involved in the great issues of early Stuart England, but uncertain of its room for manœuvre. Jacobean censorship, in which permissiveness was interspersed with punitive pounces, has, and had, the appearance of arbitrariness. The brinkmanship of professional playwrights is, in the circumstances, not surprising, but the sheer bravado of Middleton's *A Game at Chess* (1624) is.

Middleton knew the dangers of writing about the political machinations of contemporary Europe. So did the King's Men. They had the recent experience of protracted negotiations with the Revels Office over *Sir John van Olden Barnavelt* (1619): and the authors of that play, Fletcher and Massinger, had been almost timorously unpartisan. Nor should we forget that Middleton had a position to uphold. He was City Chronologer. But he was also a militant Protestant who, as early as 1617, had displayed his anti-Spanish sentiment in one of the pageants he wrote for the Lord Mayor's inaugural procession. By the 1620s, many of his party were persuaded that European Protestantism was under threat of extinction, a fear fuelled by James I's pacific policies. Middleton joined the propaganda campaign that converted unease about the possible marriage of Prince Charles to the Spanish Infanta into overwhelming public hostility. As a result, the failure of the Prince and the Duke of Buckingham to secure the

match during their 'secret' visit to Madrid in 1623 became an occasion for national celebration. Middleton voiced London's delight in *The Triumphs of Integrity*, his Lord Mayor's pageant, but the city afforded him more protection from the court than the theatre did. James I's position on the right of his subjects to criticize his foreign policy was unequivocal. He had required the Bishop of London, in 1620, to inform the clergy that they should not 'meddle in their sermons with the Spanish match nor any other matter of state'. That the King's Men should, in 1624, contemplate a production of *A Game at Chess*, with its circumstantially detailed caricature of the Spanish ambassador Gondomar, the turncoat Archbishop Marco de Dominis, and, in broader outline, some of James's own courtiers, as well as a flattering but ambiguous portrait of Prince Charles, is very surprising. That Sir Henry Herbert should license it is astonishing, and the contention that the play was powerfully backed, perhaps by the Earl of Pembroke, is irresistible. The Globe became a site for propaganda and the King's Men its instruments. *A Game at Chess* is extraordinary in many ways, not least in the geometrical precision of its envisioned staging, but it was its political clout that earned it an unprecedented nineday run. The King was out of London and the voices of opposition evidently muzzled until, on what would have been the tenth day of performance, the Privy Council stepped in with a ban. We have no evidence that the warrant for Middleton's arrest was served, but the King's Men were penalized by a ten-day interdiction. Before the year's end they had staged a possibly unlicensed play called *The Spanish Viceroy*. Very rarely in theatre history has controversy damaged the box office.

While London's theatres were being caught up in the great debates over corruption and the impact on loyal (or, especially, disloyal) subjects of the doctrine of the divine right of kings, the elaborate masques at court were celebrating the serene authority of a King who had unique access upwards (to God by way of mythological gods) and downwards (to his subjects, represented, if at all, in the anti-masques). James I was more conscious than his son of conflicting ideologies, but almost equally guilty of underrating the opposition. The Civil Wars were, after all, as much the outcome of royal insouciance as of organized rebellion, and the court masques—the most significant of early Stuart contributions to the development of English theatre—abetted the delusion of security. Without the presence of the King, the complex game of the court masque could not be played, and whilst there is scant evidence that courtiers took these masques seriously as tools of statecraft and a surprising amount that James's coarse humour was excited by them, there is every indication that they were occasions of great moment for James's Queen and for Charles I and Queen Henrietta Maria. English theatre was advanced by the theatrical conduct, not by the dramatic content of the masques, whose contemporary impact is irrecoverable from the printed page. They took place in the *absolute* present, and the King's reaction, his facial, gestural, and sometimes verbal response, was part of the event. So, too, were the personalities of the individual noble masquers, revealed in mythological roles but opening to gossip and debate their status in the here-and-now of courtly intrigue. To be present was to participate in a courtly ritual—it has been described as a 'liturgy of state'—with the masquers as celebrants and their obeisance solemnized by the presence of God's vicegerent. The setting out of a masque is of intense interest to theatre historians. At one end of the hall is a raised stage, at the other, pinpointed to be opposite the stage's centre, is the King's canopied chair of state. His is the only perfect view, and the disposition of those in attendance on him is a visual declaration of hierarchy. Between stage and throne is a large, carpeted arena, the dancing-place. Its equivalent in the public theatres is the yard where we suppose the groundlings stood. De Witt, in the drawing of the Swan, labels it *planities sive arena* (a flatness or arena), after the Roman style. In Rome, too, the arena was the site of action. The stage, as a scenic space, was inhospitable to performance. The presence of masquers anywhere other than towards the front would have exposed the artifice of the diminishing perspective, designed to gratify the King's eye. For *Tethys' Festival*, for example:

THE MASQUE OF *OBERON* (1611)

William Trumbull the Elder's eye-witness account supplies some visual details lacking in the published text.

1611, Jan. 11—The new hall of the palace was furnished as usual with its galleries round about, a green carpet on the floor, a dais at the top for the king and queen. At the bottom a very large curtain painted with the kingdoms of England, Scotland and Ireland, with the legend above *Separata locis concordi pace figantur.* When their Majesties entered accompanied by the princess and the ambassadors of Spain and Venice, flageolets played and the curtain was drawn discovering a great rock with the moon showing above through an aperture, so that its progress through the night could be observed. Old Silenus mounted on this with some dozen satyrs and fauns who had much to say about the coming of a great prince to be followed by a thousand benefits, in the hope of which the fauns danced about joyfully, exciting great laughter. They then danced a ballet, with appropriate music with a thousand strange gestures, affording great pleasure. This done the rock opened discovering a great throne with countless lights and colours all shifting, a lovely thing to see. In the midst stood the prince with thirteen other gentlemen chosen as famous dancers of the Court. Before passing into the hall ten musicians appeared each with a lute and two boys who sang very well some sonnets in praise of the prince and his father. Then ten little pages dressed in green and silver with flat bonnets *à l'antique* danced another ballet with much grace. During this a cock crew ten times, standing on the rock, and then, according to the prophecy of Silenus, there came the gentlemen in short scarlet hose and white brodequins full of silver spangles coming half way to the calf, some wearing jackets with wide folds, as the Roman emperors are represented and the sleeves the same, all in gold and silver cloth, white and scarlet feathers on their heads and very high white plumes, and black masks. Each one wore a very rich blue band across the body, except that of the prince, whose band was scarlet to distinguish him from the rest. They entered dancing two ballets intermingled with varied figures and many leaps, extremely well done by most of them. The prince then took the queen to dance, the Earl of Southampton the princess, and each of the rest his lady. They danced an English dance resembling a pavane. When the queen returned to her place the prince took her for a coranta which was continued by others, and then the gallarda began, which was something to see and admire. The prince took the queen a third time for *los branles de Poitou,* followed by eleven others of the masque. As it was about midnight and the king somewhat tired he sent word that they should make an end. So the masqueraders danced the ballet of the sortie, in which the satyrs and fauns joined. With vocal instrumental music the masqueraders approached the throne to make their reverence to their Majesties. The masques being laid aside, the king and queen with the ladies and gentlemen of the masque proceeded to the banqueting hall, going out after they had looked about and taken a turn round the table; and in a moment everything was thrown down with furious haste, according to the strange custom of the country. After this their Majesties withdrew and the ambassadors took leave.

The scene itself was a port or haven, with bulwarks at the entrance, and the figure of a castle commanding a fortified town: within this port were many ships, small and great, seeming to lie at anchor, some nearer, and some further off, according to perspective: beyond all appeared the horizon or termination of the sea, which seemed to move with a gentle gale, and many sails, lying some to come into the port, and others passing out.

Samuel Daniel composed this masque for the Queen on the occasion of Henry's being created Prince of Wales. Queen Anne masqueraded as Tethys, Prince Charles as Zephyrus, Princess Elizabeth as the Thames, with eight other English rivers and four Welsh embodied by ladies of the Queen's bedchamber. The speaking was done by actors, but the Queen and ladies performed two masquing dances and led the courtiers into the dancing-place to perform their revels of corantos and galliards before themselves completing a third 'retiring' dance. We should note that, though Tethys and her river nymphs were discovered on the scenic stage and remained there in tableau for long enough to display to advantage their water-motif costumes, they descended to the dancing-place 'and so marched up with winding meanders like a river, till they came to the Tree of Victory, which was a bay erected at the right side of the [chair of] State'. We might also notice the exquisite trickery by which the transformation of the port to Tethys' caverns almost deceived the eyes of the spectators:

First at the opening of the heavens appeared 3 circles of lights and glasses, one within another, and came down in a straight motion five foot, and then began to move circularly; which lights and motion so occupied the eye of the spectators that the manner of altering the scene was scarcely discerned: for in a moment the whole face of it was changed, the port vanished, and Tethys with her nymphs appeared in their several caverns gloriously adorned.

In publishing the account and text of *Tethys' Festival*, Daniel described his as 'the least part and of least note in the time of the performance thereof', ascribing the main credit in the masque 'wherein the only life consists in show' to the creator of 'the artificial part', Inigo Jones. It was a priority that the finest masque-writer of all contested to the bitter end.

The struggle for supremacy between Ben Jonson and Inigo Jones is notorious and its posterity sometimes overlooked. If it was waged with particular ferocity at the Jacobean court, the battle continued in the post-Restoration theatre, on the pantomime stages of the eighteenth century, and in the sensation scenes and classic pictorialism of Victorian England. It goes on still, more often bankrupting theatres than enriching them. In one respect, it is a contest between the eye and the ear, with the preponderance of production costs conceding to the eye. For the masque of *Oberon* (1611), on which he collaborated with Jones, Jonson was paid the handsome sum of £40, but the exchequer discharged over £2,000, more than half of it for costumes. What he saved on royal progresses—James had none of

Inigo Jones's design for *Prince Henry's Barriers*. At the opening of this Twelfth Night event, which began at ten o'clock in the evening and lasted through the night, this scene of the Fallen House of Chivalry was revealed to the spectators assembled in the Banqueting Hall. With the arrival of Prince Henry, the scene was transformed to show the revival of chivalry.

Elizabeth's taste for self-exposure to the people—the King spent twice over on court ceremony. And there were separate, costly establishments for the unignorable Queen and for Prince Henry, who, unlike his father in most respects, shared his extravagance. It was for the Prince that Jonson wrote his elaborate commentary on the *Masque of Queens* (1609), and his pen was quite as much at the disposal of the warlike heir as of the peace-loving monarch. He prepared, with Jones, the chivalric masque that accompanied the Twelfth Night spectacle of Prince Henry's foot-combat at barriers with six companions against fifty-six adversaries. The inconsistency of glorifying peace in masques for the King and war in *Prince Henry's Barriers* (1610) excited no recorded comment. Jonson in words and Jones in scenes and costumes systematized the Renaissance tendency to think ambiguously in terms of universal analogy. To decode everything on display in one of their masques was beyond any individual other than the notionally perfect, and therefore perfectly comprehending, spectator—the King himself.

The Stuart masque, as jointly created by Jones and Jonson, was the acme of theatrical élitism. Its customary postlude in James's court was a riotous incursion of the carnivalesque. Even at court, where the masque might not end until after midnight, a play was expected to lead on to further pleasurable pastime or to release an orgiastic urge. Dudley Carleton recorded the degenerate sequel to the performance of *Blackness* (1605), the first of the Jonson/Jones collaborations:

in the coming out, a banquet which was prepared for the king in the great chamber was overturned table and all before it was scarce touched. It were infinite to tell you what losses there were of chains, jewels, purses, and such like loose wear. And one woman amongst the rest lost her honesty, for which she was carried to the porter's lodge being surprised at her business on the top of the terrace.

A bemused witness of the mayhem after *Oberon* (1611) described it as 'the strange custom of the country'. The association of drama with misbehaviour was a Stuart legacy to the theatre. But the masques of Jonson and Jones importantly passed on, above all through the mediation of Davenant, a new vision and *new techniques* for the mutual enhancement of actors and scenery. By the time he collaborated with Davenant on the last of the great court masques, *Salmacida spolia* (1640), Jones had been experimenting for over thirty years in what was virtually his laboratory, the three-dimensional scenic stage. At its first public opening in *Blackness*, Dudley Carleton saw 'a great Engine at the lower end of the Room, which had Motion'. Overwhelmed by novelty, he seems scarcely to have noticed Jones's perspective innovation. Over the years, Jones would add to the magic of machinery and motion most of the features that would facilitate the changing of scenery in the professional theatres of the later seventeenth century: the raked Serlian stage, shutters sliding closed on grooves (the *scena ductilis*) to create a new back-scene or sliding open to reveal yet another, the arch behind a narrow proscenium to give definition to the enclosed perspective and to mask machinery, parallel wings to guide the eye and hide the joins, elaborate flying equipment, lighting effects. When Jonson introduced the first anti-masque in 1609, the invitation to radical scenic transformation was ingeniously met. The anti-masque offered an image of confusion which would be overcome and harmonized in the masquing dance. Reversible scenery, though it pleased the spectators, was too clumsy to satisfy Jones. As we have seen, for *Tethys' Festival* he preferred to use visual distractions to mask the smoother transformation provided by shutter-and-groove. The Caroline court demanded more variety. Shirley's *The Triumph of Peace* (1634) has thirteen anti-masques and *Salmacida spolia* twenty. Most intriguing in this final masque is its mirroring of the chair of state along the line of perspective. With the anti-masquers dispersed, 'the scene was changed into craggy rocks and inaccessible mountains'. The masquers then appealed to the King to occupy the Throne of Honour which was placed exactly opposite him at the vanishing-point of the perspective stage. There is no account, in the printed text, of what happened next, but Charles I was somehow transported from chair of state to Throne of Honour, to appear there when 'the further part of the scene disappeared'. For the duration of the tableau, the notional earthly body of the King was gazing from his chair of state at the visible incarnation of his divine self. The dramatic emblem is of perfect concord. The political reality was very different.

Masques in early Stuart England were not confined to court. The dispersal of courtly influence around the city of London and further afield has been extensively documented. Thus, although the audience was select, it was neither tiny (Busino, chaplain to the Venetian embassy, estimated at least 600 for *Pleasure Reconciled to Virtue* in 1618) nor, of course, uninfluential. It is, then, of more than passing interest that the participation of women was accepted in masques and their bodies always decoratively and often provocatively draped. The influence of women at the highest level was greater under the Stuarts than it had been under Elizabeth I, who found her own self sufficient. The strong-willed Queen Anne surrounded herself with equally determined companions, like the Countess of Bedford and Lady Penelope Rich. Patrons of the arts, they were also participants in the Queen's masques. Henrietta Maria, nurtured in France to recognize the courtly primacy of entertainment and devoutness, took the dramatic lead in Charles I's court. After taking part in a French pastoral drama in 1626, she was referred to by a contemporary commentator as the 'principal actress', the first recorded use of the word. She was the advocate and patron of playwrights and players, notorious for the love-fashions she promoted, and suspected, with some justification, of covert proselytizing on behalf of the Catholic Church. For the Puritan William Prynne, drama at court reeked of Catholic ritual, and the Queen's cult of love (Platonism flavoured with temptation) was a Jesuitical route to religious seduction. His *Histriomastix* (1632) is a sustained polemic against the wiles of the stage, more particularly their exercise at court, and the corrupting guile of women. The Caroline court was often taxed with effeminacy, but Prynne finds fault with fashionable ladies who

A contemporary print records the punishment of William Prynne.

Mr. William Prynne, for writing a booke against Stage-players called Histrio-mastix was first censured in the Starr-Chamber to loose both his Eares in the pillorie, fined 5000ᵗʰ & perpetuall imprisonment in the Towre of London. After this, on a meer suspition of writing other bookes, but nothing at all proved against him, hee was again censured in the Starr-chamber to loose the small remainder of both his eares in the pillorie, to be Stigmatized on both his Cheeks with a firey-iron, was fined again 5000ᵗʰ and banished into yᵉ Isle of Iersey, there to suffer perpetuall close imprisonmᵗ no freinds being permitted to see him, on pain of imprisonment.

cut their hair like nuns to signify their independence of men. Actors are 'professed Papists' and dancing a 'Devil's Mass'. Worse still, it is a Mass to which 'Queenes themselves . . . are commonly most addicted'. It has been too often argued that Prynne was the spokesman for a large body of theatre-haters. In fact, the public stage was persecuted less in the 1630s than in the six previous decades. Prynne's main target was creeping Catholicism, represented by the Arminian doctrines and ceremonials promoted by Archbishop Laud, but what he actually said counted for less than what he was believed to have said. By choosing the stage as his text for a prolix sermon on the abuse of civil (meaning Protestant) rights, and by associating stage and Queen, he highlighted the courtly interest in drama. The much-quoted reference to 'Women actors, notorious whores' was read as an attack on the Queen and her ladies. It was one of many passages cited in his arraignment. Prynne's trial and later punishment made a popular hero out of a spoilsport, but the common view that the theatre aligned itself with the court against Puritan dogma is faulty on many counts. There were, to be sure, playwrights for the Queen and for the King, but there were others who openly, or by veiled analogy, associated Charles I's personal rule with tyranny. A schism between the court and the public theatres deepened during the 1630s. The King's expelling of his jester Archie Armstrong in 1638, for a gibe against Laud, might legitimately be read as a

The Execution of Charles I. The unknown artist has emphasized the awesomeness of the occasion, and the huddle of spectators below the scaffold-stage.

sign of the court's divorce from popular theatrical sentiment—a suppression of the cleansing but unruly clown. When sides were taken in the Civil Wars, the acting profession was not unanimously royalist. The ordinance that closed the playhouses in September 1642 may have chimed with the extreme Puritanism of some members of parliament, but its real objective was public safety. With the King in Nottingham and civic order threatened, the old bugbear of lawless assembly was uppermost in the mind of parliament. Similarly, the reinforcing legislation of 1647–8 took account of public tension before the outbreak of the Second Civil War, and the army raids on illicitly operating playhouses on 1 January 1649 were designed to eliminate possible sources of trouble in the days before Charles I's trial. The continuity of surreptitious performances in London's theatres during the Interregnum is less puzzling once it is recognized that their closure was not a major plank in parliamentary policy.

The most dramatic event, and among the most theatrical, of the mid-century was the execution of Charles I. J. P. Kenyon has called it 'a great baroque drama, magnificently staged by the army, superbly sustained by their chief actor'. He might have extended the cast-list to include Richard Brandon, the King's executioner and partner in his terminal double-act. 'I think the last time they pull'd down the Stage in the City', wrote Colley Cibber in his dedication to *Love Makes a Man* (1701), 'they set up a Scaffold at Court.' The association of public executions and theatre was a commonplace. Executions were staged for the edification of the audience. They were theatre-in-education festivals. At the signing of the King's death warrant, like playhouse clowns as the tragedy closes in, Cromwell and Henry Marten inked each other's faces. The old alliance of stage and nobility seemed to be broken for ever. 'What were the lords of England but William the Conqueror's colonels?' asked a parliamentarian, 'Or the barons but his majors? Or the knights but his captains?' The disruption of social hierarchy might have returned the theatre to the people. Among prominent actors whose career spanned the Interregnum, Michael Mohun was a major and Charles Hart a lieutenant in the royalist armies, where they might have fought against Colonel Okey, a tallow-chandler in civilian life, or Major Berry, clerk to an ironworks. But the mob was more likely to riot for Church and king (as happened first in 1647) than for plays and players. The theatre, when it was finally restored, was not the people's.

There is circumstantial, but disputed, evidence that perspective scenery was used on the professional stage before the closing of the theatres. Some interaction between staging conventions at court and in the public theatres is to be expected, particularly during the brief period (1640–1) when Davenant was managing the Cockpit. But it is with the staging of Davenant's *The Siege of Rhodes* at Rutland House in 1656 that speculation can rest. We know, from John Webb's designs, that perspective scenery was used, even in so restricted a space.

Frontispiece from *The Wits*. *The Wits* is a collection of the short pieces, known as drolls, for which the actor Robert Cox was especially famous during the closure of the theatres. They are excerpts from well-known plays, whose characters are represented in the seven on-stage figures.

This was the most portentous theatrical occasion of the Interregnum, and it needs to be set in context. The English revolution had passed its climacteric with the dissolution of the Barebones parliament in 1653, and Cromwell, isolated in authority but still thirsting after consensus, found it expedient to be accommodating where he could find justification. The Lord Mayor's Show was re-established in 1655. That, and the tolerance shown towards Catholics, may have tempted Davenant back to London. The move was a bold one. He had served as soldier and gun-runner for the royalist cause and associated himself with Charles Stuart's court-in-exile. Radical Puritanism was on the wane, but plays were still under interdict. It was an 'opera' that Davenant was permitted to stage in his temporary home, and his prudence was further displayed in his division of *The Siege of Rhodes* into 'entries', after the style of masques, rather than scenes. Cromwell's love of music was common knowledge, as was his preparedness to command a masque for a state occasion. It was the use of masque-like scenery to enhance a dramatic narrative on a stage so small as to force the actors back towards the vanishing-point that was Davenant's innovation. Webb, by his own testimony, had been 'brought up by his Unckle Mr. Inigo Jones'. His frontispiece and five perspective scenes with wings and cloud borders carried Jones's scenic invention towards the Restoration stage. The transfer of *The Siege of Rhodes* to the Cockpit in 1657 was a herald of the theatre to come. The questions posed and answers half-provided by this unspectacular playhouse reflect on almost all that is essential to the history of the English stage from 1620 to 1670.

In November 1652 Cromwell pondered: 'What if a man should take upon himself to be king?' He had already dismissed a proposal to restore Charles Stuart: 'He is so damnably debauched, he would undo us all.' It was a fair point, not irrelevant to the narrative of theatrical history. But the restoration of Charles II was greeted nation-wide with celebrations that bordered on the hysterically euphoric.

It promised a return to the old ways, stirring the atavistic impulse that is often decisive in English politics. Dryden would soon embody it in the character of Moody, who brings to *Sir Martin Mar-All* (1667) a preference for 'the old Elizabeth way in all things'. The impact on theatre was immediate. Thomas Killigrew, a playwright whose conversational gifts presumably dwarfed his dramaturgy, had helped to make jovial some hours of the King's exile. In awarding him a half-share in a theatrical monopoly, Charles was publicly fulfilling a private promise. The other half-share went to the abler Davenant. The warrant issued in August 1660 gave Davenant and Killigrew exclusive rights to 'erect two companies of players' and 'to purchase, build and erect or hire at their charge . . . two houses or theatres', on condition that 'they do not at any time hereafter cause to be acted or represented any play, interlude, or opera, containing any matter of profanation, scurrility, or obscenity'. As the King knew perfectly well, he might as well have been warning fish to avoid water.

Killigrew established his King's Company in a converted indoor tennis-court, and Davenant his Duke's Company, under the nominal patronage of the future James II, in the old Salisbury Court playhouse, but their monopoly was jeopardized before the end of the year, when George Jolly's company was licensed to perform in London. The monopolists' long struggle to suppress Jolly's activities at the Cockpit was resolved in 1667, when the little playhouse was turned into a Nursery for the training of future King's or Duke's men. There is no evidence that Jolly felt flattered to be, formally, the first principal of a drama school, and the untidy dealings between the two major companies and a lesser rival established a theme that would be played with variations until 1843, when the monopoly, belonging by now to Drury Lane and Covent Garden, was rescinded.

Whilst marking time at the Salisbury Court, Davenant was planning a playhouse for scenery, and it is the opening of the converted tennis-court in Lincoln's Inn Fields in 1661 that marks the beginning of a distinctive Restoration theatre. The overriding principle was symmetry. It was a principle that guided playwrights as well as architects, as is startlingly exposed in the 'improvements' Dryden and Davenant made to *The Tempest* (1667). Had Shakespeare been writing for a Restoration playhouse, he too would

Shutter-Scene for *The Siege of Rhodes.* When the curtain in the proscenium frame was raised, it revealed side-wings in the form of vertical rocks and behind them a shutter depicting 'the true prospect of the city of Rhodes when it was in prosperous estate', illustrated here. A second shutter-scene and two scenes in relief led to the culminating 'entry' of the assault on the city, depicted on a third shutter.

have been guided into symmetry. In the new theatre, there are two rooms. The first, the scenic stage, preserves the grooves and shutters (Jones's *scena ductilis*) of the masque, together with its machinery for spectacle. It is separated from the second by two pillars supporting an arch, in the middle of which is a boxed balcony for musicians. Because it juts forward from the arch, this balcony belongs to the second room, whose separateness is emphasized before the performance by the drop-curtain that shuts off the scenic stage. In front of the arch stretches the empty proscenium or forestage, replacing the masque's dancing-place as the essential site of dramatic action. It is deep and wide, flanked on either side by boxes (lattices) above the two stage-doors which provide access and egress (and scope for symmetry games) to the actors. The auditorium is part of the same room as the forestage, which is lit by candles set in hoops and suspended from the ceiling. The reliable burning-time of a candle offers playwrights a rough-and-ready guide to the appropriate length of a play's units (acts, scenes, entries, etc.). Intervals allow the replacement or relighting of candles while the musicians play from the balcony, the audience chatters or listens, eats and drinks, strolls, or makes its awkward ablutions. In their shared room, actors and spectators are in close contact, a proximity culturally distinct from the hustings of the open-air stages. Tragic action may sometimes have occurred within the scenic stage, but Restoration comedy is a forestage interaction. Dryden, in lively debate with dramatists of the past age, records 'the last and greatest advantage of our writing, which proceeds from *conversation*'.

If a conversation is to rise above the level of mere pastime, it requires equal participation and the participation of equals. Confidence that they are being understood and responded to is an outstanding quality of the best Restoration playwrights, but the confidence was achieved at some cost to the theatre. Under Charles II's permissive eye, popular playhouses were consigned to history. Three features of the early Restoration playhouse advertise its desired clientele: admission costs ruled out the majority of Londoners, the capacity (between 600 and 800 at the comparatively ambitious Drury Lane of 1674) fed exclusiveness, and the starting-time of three o'clock suited the leisured town and court better than the busy city. As the century advanced and the hold on audiences weakened, curtain-up was steadily put back until, by 1700, it might be as late as six o'clock. But even at the later time, visits to plays were habitually followed by further conviviality. For Samuel Pepys, they were part of a social routine—seventy-three in the first eight months of 1668—and whilst he was not typical, we cannot presume him unique. Playhouses were, for their minority audience, places of easy, regular, and relaxed resort.

Pepys, almost as notoriously as Charles II, was particularly attracted to the theatre by actresses. The identity of the first woman to make a licensed appearance on the professional stage is disputed, although the year was certainly 1660

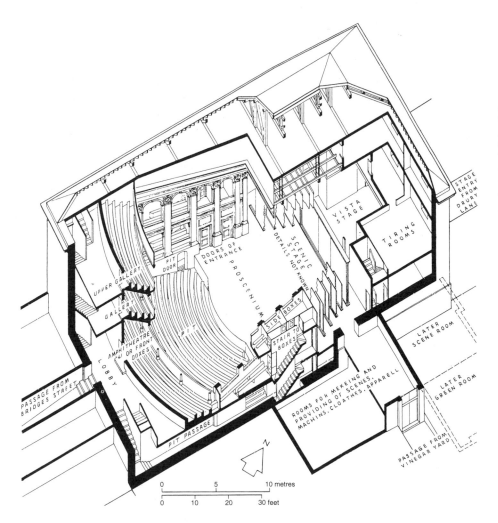

STAGE
ENTRY
FROM
DRURY
LANE

VISTA
STAGE

TIRING
ROOMS

SCENIC
STAGE
DETAILS NOT KNOWN

DOORS OF
ENTRANCE

PROSCENIUM

PIT
DOOR

UPPER GALLERY

GALLERY

SIDE BOXES

STAIR TO
BOXES

AMPHITHEATRE
OR FRONT
BOXES

PIT

LOBBY

LATER
SCENE ROOM

PASSAGE FROM
BRIDGES STREET

ROOMS FOR MEKEING AND
PROVIDING OF SCENES,
MACHINS, CLOATHES, APPARELL

LATER
GREEN ROOM

PIT PASSAGE

N

PASSAGE FROM
VINEGAR YARD

0   5   10 metres
0   10   20   30 feet

Cross-section of Wren's design, perhaps for Drury Lane (1674). Richard Leacroft's isometric drawing clearly displays the deep forestage and separate scenic stage of the Restoration theatre.

and the part probably Desdemona in a revival of *Othello*. The patents granted to Davenant and Killigrew in April 1662, in consolidation of their warrants, specified that 'all the women's parts to be acted in either of the said two companies for the time to come may be performed by women' rather than by men in women's clothes 'at which some have taken offence'. The reassessment of sexual values in post-Commonwealth England had rendered the defensiveness largely superfluous. The patents are transparently disingenuous in presenting as a social reform what was already a gentleman's pleasure. Charles II, a gentleman and libertine, set the style of the public theatres, the first monarch to mingle affably with his subjects. In St James's Park, which he opened to the public, he established the fashion for ambulatory discourse exploited in so many contemporary comedies. Conversations following encounters in parks are frequent plot-hinges, invariably involving sexual pursuit. Here again the King was a model:

Restless he rolls about from Whore to Whore,
A Merry Monarch, scandalous and poor.

The Earl of Rochester, author of that portrait, was a model, too. For King and Earl, actresses were sex objects rather than symbols of sexual equality. The reputedly virtuous Anne Bracegirdle specialized in being raped on stage, carrying into the eighteenth century the taste for such scenes that had followed hard on the arrival of actresses, and the eighty-nine breeches parts among the 400 or so new plays performed between 1660 and 1700 were designed to display female legs to prurient spectators. The case of Nell Gwyn is exemplary.

Gwyn was probably born in 1650. Claims for an earlier date have not been substantiated. She was a slum-girl, probably a child-prostitute, who sold oranges and, on occasions, herself at Killigrew's playhouse. It was there, presumably, that she first caught the eye of Charles Hart, the most glamorous of Restoration actors. She was already his mistress when she made her stage début in 1664 or 1665, shortly before the long, plague-enforced closure of the playhouses in 1665. Her progress was further delayed by the great fire of 1666, but Dryden knew her strengths well enough by February 1667 to fashion the role of Florimell in *Secret Love* around them. Florimell is a seductive madcap who dons breeches and dances a jig. This was the play that established the popularity of the 'gay couple' in comedy. Hart and Gwyn created the model for these anti-platonic lovers whose banter combines genuine affection and sexual antagonism. The proviso scene, in which Florimell and Celadon set out the rules that will enable them to preserve their individuality after marriage, was so effective that proviso scenes became almost obligatory in later comedies. *Secret Love* is significant in other ways. Dryden wrote it with the company and the audience in mind. There are seven good parts for women, only three for men. When Celadon guesses the features behind Florimell's mask, the audience hears Hart describing Gwyn: 'an oval face, clear skin, hazel eyes, thick, brown eyebrows and hair'. There is similar literalism in the jokes about the contrasting sisters, the tall, dancing Olinda, played by Margaret Rutter, and the diminutive Elizabeth Davenport (later Bowtell) as Sabina. Restoration audiences liked to be reminded of the actress behind the character. 'How many men do hover about them as soon as they come off the stage,' noted Pepys. Regulations banning backstage visits were ignored, reiterated, and still ignored. The charms of the actresses put money in the shareholders' pockets. For the revival of his play *The Parson's Wedding* (*c*.1640) in 1672, Killigrew used an all-female cast as a draw. It worked, so he followed it at once with an all-female revival of *Secret Love*. Nell Gwyn had left the stage by then. She had passed from Hart's bed to Lord Buckhurst's and then, replacing Moll Davis of the Duke's Company, to the King's. The job of royal mistress was one she took more seriously than did Charles II, by whom she had two children and for whom she

BACK-STAGE AT THE PLAYHOUSE

Nell Gwyn.
Sir Peter Lely is said to have painted this portrait of Nell Gwyn as Venus for Charles II's personal enjoyment.

The *Diary* of Samuel Pepys is liberally peppered with references to the London playhouses, actors, and, in particular, actresses; this entry is for 5 October 1667.

To the Duke of York's playhouse, but the house so full, it being a new play *The Coffee House* [by Sir Thomas St Serfe], that we could not get in, and so to the King's house. And there, going in, met with Knepp and she took us up into the tiring rooms and to the women's shift where Nell was dressing herself and was all unready and is very pretty, prettier than I thought. And so walked all up and down the house above and then below into the scene room and there sat down and she gave us fruit; and here I read the questions to Knepp while she answered me through her part of *Flora's Figary* [*Flora's Vagaries* by Richard Rhodes] which was acted today. But Lord! to see how they were both painted would make a man mad and did make me loath them; and what base company of men comes among them and how lewdly they talk! and how poor the men are in clothes and yet what a show they make on the stage by candle light is very observable. But to see how Nell cursed for having so few people in the pit was pretty, the other house carrying away all the people at the new play, and is said nowadays to have generally most company, as being better players.

forfeited a career. Just how much hers is a success story is open to debate. She was not notably more venereal than other actresses of the period. Not more than a quarter of those we know by name can be said to have lived 'respectably', though it should be said that much less attention has been paid to the periods spent by Hart, Edward Kynaston, and Cardell Goodman in the keeping of aristocratic ladies. It was precisely because actresses were not expected to be virtuous that Bracegirdle could make herself so interesting by protesting that she was.

From the beginning, Davenant outmanœuvred Killigrew, although the Duke's Company began with less experienced actors and Killigrew had obtained for the King's Company the majority of the playbooks from the pre-war repertoire. Squeezed into innovation, Davenant relished the challenge. The Lincoln's Inn Fields playhouse gave him a scenic advantage. His scenes were claptraps, and there was typical resourcefulness in his borrowing of the King's recently used coronation robes for the staging of his own *Love and Honour* (1661). The red letters customarily used on playbills to advertise special features were images of rivalry, and Davenant was generally the victor in the red-letter war. In 1663, when Killigrew's new playhouse in Brydges Street/Drury Lane was ready, the opening play was Fletcher's *The Humorous Lieutenant* (1619), part of the old repertoire on which Killigrew had hoped to base his supremacy. The Beaumont and Fletcher canon was his major resource, popular certainly, but lacking the bite of immediacy. The liveliest conversations among Restoration audiences were energized by inside knowledge. Davenant made sure that it was widely known that Samuel Tuke owed his knowledge of the Spanish original of his *The Adventures of Five Hours* (1663) to Charles II, and was rewarded with a major success. A taste for Fletcherian romance could be easily flattered by Spanish 'cloak-and-sword' plays. Tuke's deals in elevated passion expressed in mechanical verse, much of it rhyming. In a reading, its theatrical trappings seem like impositions of activity on a marmoreal debate; but it is the extraordinary in and out and roundabout use of the stage-doors that is the piece's only true innovation. By now, Thomas Betterton and Henry Harris were actors of authority to equal that of the older Hart and Mohun, and it was only the flourish of the Hart/Gwyn partnership that kept the King's Company in contention. After the success of *Secret Love*, Killigrew put Dryden under commission to write three plays a year in return for one and a quarter shares in the company. Dryden's status is, in this and other respects, unique, and the sharpness of the Duke's Company (Davenant died in 1668 but his spirit lived on in Betterton) is nowhere clearer than in the decision to open their new playhouse in Dorset Garden in 1671 with a revival of *Sir Martin Mar-All*, which Dryden had written for them before signing his contract with Killigrew. It was as if the Admiral's Men had opened the Fortune with a play by Shakespeare. *Sir Martin Mar-All* is splendidly theatrical foolery ('I never laughed so in all my life,' noted Pepys), a gift to the comedian James Nokes in the title-role and Harris as his sorely exercised servant, Warner. Critics have not found it easy to reconcile the poet Dryden with the journeyman playwright, and this play, cobbled together from two French sources in collaboration with the Duke of Newcastle, has been particularly troublesome. L. C. Knights thought it 'perhaps the stupidest play I have ever read'. The quarrel is, perhaps, between those who would think worse of Tennyson if he had written *Charley's Aunt* and those who would think better. Restoration comedy remains an object of scandal and concern, celebrated or decried for the constancy of its dialogue between fiction and contemporary reality. It plundered Molière and the French drama, Calderón and the Spanish, even incorporated set scenes from *commedia dell'arte*, but transformed rather than translated them into English to fit them for the forestage.

The opening of Dorset Garden inaugurated a new era in the rivalry between the two companies. Its scenic stage was supremely equipped, perhaps to the point where it overwhelmed drama by demanding spectacle. We get an impression of it from the engravings in the published version of Elkanah Settle's *The Empress of Morocco* (1673). The play is in the prevailing style for tragedy, written in rhyming couplets after the French model commended by the King and famously interrogated in Dryden's *Essay of Dramatic Poesy* (1668). That its scenes were both spec-

Dorset Garden playhouse. The best equipped of Restoration playhouses, Dorset Garden had also the advantage of access from the waterfront.

A scene from Elkanah Settle's *The Empress of Morocco*. Dolle's engraving of the opening scene of *The Empress of Morocco* is impressionistic rather than accurate. It shows little of the forestage at Dorset Garden, but indicates an ornate proscenium arch, a scenic stage with prison wings and arch borders, and back-shutters representing the prison wall.

tacular and admired is implicit in the publication of the engravings. The King's Company's response was to stage Thomas Duffett's travesty *Empress of Morocco*, in which Settle's kings, queens, and courtiers are transposed to pimps, prostitutes, and corn-cutters. In 1674, Dorset Garden struck gold again with Thomas Shadwell's operatic version of the Restoration *Tempest,* and the King's Company, from their newly built playhouse in Drury Lane, countered with another Duffett travesty. *The Mock-Tempest* is, in one respect, extraordinary. Whilst actresses were often hired to play prostitutes, on this occasion a prostitute was hired to play an actress. Betty Mackarel *in propria persona* holds a bawdy dialogue with the raucous Jo Haynes in an Introduction, from which we learn that she is to play Ariel, a mischievous warder in Prospero's Bridewell prison. Duffett's parodies open up the territory of the Newgate pastoral in which John Gay would find such riches some fifty years later. The last of them, *Psyche Debauch'd* (1675), aimed to score off that year's Dorset Garden spectacular, Shadwell's *Psyche*, an adaptation of the comedy-ballet by Molière and Lulli.

For John Downes, prompter to the Duke's Company, *Psyche* was an 'opera', a word regularly used to adorn spectacular plays with plenty of songs in clothing borrowed from Monteverdi and the Italians. Only Purcell's *Dido and Aeneas* (*c.*1690) among seventeenth-century English works is distinctly an opera rather than distinctly operatic. But Downes's confident use of the word should alert us to the continuing significance of music in the post-Restoration theatre. The musicians' balcony was commandingly placed, above the centre of the proscenium arch. After Charles II's reformation of the King's Musick along French lines, its string section was augmented and twelve violinists allotted to each theatre company. It was with the sound of music that the theatrical entertainment began and ended.

Comedy, as it generally has been, was more popular than tragedy, and its best examples have lasted better. In an actors' theatre, it accommodated and displayed individual personality alongside dramatic character. The late seventeenth century speaks with more assurance in prose than verse. Dryden, at ease in both, artfully combined a courtly, heroic plot in verse with a gentlemanly love-chase plot in prose. The mixture works well in *Secret Love* and *Marriage à la Mode* (1672), but weighs down Etherege's first play, *The Comical Revenge* (1664). It was his second, *She Would if She Could* (1668), that broke with verse to establish what is often seen as the characteristic tone of Restoration comedy. It is the story of a day in the life of a group of men and women about town. All the best-known comedies are set culturally and topographically in London *town*: not in the court nor in the city, often portrayed as a nest of Cromwellian vipers, and certainly not in the country (a remote place somewhere in Hampshire where there is nothing for a gentleman to do). Rambling (perambulating the town in search of sex) is an accepted pastime, and its conclusion in marriage is predictable only in terms of comic

Betterton as Hamlet. The frontispiece to *Hamlet* in Rowe's edition of Shakespeare (1709) shows a famous piece of stage business, Betterton's overturning a chair on the entrance of his father's ghost.

convention. It is almost too easy to discern an underlying melancholy in this pursuit of pleasure, undertaken by worldly-wise people who know the likely disproportion between desire and its fulfilment. Wycherley's *The Country Wife* (1675) captures in Horner the rakish image of the male who gets all and requires more, and in Margery Pinchwife the image of the randy provincial who does not yet know what 'it' is, but knows that it is in town. Marriage in such plays is not so much a happy ending as a necessary rite of passage to adultery, something which the proviso scenes of the gay couples confront but cannot resolve. The Earl of Rochester is supposed to have turned on a dog that bit him with the curse: 'I wish you were married and living in the country.'

By comparison with the known handful of comic masterpieces, the tragedies read as if translated from a foreign language and a preposterous culture, in which a gentleman will tie himself in verbal knots in order to avoid inflicting pain. A versatile actor had to speak at least two languages. Betterton is the towering Restoration figure, creator of roles as opposed as Dorimant in Etherege's finest comedy *The Man of Mode* (1676), and Antony in Dryden's blank-verse tragedy *All for Love* (1677). But Betterton was more than an actor. As Davenant's lieutenant, he had reviewed the

ACTORS OF THE RESTORATION THEATRE

Anthony Aston was a boy when Betterton and Barry were in their prime, and an old man when he recorded his memories of them.

Mr Betterton was the most extensive actor from Alexander to Sir John Falstaff, but in that last character he wanted the waggery of Estcourt, the drollery of Harper, the sallaciousness of Jack Evans. [. . .] Powell attempted several of Betterton's parts, as Alexander, Jaffier, etc. but lost his credit, as in Alexander he maintained not the dignity of a king but out-heroded Herod, and in his poisoned mad scene out-raved all probability; while Betterton kept his passion under and showed it most (as flame smokes most when stifled). Betterton, from the time he was dressed to the end of the play kept his mind in the same temperament and adaptness as the present character required. [. . .]

His favourite Mrs Barry claims the next in estimation. They were both never better pleased than in playing together. Mrs Barry outshone Mrs Bracegirdle in the character of Zara in *The Mourning Bride,* although Mr Congreve designed Almeria for that favour. And yet this fine creature was not handsome, her mouth opening most on the right side, which she strove to draw the other way and at times composing her face, as if sitting to have her picture drawn. Mrs Barry was middle-sized and had darkish hair, light eyes, dark eye-brows and was indifferently plump. Her face somewhat preceded her action, as the latter did her words, her face ever expressing the passions. Not like the actresses of late times who are afraid of putting their faces out of the form of non-meaning, lest they should crack the cerum, white-wash or other cosmetic, trowelled on. Mrs Barry had a manner of drawing out her words which became her, but not Mrs Braidshaw and Mrs Porter. To hear her speak the following speech in *The Orphan* was a charm:

> I'm ne'er so well pleased as when I hear thee speak.
> And listen to the music of thy voice. [. . .]

Parisian scenic stage, and he shared with Davenant a delight in spectacle that he was free to express as effective manager of the Duke's Company after 1668. A legal document relating to his dispute with the patentees in 1694–5 records 'the care he took as principal actor, in the nature of a monitor in a school, to look after rehearsals'. The rehearsal system he inherited was, to a modern eye, perfunctory. A revived play would be given a morning run on its day of performance. A new play would be introduced to the cast by its author, whose observations on his characters were the closest thing to directorial guidance they would receive, and who had anyway written it with their individual qualities in mind. As in the Elizabethan theatre, the disposition of actors on stage was governed by conventions of precedence, and rehearsal was restricted to scenes or episodes that were innovatory or technically complex. Betterton thought of stage speech in musical terms, with a notation based on heightened conversation rather than rhetoric. As principal actor, Betterton would often speak the prologues that so crucially set the tone of the theatrical event. Epilogues more often went to the leading actresses, who made through them their next assignation with the audience. Untouched by scandal and widely respected in a quarrelsome profession, Betterton was none the less a schemer. In 1681, with the King's Company in terminal disarray, he was party to a secret pact with Hart and Kynaston, that company's disgruntled seniors, who undertook to subvert any recovery. The following year, the two companies united under Betterton's leadership.

The London theatres had lost their initial impetus some years before the collapse of the King's Company. They were victims of the national neurosis caused by Titus Oates's invented Popish Plot. Neither company knew how to cope with an audience fiercely divided along political lines. It was over the issue of the legitimate succession or exclusion of the Catholic Duke of York that party politics became a fact of English life. The theatre was occupationally biased towards the Jacobite (Tory) group, and the King's Company did itself no favours by staging Nathaniel Lee's constitutionalist (Whig) tragedy *Lucius Junius Brutus* in 1680. The Lord Chamberlain stepped in to ban it after a few performances. In the spate of plays provoked by the Popish Plot, Settle and Shadwell also argued on the Whig side, but the preponderance was anti-Whig. Otway's *Venice Preserved* (1682) contains a slanderous caricature of the great exclusionist, the Earl of Shaftesbury (Dryden's Achitophel), and Crowne's *City Politiques* (1683) dances on the grave of Titus Oates's dreams. In sturdy times, theatres build audiences on the back of political crises, but the Popish Plot and associated Monmouth Rebellion had unnerved court and town. In the second phase of post-Restoration theatre, the wooing of the city became a necessity. The United Company was unadventurous, and the 1680s saw few advances in the art of the theatre. They did, however, see the full emergence of a remarkable actress. Elizabeth Barry may have created the discordantly passionate Mrs Loveit in *The Man of Mode*. She carried into comedy the unnerving force of felt life. But it was in Otway's tragedies that she stunned contemporary audiences. Her portrayal of Monimia in *The Orphan* (1680) created a brief vogue for she-tragedies, and in the perverse and painful *Venice Preserved* her Belvidera pushed performance closer to psychological complexity than it had ever been. She seems to have drawn something new out of Betterton, her Jaffeir. Betterton's conventional concern was with the accurate delineation of the passions, carefully separated from each other. Barry, less decorous, set out to excite passion in the audience.

It was during the 1680s that Christopher Rich began his calculated climb to management. By 1693, this hard-headed lawyer had acquired control of the United Company. Rich has had, and probably deserved, a bad press. A manager so unfamiliar with the pressures of acting and yet so ready to interfere was a novelty for the players. By 1695 most of them were up in arms. Betterton and Barry headed the list of fifteen petitioners to the Lord Chamberlain. Despite Rich's spirited resistance, their petition was granted, and Betterton became the leader of an independent company of actors. This assertion of the right of actors to control their own destiny is significant in itself, and despite their confinement to the outmoded Lincoln's Inn Fields playhouse the venture started auspiciously. The United Company under Rich had hoped to stage a new play by the brilliant new star William Congreve, and his decision to transfer *Love for Love* to Betterton and the rebels was momentous. Its opening was one of the last glittering occasions on the seventeenth-century stage.

The prologue, as so often, was spoken by Betterton, who, at the age of 60, was about to create the role of the wayward young hero, Valentine Legend. The rebels had had difficulty in tempting younger actors to join them. Cave Underhill, who played Valentine's father, was much of Betterton's age, as was the uncommonly ugly Samuel Sandford in the 'humours' role of the astrological gull, Foresight. William Smith, as the lecherous Scandal, was only a little younger. He would be dead before the end of the year. The youngest of the male leads was Thomas Doggett, nicknamed 'Aspectabund' by the company's prompter, because he was always 'wearing a Farce in his Face'. His playing of Ben Legend, Valentine's younger brother and rival for the crucial inheritance, confirmed a reputation that would last well into the next century. Prue, the impatient virgin, was played by the otherwise undistinguished Mrs Ayliff, Mrs Frail and Mrs Foresight, sisters-in-sex, by Elizabeth Barry and Elizabeth Bowman, who had acted together many times, and the witty Angelica, Valentine's beloved, by Anne Bracegirdle. Hers would be the privilege of delivering the epilogue at the end of an auspicious performance. *Love for Love* is both like and disconcertingly unlike one of Etherege's infinitely (?culpably) tolerant comedies. It opens in Valentine's lodging with the prospect of a day's rambling ahead: there is more than one love-child as a relic of ramblings past. But Valentine, despite the encouragement of the rakish Scandal and the foppish Tattle, has abandoned libertinism. What is not clear at the outset, nor quite clear enough at the end, is how far his reformation is related to the state of his pocket. Valentine has still to learn from Angelica that love cannot be bought, only returned—love for love. The point may seem trite, but it is not tritely made. There is a constant and subtle concern with the interpretation of signs, with reading and writing (three people in the play have the surname Legend). It may be—and the possibility is fraught with hope for society—that Valentine *knows*, at the play's quite bitter end, that women are not to

Facing: William III on horseback. For his equestrian portrait of William III, Kneller chose Elizabeth Barry as his model for a sturdy Ceres and the more conventionally attractive Anne Bracegirdle as the model for Flora.

be 'read' or 'written on', that Angelica can write, and is right. Congreve's purpose, then, and ultimate achievement, is to rewrite the social agenda, but *Love for Love*'s success (it had thirteen consecutive performances) owed more to its scenes of sexual fooling than either Betterton or Congreve would have liked to admit. There was, for instance, the inspired staging of Tattle's reluctant encounter with Prue. At the close of Act II, forced into action by his affectation as a rake, Tattle is dragged to Prue's bedroom. Tattle, we are fairly sure, is impotent, and the audience can spend the interval speculating on his hapless attempts to keep the predatory Prue at bay. The scene we have just witnessed was, after all, hilarious, in a way that licenses indiscreet entr'acte whispers. But Congreve is a step ahead of us. Act III opens in Prue's bedroom, with Prue and Tattle at the exact point of undress we might expect such inept fumblers to have reached in 'real' time. *They* interrupt *us*, and are at once themselves interrupted, with consequent scramblings through proscenium doors and dragging open of scenic shutters. This is the farce that Feydeau wrote, and Congreve, though he might protest that his purposes were misread, had to be wary of protesting too much.

Moral opposition to the theatre was growing, and the court, particularly after the death of Queen Mary in 1694, no longer offered assured protection. The critical scrutiny of social style is exemplified by the foundation in 1692 of the influential Society for the Reformation of Manners, and the appointment to the Chamberlainship of the Earl of Sunderland, one of its adherents, was ominous for the stage. The attack on libertinism found expression inside the playhouses, too, most impressively in such comfortless comedies as Thomas Southerne's *The Wives' Excuse* (1691) and *The Maid's Last Prayer* (1693). The emphasis here has shifted from the charms of adultery to the burdens of incompatibility within marriage. Congreve's proviso scenes, whilst looking back to Dryden and Etherege, have also to take account of the new mood. His wilfully elaborate plots and rejection of tidy pairs of characters implicitly reject the reassurance of symmetry. Jeremy Collier's powerful propaganda signalled rather than brought about the end of an era. The actors, of course, carried on: Rich with the infinitely adaptable Colley Cibber and others at Drury Lane, Betterton's company with increasing difficulty at Lincoln's Inn Fields. It was, however unjustly, the plight of Betterton and the plays of Congreve that Collier had most in mind when he wrote, 'the modern poets seem to use smut as the old ones did Machines, to relieve a fainting situation'. Vanbrugh's Loveless, as Collier knew well, relieved Berinthia's fainting situation by carrying her off to the closet, where 'there's Moon-shine upon the Couch'. 'Help, help,' says Berinthia *very softly*, 'I'm Ravish'd, ruin'd, undone.' *The Relapse* (1696) offered rich pickings to Collier.

As the seventeenth century approached its end, the London theatres were faltering and there was a spate of prosecutions of actors for 'uttering profanities'. Confidence to resist is best represented by the low comedians, descendants of

Jo Haynes at Drury Lane. Jo Haynes's cheeky epilogue excited imitators. Pinkethman was one, and at the 1704 May Fair he went one better by delivering an epilogue from the back of an elephant.

Richard Tarlton. William Pinkethman performed at the summer fairs, where, according to Downes, 'he has gain'd more . . . in Twelve Years, than those that have Tugg'd at the Oar of Acting these 50'. Thomas Doggett, whose eccentric performance as Ben in *Love for Love* had been the outstanding success of that brilliant production, took himself off on provincial tours. Jo Haynes, whose chequered career encouraged Tobyas Thomas to publish the first biography of an English actor in 1701, refused to be tied down by anyone. We have a glimpse of him, cocking a snook at Drury Lane decorum by delivering an epilogue from the back of an ass. But a move to disempower actors had begun and would continue. Cibber tells us that, after the 1696 alterations to Drury Lane, 'the Actors . . . are kept so much more backward from the main Audience than they us'd to be'. It was the beginning of an attack on the forestage as a site of interaction. In the longer run, the audience would occupy one room and the actors another.

7

FRENCH RENAISSANCE AND NEO-CLASSICAL THEATRE

WILLIAM D. HOWARTH

IT would be no exaggeration to say that the history of the French theatre through the years 1550–1715 presents a microcosm of the country's political and social history during this same period—a period so crucial for the development of the modern French state. What we witness is the transition from the disorganized and anarchic survivals of the medieval theatre, kept alive by the religious guilds and a calendar dominated by church festivals, to a highly organized, centralized state theatre which not only mirrored the political structure of Louis XIV's France, but also served in its way as an organ of absolutist policy. As long as the Religious Wars continued, with the great feudal houses fighting for the real control of the kingdom, there was little chance of a settled cultural life in the capital; but once Henri IV's reign had brought peace and the foundations of prosperity, it became possible for Richelieu to begin to implement the far-reaching plan for a France secure and united within her own frontiers, and able to play an increasingly dominant role in Europe. And to this domination in Europe, planned by Richelieu and achieved by Louis XIV, French cultural hegemony was to make a not inconsiderable contribution. The ostentatious cultivation of the arts at the court, and in the capital, of the Sun King helped to convey the message to an admiring—or envious—Europe that, just as Louis was aspiring to be the modern counterpart of Alexander or Caesar, so Paris should be looked on as the modern equivalent of Athens or Rome.

The first tentative foundations of a steady movement towards a modern national theatre in France are to be seen in the striking, if fortuitous, conjunction of two dates in the mid-sixteenth century. In 1548 the powerful Paris magistracy, the Parlement de Paris, issued a decree banning all performances within their jurisdiction representing 'the passion of Our Lord and any other sacred mysteries': the Confraternity of the Passion, the religious guild which held the monopoly of theatrical performances in the capital, was to be confined to 'other secular, honest and decent [plays] provided that there is in them no offence or insult to anyone'. For the mystery plays had become the subject of increasing criticism: from devout Catholics because their popular style debased the sacred subject-matter, and from Protestants, whose attitude tended towards a more wholesale condemnation of all theatre, especially the religious mysteries. The following year, 1549, saw the publication of an important theoretical manifesto, *La Défense et illustration de la langue française* (*The Defence and Ennobling of the French Language*) by the young humanist Joachim du Bellay. This text, in which du Bellay was acting as spokesman for the group of poets known as the Pléiade, called for a campaign to establish French as pre-eminent among the literatures of modern Europe; and, with particular regard to dramatic writing, for the replacing of the medieval genres of mystery play, miracle, and farce by those of tragedy and comedy, which would command the authority and the prestige of the drama of antiquity. These young humanists were treading the path already pioneered before them by their Italian counterparts; and to begin with imitation of Italian drama was much in evidence, and the influence of Latin writers (Plautus, Terence, and Seneca) more considerable than that of the Greek tragedians. However, by the end of the century a corpus of French tragedies had been built up, including adaptations of the Greek masterpieces alongside imitations of Seneca: a highly derivative body of texts, but one which reflected the ideals of du Bellay's *Défense*.

However, while its importance for the future development of neo-classical drama in France is by no means negligible—for the *form* of tragedy (five acts in rhyming alexandrines) and its rhetorical and declamatory *manner* were established in this way—this was something that took place almost entirely outside the professional theatre. The Pléiade poets, their contemporaries, and their successors constituted a self-contained coterie of scholars, and their plays were destined for performance either at the royal or princely courts—until their patrons demanded something more entertaining as a relief from the horrors of the civil wars—or in the colleges. The fortunes of the professional theatre in Paris were at a very low ebb: the Confraternity of the Passion continued to enjoy the monopoly of theatrical performances at the Hôtel de Bourgogne, into which they had moved in 1548, and from time to time, from 1578 onwards, they would lease this theatre to a succession of itinerant companies. There are records of a number of such companies in the provinces at this time, where they played (especially in university cities) the new humanist tragedy, but also, to more popular audiences, the religious plays now forbidden in Paris.

Italian comedy travels to France. This painting almost certainly represents a scene from a scenario performed by the first professional company of Italian players in Paris, I Gelosi. Isabella and Pantalone are recognizable in the foreground.

By 1600, then, the form of French neo-classical tragedy was fairly well developed, as well as its general subject-matter. The matter for tragedy was to be found in mythology (e.g. Robert Garnier's *Hippolyte*, 1573); in ancient history (e.g. Étienne Jodelle's *Cléopâtre*, 1552) and exceptionally in the history of more modern periods (*L'Écossaise*, on the subject of Mary Queen of Scots, by Antoine de Montchrétien, 1601); and in the Bible (e.g. Garnier's *Les Juives*, 1583) or the lives of the saints. The episodic, all-embracing, and spectacular manner of the mysteries had given way to a kind of writing which already possessed the unity of tone and the clear focus on a single event that were to be features of seventeenth-century tragic form—but without the latter's dramatic impact; for Renaissance tragedy is static, oratorical, and didactic, consisting all too often of elegiac lamentation about an event that has already taken place.

Comic drama, though borrowing from Plautus and Terence, and frequently based on Italian models, did retain features of character and plot from the medieval farces; and Renaissance comedy is certainly a more lively affair than the tragedy of the period. The construction varies from the simple to the highly complicated, depending on the model chosen: in a unique imitation of Aristophanes (*The Birds*, for instance) Pierre Le Loyer presents a cornucopia of visual as well as verbal delights (*La Néphélococugie*, 1578), while Jodelle's *L'Eugène* (1552) retains the simple directness (and the bawdy flavour) of medieval farce; Pierre de Larivey's *Les Esprits* (1579) is more typical of the genre, in that it follows Lorenzo de' Medici's *Aridosia*, itself a fusion of source-material from Plautus and Terence. Where, and how, these plays were performed—and in some cases, whether they were performed at all—is, as with the tragedies of the period, largely a matter for conjecture rather than certainty; and although it is safe to assume that many of them were performed in the colleges and the private houses of rich patrons, documentary (and particularly visual) evidence is entirely lacking.

The situation is different, however, once we turn to consider the distinctive contribution of the Italian companies which performed in Paris from the 1570s onwards. The original company was that of Alberto Ganassa, famous as an early Arlecchino, who was invited to Paris in 1571 to perform at Charles IX's wedding, but was not allowed to stay in France because the Parlement found his admission charges too high. However, he returned in 1572 to perform at the wedding of Henri IV and Marguerite de Valois; and this was followed by repeated visits by the company of I Gelosi under the direction of Flaminio Scala (who himself played Flavio, the young lover, and was a noted author of scenarios). From now on, it

La bonne mere Guillemette.   Agnan Magifter.   Peronne.

Mon fils tu ès ja grãd, tu deurois te cognoiftre, | Ma mere laiffez moy, ie veux baifer Perône, | Vous ne me tenez pas Magifter de village,
Et fage te monftrer pour vn iour te pouruoir, | Car elle m'a donné vn foufflet pour faueur, | Puis que ie vous ay peu le dernier coup donner,
Ne cours dõc point apres la fille de ton maiftre, | Aproche mon foulas, ne t'enfuis ma mignône, | Ma foy vous eftes trop éueillé pour voftre aage,
Il la faut courtifer autrement pour lauoir. | Par la mort d'vn étron tu as raui mon cœur. | A dieu dõc Magifter, il m'en faut retourner. iij

Native French farce. The farce-player Agnan Sarat in one of his stock roles as a lecherous pedant, pursuing the fleeing Péronne (played by a man). Little is known of Agnan's repertoire apart from this series of engravings, some of which suggest characters and subject-matter of mixed French and Italian origin.

becomes possible to think in terms of a form of theatre with a strong visual impact; we can talk with some confidence of the costumes and masks distinguishing the Arlecchino, Pantalone, and other stock characters; and we can obtain a reliable idea of stage-setting, and an occasional clue as to gesture and comic effects.

A precious source of iconographical evidence regarding the roles and costumes of the Italian comedy in its earliest form in France is the *Recueil Fossard*, a collection of engravings and woodcuts discovered during the present century at Stockholm. An interesting feature of the same collection is that it includes, alongside the stock characters of the *commedia dell'arte*, a number of engravings of the French actor Agnan Sarat, who was one of the first professional performers to leave a name to posterity, and whose activities are recorded in a contract signed by the Confraternity of the Passion, leasing the Hôtel de Bourgogne to Sarat and his colleagues in 1578. These engravings suggest farcical subject-matter similar to that of the Italians; but there is the barest hint of a more varied repertory in the doggerel verse (*c.*1600) addressed to a courtier whose fine clothes had ended up on the stage, via an old-clothes dealer:

> How many times did your silken finery
> Clothe the King of Troy, or Amadis and his knights,
> When Agnan with his ugly face
> Played at the Hôtel de Bourgogne some tale of long ago?

Farce at a fairground theatre. The open-air stage survives in the quack doctor's booth at the seasonal fairs. The characters appearing in this sketch show how Italian influence had merged with native farce tradition: Polichinelle derives from the Neapolitan Pulcinella, while the character here named Brigantin is dressed as an Arlecchino (though he is not masked).

Otherwise, nothing; and the first French actor to have left anything approaching a coherent record of his professional career is Valleran Lecomte. Valleran, whose company is first recorded as playing in Paris in 1598, was already an established actor when he performed at Bordeaux in 1592; he became *chef de troupe* soon after that, and spent several more years in the provinces before appearing at the Hôtel de Bourgogne in 1598, when his company is designated 'troupe du Roi'. From now until 1600, when financial hardship forced him to take to the road again, and from 1605 until 1612, when he finally left the capital to play in Holland, Valleran's Paris career is well documented by the signing of repeated contracts with the Hôtel de Bourgogne; acts of association with other French companies, or with visiting Italian troupes; and the hiring of young actors whom Valleran helped to train. Essentially, his is a story of a precarious struggle to establish a mixed repertory of tragedy and farce, and it is clear that competitors fared better when they offered the less cultured elements of their Paris public a diet of farce alone. Nevertheless, Valleran and his company can be credited with supporting the early career of Alexandre Hardy, who as well as a small-time actor was also the first professional playwright attached to a theatre: it is no doubt Hardy's plays that are referred to in a document of 1599 as the 'romances, comi-tragedies, pastorals and comedies' that were to be executed by Valleran's company. Documentary evidence from these years attests the first known presence of actresses on the Paris stage; while the two outstanding actors of the next generation—Pierre Le Messier (stage name Bellerose) at the Hôtel de Bourgogne and Guillaume des Gilleberts (stage name Mondory) at the Théâtre du Marais—were started on their career by Valleran during this period.

Hardy was to supply the Hôtel de Bourgogne with plays until 1630 or so; he was a prolific playwright, in the mould of his contemporary Lope de Vega, and the thirty-odd plays he published form a small fraction of the total he wrote. Compared with the Pléiade poets he lacked finesse and literary skill; but he had a feel for the theatre as well as some classical education, and was able to bring dramatic action and spectacle to serious drama in place of the elegiac laments of

Tennis-court into theatre. The origins of the French public theatre in the *jeu de paume*, or tennis-court, were to determine the elongated rectangular shape of theatre design until well into the eighteenth century. Throughout the seventeenth century there was an abundant supply of such buildings, and conversion was easy and cheap.

Standing spectators in the *parterre*. This engraving by Abraham Bosse of the standing *parterre*, flanked by the seats in the raised *amphithéâtre*, gives an excellent impression of the atmosphere in a mid-seventeenth-century theatre, with its small box stage and a long, narrow auditorium. It has been suggested that the engraving represents the Hôtel de Bourgogne: if so, there is a considerable foreshortening, since contemporary records show that the *parterre* there could contain nearly 1,000 spectators.

his predecessors. With Hardy and his contemporaries and immediate successors (Théophile de Viau, in his tragedy *Pyrame et Thisbé*, 1622; the pastorals of Honorat de Racan and Jean Mairet; and Pierre Corneille's early works, the tragicomedy *Clitandre* (1632) and the comedy *L'Illusion comique* (1636)) we see the flourishing of the style we now call 'baroque': a vigorous, exuberant, and imaginative French counterpart to the English Elizabethan and Jacobean theatre and the *comedia* of Lope and Calderón. Elements included plot-devices such as disguise and ambush; violent action—rape and murder—on stage; striking changes of tone from the elevated to the familiar; sprawling construction; a fondness for the complications of the play within the play; self-indulgent literary embellishment and extravagant poetic conceits—and subsuming all this, an evident delight in theatricality. If these characteristic features of the drama of Shakespeare and his contemporaries were also to be found in the representative theatre of a whole generation in France, how did it come about that they were so swiftly abandoned in favour of a much more chastened, disciplined, and intellectualized form of dramatic composition?

First, we can point to factors concerning the sociology of the theatre, and its relation to the general political situation. Changes in the make-up of theatre audiences in terms of class and sex reflected the greater importance of women in the social life of court and capital, which brought a refining influence into the arts; the fact that theatre-going became more respectable throughout the reign of Louis XIII; and above all, perhaps, the deliberate policy of Richelieu. In his determination to consolidate the stability of the monarchy by encouraging orthodoxy in every aspect of French life, the Cardinal was very much aware of the important contribution to be made by literature and the arts. He founded the French Academy in 1635; and the particular interest he took in the theatre, together with his ambition to make a personal mark in this area, led to the creation of a 'Company of five authors' (including Corneille) who worked under his direction. The fact that Richelieu favoured a certain type of play inevitably influenced the evolution of dramatic genres in the 1630s; and at the time of the contro-

versy over Corneille's *Le Cid* he made a decisive intervention when he referred the play to the judgement of the Academy.

Specifically literary factors are easier to gauge; and here, the move towards refinement and self-discipline is reflected in the development of a more flexible, less self-consciously 'poetic' verse-form which was to become an ideal vehicle for rational debate, analytical soliloquy, and the vigorous exchange of points of view: in other words, for the tragedies which were to be characteristic of Corneille and other playwrights from 1640 onwards. The desire for a more disciplined kind of dramatic composition is seen too in the debate between 'irregular' and 'regular' dramatic forms, and in the triumph not only of the three unities of time, place, and action, but also of obedience to an equally important, if unwritten, fourth unity: that of tone.

However, the unities were not—as used to be suggested—accepted unwillingly by reluctant playwrights at the behest of their influential patrons; nor did these formal changes take place in a theoretical vacuum removed from the practicalities of the real theatre. On the contrary: they were the subject of willing experimenta-

Richelieu's private theatre. The first theatre to be purpose-built was Richelieu's private theatre inaugurated in 1641 in the Palais-Cardinal; later, under the name of the Palais-Royal, it was to become the home of Molière's company. This painting shows, in spite of the theatre's elegance, the same basic shape as that of the tennis-court.

tion by a group of ambitious young dramatists—the avant-garde of their day, one might say—all eagerly competing for box-office success; and the development of a theory of regular drama based on the three unities was accompanied by a parallel evolution in staging practice. The one complements the other: the emergence of a 'classical doctrine', as a theoretical basis for a form of serious drama capable of satisfying the rational needs of an educated audience, corresponds to the transition from the multiple set bequeathed by the medieval theatre to the single perspective set imitated from Italian models and which, once accepted, would remain in force for over 200 years.

In this respect as in others, *Le Cid* of 1637 can be seen as the 'hinge' between the old and the new. A modified form of the conventional multiple set, comprising about five compartments, had survived into the 1630s, and this can be seen as the

---

A RATIONAL DEFENCE OF THE UNITY OF PLACE

The *Pratique du théâtre*, published in 1657 by the abbé d'Aubignac, who had been a protégé of Richelieu's, puts forward a defence of regular drama based on the unities of time and place which is founded not, as with many theorists, on the authority of Aristotle and others, but on common sense and reason. It was this recourse to rational critique which hastened the demise of the multiple set in the 1630s.

If [the poet] requires his characters to appear in different locations, he will make his play absurd by the lack of that verisimilitude which should be its principal foundation . . . We can see from this how ridiculous that wall was in Théophile's *Pyramus and Thisbe* which came out on to the stage to enable the lovers to speak to each other through it, and then disappeared when they went off-stage so that the rest of the characters might see one another. For not only did the two compartments which were on either side of this false wall represent the private apartments of Thisbe and Pyramus, so that it was against all reason that the King should arrive in the same place to address his confidants—and how much more contrary to reason that a lion should make its way there to frighten Thisbe!—but I should like to know how, according to the detail of the narrative, this same wall could become both visible and invisible? by what magic it was able to prevent the two lovers from seeing each other, but did not prevent the other characters? or again, what extraordinary power it possessed, so that it could both be present, and cease to be present?

Facing, above: stage design for a tragedy. One of the forty-seven sketches left by the Hôtel de Bourgogne's *décorateur*, Laurent Mahelot, recording this theatre's stage designs in the 1630s, this illustration for Théophile de Viau's *Pyrame et Thisbé* (first performed in 1622) shows the multiple 'compartmented' set at a relatively late point in its evolution. 'Town' elements (the King's palace and the homes of Pyramus and Thisbe (with the famous wall apparently sliding into position when required)) are flanked by out-of-town components: mulberry tree, lion's cave, fountain, and Ninus' tomb.

Facing, below: stage design for a comedy. Here we have a more advanced example of the compartmented set, showing the separate (but inter-related) elements harmoniously composed by the use of perspective into a homogeneous whole. Comparison with the *Pyrame et Thisbé* sketch indicates, as do several other of Mahelot's drawings, the reuse of stock elements from the Hôtel de Bourgogne's store.

basis of the forty-seven set designs illustrated in the priceless sketch-book of Laurent Mahelot, the *décorateur* of the Hôtel de Bourgogne. As long as the convention of the multiple set—symbolic suggestion, rather than realistic representation of 'place where'—was accepted by playwrights, designers, and audiences, it was perfectly viable; but by 1637 various factors were combining to bring it under attack. Some of Mahelot's sketches bear witness to a tendency to 'compose' the elements into a coherent whole; the better-informed spectators were no doubt aware through published engravings of the advances made by the Italians in the field of perspective design, and were looking for a more representational set; while the growing acceptance of unity of place (as a corollary of the twenty-four-hour rule) again pointed towards a unified design.

Le Cid was not performed at the Hôtel de Bourgogne, but (like Corneille's other early plays) by the rival company under Mondory which occupied the Théâtre du Marais. Mondory's outstanding success as an actor-manager (he had become *chef de troupe* in 1634) depended partly on his own talents, partly on the quality of the play-scripts he was offered by Corneille and others. Another of his great successes was as Herod in the tragedy *Mariane* by Tristan l'Hermite (1636), and some idea of the energy and passion of his acting style may be gained from the fact that he suffered an apoplectic seizure in a revival of this play in 1637, and had to retire prematurely from the stage. Although the Marais is less well documented than the Hôtel de Bourgogne, it is known that it shared the long, narrow shape of other early theatres converted from tennis-courts; and in the absence of evidence to the contrary we may be confident that staging practice was similar to that illustrated by Mahelot at the rival theatre. Confusion about the precise location of the action of each scene within the multiple set may have contributed to the controversy about *Le Cid*, but this was largely fuelled by the jealousy of rival playwrights. The real subject of the controversy was the clash between irregular and regular theatre, between tragicomedy and tragedy, in a word between old and new. In its 1637 form Corneille's play conformed to the unities, but at a cost, for its action-packed plot fits uneasily into twenty-four hours. It remains the masterpiece of baroque tragicomedy, even though Corneille was later to respond to the critical furore by trying to bring it into line with the precepts of classical tragedy.

In one other way *Le Cid* made theatrical history in 1637. For the play was so phenomenally successful that, as a temporary expedient, room was found for spectators on the stage itself. The temporary expedient was to last for well over 100 years. These highly prized (and highly priced) seats became a vital source of income: theatre companies could not afford to discontinue the custom, and new theatres even incorporated stage seats as a permanent feature of their design. The presence of exhibitionist young noblemen on stage can hardly have been conducive to realism, even in the case of a single representative set; but this feature must certainly have helped to kill off the multiple stage, since there could be no future in a compartmented set once spectators were seated between the acting area and the downstage compartments to which it related.

As a result of the controversy over *Le Cid*, Corneille turned to tragedy, with very few exceptions choosing subjects from Roman history. Most of his contemporaries followed suit; and tragicomedy, from having been by far the dominant dramatic genre in the 1630s, was rapidly overtaken by tragedy in the following decade, and continued to suffer a steady decline in the second half of the century, when publication figures show it to have been virtually eliminated by the more regular genres of tragedy and comedy. If, until *Le Cid*, Corneille had been one of a group of half-a-dozen aspiring young dramatists, there was no doubt at all, after *Le Cid*, of his pre-eminence. Although he was now prepared to adopt a more restrained and sober manner, he continued to create, in tragedies like *Horace* (1640), *Cinna* (1640), *Polyeucte* (1642?), *Pompée* (1643), *Rodogune* (1644), or *Nicomède* (1651), a form of heroic drama whose central characters are endowed with the strength of will to meet the challenge of adversity rather than, as in more conventional tragedies, being overwhelmed by a malign fate; indeed, he was to assert his preference for the admiration inspired by his heroes, over the tragic emotions of pity and fear prescribed by orthodox Aristotelian theory.

Neo-classical tragedy as we know it had very quickly—in no more than a dozen years at most—been accepted by playwrights and their public as the form of serious drama best designed to offer refined and rational aesthetic pleasure, while at the same time being capable of conveying a political or moral message. In the case of comic drama, however, examples of the genre were spread over a wide spectrum, from the five-act verse formula established by Corneille in the 1630s, and which was to be the vehicle for Molière's masterpieces, to the one-act plays given as curtain-raisers or afterpieces at the Hôtel de Bourgogne, and indeed to the sketches performed by mountebanks at the fairground theatres. Popular farce was kept alive at the Hôtel de Bourgogne by a trio of talented actors who, performing tragedy under their 'literary' stage names (La Fleur, Fléchelles, and Belleville), made their real reputation in the company's repertoire of farces under the names of Gros-Guillaume, Gautier-Garguille, and Turlupin. All three died in the mid-1630s; but not before the young Jean-Baptiste Poquelin, whom we know as Molière, had been taken to see them by his grandfather, and had conceived a precocious taste for this form of entertainment. Julien Bedeau, whose stage name was Jodelet, was a talented comic actor celebrated for his floured face who kept farcical comedy going at the Marais. In 1659, shortly before his death, he was to be taken into Molière's company.

If we bear in mind the extent to which the history of comic drama in Europe is characterized by critical, and creative, tension between two competing theoretical views of comedy—the mimetic (comedy as representation of real life) and the ludic (comedy as exaggeration or distortion of reality)—it can be seen that the

Les noms de Turlupin, de Gros Guillaume, & de Gautier Garguille, ne font que des noms de Théatre. Le premier se nommait Henri Legrand, le deuxième Robert Guérin, & le troisième Hugues Guéru. Tous les trois anciens garçons Boulangers, quittèrent cette profession pour celle de Comédien. Ils jouèrent au Théatre de l'hôtel de Bourgogne des farces qui urent un tres-grand succes. Quant au Capitaine fracasse, ce role d'Espagnol était celui d'un figurant.

Farce at the Hôtel de Bourgogne. The three celebrated farce-players of the Hôtel de Bourgogne company in the 1630s are here accompanied by Captain Fracasse, one of several popular embodiments of the braggart soldier, frequently portrayed as a Spaniard. This role, of Italian origin, was exploited with success in France; cf. Corneille's Matamore in *L'Illusion comique* (1636).

mid-century theatrical scene on which Molière made his début was a particularly fruitful one for a comic dramatist. On the one hand we have the evidence of the success of the farce-players in the popular theatre, and of the continued visits to Paris by the Italian players. Louis XIII had a particular regard for the Italians, and a sustained correspondence survives between the King and the Duke of Mantua, their patron. After visits in the 1620s and 1630s, there was a longer stay from 1639 to 1648 by a company which included Tiberio Fiorilli, the celebrated Scaramouche; a return visit took place in 1653, and in 1660 the Troupe de la Comédie Italienne took up residence in Paris, under the leadership of Scaramouche, on a permanent basis. For the most part, their repertory continued to be the 'pièces à canevas', or scenarios, in which the *commedia dell'arte* players had been trained on their home ground; playing before a foreign audience, they inevitably emphasized the visual features of their performance, and such comments as are extant in correspondence and memoirs speak admiringly of their traditional *lazzi* and acrobatic feats.

The contribution made by the Italians to the development of French comedy in this period was of a kind which complemented the traditional material inherited from the indigenous French farce. While the latter characteristically offered a simple study of a relationship, frequently in a domestic context which set the henpecked husband against the shrewish wife, a typical Italian scenario would present a more complicated plot revolving round the obstacles to the marriage of a couple of young lovers, helped by their servant to outwit the traditional father-figure. Molière was to draw on popular farce in both its French and Italian forms; but he was also to adapt to his own use the five-act verse framework of mimetic comedy, set in the context of contemporary Parisian social life by Corneille, who had defined his purpose in such comedies as being to paint 'a portrait of the manners and conversation of persons of good breeding'. The exclusion of what we should call genuinely 'comic' material from this definition was entirely in keeping with the firm distinction between comedy and farce established by the Renaissance theorists, for whom the imitation of life, not the arousing of laughter, was to be the comic dramatist's goal. Moreover, there is abundant evidence that this distinction was one which was observed in practice by the purists among the educated theatre-goers of mid-century Paris, who affected to look down on laughter as a concession to the more vulgar elements in a person's make-up. This is no doubt what Molière had in mind when he wrote that 'it is a strange undertaking to try to make the gentry laugh'; however, it was his achievement to bring about a fusion of literary comedy and popular farce, as framework and animating principle respectively of a totally original form of comic drama.

After a good education, Molière turned his back on the professional security of his bourgeois family in 1643, and threw in his lot with the Béjarts, a family of actors, founding the Illustre Théâtre with them. Financial failure obliged the company to leave Paris, however, and from 1645 to 1658 they toured the provinces, especially (to judge from surviving records) the south and west of the country. At the end of this period they felt confident enough to try their fortunes in Paris once more, playing at first in the Théâtre du Petit-Bourbon (a part of the Louvre) by the King's favour, and later transferring to the Palais-Royal (formerly the Palais-Cardinal, where a theatre had been constructed by Richelieu). In the incredibly short span of fifteen years up to his death in 1673 at the early age of 51, Molière achieved recognition by his contemporaries as an outstanding man of the theatre, and was soon to be acknowledged by posterity as the greatest comic writer of all time. In the provinces, and to begin with after the return to Paris, Molière and his

Molière and the Italians. A satirical portrait of Molière in the role of Sganarelle, taking lessons from Scaramouche, with whose company he shared a theatre for a time. The caption reads: 'What will he be like, with such a great master to teach him?'

Scaramouche enseignant.   Elomire estudiant.

Qualis erit? tanto docente magistro.

I. Weyen sculp.

company offered a conventional mixed repertory of tragedy, tragicomedy, comedy, and farce. As an actor, he had a personal ambition to shine in tragedy as well as comedy, but this was not where his talent lay—or rather (to judge by comments from objective sources, as well as by the lampoons to which he was subjected by spokesmen for the rival company at the Hôtel de Bourgogne) his manner as a tragic actor was not sufficiently stylized and declamatory to suit current taste. In comedy, it was a very different matter. It is often suggested that he developed his comic gifts by modelling himself on Scaramouche, with whose company he shared his theatre, and with whom he enjoyed an excellent relationship. At all events, the proportion of tragedy to comedy in his repertory steadily diminished, as did that of comedy by other authors. As both actor and playwright were confirmed in the favour of the public, so Molière's company gradually became a vehicle for his own plays, written to cater for the talents of his colleagues, and in particular for his own comic skills.

He began by specializing in the Italianate role of Mascarille, an extrovert, scheming, and resourceful valet; but in 1660 this type was permanently replaced (with very few exceptions, such as the Scapin of *Les Fourberies de Scapin* (*Scapin's Tricks*, 1671)) by a role owing more to sources in indigenous farce, as well as to the playwright's observance of his fellow-men. Sganarelle, a more passive, introspective character, prone to self-delusion even when he is not the victim of the wiles of others, was to offer rich scope for Molière's talents; witness this tribute from a contemporary commentator:

Never was anything seen so diverting as the posturings of Sganarelle behind his wife's back. His facial movements and gestures give such a convincing representation of jealousy that even without speaking he would be recognised as the most jealous of men . . . One would need the brush of Poussin, Le Brun or Mignard in order to do justice to his admirable antics. You never heard such a simpleton, nor saw such a foolish face; and one doesn't know whether to admire the author more for the way he has written the play, or the way he acts it.

More important, this simple figure from the farce tradition contained new possibilities for development; and, as well as appearing under his own name in half a dozen plays, Sganarelle provides the basis for a series of distinctive creations appearing throughout Molière's mature theatre, as the comic dramatist exploits the skills of the gifted comic actor. It has been said that the whole of Molière's comedy is characterized by the interplay between two kinds of imagination: whereas the Mascarille/Scapin type represents the imaginative artistry of the consummate trickster, the derivatives of Sganarelle are characters who try to impose on those around them a subjective fantasy of their own imagining. Of this character-type, the best-known examples are no doubt Arnolphe in *L'École des femmes* (*The School for Wives*, 1662), Orgon in *Tartuffe* (1664), Alceste in *Le Misan-*

Facing: Molière as tragic actor. Molière in the role of César in Corneille's *Pompée*. As a tragic actor he was not a success, and was the subject of some forthright criticism: a contemporary text alleges that 'when Molière is playing, we can laugh both in comedy and in tragedy'.

thrope (1666), Harpagon in *L'Avare* (*The Miser*, 1668), Monsieur Jourdain in *Le Bourgeois gentilhomme* (*The Would-be Nobleman*, 1670), and Argan in *Le Malade imaginaire* (*The Imaginary Invalid*, 1673).

Molière's success was not achieved without making enemies among rival actors and playwrights; and several of his plays, especially *Tartuffe* and *Dom Juan* (1665), were subjected to fierce opposition by the church authorities, the former being banned from 1664 until public performance was finally authorized in 1669. Throughout these years, however, Molière was able to count on the King's protection; and his company was in regular demand for performances at court or at one of the great princely houses. As a result of the King's patronage Molière quite early developed, alongside the series of plays written for his town theatre, a genre specifically conceived in response to the requirements of court entertainment. This new art-form, the *comédie-ballet*, bore some relation to the court ballet (and the English court masque) popular in previous reigns, but its originality was the integration of the interludes of music and dance into the narrative framework of comic drama. The King's Italian master of music, Jean-Baptiste Lulli, collaborated with Molière in a number of such works, of which *Le Bourgeois gentilhomme* is the undoubted masterpiece; while Molière's last play, *Le Malade imaginaire*, was a *comédie-ballet* created in collaboration with Marc-Antoine Charpentier.

If the development of this mixed genre shows Molière's ability to exploit the tastes of Louis and his courtiers to positive artistic effect, he was no less able at the same time to exert a decisive influence on the development of main-line comic drama, to mould the taste of Paris audiences, and to impose on them an equally original conception of comedy—and this in spite of the active opposition of jealous rivals, and of the pedants, the *précieuses*, and the prudes whom he satirizes in his brilliant conversation-piece *La Critique de l'École des femmes* (1663). Here, he defends himself through his spokesman Dorante against the criticism that *L'École des femmes*, despite its success, was an unsatisfactory mixture of two styles. Either it was a farce, says his pedantic critic, in which case Arnolphe should not have been shown with the sympathetic traits that make him a rounded character; or else it was a comedy 'according to the rules', in which case the character should not have been portrayed in such an exaggerated fashion, 'with ridiculous sighs and rolling his eyes in an extravagant manner' (valuable first-hand testimony as to the way Molière played the character of the jealous lover). Certain aspects of the play, in other words, were judged too 'comic' to deserve a place in a 'comedy'. Molière's reply, expressed by the enlightened Dorante, is simple but far-reaching: there is no reason, he says, why a character should not be a sympathetic representative of ordinary men and women in some respects, and a figure of fun in others. With this basic formula, at an early stage in his career, Molière consciously inaugurates the fusion of the refined literary comedy then in vogue with the aesthetic principle of popular farce. In Arnolphe's case, everything is subordinated to his obsession

Facing: comedy moves indoors. Despite its importance in other respects, *L'École des femmes* (1662) is backward-looking in that it retains the conventional outdoor décor which goes back to Plautus. However, for his next play, *La Critique de l'École des femmes* (1663), Molière adopted the representational indoor set which was to make such a vital contribution to plays like *Le Misanthrope* and *Les Femmes savantes*—and which he had already used, as a considerable innovation, for *Les Précieuses ridicules* in 1659. Frontispiece engravings for comedies of the period do possess a genuine documentary value (unlike those for tragedy, which are invariably inspired by the illustrator's fantasy).

with cuckoldry and the elaborate precautions he has taken to guard against conjugal misfortune: he is the deceived husband of the farces, transformed into a respected, worthy citizen whose one *idée fixe* makes him a figure of fun. And, like the rest of Molière's heroes, he provides an admirable illustration of Bergson's theory of laughter: such characters are comic because of the rigidity of their outlook on life, the mechanical inflexibility of their behavioural responses. They are caricatural portraits, it is true, and sufficient imagination has gone into their creation to make them ready targets for our laughter; but the reality they caricature is never so remote that they become mere creatures of fantasy.

The critical problem posed by Molière's formula for comedy based on character—how far can the playwright go in presenting a 'rounded', sympathetic figure and still make him the mainspring of the play's comic effect?—is seen at its most

L'ESCOLE DES FEMMES

LES PRECIEUSES RIDICULES

Spectacular entertainment at court. The genre of court ballet, which brought together the arts of music, dance, and spectacle, flourished throughout the seventeenth century under the patronage of successive kings. Many examples, like the one illustrated here (performed both at the Louvre and at the Paris Hôtel de Ville), introduced elements of the exotic and the burlesque.

acute in the case of *Le Misanthrope*. Here, the comic pattern of the character whose practice contradicts his doctrinaire theorizing is so intellectualized and refined that its affinities with the mechanism of farce are often overlooked. When we laugh at Alceste it may be, to use a contemporary's felicitous phrase, with 'the laughter of the mind'; but that should not obscure the family likeness between this opinionated, irascible character, living in an unreal fantasy world where he is right and everybody else wrong, and the Sganarelle figure as he appears in Molière's other comedies. There can be little doubt that a production that brings out this family likeness will be nearer to the spirit in which the play was originally conceived than the nineteenth-century interpretation which saw Alceste as the noble victim of a heartless and hypocritical society.

A feature of Molière's dramatic writing that should not be overlooked is the important service he performed in 'bringing comedy indoors'. Comedy in the ancient world had used an outdoor setting, and its action—love-making, quarrelling, plotting, bargaining—had been carried on in the market-place. Molière had been content to follow this convention in a play like *L'École des femmes*; but in the major comedies of his maturity he broke new ground and set his action indoors. This was not a total novelty (Molière had used an indoor setting himself for one or two of his one-act plays); but there can be no doubt that its systematic adoption in a five-act format reflects an awareness of the suggestive potential of the representational set. The portrait of a salon society in *Le Misanthrope* or in *Les Femmes savantes* (*The Learned Ladies*, 1672) certainly gains immeasurably from the indication of a precise social milieu; while the impact of *Tartuffe* would surely have been much less controversial if the dialectical confrontation between the religious hypocrite and his opponent Cléante had not been reinforced by the visual evocation of a bourgeois household.

Although Molière enjoyed the patronage of the young King on his return to Paris; although his company was granted the title 'la troupe du Roi au Palais-Royal' with a reasonably generous subsidy; and although there was every reason to be grateful for Louis's support in the case of *Tartuffe*, royal favour was always precarious. The King preferred musical entertainment and spectacle to straight drama: partly because such court spectaculars could more easily be presented as a form of sycophantic homage to the Roi Soleil, partly because he himself genuinely enjoyed dancing in the court ballets which formed a central part of these lavish entertainments. The performance of *Le Bourgeois gentilhomme* at Chambord in 1670 brought Molière's series of *comédies-ballets* to a peak of success; but it also marked the end of the collaboration between the 'two Jean-Baptistes' after they had worked together in no fewer than ten plays. Lulli now captured the King's favour, and it was the foundation of the Paris Opera, combined with intrigue on the Italian's part, which brought this about. In 1669 the Finance Minister Colbert had been persuaded to create the Académie Royale de Musique (the Paris Opera's official title), and a second-rate poet and librettist, Pierre Perrin, was appointed director. An entrepreneur, the marquis de Sourdéac, built a new theatre, and it was here that Perrin put on the first French opera, *Pomone*, in 1671. But he was no businessman, and when he was imprisoned for debt in 1672 Lulli was granted the privilege—having, according to one account, agreed with Molière to make a joint bid and then gone behind his colleague's back. The privilege carried a monopoly with it, and an injunction was at once served on the other Paris theatres, preventing them from employing a full-scale orchestra or chorus. Lulli continued to perform without acknowledgement works he had composed in collaboration with Molière; and his triumph was complete when, on Molière's death in 1673, he persuaded the King to make over the Palais-Royal theatre as the home of the Paris Opera. The preamble to the letters patent creating the Opera observes that 'for many years now the Italians have established various *Academies* in which musical plays are performed, which are called *Operas*'; not only will the foundation of a counterpart in Paris enable the French to catch up with the Italians, the Germans, and the English, but the new establishment is to enjoy quite a different standing from that of the non-musical theatres, for 'it is our good pleasure that all gentlemen, ladies and other persons may sing in the said Operas without derogating from their noble rank, or from their privileges, offices, rights and immunities'.

This text has very far-reaching implications, suggesting as it does that the profession of actor in non-musical theatre should continue to attract the social stigma of 'derogation', with all that it involves in terms of loss of civil rights; this was of course in addition to the penalty of excommunication laid down by the Church, and which was highlighted in notorious fashion when attempts were made, on Molière's death, to prevent his being buried in consecrated ground. An edict published by Louis XIII in 1641 had enshrined the doctrine that actors were not to be discriminated against:

insofar as the said actors exercise such control over their actions in the theatre that these remain wholly exempt from impurity, we desire that their occupation, which is capable of providing innocent diversion for our people from certain blameworthy activities, shall not be held to their discredit, nor prejudice their reputation in public intercourse.

In fact, certain civil penalties did remain in force; but it was the attitude of the church authorities towards the theatre that constituted the biggest threat to the acceptance of actors as full members of the community. There were considerable variations to be seen in ecclesiastical opinion: some bishops were quite ready to attend performances of plays, especially at court, and the most notorious of the published attacks on the theatre, Bossuet's *Maximes et réflexions sur la comédie* of 1694, was a reply to a pro-theatre pamphlet, the *Lettre d'un théologien illustre . . .* written by a cleric, Father Caffaro, in the same year. However, the moral rigorism of the 1690s—illustrated at court as well as elsewhere, where the influence of Madame de Maintenon led to a new climate of austerity quite hostile to theatrical entertainment—put increased pressure on members of the profession; and when the actor Michel Baron retired in 1691 before he was 40, this was hailed by the duchesse d'Orléans as a victory for the bigots:

I must tell you that our leading actor is leaving the theatre. The zealots have put the fear of God into him by saying they were going to outlaw all the profession, and that the first to renounce it would get a pension . . .

In spite of the edict of 1641, and whatever more enlightened thinkers might say, the civil and religious disqualification of actors was to remain in force until the Revolution.

For most of the century, the Hôtel de Bourgogne enjoyed a pre-eminent position among the theatres of the capital. Even when Mondory's company was granted the title of 'Troupe du Roi au Marais' in the 1630s, and when a similar title was bestowed on Molière's company at the Palais-Royal some twenty-five years later, it was nevertheless the case that the 'Troupe Royale', the 'Grands Comédiens' of the Hôtel de Bourgogne, came first: a pre-eminence that was underlined by the larger subsidy given to this theatre. After the disappearance of its famous trio of *farceurs* in the 1630s, the company continued to prosper under Bellerose, who succeeded Gros-Guillaume as its director—a position he held for twenty-five years—with a reputation based above all on its playing of tragedies: those of Corneille's contemporary Jean de Rotrou, and subsequently Corneille's own later plays, as well as his earlier masterpieces once they had passed into the 'public domain': that is to say, once they had been published, since there were no copyright restrictions in force.

Molière was the consummate man of the theatre. Corneille, though primarily a man of letters, enjoyed a long career of considerable variety and innovation as a playwright. In addition to the early comedies, and the tragedies for which he is best known, he wrote the highly imaginative *L'Illusion comique*, an apologia for the theatre in the form of a play-within-a-play (indeed the structure turns out to involve a play-within-a-play-within-a-play); and it was his mythological tragedy *Andromède* which provided the Italian Giacomo Torelli with the subject for the most successful machine play of the mid-century (Théâtre du Petit-Bourbon, 1650). With Jean Racine, however, we come to a playwright who was almost exclusively a tragic *poet*: many would say the greatest tragic poet of all time. In a singular departure from his characteristic manner, he did compose an amusing satirical comedy after Aristophanes, *Les Plaideurs* (*The Litigants*, 1668); he is known to have coached the actress Marie Champmeslé, his mistress, in the speaking of his verse; and he was sufficiently ambitious for the success of his plays to transfer his *Alexandre le grand* (1665) from Molière's company at the Palais-Royal to the 'Grands Comédiens' at the Hôtel de Bourgogne. But otherwise, he kept a certain distance between the practical theatre and himself; and after a relatively short period of almost unqualified success with his tragedies (from *La Thébaïde* (1664) to *Phèdre*, his masterpiece (1677)) he withdrew from dramatic writing. Disillusion at the cabal mounted against *Phèdre* in favour of the rival *Phèdre et Hippolyte* by Nicolas Pradon, and his own religious conversion, were factors which contributed

AN ELOQUENT APOLOGIA FOR THE THEATRE

In Corneille's *Illusion comique* (1636) a bourgeois father seeks his missing son. Though Pridamant is mortified when the magician Alcandre reveals that Clindor has become an actor, this eloquent apologia in Act v conquers his anti-theatre prejudice. Mondory, who played Clindor, had similarly left a good family to go on stage.

ALCANDRE. Stop your lamenting: the theatre is an object of idolatry on all sides nowadays; and what was once looked on with scorn has become the delight of all men of taste. It is the talk of Paris and the envy of the provinces, the favourite entertainment of the princes of the blood, the pastime of the common people and the quality alike. It is quite the most popular of all recreations, and those to whose wisdom and care the well-being of the state is entrusted look to the delights of theatrical entertainment as relaxation from the burdens of high office. Even our great King, that renowned captain whose martial exploits are feared in all corners of the earth, condescends from time to time, crowned with a victor's laurels, to grace the theatre with his presence. There, Parnassus displays its wonders; the choicest wits forgo their nights' rest in its service; and Apollo's favourites all devote a portion of their learned labours to the theatre. If men are to be judged by their wealth, the theatre is a domain which brings in handsome rents; and in the exercise of this agreeable profession your son enjoys more honour and greater riches than if he had stayed at home. Desist then from this all too common error, and complain no more of his good fortune.

PRIDAMANT. I dare protest no longer: it is only too clear that the profession he has adopted is much to be preferred to my own . . .

to this decision; but equally strong, without a doubt, was his social ambition, and the desire to abandon a career which involved a degree of 'derogation' for that of courtier-poet (he was appointed Historiographer Royal in 1677) in the King's close entourage. However, he was later persuaded by Madame de Maintenon to compose the religious dramas *Esther* (1689) and *Athalie* (1691) for private performance by the young ladies of the convent school of Saint-Cyr.

There have been those who have regarded Racine's tragedy as a form of 'closet drama' more suitable for the study than the stage. Such a view could not be more erroneous; and it is belied by the continuing success of his plays with succeeding generations of spectators at the Comédie-Française and elsewhere. In particular, certain of his female roles—Hermione in *Andromaque* (1667), Agrippine in *Britannicus* (1669), Monime in *Mithridate* (1673), and above all the eponymous heroine of *Phèdre*—have always been seen as providing the supreme challenge for a French actress. Racine's portraits of women are indeed the most striking, and his studies of sexual jealousy as a predatory and destructive force (Hermione, Roxane in *Bajazet* (1672), Ériphile in *Iphigénie* (1674), and Phèdre) have an intensity that is diluted in their male counterparts (Néron in *Britannicus*, Mithridate), whose motivation is determined by political as much as sexual factors. His *Phèdre*, based on the Hippolytus plays by Euripides and Seneca, must rank as the outstanding modern play on a Greek mythological subject; and in his 'Greek' plays particularly, Racine shows an unrivalled understanding of the nature of Aristotelian tragedy.

However, it would be wrong to think of Racine's tragedy as being exclusively based on subjects from Greek mythology. His prefaces show a remarkably combative nature; and competition with Corneille soon led him to challenge his older rival in the field of the Roman history play which the latter had made his own. This he did with great success in *Britannicus* and in *Mithridate*; while as for his *Bérénice*, this play appeared at the Hôtel de Bourgogne in 1670, exactly a week before Corneille's *Tite et Bérénice* was put on at the Palais-Royal. Despite radical differences in treatment, the subject of the two plays was manifestly the same, and a legend was subsequently given currency to the effect that the two dramatists had been engaged to tackle the theme by the duchesse d'Orléans; however, it is more likely to have been a case of rivalry between the two theatres—there were several other instances of this in the period—encouraging one playwright (in this instance, one presumes that it was Racine) to appropriate the subject on which the other was working. In any case, there could be little doubt as to which play was the more successful. Tragedy, as has been seen, fared better at the Hôtel de Bourgogne, where Floridor (Josias de Soulas), who had followed Bellerose as *chef de troupe*, and Mlle de Champmeslé gave powerful support to Racine's simple, elegiac manner.

Racine was to retire from the theatre in 1677, while Corneille's last play, *Suréna*, had been performed in 1674; in the ten years of their direct rivalry, both the judgement of theatre audiences and that of the salons had given victory to Racine—though Corneille retained the loyal support of many followers of an older generation. The most successful tragedy of the period, however, was by neither: it was *Timocrate*, by Pierre Corneille's younger brother Thomas. This play had an initial run of eighty performances at the Marais in 1656–7, not to speak of simultaneous performance—unauthorized—at the Hôtel de Bourgogne. In fact, Thomas Corneille could well qualify as the most truly *popular* dramatist of the second half of the century. A prolific writer, skilful at gauging literary and theatrical fashions, he followed in *Timocrate* the vogue for the adventure-novels of the 1650s; the play has a very complicated, and completely fictitious, plot, while the high-minded quest for personal glory which inspires the heroes of the elder Corneille is here reduced to a trivial pursuit of gallant conquests. Not that he did not compose more orthodox tragedies (some, like *Ariane* (1672), of real distinction and others, like *Le Comte d'Essex* (1678), striking out in comparatively new directions); and he also wrote a great many comedies, from those with an involved plot of Spanish inspiration to the more sentimental type with a *précieux* flavour. In addition, he wrote libretti for operatic works by Lulli, and, in collaboration with Jean Donneau de Visé, he also composed a number of elaborate machine plays. Altogether, Thomas Corneille was certainly the most versatile playwright of his age, even if the various formulas which brought him popular success proved for the most part to have little appeal for later generations.

Molière's death in February 1673 was followed not only by the transfer of his Palais-Royal theatre to Lulli, but by the threat of forcible dissolution for his company. In fact, a number of his former colleagues did leave in order to join the Hôtel de Bourgogne; however, their place was at once taken by actors from the Marais theatre, with the consequence that it was the Marais troupe which was wound up by official edict and amalgamated with Molière's company. The theatre constructed by the marquis de Sourdéac in the Rue Mazarine was now purchased, and this was to be known as the Hôtel de Guénégaud. The Italians followed the new company there, paying them a rent for the right to perform on four days of the week. With the demise of the Marais, theatre-going in Paris became concentrated in the more fashionable central areas: its remote situation had always been a disadvantage for the Marais theatre; and Pierre Corneille is said to have remarked, when one of his late plays was accepted there after having been turned down by both the other theatres, that he was bringing life back to the 'desert', where people had forgotten that a theatre existed. The Hôtel de Bourgogne continued to enjoy its superior reputation for tragedy (as well as the King's support); while the Guénégaud theatre was to be best known for its comedy and for the machine plays which had been a speciality of the Marais company.

But in July 1680 the King's interventionist policy of centralization was to be carried a stage further, when the death of La Thorillière (François Le Noir), the leader of the Hôtel de Bourgogne company, who had migrated from the Palais-Royal in 1673, provided a pretext for the amalgamation of the two theatres. Against considerable opposition from the actors—among other things, the decision necessitated some compulsory retirements—it was decreed that, henceforth, the joint company should play at the Guénégaud on every day of the week; while the Italians would move to the Hôtel de Bourgogne. This second 'jonction des troupes', or amalgamation by royal edict, was effectively the foundation of the Théâtre-Français, or Comédie-Française, in very much the form in which we know it today; though the epithet was not intended to indicate its status as a *national* theatre in the modern sense, but simply to distinguish it, as a company playing in French, from the Théâtre-Italien, whose company played in their own language. Both were 'royal' companies, their members servants of the King; and in return for this privilege, and a not ungenerous subvention, they were never allowed to forget that priority in their calendar must be given to the service of King and court. This involved (particularly for the French company) a burdensome schedule of continual visits to Versailles, Saint-Germain, Fontainebleau, Chambord, and Marly, with an inevitable impoverishment of the resources available for performance in Paris, and a consequent limitation of the possible repertory.

Organization of the two theatres under the new dispensation was shortly afterwards handed over to Madame la Dauphine; and her published regulations for the

Comédie-Italienne date from 1684, those for the Comédie-Française from 1685. Although a similar set of regulations for the Opera was not to follow until 1713, to all intents and purposes the administrative machinery for the 'state' theatres was in position well before the end of the century. The overall responsibility for the running of the theatres was to be in the hands of the First Gentlemen of the Bedchamber (Premiers Gentilshommes de la Chambre), who would exercise it through the intermediary of the Superintendent of Royal Entertainment (Intendant des Menus Plaisirs). It was the First Gentlemen (or the particular one of the four whose duty it happened to be at the time) who decided on the admission of an actor or actress, terms of probation, the allocation of a full, half-, or quarter-share, and questions of retirement and pension, as well as all sorts of disciplinary matters. With male actors this was often delicate enough, and cabals and pressure-groups could easily develop; but when—as sometimes happened—decisions had to be taken concerning an actress who was the protégée of one of the First Gentlemen, there were obvious opportunities for considerable dissension in the company between rival actresses and their groups of supporters. Such, however, was the external administration of a system which was to remain almost without change up to the outbreak of the Revolution; while the internal government of the Comédie-Française with its company of *sociétaires*, each with a say in its running by means of regular general assemblies and elected committees, was to have an even longer life.

But its status as a privileged national institution did not guarantee the new theatre an untroubled existence; and in 1687, as a bolt from the blue, there came an order to move out of the Guénégaud theatre. The completion of the Collège des Quatre Nations was imminent, and it was held that it would be an affront to the dignity of this splendid building (now the Institut de France) if it had to suffer the immediate neighbourhood of a public theatre. Negotiations to secure an alternative site began at once; but such was the extent of anti-theatre prejudice, encouraged by the Church and various monastic orders, that more than once arrangements were apparently concluded, including the drawing-up of plans, when the King yielded to pressure and vetoed the move. Finally a site was found in the present-day Rue de l'Ancienne Comédie, where a tennis-court was adapted by the eminent architect François d'Orbay, a warm-hearted supporter of the theatre who refused to accept any payment for his work. Here, the presence of spectators on stage was recognized by giving them a permanent seating area surrounded by a balustrade; but, in spite of the opportunity to break with the past in other respects as well, there were no seats provided in the pit, and the standing *parterre* was to remain a distinctive feature of French theatres until the latter half of the eighteenth century.

This wholly male section of the audience has often been given a bad name on account of its volatile nature, and indeed serious disturbances did occasionally break out there. These seem to have been due principally to the military and to liveried servants; and one continual grievance of the theatre authorities concerned members of these groups who habitually entered without paying and who, when challenged, would cause trouble. A royal edict of 1674 had addressed the problem:

His Majesty, having been informed of the disorderly acts that are committed daily on the premises in which the actors of His Majesty's second company perform in the rue de Guénégaud, which curtail the liberty of the public and interfere with their entertainment; and seeking to remedy this situation, expressly forbids all persons, of whatever quality and condition they may be, even including officers of the Royal Household, to enter the said theatre without paying. His Majesty moreover requires the sieur de Francini, Lieutenant-général of police, to install the number of guards he may deem necessary at the entrance to the said theatre, to ensure the execution of the present ordinance.

The fact that, even after the *jonction des troupes*, a virtually identical form of words recurs time and again with reference to public order at both the Comédie-Française and Comédie-Italienne suggests that any edict on this subject was likely to remain a dead letter. However, in spite of the restlessness inevitable in a body of standing spectators, the frequent presence of pickpockets, and the possibility of disruption from a small minority of troublemakers, the *parterre* was a valued section of the audience. There is no reason to disbelieve Molière when he pays tribute to the soundness of its judgement, and numerous anecdotes testify to the presence among its *habitués* of men of wit and culture. If there were some who adopted (until the authorities succeeded in stamping it out) the mindless habit of signifying disapproval of play or performance by piercing whistles, others preferred to demonstrate their knowledge of the text by ingenious emendations or clever sallies. A story is told concerning an actress, noted neither for talent nor for beauty, who was making her début in *Andromaque*:

One spectator kept grumbling beneath his breath at hearing her mangle the poetry of 'le tendre Racine', for whom he had an ardent admiration; and in the passage in which Andromaque says to Pyrrhus: 'My lord, what can you mean? What will be said in Greece?' our friend replied spontaneously to her question: 'That you, Madame, are hardly an artist's masterpiece!'. With that he strode out, leaving the *parterre* applauding his impromptu verse, and the actress most embarrassed at her want of comeliness.

Altogether, we may perhaps regard the *parterre* as possessing at best something of the character of an audience of promenaders at the Albert Hall; but it is difficult to be equally charitable about the behaviour of the young fops who occupied the seats on stage. In the following passage from *Les Fâcheux* (*The Society Bores*, 1661), Molière describes one of the nuisances who prevent the hero from keeping a rendezvous with his mistress:

I was seated on stage prepared to enjoy the play, of which I had heard good reports. The

ô bien don Ie fuis ton feruiteur

Spectators on stage at a comedy. This engraving shows fashionable male spectators both seated and standing on stage at the Hôtel de Bourgogne in 1666. Extrapolation beyond the picture-frame towards the front of the stage suggests a total of about forty on-stage spectators, though at this time only thirty-two were officially allowed.

actors were just beginning, and everyone was quiet, when a young man of fashion covered in lace made a noisy entrance, shouting: 'Come on! Let's have a seat at once!', taking the whole audience by surprise and disturbing a particularly good passage . . . The actors tried to carry on with their performance, but the intruder made another fuss about his choice of seat: although there was plenty of room at the side of the stage, he strode across, planting his chair in front of the actors, and had the insolence to block the view of the stage from three-quarters of the *parterre*.

A satirical caricature, of course; but feelings did run high, and this is far from being the only protest from practising dramatists. Nor was it only a question of antisocial behaviour by young exhibitionists: as a journalist pointed out, when Tartuffe looks round carefully to see whether anyone might surprise his attempted seduction of Orgon's wife, he is surrounded by 'a hundred witnesses of his tête-à-tête'.

It appears that, at Molière's Palais-Royal, stage seats had been limited to thirty-two, but by the turn of the century the new Comédie-Française had been adapted to provide 140 places on stage, while there is evidence to show that on occasion even this number was greatly exceeded. Like the *parterre*, the stage was a male preserve, the only reported exception being a performance of Claude Boyer's *Judith* in 1695, a play remarkably popular with the ladies:

Men had to give up their seats on stage and stand instead in the wings. Can you imagine two hundred women sitting on the benches where men alone are normally seen, with handkerchiefs at the ready on their knees, to wipe their eyes in the more touching passages? One particular scene in Act IV, where they wept copiously, was called on that account 'the scene of the handkerchiefs'. The *parterre*, instead of weeping with them, had a good laugh at their expense.

Boyer's play was one of the more successful contributions to a vogue for religious tragedy, both at the Comédie-Française and at court, of which Racine's two late plays are the best-known examples, and which no doubt represents an attempt to placate the anti-theatre lobby. In general, the 1690s seem to have witnessed a considerable falling-off in theatrical activity at court, largely due to the prevailing climate of piety and moral austerity. At the town theatres, however, there is little sign that the hostile attitude of the Church acted as a deterrent either to dramatists or to their patrons. Comedy in particular flourished in the closing decades of Louis XIV's reign, and although there were complaints that Molière's success had made it too difficult for those who followed, there is plenty of evidence in practice of innovation and independence. Florent Carton Dancourt, an actor in the Comédie-Française company, kept his theatre supplied with a series of comedies of manners, exploiting the social fashions of the capital in an entertaining way. Alain-René Lesage's satire *Turcaret* (1709) could be called a black comedy: it is a mordant attack on the financial practices and the moral corruption of the time; but after *Turcaret* he quarrelled with the Comédie-Française, and devoted his talent as a dramatist to catering for the fairground theatres. Jean-François Regnard came nearest to being a successor to Molière, though his idiosyncratic brand of verbal comedy, seen in plays like *Le Légataire universel* (1708), is more to modern taste than his closer imitations of Molière's manner.

Both Regnard and Charles-Rivière Dufresny had begun by writing for the Italians. By 1690, the Comédie-Italienne had established itself as an alternative French-language theatre (in spite of opposition from the Comédie-Française, which had been overruled by the King), though their manner remained very distinctive. Their traditional extempore performance, based on scenarios or plot synopses, had largely given way to fully scripted dialogue; but the conventional roles survived, as did the characteristic *lazzi* and the emphasis on a very physical form of comedy. This was popular theatre in both senses of the word, offering light-hearted relief from the uniformity and regularity that play-goers already saw as features of too many of the comedies, as well as the tragedies, offered at the Théâtre-Français. Ludic entertainment *par excellence*, not unlike a series of music-hall sketches, the Italians' repertory was often vulgar, even obscene. This produced warnings from the authorities in the mid-1690s; and in 1697 it was judged that they had gone too far, and the company was disbanded and its members sent into exile. The most likely reason for this was the one later cited by Saint-Simon:

The King expelled the entire company of Italian actors quite precipitately, and would not consider engaging another. As long as all they had done was to overflow in filth and sometimes in impiety on their stage, people had done nothing but laugh; but they took it into their heads to perform a play called *La Fausse Prude*, in which

Spectators on stage at a tragedy. The disturbing effect of on-stage spectators at performance of a tragedy is well suggested by this engraving of a scene from a revival of Corneille's *Cinna* towards the end of the century: in dress and posture the young fashionables vie with the actors, while in volume they easily outnumber them.

Madame de Maintenon was easily recognised. Everyone rushed to see it, but after three or four performances, given consecutively because their finances demanded it, they were ordered to close their theatre and leave the kingdom within the month. This caused a great commotion, and if the actors lost their establishment through their blindness and folly, she who had them expelled did not gain by it, on account of the freedom with which this ridiculous event caused people to speak of it.

There were, however, those at the time who saw this as a pretext for what was primarily an economy measure which would save the royal treasury a large annual sum, at a time when ruinous foreign wars had emptied the coffers. The most obvious effect of the expulsion of the Italians, and the closure of their theatre from 1697 to 1716, was the strengthening of the monopoly of the Comédie-Française by eliminating its only official rival in the field of non-musical theatre. On the other hand, there was an immediate response by the fairground theatres, who took the opportunity to fill the gap left by the Italians' departure with entertainment of a popular nature; and since this ultimately led to the development of the Opéra-Comique, the long-term effect of the expulsion must be seen, paradoxically, to have been a weakening of the monopoly situation.

The fairground theatres had formed part of the attractions of the annual Foire Saint-Germain (in winter) and Foire Saint-Laurent (in summer) since the end of the sixteenth century. Their booths, originally makeshift constructions housing tightrope-walkers and acrobats as well as simple comic sketches, had gradually developed into something more permanent and presentable, as the Foires themselves evolved into a highly prized part of the Paris scene, frequented by members of all classes. Fairground performers possessed some official status—they had to be licensed by the authorities—but they were not to begin with seen as constituting a threat to the monopoly theatres: their activities were too varied, and their presence too ephemeral. In 1697, however, when the fairground operators decided to fill the gap left by the exiled Italians, using some of their material, this was perceived as direct competition; and by the same token, their performance of sung sketches was interpreted as infringing the Opera's monopoly. For the closing years of the old King's reign, their story becomes one of continual persecution: not only were they subjected to repeated attempts to restrict the number of performers, or the use of spoken or sung dialogue, but contravention of these restrictions was dealt with by the most violent means. In 1709, backed by an injunction from the Parlement, the Théâtre-Français caused two theatres at the Foire Saint-Germain to be broken up and destroyed; although the operators were back in business the next day, a more thorough destruction followed: action which was supported, on appeal, by the King's Grand Conseil. As regards musical theatre, things changed in 1712 when the Opera, short of money, sold a form of privilege to a fairground proprietor, Catherine Vondrebeck, thus paving the way for the creation of the

Opéra-Comique (the term first appears in 1715); but the Comédie-Française was more tenacious, and harassment of the fairground theatres continued well into the Regency.

These skirmishes were little more than an irritant to the privilege theatres, safe in the possession of their monopoly, and it was to be long enough before the Opéra-Comique would be sufficiently well organized to pose a serious threat. When Louis died in 1715 the privilege theatres, like the absolute monarchy, must have looked established for all time: a tangible expression in the field of the performing arts of the superiority of French culture, itself a reflection of the political supremacy won by Louis. No matter that its major creative achievements belonged to the 1660s and 1670s (not unlike the most spectacular glories of the Roi Soleil); no matter that there was a widespread feeling that (again like the monarchy) the present regime in the theatre was becoming over-conventional and dull; no matter that the driving out of the Huguenots and the military suppression of peasant riots could be matched, *mutatis mutandis,* by the expulsion of the Italians once they had stepped too far out of line and the uncompromising severity shown to the fairground performers. Such were the prices that had to be paid for cultural, as for political, orthodoxy in an absolutist regime. On the credit side, however, must surely be reckoned the founding of what would develop, whether wholly predictably at this date or not, into Europe's most prestigious theatre.

PART THREE

European and Western Theatres from 1700

8

EIGHTEENTH-CENTURY THEATRE

PETER HOLLAND AND MICHAEL PATTERSON

AT first sight the eighteenth century appears to be the least interesting and significant period of theatre history since the Middle Ages. Some histories of theatre virtually omit it, while others treat it as some sort of connecting corridor from the splendours of the Renaissance to the innovations of the nineteenth century, essential but not worth lingering in.

Indeed the eighteenth century produced few great dramatists; several comic talents perhaps: Sheridan and Goldsmith, Marivaux and Beaumarchais, Goldoni and Gozzi, Holberg and Lessing. But an anthology of world drama could legitimately be published without including the works of any of these. Only in the emergent theatre of late eighteenth-century Germany can one point to the major dramas of Goethe and Schiller. Nor could the eighteenth century boast of important innovations in theatre technology, except towards the end of the century with the replacement of candle-light by oil-lamps.

What was significant about the theatre of the eighteenth century, however, is that it developed in Continental Europe a function in society unparalleled since its role in ancient Greece. From being an entertainment at court or in the marketplace it became a political forum for the bourgeoisie, a focus for national identity and even revolution. It moved from being formal and stylized, or from being vulgar and coarse, to a new level of realism; the stage began to search for authenticity and newly to mirror the everyday lives of the spectators. The theatre began to analyse its own aesthetic and to develop from a craft into an art. Actors and actresses rose from the social level of prostitutes and jugglers to become favoured members of society, and playwrights began to be paid properly for their work. In short, the theatre reflected the philosophical glory of the century, the questioning, the tolerance, and the democratic thinking of the period of Enlightenment.

But in the relative security and unquestioned national identity of eighteenth-century England, the theatre was less concerned with philosophy than with profits, and its profits depended on appeasing new social forces of bourgeois morality and new impositions of governmental control. In April 1698 Jeremy Collier, a dissenting clergyman, published *A Short View of the Immorality and Profaneness of the English Stage together with the Sense of Antiquity upon this Argument*. For a short view the book was not particularly short; running to over 280 pages, Collier mounted an extraordinary, vitriolic attack on contemporary drama. His microscopic examination of the language of Congreve, Vanbrugh, and others was designed to prove his case that, though drama *ought* to be moral ('The business of plays is to recommend virtue and discountenance vice'):

Our poets write with a different view . . . Their liberties in the following particulars are intolerable, viz., their smuttiness of expression, their swearing, profaneness and lewd application of scripture, their abuse of the clergy, their making their top characters libertines and giving them success in their debauchery.

In the aftermath of *A Short View* dozens of pamphlets appeared, attacking and defending the stage; Vanbrugh replied in a pamphlet in June, Congreve responded with another in July. It was not only a pamphlet war: Congreve in *The Way of the World* (1700) provided the best answer to Collier, proof that Restoration comic drama could deal seriously with evil, and Farquhar, in *The Recruiting Officer* (1706) and *The Beaux' Stratagem* (1707), showed that comedy could move beyond London to explore issues as serious as recruiting or divorce without offending Collier. But more worryingly for the theatres, the controversy moved to the courts: actors found themselves prosecuted, with admittedly limited success, for speaking particular lines noted down by informers placed in the audience by the Society for the Reformation of Manners. Adding 'Egod' to a line was now a dangerous business; even speaking lines previously authorized could result in a prosecution.

Something about Collier's invective was clearly timely. Changes in social power, the increasing significance of bourgeois values in opposition to the aristocratic world of the Restoration, found in Collier a suitable champion for a morality that this newly important segment of society wished to see enshrined at the centre of public behaviour. The theatre was the most public forum for such a debate. Apart from anything else, the Collier controversy pushed theatre into the centre of debate and established writing about the theatre as an energetic field of literary and moral argument. Sir Richard Steele, the most vigorous exponent of the attempt to reconcile the new dominant morality with drama, may have succeeded in creating a moral and sentimental drama, particularly in *The Conscious Lovers* (1722), but his efforts were mostly concentrated on essays about drama and theatre in the widely-read periodicals, like the *Spectator* and the *Tatler*, which he published.

While Steele may have transformed the nature of high comedy, the most popular changes in comedy were to happen elsewhere. On 29 January 1728 at the Lincoln's Inn Fields Theatre a packed house watched the first night of John Gay's *The Beggar's Opera*. Gay's ballad-opera had been turned down by Colley Cibber, one of the managers at the rival patent theatre, Drury Lane, the only other company licensed to act in London. The style of Gay's opera was too novel for that staid company, whose major dramatic success was to be George Lillo's tragedy of an apprentice led astray, *The London Merchant* (1731), a play whose huge impact on the development of European drama was out of all proportion to

SOCIAL SATIRE AND POPULAR SONG

John Gay's *The Beggar's Opera* (1728) created a triumphantly successful new genre by adding new satiric words to the tunes of popular ballads. Peachum, the master criminal, still recognizes that a lawyer is an even more efficient thief than himself.

MRS PEACHUM. I am very sensible, husband, that Captain Macheath is worth money, but I am in doubt whether he hath not two or three wives already, and then if he should die in a session or two, Polly's dower would come into dispute.

PEACHUM. That, indeed, is a point which ought to be considered.

> AIR XI, *A Soldier and a Sailor*
> A fox may steal your hens, sir,
> A whore your health and pence, sir,
> Your daughter rob your chest, sir,
> Your wife may steal your rest, sir,
>   A thief your goods and plate.
> But this is all but picking,
> With rest, pence, chest, and chicken;
> It ever was decreed, sir,
> If lawyer's hand is fee'd, sir,
>   He steals your whole estate.

The lawyers are bitter enemies to those in our way. They don't care that anybody should get a clandestine livelihood but themselves.

its minimal success in influencing English tragedy. Gay's friends persuaded John Rich, the manager at Lincoln's Inn Fields, to take on his startlingly new combination of popular song, underworld comedy, and political and social satire. Rich was ready to give up after the first rehearsal but was encouraged to persevere. On that first night the audience numbered over 1,200: 250 in the boxes, 300 in the pit, over 600 in the galleries, and a few people sitting on the stage. Initially the audience was bemused. As Alexander Pope, Gay's friend, reported later,

We were all at the first night of it, in great uncertainty of the event, till we were very much encouraged by overhearing the Duke of Argyll, who sat in the box next to us, say 'It will do,—it must do!—I see it in the eyes of them'. This was a good while before the first act was over.

By the end the applause was tumultuous and the new form was an unprecedented success. Before the season was over, the play had been performed sixty-two times. The popular tag claimed the work 'had made Rich gay and Gay rich'—indeed Gay told his friends he earned over £600 from the play.

Rich's audience was used to innovation. From 1716 he had experimented with pantomime, his own adaptation of *commedia dell'arte*, starring himself as Harlequin under his stage name John Lun, and the experiment of a new annual pantomime continued till his death. He also tried adding other entertainments to the main bill; his development of the 'whole show' firmly established the practice of adding short farces, musical entertainments, processions, rope-dancers, contortionists, indeed almost anything to create a varied evening's entertainment for the audience. Some would complain about the unseemliness of this rag-bag; no one could dispute its success.

The extraordinary triumph of *The Beggar's Opera* not only spawned dozens of imitations but also confirmed Rich in the plan of moving to a larger theatre. In 1730 he began to design a theatre in Covent Garden, completed in 1732. Others too perceived the implications of the work's success: in 1729, in the teeth of city opposition, Thomas Odell opened a new theatre in Goodman's Fields in the East End of the city, at the opposite end of London from normal theatrical activity. Odell had recognized that a sizeable part of the audiences filling the theatre to see Gay's work came from the eastward spread of the expanding city. The new audience deserved its own local theatre. Odell's plan was helped by the creation in 1720 of the Little Theatre in the Haymarket, mostly used by visiting companies at first but a wedge driven into the near-monopoly of the patent houses, Drury Lane and Lincoln's Inn Fields.

In spite of the reformers' use of the law-courts, the moral and clerical concerns of Collierism did not result in new legislation. Government only uses moral anxiety for its own ends. The Licensing Act of 1737, by far the most important governmental control of theatrical activity in the century, was rushed through parliament

THE LICENSING ACT 1737

Pushed through Parliament at high speed, Walpole's legislation re-established censorship of plays through the Lord Chamberlain's office. The Examiner of Plays had to approve the texts of all spoken drama in British theatres until the office was abolished in 1968.

And be it further enacted by the authority aforesaid, that from and after the said twenty-fourth day of *June*, one thousand, seven hundred and thirty seven, no person shall for hire, gain or reward, act, perform, represent or cause to be acted, performed or represented any new interlude, tragedy, comedy, opera, play, farce, or other entertainment of the stage, or any part or parts therein; or any new act, scene or other part added to any old interlude, tragedy, comedy, opera, play, farce or other entertainment of the stage, or any new prologue or epilogue, unless a true copy thereof be sent to the Lord Chamberlain of the King's household for the time being, fourteen days at least before the acting, representing or performing thereof, together with an account of the playhouse or other place where the same shall be and the time when the same is intended to be first acted, represented or performed, signed by the master or manager, or one of the masters or managers of such playhouse or place, or company of actors therein.

by Walpole: first proposed on 20 May it became law on 24 June. As Lord Chesterfield noted in his speech in the House of Lords, 'It seems designed not only as a restraint on the licentiousness of the stage; but it will prove a most arbitrary restraint on the liberty of the stage.' Two separate sets of interest combined to ensure the Bill's passage. The first was Walpole's anxiety at the mounting use of the stage for political satire; Gay's political satire in *The Beggar's Opera* had stung Walpole but the sequel, *Polly* (1729), was so direct in its mockery that Walpole was able to use the existing powers of the Lord Chamberlain to have the play banned completely. Yet by the middle of the next decade, in Henry Fielding's exuberant satiric plays like *The Historical Register for the Year 1736* (1737) at the Little Haymarket Theatre, the theatres were using an unprecedented freedom to attack government policy. Walpole's pretext for the Bill was an anonymous and now lost play, *The Golden Rump*, which mocked the King himself as an animate idol in need of the Queen's assistance with enemas to help his golden bowels. The obscenity and ridiculing of the monarch were sufficiently outrageous to guarantee the Bill's passage. The play was passed to Walpole by Henry Giffard, the manager of Goodman's Fields; Fielding for one suspected that the play had been planted by Walpole himself.

Walpole's interests neatly merged with the anxieties of the managers and financial backers of the two patent theatres. The threat of the new London theatres was substantial and frightening. Colley Cibber, one of the triumvirate of actors who ran Drury Lane, was worried:

How could the same stock of plays supply four theatres which (without such additional entertainments as a nation of common sense ought to be ashamed of) could not well support two? Satiety must have been the natural consequence of the same plays being twice as often repeated as now they need be, and satiety puts an end to all tastes that the mind of man can delight in.

The Licensing Act not only restricted the production of legitimate drama to the two patent theatres but also required all plays to be censored by the Lord Chamberlain, a task he delegated to his Examiners of Plays. But few plays were banned outright; most were slightly altered, generally by that process of self-censorship that ensured a play was acceptable long before the manuscript was submitted for licensing.

There were many ways round the first restriction. Plays were, for instance, offered free to audiences paying, notionally, for a concert or for food and drink. The irrepressible Samuel Foote invited his audience to tea, then performed his satirical revue playlets as a noon matinée and was finally appeased by being granted a personal patent in 1766, in part an act of sympathy for his loss of a leg after a riding accident—though, typically, Foote carried on acting, starring in his satire on doctors, *The Devil upon Two Sticks* (1768). Equally typically, Foote managed to sell on the patent, which was explicitly restricted to his own lifetime.

What the highly competitive conditions of London theatre seemed to be waiting for was a star performer, somebody who could transform both acting and the social status of theatre. On 15 September 1747 the new season at Drury Lane opened with a performance of *The Merchant of Venice*, starring Charles Macklin as Shylock, a role he had begun to play in 1740 and would carry on playing almost till his retirement from the stage in 1789. Macklin's brilliant performance, changing Shylock from the comic tradition into a fierce and powerful figure, was already a known quantity. But Drury Lane was under new management and the prologue to the season, written by Dr Johnson, set out the policy of the company.

> Ah! let not censure term our fate our choice,
> The stage but echoes back the public voice.
> The drama's laws the drama's patrons give,
> For we that live to please, must please to live . . .
> 'Tis yours this night to bid the reign commence
> Of rescued nature and reviving sense;

> To chase the charms of sound, the pomp of show,
> For useful mirth and salutary woe.

The audience was already being asked to accept one major innovation: gentlemen were now banned from taking their seats on the stage or wandering backstage during the performance. The play was clearly to be more important than the social activity of the audience. One of the managers, James Lacy, had been running Drury Lane for three years; he had now taken on as co-manager the newly established star of the theatre, the greatest actor of the century, David Garrick. Indeed Garrick seems to have been the man who encouraged the use of the word 'star' to describe a famous actor. Benjamin Victor, writing in 1761, could look back on Garrick's appearance as the arrival of 'a bright luminary in the theatrical hemisphere . . . [which] soon after became a star of the first magnitude and was called Garrick', the first recorded application of 'star' to the theatre.

Garrick, an unsuccessful wine merchant, had his first play performed in 1740. He began acting in 1741, out of town, at Ipswich. Ipswich was one of the permanent theatres on the increasingly important and established provincial circuit as temporary fit-up stages began to be replaced by permanent purpose-built theatres in, for instance, Bristol (1729), York (1734), and Ipswich (1736). His London début as an actor, as Richard III at Goodman's Fields in October 1741, was an immediate and spectacular success. He soon followed it with the other major Shakespeare tragic roles, defining his success from the beginning in terms of his unswerving allegiance to Shakespeare as the centre of the national culture.

Garrick was Lacy's natural choice and soon dominated the partnership. His reign at Drury Lane lasted till 1776. Whatever else he achieved, his career transformed the social status of acting: as Dr Johnson said, 'his profession made him rich and he made his profession respectable'. He was a friend of the high and mighty in London society, fully accepted as a gentleman, living in his riverside villa at Hampton with its gardens by Capability Brown and its temple dedicated to Shakespeare. At his death, he was accorded a grand funeral and burial in Westminster Abbey. His magnificent library with its unrivalled collection of English drama was a resource for scholars and he bequeathed it to the British Museum at his death, where it still forms the cornerstone of the British Library's holdings of early drama. He corresponded with the major figures of European theatre, particularly in France, where his visits had a profound influence on Diderot's revolutionary thinking about the nature of acting in *Le Paradoxe sur le comédien* (*The Paradox of the Actor*), which contrasts the tradition of feeling with Garrick's virtuoso demonstration of the application of intellect and observation in the creation of emotional intensity in a role. He was also the most painted actor ever, the subject of hundreds of paintings and engravings, a very visible icon of acting.

Garrick's management hardly produced any major new plays for the repertoire. New comedies and tragedies, new farces and afterpieces were regularly produced

Stage scenery, 1781. This set, for a sea-coast scene, was designed by Philip de Loutherbourg. Its representation of landscape and its careful exploration of depth through the wing-pieces were innovatory strengths of de Loutherbourg's work.

and Garrick corresponded encouragingly with some dramatists and dealt with the antagonism of the many disgruntled would-be playwrights whose prized work he turned down. But none of the new work has kept its place as an established part of the performed or read drama. His own plays were pragmatic, often very popular, responses to the audience's taste. The Drury Lane repertoire came increasingly to be built around the concept of a classic repertory, a stock body of plays revived each season.

Always interested in production, Garrick experimented with new effects and spectacle; in 1771, for instance, he attracted Philip de Loutherbourg to London to develop new styles of stage lighting and scenery, new ways of representing place and landscape, and new methods of using light to brilliant effect in night scenes, culminating in his development of the picturesque and topographically exact scenery for *The Wonders of Derbyshire* (1779).

Garrick was prepared too to explore the possibilities of costuming. Where, as Steele wrote in 1711, 'the ordinary method of making an hero is to clap a huge plume of feathers upon his head', Garrick was intrigued by the opportunities for a more 'authentic' costuming, following the lead of Aaron Hill, who designed a set of 'old Saxon habits' for his play *Athelwold* in 1731, though it was Macklin who achieved a 'Scottish'-styled *Macbeth* in 1773.

In all Garrick worked towards a coherence of style and an integrity of effect in production. If hardly a modern-style director he was trying to create a unity in performance to an extent previously unknown. The attitude extended to his work with the acting company: under his management, rehearsals were taken with great seriousness and actors' temperamental displays were firmly dealt with. Garrick may have been the leading actor in the company but he collected an ensemble of fine performers around him: Spranger Barry and Kitty Clive, Peg Woffington and Hannah Pritchard, all the best actors of the age worked with Garrick.

His own acting was as daring and virtuosic as possible. For his performance as Hamlet he had a wig-maker produce a trick wig whose hair he could make literally stand on end in his confrontation with his father's ghost. But such tricks were far less important than the intensity and innovatory naturalism of his effects. Above all, he seemed to enjoy acting and audiences enjoyed watching him. He also depended on a rapidity and vivacity of effect, allowing his mobility to generate emotional intensity without ever being trapped by a single overarching mood. Where the power of Betterton in the Restoration or his successors like James Quin lay in declamatory style and heroic force, Garrick's success lay in the exploration of character and a form of genius that allowed him to go beyond the rules. In a period when more and more books and pamphlets were appearing on the actor's art (by Aaron Hill, John Hill, and others), Garrick redefined the terms of acting. In tragedy, where stance played a major part, Garrick added a quality of stillness— he was notorious for his use of pauses for effect—and an imagination that went beyond any pattern or expectation. As he wrote to a friend, 'I pronounce that the greatest strokes of genius have been unknown to the Actor himself, till circumstances and the warmth of the scene has sprung the mine as it were, as much to his own surprise as that of the audience.' He did not succeed in every role; when Spranger Barry was playing Romeo at Covent Garden in 1750, Garrick insisted on playing the same role at Drury Lane on the same nights, but popular opinion, though annoyed at the way this actors' squabble proved boring, acknowledged Barry's superiority, particularly in the first half. As Lear, though, his emotional power was unquestionable; as a contemporary rhyme judged:

> The Town have found two different ways
>   To praise the different Lears:
> To Barry they give loud huzzas,
>   To Garrick only tears.

In comedy there was no contest: the detail of his characterization was coupled to

Mr. Garrick delivering his Ode, at Drury Lane Theatre, on dedicating a Building & erecting a Statue, to Shakespeare.

To him the Song, the Edifice we raise, | *He merits all our Wonder, all our Praise.*

The Ode, page 2.

DAVID GARRICK AS HAMLET, 1775

Above: David Garrick delivers his ode in praise of Shakespeare. After the rains washed out some of the events of the Stratford Shakespeare Jubilee in 1769, Garrick repeated his brilliant recitation of his ode on stage at Drury Lane, complete with orchestra, chorus, and a statue of Shakespeare.

Lichtenberg, a German visitor to London, wrote finely detailed accounts of the actors he admired. His description makes no mention, however, of the trick wig, whose hair could stand on end, that Garrick used at this moment in his performance.

Hamlet has folded his arms under his cloak and pulled his hat down over his eyes; it is a cold night and just twelve o'clock; the theatre is darkened, and the whole audience of some thousands are as quiet, and their faces as motionless, as though they were painted on the walls of the theatre; even from the farthest end of the playhouse one could hear a pin drop. Suddenly, as Hamlet moves towards the back of the stage slightly to the left and turns his back on the audience, Horatio starts, and saying: 'Look, my lord, it comes,' points to the right, where the ghost has already appeared and stands motionless, before any one is aware of him. At these words Garrick turns sharply and at the same moment staggers back two or three paces with his knees giving way under him; his hat falls to the ground and both his arms, especially the left, are stretched out nearly to their full length, with the hands as high as his head, the right arm more bent and the hand lower, and the fingers apart; his mouth is open: thus he stands rooted to the spot, with legs apart, but no loss of dignity, supported by his friends, who are better acquainted with the apparition and fear lest he should collapse. His whole demeanour is so expressive of terror that it made my flesh creep even before he began to speak. The almost terror-struck silence of the audience, which preceded this appearance and filled one with a sense of insecurity, probably did much to enhance this effect.

the most careful naturalism. Arthur Murphy, describing Garrick, as Sir John Brute in Vanbrugh's *The Provoked Wife*, falling asleep in a chair, noted that 'sleep comes upon him by the most natural gradations. Not the minutest circumstance about a man in this situation escapes him. The struggle between sleep and his unwillingness to give way to it is perfectly just.' Over and over again, critics noted how Garrick could control the audience, a necessary talent at a time when audience complaint was likely to be noisy and dangerous: Garrick's experiment of bringing over a troupe of French dancers under Noverre in 1755 resulted in riots of anti-Gallic feeling lasting six days. Even Garrick could not control the audience's patriotic violence: coming onto the stage to try to quiet the hissing, he was greeted by ironic jeers of 'Monsieur'.

In 1767 the town council of Stratford-upon-Avon invited Garrick to contribute to a suitable monument for the refurbished town hall in exchange for the freedom of the town. Garrick's imagination was fired; everything in his work and in his idolatry of Shakespeare seemed to have led up to this moment. He set out to celebrate the Shakespeare bicentenary in Stratford, slightly belatedly, in September 1769 with a jubilee, a grand series of events in praise of Shakespeare. Pavilions were constructed on the banks of the Avon for the programme. There were to be processions through the town, balls and masquerades, concerts and horse-racing; the climax was to be his own ode on Shakespeare performed by himself at the peak of the festivities, accompanied by orchestra and chorus. Predictably enough torrential rain blighted everything and the irritation of London society transported to the depths of the provinces grew as they were asked to pay outrageous prices for bed and board. It was easy for mockers to ridicule the jubilee as an over-ambitious piece of Garrick's self-aggrandizement. Even George Colman the Elder, often Garrick's collaborator, wrote a satiric play about it for Covent Garden. Garrick's response was, as usual, both pragmatic and successful. He wrote his own play about the jubilee, mocking provincial naïvety (always likely to appeal to a London audience), and built into the play the spectacular procession of characters from Shakespeare that had been a damp squib in Stratford. Running for dozens of performances that season (though quickly forgotten thereafter), *The Jubilee* easily recouped Garrick's financial losses.

If in Garrick's whole career he failed to generate any significant new drama, his achievement was focused on his deification of Shakespeare. While the plays continued to be adapted and reworked, Garrick was also concerned to put back as many lines as he could and to extend the range of Shakespeare plays performed, always placing Shakespeare at the centre of his company's repertory. In the division between study and stage that was deepening throughout the century, Garrick tried to bridge the gap. National pride in its greatest writer, literary admiration and scholarship, and theatrical viability combined to tie Garrick and Shakespeare together in the forefront of the culture.

It used to be the standard critical view that the aftermath of the success of Collier and Steele, the ending of the era of Restoration dramatists, established the triumphant dominance of sentimental drama, comedy where tears outweighed laughter and moral orthodoxy, and reform overcame any subversive energies. But the acceptance of a dramatic form by a literary culture is not the same as its acceptance in a theatrical one. Very few works managed to be both; Richard Cumberland's *The West Indian* (1771) was one exception, an intriguing redefinition of the title-character from the stereotype comic butt into a naïve but thoroughly moral man finding his way through the dangers of London society. Cumberland's humane sensitivity suggests a modern liberalism, as in *The Jew* (1794), a reasoned argument against anti-Semitism. The fashion for such serious comedy meant, as Oliver Goldsmith complained, that 'while the comic poet is invading the province of the tragic muse, he leaves her lovely sister quite neglected'. But Benjamin Hoadly's *The Suspicious Husband* (1747), the most performed comedy at Drury Lane under Garrick, allows the conventions of sentimentalism to exist in uneasy alliance with the staple features of Restoration comedy. By the 1770s, dramatic comedy found a new lease of life, a new excitement, precisely by rediscovering the possibilities of Restoration form, just as Richard Brinsley Sheridan found the source for his satire on contemporary opera and tragedy, *The Critic* (1779), in Buckingham's *The Rehearsal* (1665). In Sheridan's *The School for Scandal* (1777) sentimentalism is displayed most completely as a mask, Joseph Surface's ingratiating device to cover his hypocrisy. Goldsmith's anxiety about comedy is part of his distinction between 'laughing' and 'sentimental' comedy in an important essay in 1773; in the latter 'almost all the characters are good and exceedingly generous; they are lavish enough of their tin money on the stage, and though they want humour have abundance of sentiment and feeling'. Goldsmith's battle-lines are perhaps too rigidly drawn but there is no question that his *She Stoops to Conquer* (1773) celebrates laughter over sentiment as well as the honest virtues of the country over the snobbery of the town.

Sheridan and Goldsmith may have found their way back to a richer seam of comedy but just as important to the evening's performance was by now the afterpiece, pantomimes like Theobald's *Harlequin Sorcerer* (performed 337 times at Covent Garden between 1747 and 1776, far and away the most performed play of the period), farces like Garrick's own *Miss in her Teens*, short musicals like Issac Bickerstaffe's *The Padlock*. If the theatrical bill was now less extreme in its multiplicity than earlier in the century, the comic energies of the afterpieces go a long way to balance any sentimental, moral dullness in the main play. Farce, not sentimental comedy, is really the dominant dramatic form of the period, a genre hardly explored before but now richly developed, enthusiastically enjoyed, and, especially, given a particular position in the structure of the evening's entertainment. The audience's demand for amusement rather than, as the literary culture would

have approved, instruction meant that it was hardly surprising that both the patent theatres suffered from damage when the audiences rioted at the managements' attempts in 1762 to abolish half-price admission after the third act of the main play.

Theatre audiences in London were, in any case, growing throughout the century at an even faster rate than the population growth of the city. The changes in theatre architecture reflected this. Where Drury Lane at the beginning of the century had a capacity of no more than 1,200, by 1794, after a number of rebuildings, it could hold over 3,600. Covent Garden expanded from 1,330 in 1732 to 3,000 in 1782. The other new theatres in London now functioning semi-legally, performing pantomimes, equestrian spectacles, and unlicensed plays, were similarly large: for instance, Sadler's Wells (from 1765), with a capacity of 2,600, or the Royal Amphitheatre run by Philip Astley (from 1788), with a capacity of 2,500 after 1803, or the Royal Circus, run by Astley's rival Charles Hughes (from 1782). These colossal auditoriums cried out for spectacle and tragedy. Throughout his management Garrick looked for new tragedies. He rejected the Scottish playwright John Home's *Douglas*, only to find it a phenomenal success in Edinburgh in 1756 (when a chauvinist spectator called out, 'Whaur's yer Wullie Shakespeare noo?') and equally successful at Covent Garden. Home's play, a full-blown

Covent Garden Theatre, 1763. This engraving of the theatre shows members of the audience clambering over the tiny orchestra pit during a theatre riot. Note the chandeliers over the stage to light it and the on-stage boxes on the side of the forestage, with an entry-door on either side downstage of the proscenium arch.

A provincial theatrical performance in England, 1788. James Wright's engraving of *Macbeth* scene 1 is wickedly mocking of theatre outside London but the energy of the scene and the audience excitement is also portrayed with sympathy. Such fit-up provincial theatres were gradually replaced by purpose-built well-equipped theatres.

romantic tragedy, looked forward. Arthur Murphy, later an efficient writer of comedies, had his first major success with *The Orphan of China* (1759), adapted from Voltaire, looking sideways at France for a tragic form. But the effective search for a neo-classical style needed to find a grandeur of scale in performance.

Garrick's share in the Drury Lane Theatre was bought out by Sheridan but, realizing that the chores of management did not suit him, in 1788 Sheridan passed the day-to-day work on to John Philip Kemble. Kemble had made his London début in 1783. His sister, Sarah Siddons, had returned to Drury Lane in triumph the previous year. Siddons had had a disastrous start with Garrick's company in 1775 and had soon left for the provincial circuit. She learnt her craft first on the important Yorkshire touring circuit, run for over thirty years by Tate Wilkinson, which was also her brother's training ground, then playing at the various Theatres Royal, many of which had now been given royal patents in their own right (Bath in 1768, York in 1769)—before the end of the century the numbers and size of theatres outside London increased at a prodigious rate. Siddons's return to London, playing Isabella in Garrick's version of Southerne's tragedy *The Fatal Marriage*, marked the beginnings of a new style of acting.

Kemble and Siddons gave the public the kind of tragic performance the theatres demanded: monumental, aristocratic, full of heroic proportions and emotional intensity. They established the standard for an English classical style. Both worked on stage with an unremitting concentration. William Hazlitt the critic summed up Kemble's style perfectly:

In short, we think the distinguishing excellence of his acting may be summed up in one word—*intensity*; in the seizing upon some one feeling or idea, in insisting upon it, in never letting go, and in working it up, with a certain graceful consistency and conscious grandeur of conception to a very high degree of pathos or sublimity.

Clearly he was never going to succeed in comedy; his performance as Charles Surface in *The School for Scandal* was 'as merry as a funeral and as lively as an elephant'. But, where Garrick had succeeded through lightning changes, Kemble's single-minded conception of a role's through-line had a vastness of scale that suited Shakespearian tragedy in the vast spaces of the new Drury Lane and Covent Garden. His productions—and he, even more than Garrick, worked to create a unified, integrated style for a performance—were full of spectacle, especially in plays, like Shakespeare's *Henry VIII* and *Coriolanus*, that gave him a space to create enormous processions as well as opportunities to explore his concept of antiquarian realism in sets and costumes.

The streak of the showman, conspicuously absent from his own acting style, as well as the competition from the rival establishments unable to perform 'legitimate' drama, the legal prerogative of the patent theatres, led to Kemble's encouragement of the fashion for melodrama, of putting animals on the stage (including an elephant), of allowing the craze for Master Betty, the 'Infant Roscius', a 13-year-old lionized for his performance in all the classical roles in 1804–5, to run its course. As one annoyed newspaper complained, Kemble ignored any play that could not 'be

John Philip Kemble as Coriolanus, 1797. The nobility and dignity of Kemble's acting was most completely demonstrated in his patrician performance as Shakespeare's Coriolanus. The setting carefully underlines the grandeur of the actor with the martial statue contrasting with the cloaked actor. Painting by Francis Bourgeois.

converted into a pageant but brings forward with much pretence any drama that has its proper capabilities of ostentatious spectacle'. The patent houses, though the homes of the traditional repertory, were by no means staid in their productions as they fought for audiences.

Often criticized as an actor for his artificiality and stiffness, his idiosyncratic pronunciation, and his excessive reliance on pregnant pauses (a consequence of his breathing difficulties caused by asthma), Kemble was unequivocally noble in manner, ideal for the patrician coldness of Coriolanus. As Leigh Hunt noted, 'it is in characters that are occupied with themselves and with their own importance . . . that Mr Kemble is the actor'. Kemble's studied effects complemented perfectly the naturalness that was the key contemporary perception of Mrs Siddons:

Mrs Siddons has the air of never being the actress; she seems unconscious that there is a motley crowd called a pit waiting to applaud her, or that there are a dozen fiddlers waiting for her exit.

Praised for her sweetness and pathos, praised for her terror and dignity, she was praised above all for her identification with her role: as an early biographer commented, 'When Mrs Siddons quitted the dressing-room, I believe she left there the last thought about herself.'

Kemble and Siddons left Drury Lane and the squabbles with Sheridan in 1802 and moved to Covent Garden. After the theatre burnt down in 1808, Kemble rebuilt it on an even larger scale than before. New theatres and grand productions are expensive, even with a huge capacity. Kemble's attempts to put up the admission prices sparked off the 'Old Price' riots that lasted in the theatre for sixty-six nights till Kemble was forced to give in.

The demonstrations and celebrations of the Old Price riots proved conclusively that the audience in England now ruled the theatre and that, as Dr Johnson had predicted in 1747, 'the drama's laws the drama's patrons give'. But to provide the kind of entertainment the audiences expected, the theatre in England had developed, to an extent unimaginable at the beginning of the century, its skills and machinery, its sets and costuming, its showmanship and excitement. Whatever else their work may be about, the huge theatres of 1800 were unequivocally always celebrating their own artistic and spectacular triumphs.

In those countries of Continental Europe that had enjoyed a strong theatrical tradition in the Renaissance, the development of theatre in the eighteenth century was determined largely by the need to create a stage that reflected the changing make-up of society. Against the background of declining aristocratic power accompanied by growing rationalism, high tragedy and formalized acting no longer seemed relevant to the concerns of the expanding middle classes. Even the comedies of the previous century, like those of Molière, presented the image of a stable society, one in which impending disaster may be averted by the intervention of the all-powerful Sun King (as in *Tartuffe*), comedies in which we are invited to laugh at the follies of individuals who fail to conform to the behaviour of their peers. By the time we come to Beaumarchais's comedies over a century later, it is the extraordinary individual that wins our sympathy and we laugh instead at the follies of society.

This change began with the death of Louis XIV in 1715. When the Sun King set, the austerity of the recent past was rapidly repudiated. The aristocracy threw itself into a bout of elegant self-indulgence, celebrating the sensual delicacy of the rococo in opera, ballet, and masked balls. The Regent invited a leading Italian theatre personality, Luigi Riccoboni, to reopen the Comédie-Italienne in Paris, and in 1716 it returned to the Hôtel de Bourgogne, from which it had been banned in 1697, its front-curtain now boasting an impressive phoenix with the motto: 'Je renais' ('I am born again'). Here, as he had tried in Italy, Riccoboni attempted to educate his public to a more serious taste, but after the gloom of the final decades of Louis's reign audiences demanded spectacle, music, dance, and laughter. Riccoboni had to concede to public taste and his French adaptations of the *commedia dell'arte* were so successful that, on the accession of Louis XV in 1723, he was granted a generous annual pension. While these harlequinades, the forerunners of comic opera, were light, they were not mindless. In one of them Socrates instructs Harlequin:

It is essential to give witty expression to the voice of reason and to useful truths for the correction of manners . . . to avoid above all trivial jokes, empty pleasantries, puns and all such licence which damages morals and offends common decency.

This progressive and serious conception of the role of comic theatre was further developed at the Comédie-Italienne by Marivaux in his *comédies gaies*. Marivaux displayed a remarkable ability to enter into the mind of his heroines, helped no doubt by his close working association with one of Riccoboni's leading actresses, Silvia. His recurrent theme of sensitive individuals out of joint with an uncaring society paved the way for the domestic dramas for which the eighteenth century was to be so famous.

This move away from the stylization and formality of the classical French stage towards more domestic forms did not occur only within the elegant surroundings of the Comédie-Italienne. Even before the death of Louis XIV the monopoly of the Comédie-Française was being challenged from a much more populist quarter: *le théâtre de la foire* (fairground theatre). For many years this form of theatre was refused permission to use speech, song, or dance; so the performers mounted spectacular dumb shows, often with striking stage effects (waterfalls, live flying animals, etc.), which were accompanied by scrolls of text in couplets, sung by actors planted in the audience and by the more literate members of the public.

The content of these couplets was often felt to be so subversive that one leader of such a troupe, the unfortunate Octave, was thrown into prison for offending the court.

Meanwhile, the Comédie-Française continued to build on its tradition of being the home of fine tragedy. The plays of Racine and Corneille were revived (including the first professional production of Racine's *Athalie* in 1717). It also discovered in Voltaire not only a prolific playwright (he wrote his fiftieth play at the age of 84) but a true man of the theatre: actor, director, and critic. Though impressed by Shakespeare, he felt that his apparent wildness was unsuitable for the French stage. So Voltaire continued to write his plays according to the neo-classical rules but adopted some of the more spectacular effects of Shakespeare, e.g. ghost-scenes. As he wrote in the Preface to his *Tancred*: 'il faut frapper l'âme et les yeux à la fois.'

Lekain in Voltaire's *Orphan of China*. The splendidly elaborate costume clearly owes more to eighteenth-century notions of Genghis Khan than to historical accuracy.

('One must make an impact on the soul and the eyes at the same time.') His plays, too, while cast in the classical mould, contained much of his own contemporary humanist ideas: thus his version of *Oedipus* (1718) contains a biting attack on the priesthood, and in *L'Orphelin de la Chine* (*The Orphan of China*) (1755) the characteristically sentimental eighteenth-century exercise of virtue by the heroine overcomes the lustful desires and menaces of Genghis Khan.

In many respects Voltaire's tragic figures were in fact contemporaries in classical costume, and by the mid-century even the tradition-bound Comédie-Française had to acknowledge the new taste of the public for a theatre which reflected their own society. It was above all Denis Diderot, as both playwright and theoretician, who established the new vogue for 'bourgeois drama' or *comédie larmoyante* ('lachrymose comedy', a sentimental piece with a happy outcome) on the model of Lillo's *The London Merchant*. Most important of Diderot's plays was *Le Père de famille* (*The Head of the Family*). Presented at the Comédie-Française in 1761, the piece preserved, however improbably, the three unities, but was innovative in placing high emotion and potentially tragic conflict in a contemporary domestic setting. The realism of the piece was reinforced by the prose dialogue, the precise stage directions encouraging full use of the stage, and the technique of cross-cutting two conversations, which sacrificed theatrical focus for the sake of naturalness.

The actors, unused to performing in prose, were, according to Diderot, 'trembling as they went on stage as if it were their first time'. More importantly perhaps, and as a mark of royal approval, it was recorded that Louis XV, one of the 'most hardened egoists of the day', wept copiously at this story of a son whose love for a beautiful but impoverished girl drives him to disobey his father. Fortunately for both Louis and the outcome of the play, the girl turns out to be a cousin, and father and son indulge in a sentimental reconciliation. For without this *deus ex machina* the evil uncle might have made arbitrary use of a warrant of arrest to incarcerate his young niece; the son might indeed have rebelled against authority, and the play would have acquired a much more revolutionary slant.

This new-found acknowledgement of eighteenth-century theatre as performing an important role in examining the nature of society and allegedly in improving the quality of that society (later most clearly asserted in Schiller's essay 'The Theatre Considered as a Moral Institution' of 1784) had a predictable effect on theatre practice. Actors who developed a more natural style of delivery became popular. Already at the Comédie-Française Adrienne Lecouvreur had enchanted audiences with her natural charm in place of the mannered gestures of the older actors, and had reinforced this by adopting more realistic costume. So, when she performed the role of Queen Elizabeth in 1721, in place of the conventional Versailles gown and high wig, she wore an English court dress with a sash of the Order of the Garter. Later, another star of the Comédie-Française, Mlle Clairon, similarly shocked her audiences by appearing with 'half-naked arms' and even, awakened from sleep in a version of *Dido*, in a plain shift. Such authenticity in costume reached a turning-point when in 1789 Talma, who significantly had begun his theatrical career in London, sought advice from the artist David and appeared in Voltaire's *Brutus* wearing a Roman toga. After recovering from the shock of seeing bare arms and legs on stage, the public were soon won over to this new style of presentation, so much so that when two years later Talma wore a Roman hair-style for the role of Titus, he inaugurated a fashion which was to become the rage for modish French revolutionaries.

Stage-sets also developed a new realism, necessitating the removal of spectators from the stage, which was finally achieved at the Comédie-Française in 1759. Perspective scenery created by sliding wing-flats and a backdrop with objects painted on it were still the norm, but there was now greater care taken with the authenticity of setting and the realism of the furniture and props. No longer were the French public prepared to tolerate absurdities like the appearance of the 70-year-old actor Baron in the role of a child in La Motte's *The Maccabbees* (1721). No one was fooled by the boy's cap perched on his head, and when two actors had to assist him to his feet again after he had embraced the knees of the King, the merriment of the audience was loud and sustained.

Middle-class audiences now preferred a theatre that set out to explore their problems in a realistic even if sentimental manner, but they soon went further. The bourgeoisie could boast ever-growing wealth and intellectual supremacy over

Performance in the Hôtel de Bourgogne, 1769. By the latter half of the eighteenth century the Parisian public were demanding greater realism on stage; hence the 'authentic' spinning-wheel and stool and the detailed, even if painted, back-drop.

the aristocracy but were still excluded from political power. So, in addition to a theatre that reflected their world, they demanded one that would give voice to their aspirations. These aspirations in the French theatre were to focus on the unlikely figure of a Spanish barber, created by a writer and adventurer who had only recently begun to write for the stage.

Beaumarchais's first Figaro play, *The Barber of Seville* (1775), presented the unoriginal but for the times provocative view of the servant as being cleverer and, despite a certain pardonable roguishness, essentially more honest than his master; but it was the much more aggressive stance of Beaumarchais's next comedy which was to cause a furore in Paris and firmly to establish theatre as a major influence in French political life. Beaumarchais read his *Marriage of Figaro* to the members of the Comédie-Française in 1781, but it was to be some years before it could be seen by the Parisian public. No less an admirer of Beaumarchais than the Queen Marie Antoinette arranged for the play to be read to the King: 'Louis XVI accompanied the reading with comments of praise or disapproval; more and more, however, he was moved to utter: "That goes too far! That is indecent! etc."' On hearing Figaro's fifth-act monologue attacking the aristocracy, he leapt to his feet and cried with prophetic insight: 'That's terrible! It will never be performed: for this play not to be a danger, the Bastille would have to be torn down first.'

Despite this royal condemnation, a private performance of the play was arranged. However, on the appointed evening in June 1783, word came from the King that even the aristocracy were not permitted to see this notorious piece. Predictably, the demand to see it

Beaumarchais's *Barber of Seville*, c.1775. Note the realistic furniture and props and the natural and flowing postures of the actors, a contrast with the stiff formality shown in Watteau's painting of half a century earlier.

staged grew even stronger, and a further private performance was organized to take place on 26 September 1783 at the château at Gennevilliers. The hall was so packed with gentry that Beaumarchais felt compelled to break a few windows to let air into the stuffy auditorium—an act that was recognized as being symbolic even then. Such was the success that the pressure to release it for public performance grew even greater.

This was achieved remarkably simply: Beaumarchais was permitted to read his play to Breteuil, the Royal Minister, and an assembled company of leading

THE MARRIAGE OF FIGARO

Figaro attacks the aristocracy in a monologue which initially convinced Louis XVI that the play must be banned, but which was greeted with thunderous applause at the Parisian première in 1784.

No, Monsieur le Comte, you shall not have her . . . . Because you are a great lord, you imagine you are a great genius! . . . Nobility, wealth, titles, and appointments, they all make you so proud! And what have you done in return for so many favours? You took the trouble of being born and nothing more. For the rest, you are a fairly ordinary man; while, as for me, my God! Lost in a dark mass, I have had to use more knowledge and strategy just to stay alive than was needed to rule the whole of Spain for a century.

arbiters of literary taste. The wit of the piece and Beaumarchais's own charming delivery won over these gentlemen, and the King grudgingly allowed a production to go ahead, quietly hoping that it would be a flop. So, just eight days after the Comédie-Française had moved back into their renovated theatre, on 27 April 1784, the most important first night in eighteenth-century France took place. The alternative title of the piece, *La Folle Journée* (*The Crazy Day*), could not have been more appropriate. Already the previous night ladies of the nobility and women of the bourgeoisie had, at the expense of both protocol and comfort, shared actresses' dressing-rooms in order to ensure that they would get a seat. Others took lunch in the auditorium. As the time of the performance approached, the throng outside the theatre swept aside the guards and forced the gates, causing several ladies to faint. Less than half those pressing to get in managed to get tickets. Every single minister was present, as were all the brothers of the King. Unsurprisingly, the King stayed away. Beaumarchais himself had the prudence to arrange for two *abbés* to sit either side of him to indicate the seriousness of the moral intent in his piece, although more cynical observers commented that they were there to provide the spiritual guidance that he so obviously needed.

The performance lasted from half-past five to ten o'clock, interrupted by tumultuous laughter and applause, with only occasional whistling and hissing. Dazincourt as Figaro rapidly won over the public with his display of native cunning, and this was reinforced by the pert charm of Louise-Françoise Contat as Suzanne; but the greatest favourite of the evening was Jeanne-Adelaïde Olivier's Chérubin, in whose youthful yearning the French public perhaps saw something of their own longing for change. The quality of performance was further heightened by the use for the first time of oil-lamps, which gave the stage an unprecedented brightness and avoided the usual distraction of stage-hands having to trim candle-wicks. No doubt to Louis's chagrin, the piece was repeated over seventy times.

It is an irony of course that this piece, which Napoleon later described as 'the Revolution in action', and which at the time indirectly led to Beaumarchais's arrest, should have been mounted not just with the acquiescence but with the full support of the aristocracy, whose right to govern was so seriously challenged in the play. But, however much the effectively suicidal connivance of the nobility was required at the time, the tide of history and with it the new political role of theatre in France would anyway sooner or later have broken through the banks of censorship. A few years later, in 1789, Danton himself as leader of the Revolutionary Council ordered the Comédie-Française to perform Chénier's *Charles IX* with Talma in the title-role. The scene in which a dagger is blessed before being used to strike the mortal blow on the King was greeted in performance by ten minutes of unbroken applause. The stage was being used quite manipulatively to prepare the public for the execution of Louis XVI, and, in recognition of the value of the theatre to the Revolution, a directive of January 1791 permitted anyone to open a theatre in Paris.

So from the virtual monopoly of the Comédie-Française in Paris at the start of the century, with its classical repertoire and formal playing style, French theatre now took place in many different venues, offering middle-class dramas and comic operas, played in a lively, natural style. What began the century as aristocratic entertainment had now become the forum of the people.

In Spain, the combination of the brilliance of the legacy of Golden Age drama and the lack of any genuinely talented new dramatists proved stultifying. The achievements of the previous century had become a rigid form, hindering innovation. Even the political change at the accession of the Bourbons failed to have any impact on the development of drama. Instead earlier plays continued to be performed, usually heavily adapted to accommodate the increasing audience demand for stage spectacle, the one sure way to achieve success. By the 1760s the government sought to reform the poor state of the theatres by attempting to ensure adequate rehearsal, a less stilted performing style, and a ban on ad libs.

But the major source for significant change was, inevitably, the influence of the new possibilities being explored in French drama. Nicolás Fernández de Moratín tried to combine the traditional style with French neo-classical tragedy to resist the trivialization of drama. But beyond the observation of the unities and a certain new concentration on the triumph of love over honour his tragedies did little to alter the prevailing style.

His son Leandro Fernández de Moratín sought, with far greater success, the reconciliation of national traditions with neo-classical and Enlightenment forms in his five comedies. *La comedia nueva* (*The New Play*, 1792) mocked earlier plays for using a neo-classical façade to mask their interest in superficial spectacle. Instead he explored, particularly in his best play *El viejo y la niña* (*The Old Man and the Young Girl*, 1786), new areas of concern for drama: bourgeois morality over aristocratic codes of honour, material conditions rather than social structure, free choice against arranged marriages, and the concerns of the significant social group of the untitled nobility. Profiting from the *comédie larmoyante*, his plays had a new intensity of emotion. This new centring on the social interests of a bourgeois theatre provided the theatre with the means to exchange the patronage of the court for the influence of popular taste without kowtowing to the demand for spectacle; it also combined the strengths of the national tradition with the opportunities offered by the drama of the rest of Europe.

While Italy could not look back to a Golden Age, its native tradition, *commedia dell'arte*, for all its aristocratic patronage, brought its own problems. Though the materials of the form seemed increasingly exhausted and noble support fell away, it was far from clear what could replace it. The development of *opera buffa* (comic opera) seemed to divert energies from the dramatic comedy. Again and again the best talent (actors, playwrights, and designers) was also sucked out of the country, lured to France, where success seemed easier.

In tragedy and scripted comedy, the model of French neo-classical forms was too quickly and easily dominant to allow the development of a national alternative. Luigi Riccoboni, son of a famous Pantalone, attempted to renew the Italian repertoire and create a popular Italian serious theatre, gradually replacing work by Corneille and Racine with plays by Trissino and Tasso and new work by Pier Martelli. But only with *Merope* by Scipione Maffei did Riccoboni find a major success. *Merope*, a weighty verse tragedy, was premiered in Modena in 1713 and triumphed in Venice the following year as well as being toured by Riccoboni. But though admired, Riccoboni's troupe had comparatively little impact and, when his attempts to reform comedy were signally unsuccessful at Venice's Teatro San Luca, he headed off to Paris, where, as we have seen, he Gallicized *commedia*, the reverse of his resistance to the French influence in Italian theatre.

The best Italian tragedy of the mid-eighteenth century was produced for opera, particularly in the libretti of Pietro Metastasio, whose work, from *Didone abbandonata* (*Dido Abandoned*, 1724) onwards, was hugely in demand. Serious dramatic tragedy had to wait for the end of the century for a significant new impetus. In 1775 Count Vittorio Alfieri wrote his first tragedy, *Cleopatra*, which succeeded in Turin, Alfieri's home city, with Girolamo Medebach's troupe. However, he soon recognized, by bitter experience, that the public theatres were likely to butcher his plays to please audiences. In Rome in 1781 he wandered into a performance of his *Orestes* to find that the play had been given a happy ending, with Orestes now reconciled to his mother rather than killing her. Leaping on to the stage and arguing with the actors, Alfieri managed to secure an apology from the theatre's management. Appalled by the actors' lack of professionalism he had most of his later plays performed in private houses before carefully selected audiences by a group of amateur actors under his control. The literary success of Alfieri's work had little effect on the popular theatre. Not the least of his problems was the need to develop an Italian dramatic language, rather than the local, regional forms used for most drama; he had both to learn Italian and use it for his own drama. His own nationalism found a response in the social unrest that would lead to the Risorgimento, the search for a national unity.

In 1750 in the closing speech of the season delivered by the leading actress of the Medebach company in Venice, Carlo Goldoni announced that he would provide no fewer than sixteen full-length new comedies for the company's next season. As a boast it was outrageous; astonishingly he fulfilled it. Venice was now the centre of Italian cultural life, with numerous theatres and a vibrant intellectual society. Though long interested in writing for the stage, Goldoni had produced plays only intermittently while practising law. Initially he wrote the scenarios required by *commedia* style: in 1745, for instance, he developed *Arlecchino, servitore di due padroni* (*The Servant of Two Masters*) in scenario form for the brilliant actor Antonio Sacchi. It was only in 1753 that he combined his original text and Sacchi's interpretation of its *lazzi* into the fully scripted form in which it is now played. But Goldoni moved beyond this mixture of written and improvised drama, aiming to overturn the crudities and sheer self-indulgence that had come to dominate *commedia* by developing a scripted comedy, using the traditions but abandoning the masks and improvisational techniques central to its performance.

In 1748 Goldoni became a full-time dramatist working for Medebach's new company based at the Teatro San Angelo, one of three theatres specializing in comedy. *I due gemelli Veneziani* (*The Venetian Twins*) was his first major success, a comedy that, building on convention, found a new possibility of seriousness, social realism, and decisive social comment even as it exploited the virtuoso skills of Cesare D'Arbes, who played both twins.

Frighteningly prolific, Goldoni did not seek immediate reformation but rather a gradual transformation of Venetian theatre. Inevitably it was not an uncontested change. Initially the conservative opposition was focused on the parodies of Goldoni by Chiari produced at rival theatres: Goldoni's *La vedova scaltra* (*The Cunning Widow*, 1748) at the San Angelo was mocked in Chiari's *La scuola delle vedove* (*The School for Widows*) at the San Samuel. Goldoni, who watched Chiari's play disguised in cloak and mask, found it little more than plagiarism with added comments ridiculing his jokes.

Chiari was hardly likely to stop the flow of reform, a programme for which Goldoni outlined in the first play of the sixteen in the 1751–2 season, *Il teatro comico* (*The Comic Theatre*), a polemical manifesto masquerading as a play. In 1753, after arguments over royalties, Goldoni parted company with Medebach, moving to the larger Teatro San Luca. Over the next decade, his work developed in subtlety and ambition, exploring Venetian society in depth. As early as 1748 in *La putta onorata* (*The Respectable Girl*) Goldoni had included scenes of Venetian gondoliers, portraying their language and customs; he arranged for gondoliers to attend the theatre free and, as he recorded in his *Memoirs*, 'they were enchanted to see themselves represented on stage'.

But by this time his reforms had attracted a more serious opponent than Chiari. Carlo Gozzi was a leading member of the Accademia dei Granelleschi, a literary society founded in 1747 with the express aim of preserving purity of style; Goldoni's fascination with regional dialect was bound to be offensive to him. Initially basing his attack in pamphlets and parodies, Gozzi was stung by Goldoni's mockery of critics who do not write into producing his own plays. Working with Sacchi's company, Gozzi wrote a series of plays which sought to regenerate *commedia* and re-establish what he viewed as its aristocratic and

emphatically conservative political and social values through fables, the *fiabe*. He repudiated any taint of Francophile radicalism that Goldoni was seen as sharing. The best of the *fiabe*, *L'amore delle tre melarance* (*The Love of Three Oranges*, 1761) and *Turandot* (1762), combined fantasy derived from folk-tales with exotic settings and stage spectacle, denying at every step the realism, scenic simplicity, and contemporary idiomatic language of Goldoni. Part scripted and part improvised, Gozzi's fables also made Aristophanic use of the opportunities for direct satire of Goldoni and topical events.

Initially successful, the vogue for the *fiabe* passed (not to be rediscovered until the twentieth century in adaptations into opera) and Gozzi eventually abandoned the stage, frustrated by the endless quarrels with Sacchi's actors. But Gozzi's initial success, coupled with the temptations of higher salary and reasonable security, was enough to cause Goldoni to follow the high road to Paris, becoming house dramatist at the Comédie-Italienne. Goldoni worked in Paris until his death without ever really adapting to the different circumstances, exiled from the language and society he had so exhilaratingly examined, the very sources of his theatrical inspiration.

On turning from the established theatre nations of England, France, Spain, and Italy to the rest of Europe, a quite different picture is encountered. With the exception of the relatively stable imperial thrones in Vienna and Petersburg, the situation of the German-speaking countries and of the nations of northern and eastern Europe was not conducive to the creation of national theatres. What is now called Germany was divided into over 360 different states, each with its own laws, currency, and measurements; the Netherlands were recovering, albeit with growing prosperity, from the Wars of Spanish Succession (1702–13); Sweden's role as a European power ended when Charles XII fell in battle in 1718; Denmark was politically strongly influenced by Germany; and Norway was little more than a satellite of Denmark. Bohemia was under the dominance of German culture, Hungary owed allegiance to Austria, and Poland had to endure a succession of 'Partitions'. Moreover, every European nation outside England looked for cultural guidance towards France. The French language dominated in court circles, the French taught one to dance and to cook well. Leibniz, Germany's leading Enlightenment philosopher, wrote in French, and the Prussian king, Frederick the Great, went on record as saying that German was a language fit only to use for speaking to one's horse.

Amongst this adulation of the French, it was the little country of Denmark that tried to establish the first national theatre of northern and eastern Europe. Ironically, the initiative came from two French theatre-practitioners who opened the first theatre for Danish productions, the Grønnegade Theatre in Copenhagen. Two days after the predictable translation of Molière with which the theatre

Facing: *Theatro Regio*, Turin, *c*.1740. Italian theatre used extremely elaborate stage sets. The experience of watching the play was no doubt helped by waiters serving drinks during the performance. The armed guard kept order as in theatres elsewhere in Europe.

Holberg's *Jeppe from the Hill*, c.1740. Note the wing-flats, the chandeliers and footlights, and the spectators on the stage.

opened, Denmark was to acknowledge its first and immediately successful national playwright, Ludvig Holberg. Born in Norway, but obliged to write in Danish, Holberg was a university professor, but fortunately one with a strong sense of theatre. His first play, *Den politiske Kandestøber* (*The Pewterer who Wanted to be a Politician*), owes a great deal to Molière in ridiculing the vices of an extreme comic type, in this case a simple artisan who has pretensions to becoming mayor but is exposed as entirely inadequate to the task. But Holberg goes beyond mere imitation of his French model in the authenticity of the Danish setting and in establishing a testing situation for his comic figures, a style of serious comedy which raises social and philosophical questions in a manner that anticipates Kleist's comedies.

The opening night of *The Pewterer* was a huge success, with many of the would-be audience obliged to attend the event in the yard outside the theatre. There were rumblings amongst the city fathers that the play was mocking them, but Holberg could claim that his intention was on the contrary to add lustre to the image of the authorities. However tongue-in-cheek this defence of his comedy was, he clearly subscribed to the characteristic Enlightenment view that comedy was socially useful in the correction of vices, whether those revealed in the behaviour of the comic protagonist or in that of society at large.

Despite the initial success of Holberg's plays and his own prolific output (in 1723 alone he wrote ten comedies, and by 1727 had written twenty-six), this first attempt to create a new national theatre was dogged by problems, above all the continuing domination of French culture. The Grønnegade theatre was obliged to alternate the comedies of Holberg with the more polished products of Molière, and, while the King supported the enterprise, the robust knockabout humour of the native Danish performances was felt to be too vulgar for actual royal attendance. So, when the theatre was invited to perform at the palace, they offered two French pieces rather than anything by Denmark's own national playwright. Already in 1727 Holberg's *Funeral of Danish Comedy* was performed, and attempts later in the century to revive a Danish national theatre were directed to the performance of operettas and festive operas.

A similar story can be told of Denmark's neighbour, Sweden. An attempt in 1737 (undertaken by a French actor) to found a Swedish national theatre had failed by 1754, and it was only with the accession of Gustav III in 1771 that the theatre in Stockholm began to receive whole-hearted royal support. Gustav was a wholly theatrical personality: he learnt of his accession to the throne during a theatre performance in Paris, and his life ended at a masked ball, when he was shot dead by aristocratic conspirators. He wrote his own plays, one of which was performed in Paris and Vienna, and he loved to act himself, on one occasion taking the lead role in five tragedies and a number of comedies within two weeks. By the time of his death in 1792 there were four permanent theatre companies in Stockholm, a city of not more than 75,000 inhabitants.

However, despite all this encouragement, the foundation for a Swedish-language theatre was not created. No national playwright emerged, and so, apart from Gustav's own plays, most productions were translations of Holberg or of French originals. As in Denmark, it was the musical theatre which proved more popular and led to the building of splendid theatres. Most famous of these today, because it is preserved in its original state, is the court-theatre of Drottningholm, built in 1766. The harmoniously proportioned auditorium is reflected in the generous proportions of the stage, and the sets were not permitted to disturb this aesthetic balance. Thus, whether the setting was a royal palace or a peasant's hut, the dimensions of painted wing-pieces and backcloth remained the same. Here, as in Goethe's work at Weimar, we see, perhaps for the last time in mainstream European theatre, the pursuit of beauty at the expense of authenticity.

This tension between what was aesthetic and what was authentic, characteristic of eighteenth-century theatre, is well illustrated by the development of Dutch theatre. Here, perhaps owing to their secure and prosperous society in the latter half of the century, there seemed less pressure to create a self-consciously national

Drottningholm Theatre with palace set. The lines of the auditorium are continued on to the stage, creating a shared harmonious space, free of much concern with authenticity.

theatre. The splendid Schouwburg in Amsterdam, originally founded in 1638, rebuilt after the fire of 1772, attracted a regular audience of burghers, whose casual behaviour shocked visitors from other countries: rowdiness, eating, drinking, talking, and wearing hats were all commented on. The two leading actors at the theatre, who often played the same role on alternate evenings, in many ways incorporated the ongoing debate of the century. Jan Punt was a devoted follower of the Comédie-Française style, concentrating on graceful gesture and pleasing vocal delivery; Marten Corver was much more committed to discovering the truth of performance, adopting a more natural style of acting and wearing historically accurate costumes. Their rivalry led to a spate of vituperative pamphlets in the early 1760s and eventually to Corver's leaving the Schouwburg to form his own ensemble, with whom he might pursue authenticity. Perhaps he should have been warned: it was just such an attempt at realism that had led to the fatal Schouwburg fire of 1772. In order to play a dungeon scene with appropriate gloom, most of the candles had been screened off; the screens caught fire and led to the catastrophe.

In what would become Czechoslovakia too the development of a national theatre was closely linked to the linguistic separatism from the power of German, the language which dominated Czech in Prague; the need was defined by the national revivalists as one for a 'patriotic theatre'. Only in 1786 with the founding of the Bouda Theatre were Czech plays on Czech matters written.

In Poland, where a much stronger popular theatrical tradition had existed, the National Theatre, a professional Polish-speaking public theatre, was founded in Warsaw in 1766 by Poland's last king, Stanislas August Poniatowski, as part of an extensive cultural reform in which stage-plays were intended to be the means for the widespread dissemination of Enlightenment ideas. After a sticky start the company flourished under Wojciech Boguslawski, its director from 1783 to 1814, though much of its repertoire was, predictably, adapted from French and German theatre. As Poland vanished as an independent state, Boguslawski's nationalism ran into censorship, especially after Russian objections to the patriotism of his comic opera about peasant disputes, *Cracovians and Highlanders* (1794).

Russian theatre lacked a substantial earlier dramatic tradition. Though, under orders from the Tsar, Pastor Gregory wrote *The Comedy of Esther*, produced at court in 1672 with a cast of sixty-four and lasting ten hours, this hardly inaugurated a continuous dramatic tradition. A later tsar, Peter the Great, tried to found

Van Elvervelt's *Head of the Family* (after Diderot), Amsterdam. Note the bourgeois costumes, the emotionally charged gestures, and the use of the full depth of the stage, which lends a feeling of realism to the scene. Meanwhile in the auditorium the notoriously rowdy Dutch burghers sit with their hats on or totally ignore the play, as at bottom right.

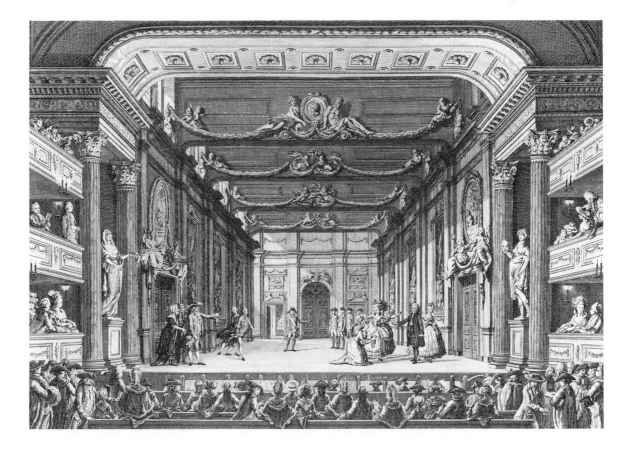

a theatre in Moscow. Much opposed by those who saw theatre as irredeemably anti-Russian, the theatre was built opposite the Kremlin and survived four years (1702–6). While attendance was high when the Tsar was present, at other times the audience averaged twenty-five; the Russian aristocracy did not take to theatre-going. The Empress Elizabeth (ruled 1741–62) made theatre attendance compulsory for her court to ensure an audience. She founded a permanent professional theatre in St Petersburg in 1756, formed by merging the amateur dramatic society of the cadet corps with Fyodor Volkov's semi-professional company which included the actor Ivan Dmitrevsky, later known as 'the Russian Garrick' after his success in Russia led to trips abroad to examine the work of Garrick, Clairon, and Lekain. Volkov's repertory included the plays of Alexander Sumarokov, an aristocrat who saw in drama the crucial means of social change; his adaptation of *Hamlet* in 1748 oddly makes Polonius, not Claudius, the murderer of Hamlet's father. When a

Cracovians and Highlanders, 1794. According to a contemporary commentator, Boguslawski's Polish comic opera 'aroused tremendous enthusiasm. The play is on a national subject and presents, with much talent, a dispute of peasants . . . The author [is] an expert at playing on human feelings, and at the same time as good a patriot as a playwright.'

play of Sumarokov's was a great success in an amateur production in 1749 the Empress encouraged a repeat performance at court. In 1752 Elizabeth brought the Volkov company to St Petersburg and the Russian Patent Theatre of 1756 was initially run by a triumvirate of Volkov, Dmitrevsky, and Sumarokov.

Catherine the Great, herself a prolific though undistinguished dramatist, continued her predecessor's interest in drama, founding the Imperial Theatre School in 1779 for the training of actors and dancers. Much more significant for the spread of theatre through Russia were her charters of 1762 and 1785 which effectively released the nobility from most of their obligations of state service, enabling them to return to their estates and recreate metropolitan pleasures. The result was the emergence of 'serf theatres'. Over 170 of these were created, of widely differing quality and achievement and with widely differing facilities; the finest had elaborately equipped theatres to rival those in many European cities. The performers were exploited and abused, humiliated and rewarded for their work by their masters.

Sumarokov's tragedies, like *Khorev* (1747), spawned further patriotic, historical epic dramas to a neo-classical formula and others copied the formula of his comedies, firmly based on Molière. Only in the plays of Denis Fonvizin did comedies emerge that would last. Distinctively Russian in their concerns, Fonvizin's comedies gently satirize contemporary manners. Like Sumarokov he mocked Russian Gallomania and the brutality and smugness of provincial gentry. He even, though cautiously, questioned the institution of serfdom itself. His antipathy to the dominance of French culture in aristocratic education is part of his celebration of Russia. It was, in effect, a reaction to the tendency of other dramatists to try simply to Russify foreign drama. Fonvizin's best comedy, *The Minor* (1781), was immediately recognized as a masterpiece, with forty-six productions in twenty years; in it he created numerous character-types that were to be the staple of Russian dramatic comedy for the next century. Only in the 1780s was the developing Russian drama able to explore Enlightenment ideas without using Enlightenment forms.

By contrast with the sometimes tentative efforts of these nations to establish native theatre traditions, Austria found itself blessed with a capital that in Continental Europe was second only to Paris as a cultural centre. The major achievements of Vienna were in music and opera, and the city was to host many of the premières of Mozart's works later in the century. Grand opera and lavish court entertainments attracted stage and costume designers from all over Europe, most

Private theatre, aristocratic splendour. The extraordinary castle theatre at Cesky Krumlov in Czechoslovakia shows the sophistication and splendour of an aristocrat's private theatre. Its ornate decoration and elaborate sets imitate and realize the imaginings of the Bibienas' theatres in Italy.

Design by Giuseppe Galli-Bibiena, 1740. It is impossible to know how much of the extraordinary feeling of space and the receding perspectives could in practice have been created on stage, but such monumental designs established an ideal for the eighteenth century and beyond.

notably the Italian Galli-Bibiena family, four generations of gifted architects and designers, who were responsible for the opera-houses in Dresden and Bayreuth as well as in Vienna. Their soaring stage designs, created with a freer arrangement of painted flats than the conventional symmetrical arrangement of wings and backdrop, offered intriguing angles (the *scena per angolo*) and a monumental quality that was to live on into the nineteenth century and would re-emerge in the spectacular designs of Edward Gordon Craig.

Serious spoken theatre, on the other hand, did not develop with the same vigour in Vienna: Austria was to wait until the next century for its major tragedian, Grillparzer; there was a standing French company in Vienna; Italian troupes were frequent visitors; and even though in 1776 the Burgtheater was declared to be the Austrian National Theatre, it continued mainly to stage imports from Germany, France, Italy, and England, when it was not devoted to opera. The strongest native tradition was in the popular theatre, in the knockabout farce and improvisation of performers like Stranitzky, Prehauser, Kurz, and Laroche. Their comic figures, scatological versions of the *commedia dell'arte* Harlequin, whether called Hanswurst, Bernardon, or Kasperl, entertained generations of Viennese and were of such quality that they were allowed to perform in Vienna's most reputable theatres. However, in 1753 the Empress Maria Theresa, determined to raise the quality of theatrical entertainment in her capital, issued a ban on improvisation and coarse knockabout pieces. Actors were to refrain from 'all indecency and nonsensical expressions'. A first offence would attract a warning; a second offence two weeks' prison; and a third a life sentence. Despite this the popular tradition lived on and indeed flourished in the so-called *Vorstadttheater* (suburban theatres), eventually bearing fruit in the delightful comedies of Johann Nestroy and Ferdinand Raimund and so providing Austria with a genuine native theatre.

It was Austrian domination that doused the brief flickering of an independent Hungarian theatre in the sixteenth century. The proper establishment of a modern theatre derived, inevitably, from the imperial court in Vienna when György Bessenyei, a member of the Empress Maria Theresa's Hungarian Guards, set out a programme to create a national culture, not least through the re-establishment of Hungarian as a valid literary language. A professional theatre in the country was bound to emerge in Buda and Pest as the twin cities acquired greater and greater significance as centres of government.

It was in Germany that the theatre proved to be the strongest focus for national sentiment in the eighteenth century. Here too there was a lively theatre tradition of wandering players, who, deprived of theatre buildings in which to perform, were obliged to stage their plays in inn-yards and market-places. They were constantly on the move in search of new audiences, and had to act in a coarse

Strolling players in Munich. Despite the crude conditions in which they had to perform, the costumes are elegant versions of the court dress of the day, an important visual enticement when the market-place offered so many distractions.

flamboyant style to compete with the other attractions of the fairground or tavern. The favoured presentation was the *Haupt- und Staatsaktion*, in which a historical event was re-enacted, interspersed with comic scenes, usually dependent on the figure of the Hanswurst. Hanswurst was so strongly identified with the crudities of this popular theatre that, in his attempt to reform German theatre, Gottsched arranged for his theatrical colleague, Karoline Neuber, to stage a play in 1737 in Leipzig in which Hanswurst was ceremoniously banned from the stage—with as little success as Maria Theresa had had in Vienna.

Beside this popular tradition the only major source of theatre was in the spectacle of Jesuit dramas, mounted in schools and performed by amateurs in Latin. Without any native German tradition other than that of the strolling players, it is understandable that the first attempts by Gottsched to create a national German theatre led to a slavish imitation of French models, even to the wholly inappropriate adoption of the French alexandrine as the metrical form for his dramas, as in his insufferably tedious *Der sterbende Cato* (*The Dying Cato*, 1731).

Discovering a new impulse for German theatre that did not derive from France was but one of the achievements of Gotthold Ephraim Lessing. In 1759 he published a series of critical essays entitled *Briefe die neueste Literatur betreffend* (*Letters on Modern Literature*), in one of which he denounced Gottsched for his

Comedy with German players. Pantalone (left) reflects the influence of the *commedia dell'arte*, while Hanswurst (right, with his cap and slapstick) is there to make ribald comments to the audience. The small elegant set would have been created by wing-flats and backdrop but anticipates the realism of the box-set.

LESSING'S 17th *LETTER ON MODERN LITERATURE*, 1759

Lessing's early use of 'Genius' is interesting: the ebullient and creative period of the Storm and Stress became known as the *Geniezeit* (Age of the Genius), and the notion that artists are 'geniuses', recipients of special inspiration, was a major feature of the Romantic movement in Europe.

When Karoline Neuber flourished, and many followed the calling to serve her and the German stage, our dramatic literature was indeed in a miserable state. Rules were unknown; there were no models to follow. Our *Staats- und Helden-Aktionen* were full of nonsense, bombast, filth, and vulgar humour. Our comedies consisted of disguises and magic, and slapstick fights were their wittiest moments. . . . And how did Herr Gottsched go to work on this? He knew a little French and began to translate; he encouraged anyone who could rhyme and understand Oui Monsieur to translate as well; . . . he laid his curse on improvisation; he had Harlequin solemnly banished from the stage, which itself was the greatest Harlequinade ever performed; . . . in short, he did not wish merely to improve our old theatre but to be the creator of an entirely new one.

Had he translated the masterpieces of Shakespeare for us Germans—with a few minor changes—I am certain that it would have had better results than being made so familiar with Corneille and Racine. For one thing it would have been much better suited to the taste of the common people; . . . for another Shakespeare would have awakened quite different talents amongst us than the French have succeeded in doing. For a Genius can be set on fire only by another Genius.

clumsy attempts to force German theatre into a Gallic strait-jacket and pointed instead to Shakespeare as a model, a writer who succeeded in combining the depth of thought and beauty of expression of Racine and Corneille with the robust vitality and exciting theatricality of the popular tradition. This recommendation had a decisive effect on two aspects of German playwriting. First, it led to the adoption of blank verse as the preferred medium for serious drama, and Lessing's own *Nathan der Weise* (*Nathan the Wise*, 1779) was the first major German play to be written in blank verse. Secondly, it encouraged writers to discard the constraints of the neo-classical rules and to risk the portrayal of wider historical themes, using a multiplicity of settings and a wide range of characters. This *Shakespearomanie* (Shakespeare mania) was to affect Goethe, Schiller, the writers of the *Sturm und Drang* (Storm and Stress), and indeed virtually all German-language playwrights until the present day.

The director reprimands an actor for improvising. Note the sliding wing-flats and candles, and the casual presence of people in the wings, although they would almost certainly be visible to some members of the audience.

Lessing himself wrote plays set in contemporary Germany, most notably his comedy *Minna von Barnhelm* (1767), in which, in the guise of a love affair between the Saxon title-figure Minna and the Prussian officer von Tellheim, he appeals for reconciliation between two of the hostile states of the Seven Years War (1756–63). At one significant point in the play a foppish French soldier comically attempts to converse with Minna in German, and when in desperation he appeals to her to speak French, she counters: 'Sir, I would seek to speak it in France. But why here?'

German language then was acknowledged not only as a suitable medium for a literary theatre but also as the rallying-point for the German 'nation'. It was in fact the only objective source of cultural identity, and there was, outside the theatre, no national forum where German was spoken—no national parliament, no central court, no equivalent of the Académie Française. The role assumed by the German stage in the eighteenth century is now enshrined in the word *Bühnensprache* (stage-language), the approximate equivalent of the English 'Queen's' or 'Oxford English'. Moreover, the rising class of the bourgeoisie also had no public forum in which they might effect social change or influence the course of political life; so the theatre could be looked on as the one place where moral and social issues might be debated in public.

A national theatre was therefore thought of not merely as a means of raising the quality of German theatre but also as a way of promoting German identity and values. To this end, in 1767 a number of wealthy burghers in the free city of Hamburg embarked on the 'Hamburg Enterprise', the establishment of the first German National Theatre. Lessing, in the role of 'theatre poet', assisted in the undertaking and produced regular articles on the theatre, the *Hamburgische Dramaturgie* (*Hamburg Dramaturgy*), still one of the most important discussions of theatre in the German language. However, the experience of Denmark and Sweden was repeated here. Despite attempts to make the National Theatre popular and accessible (when Lessing's *Minna von Barnhelm* was staged, dancers and acrobats entertained the public between the acts), there was too little public support, and within two years it was forced to close.

The only viable support for theatre came, as with most of the arts, from the many courts of Germany. Initially, German acting troupes were offered a home for the lean winter months and were permitted to tour in the summer. This not only offered performers financial security but also created an environment where they could approach their art with greater seriousness. So in 1753 the great German actor-manager Konrad Ekhof was able to form an academy of actors at the court of Schwerin, at which for the first time plays were read and discussed before being rushed into rehearsal.

Two court theatres were particularly associated with this development of German drama, which was to establish Germany as a major theatre nation in Europe. In 1777, when Duke Karl Theodor moved his court from Mannheim to Munich, he created for his disappointed townspeople the Court and National Theatre of Mannheim. The director of the new theatre, Freiherr von Dalberg, attracted to it a number of leading actors, including the man destined to become Germany's greatest classical actor, August Wilhelm Iffland. He also appointed as writer in residence a poet and playwright of great promise, Friedrich Schiller. While still at school Schiller had written a wordy but exciting Storm and Stress melodrama about two brothers, *Die Räuber* (*The Robbers*). One brother, Franz Moor, is an evil hypocrite who convinces his father that his other son is unworthy of paternal love. The rejected son, Karl Moor, in an act of defiance forms a band of robbers and indulges in wild escapades in the Bohemian woods, finally arriving home to find his father incarcerated by the wicked brother. Karl is reconciled with his father, who dies in his arms, he murders his fiancée from a perverted sense of honour, and finally leaves to find a poor man who will benefit from the reward for his arrest.

The play contains obvious echoes of Shakespeare, especially the rejection and resulting feigned madness of Edgar, and Gloucester's

'Mirror-scene' from Klinger's *Twins*, 1776. Note the flamboyantly theatrical 'Storm and Stress' pose and the dramatic use of off-stage lighting.

treatment at the hands of Edmund. As with Shakespeare and other Storm and Stress writers (Lenz, Klinger, Heinrich Leopold Wagner) who were so under Shakespeare's spell, Schiller filled *The Robbers* with dramatic incident and frequent changes of setting. His play has an episodic quality which can now be traced in a line of descent through Büchner and the expressionist playwrights to Brecht's 'Epic Theatre'. It was Dalberg at Mannheim who, while insisting on cuts and on setting the piece in the Middle Ages instead of in contemporary Germany, was bold enough to stage this adolescent firework. The première took place on 13 January 1782. It began at five o'clock and did not end until a quarter past nine. Many members of the audience occupied their seats from one o'clock to make certain of not missing a piece that, since its anonymous publication the previous summer, had already gained considerable notoriety. An eye-witness reported:

The theatre resembled a madhouse: rolling eyes, clenched fists, stamping feet, hoarse shouts in the auditorium! Complete strangers embraced each other in tears, women staggered almost fainting towards the exits. It was a general dissolution as in the time of Chaos, from whose mists a new creation springs forth!

After this controversial beginning Schiller later formed a productive alliance with the leading author of Germany, Johann Wolfgang von Goethe. Goethe wrote poetry, novels, critical articles, scientific papers, and was a Privy Councillor at the court of Weimar; yet he still found time to write plays and to work as theatre director at the court theatre. Schiller's move to Weimar in 1798 initiated a period of intense theatrical activity, cut short by his premature death in 1805. As Goethe reminisced in 1825:

Just think that the boring period of French taste had only just ended, that Shakespeare still seemed fresh, that Mozart's operas were new works and that Schiller's plays appeared here year after year directed by the author himself . . . I can't deny, it was really something.

In terms of playwriting Goethe contributed little to the development of German theatre. His medieval piece about the robber-baron Götz von Berlichingen of 1773, written in a Storm and Stress intoxication with Shakespeare, was episodic and unwieldy, as was his life's work, *Faust*; neither was suitable for a conventional stage of the time, whatever their other merits. By contrast, his neo-classical pieces like *Iphigenie auf Tauris* (*Iphigenia on Tauris*) (1779, rewritten 1798) and *Torquato Tasso* (1790) were much simpler to perform but, in their observance of the unities and lack of stage action, looked back to French neo-classicism rather than forwards to any innovative dramatic style. Significantly, Goethe seemed to regard his own plays as more suitable for reading than performance and only reluctantly directed them at his own theatre in Weimar, giving preference instead to much more popular and accessible works by playwrights like Iffland and Kotzebue.

Goethe's major contribution to the theatre lay rather in the care with which he mounted productions. Though a strict disciplinarian, he was generous and supportive to his actors, offering them a reasonable remuneration and treating them with a respect unusual at a time when actors were often still regarded as social pariahs; above all, he fostered a genuine ensemble spirit within the company. Goethe would begin work on a play with careful read-throughs of the text and would then, after lines had been learnt, direct his actors with particular attention to the delivery of lines and overall stage-picture. In the case of important pieces like Schiller's tragedies, weeks or even months of discussion took place before rehearsals were embarked upon.

At a time when the growing popularity of the theatre meant that actors hardly had time to rehearse, noting only their entrances and exits, and often did not even succeed in learning their lines, depending on the prompter or on improvisation to carry them through the performance, such discipline and concentration made the theatre work of Weimar a model for the rest of Germany. The formality and overriding aesthetic concerns of Goethe's style found many critics, especially when matched against the realism of actors like Friedrich Ludwig Schröder in Hamburg, famous for the psychological truth of his Shakespearian roles. But, by engaging in a process of education for both his audience and his actors, Goethe made it possible to stage serious verse-drama in German, not only his own plays, but also those of Schiller, and later of Kleist and Hebbel, and the new verse translations of Shakespeare. To have established such an excellent model of theatre practice was no mean achievement in a nation which half a century earlier had been struggling to create a theatre worthy of the name.

It is curious how often facets of the European experience reappear in the development of theatre in America through this century. As in Russia, amateur theatricals led to the establishment of the first theatre, a servant-run theatre in Williamsburg in 1716. Others appeared in Philadelphia (1724), Charleston (1736), and New York (in the 1730s). Soon professional companies emerged: Walter Murray and Thomas Kean formed the first in the country, starting in Philadelphia but performing in New York in 1750. Their initiative was copied by the London Company of Comedians, who dominated American theatre for fifty years, founded in 1752 by Lewis Hallam, Sr., and run from 1758 by David Douglass. As in Europe, this company too responded to growing national pride, renaming themselves the American Company of Comedians in 1763. Sent from London they toured New York, Philadelphia, Charleston, and other towns, establishing and converting theatres up and down the East Coast, including Southwark and the John Street Theatre, later the Theatre Royal, in New York.

But theatre in America found itself in the crossfire of the struggle for independence. The frequent attempts to ban plays, including the 1774 decision by the

Continental Congress, were as much part of the opposition to British imports as moral objections to drama itself. Only in the aftermath of independence did theatre begin to flourish again as a national cultural form. Restrictive legislation was soon repealed and theatres opened across the country, including New England. American drama too had a new beginning: Royall Tyler's *The Contrast* (1787) was the first American comedy professionally performed. From its tentative beginnings, American theatre ended the century as a rapidly growing, socially acceptable art-form.

Given the growing social importance of eighteenth-century theatre, it is unsurprising that one of the major achievements of the century was in raising the status of theatre practitioners. It may have been predictable that strolling players should have been constantly harassed by the authorities. But the Licensing Act in England in 1737 was

An act to explain and amend so much of an act made in the twelfth year of the reign of Queen Anne, entitled *An act for reducing the laws relating to rogues, vagabonds, sturdy beggars and vagrants into one Act of Parliament; and for the more effectual punishing such rogues, vagabonds, sturdy beggars and vagrants* . . . as relates to common players of interludes.

It was not only in hasty legislation that actors were treated as social outcasts. For over a decade Adrienne Lecouvreur had been the brightest star at the Comédie-Française; when she died aged 38 in 1730, public tributes and eulogies poured in from all over France. But the Church would not allow an actress proper burial: her body was taken by the police at night to a piece of waste ground beside the Seine, thrown into a pit, and covered with lime with the earth trampled flat to hide the grave. When, thirty years later in Germany, Karoline Neuber, who had done so much for German theatre, was carried to her grave, the coffin had to be lowered over the graveyard wall, since the clergy would not allow it to pass through the church gate.

In some countries actors were well rewarded. A proposal for an English company in 1703 would have paid the senior male and female actors up to £150 a year while others were to be paid a guinea a performance. But elsewhere actors' wages, except in opera, were only just enough to survive on: the leading actor Ekhof received a weekly salary just enough to buy a pair of shoes. Many female actors, who were usually required to supply their own costumes and accessories, were virtually driven to prostitution, even though a few, like Mlle Clairon, exploited their own exploitation and acquired wealth and influence. In the 1760s London theatres began to create contributory pension schemes and in 1776 an Act of Parliament formalized Drury Lane's benevolent fund to provide some financial security for actors unable to work through illness, accident, or old age.

The situation for playwrights was far worse. In the mid-century shareholders of the Comédie-Française could expect an annual dividend of twenty times the usual payment to authors. In England the system of giving authors shares of the profits on the third and sixth nights meant that a poor play which did not survive that far earned the playwright nothing. Even the great Voltaire sometimes had to waive any fee in order to see his plays performed. It was Beaumarchais who led the playwrights' campaign for fair royalties in France, warning that authors would stop writing for the theatre unless they were properly rewarded. A breakthrough was achieved in 1780 when it was agreed with the Comédie-Française that writers should receive one-seventh of the box-office income, thus creating the basis for the modern system of royalty payments. In England by the end of the century the best hope for substantial payment was selling the play outright either to the theatre company or to a publisher; by the 1790s the theatres were paying anything up to £500 for a full-length play, though the more normal payment was closer to £100.

Brockmann as Hamlet. The immensely popular Brockmann assures himself of attracting attention by seeking a prominent position on stage while the players in the 'mouse-trap' scene in the background are virtually ignored.

As the century progressed actors came generally to be treated with much greater respect and generosity. In Weimar in the 1790s Goethe insisted that actors should be reasonably paid and his performers were treated as acceptable members of court society. In 1779 Garrick was buried in Westminster Abbey; in revolutionary France Talma was a public hero; while in Germany and Austria Franz Brockmann's performance as Hamlet conferred on him the kind of public adulation we would now associate with a pop star, with his image being reproduced on tobacco tins and playing cards.

Perhaps the most remarkable change was the establishment of theatre as an art with its own history. Throughout the century and throughout Europe, writers started to publish histories of theatre, manuals of actor training, descriptions of sets and costumes, biographies of actors and managers, analysis of plays in performance: in England, for instance, works like Charles Gildon's *The Life of Mr Thomas Betterton* (1710), John Hill's *The Actor* (1755), Benjamin Victor's *The History of the Theatres of London and Dublin* (1761), and Francis Gentleman's *The Dramatic Censor or Critical Companion* (1770) are the tip of this iceberg. The century's great theatre theorists like Diderot and Schiller belong in this context. Theatre now commanded serious attention as an artistic and cultural form, a specific art in a culture flowing often awkwardly across the whole of Europe; it mattered to the intellectual community of Europe as never before.

9

NINETEENTH-CENTURY THEATRE

MICHAEL R. BOOTH

[*The increasing storm of lightning, thunder, hail, and rain, becomes terrible. Suitable music. Enter* ROMALDI *from the rocks, disguised like a peasant, with terror, pursued, as it were, by the storm.*]

ROMALDI. Whither fly? Where shield me from pursuit, and death, and ignominy? My hour is come! The fiends that tempted, now tear me. [*Thunder*] The heavens shoot their fires at me! Save, spare, oh spare me!

THE clap of thunder from *A Tale of Mystery*, first performed at Covent Garden Theatre in 1802, peremptorily announced to the new century a new direction in theatre. Thomas Holcroft's play is an adaptation of Guilbert de Pixérécourt's piece for the Ambigu in Paris, *Coelina; ou, l'Enfant du mystère* (1800); the new melodrama was an international phenomenon. Eighty years later Henrik Ibsen, writing to the Norwegian director of *An Enemy of the People*, told him that its staging should reflect truth to nature, 'the illusion that everything is real and that one is sitting and watching something that is actually taking place in real life'. Between these poles of the seemingly ordinary and the undoubtedly extraordinary revolved the theatrical world of the nineteenth century.

Not that this world was created on 1 January 1800, despite the close proximity of *Coelina* and *The Tale of Mystery* to the very beginning of the century. Gothic tragedy with all kinds of romantic and supernatural effects flourished on the London stage in the 1790s, and the proliferation of Parisian theatres after the Revolution had encouraged the rise of a vigorous popular theatre which also contained a strong element of melodrama, spectacle, and Gothic darkness. Even the sacred stage of Weimar was invaded by Pixérécourt's drama of bandits and a heroic dog, *Le Chien de Montargis*, much to Goethe's disgust: a confrontation between the 'elitist' and the 'popular' in a nutshell.

In the nineteenth century these earlier tendencies came to fruition, and for forty years theatre was marked by a strong romanticism, a plethora of scenic effects, an intensely emotional but codified acting style, and a developing stage technology that advanced the art of theatre in the direction of the strikingly visual and grandly spectacular. This theatre—like all theatre—was constantly changing, adapting to new social contexts, new audiences, and new cultural influences. After 1840 it was different in character, but it is necessary to start with melodrama and the means of production because they defined and contained all nineteenth-century theatre; for a century melodrama remained the most popular form of entertainment.

For the art of theatre, melodrama meant that acting style, staging methods, and lighting had to express dramatic content appropriately. The increasing importance of the scene-painter in creating a beautiful and impressive visual environment for the stage action—a phenomenon by no means confined to melodrama—meant that the painterly and pictorial were of marked significance early in the century, a significance that was spectacularly elaborated toward the end of it.

Excluding city streets and, as the century wore on, the growing use of familiar urban topography in a new urban drama, there were basically two kinds of dramatic landscapes in melodrama. One was the romantically sublime, extensively employed in the supernaturally inclined tragedy of the late eighteenth century and the Gothic melodrama of ruins, mountains, forests, rocky gorges, and bleak moors. This kind of landscape expresses its Gothicism through painting, light, and sound. C. E. Walker's *The Warlock of the Glen* (1820) depicts '*The moor at midnight. In the background are the ruins of an abbey surrounded by a few withered trees—the wind is heard at intervals, and the thunder is dying away in the distance. Stage dark.*' The other

David Osbaldiston in *Esmerelda*. This English melodrama was first performed in 1834, and the typical 'penny plain' print emphasizes the posture and gesture of the actor. Though stylized, 'penny plains' and 'twopenny coloureds' provide good visual evidence for the acting of melodrama.

Mr. OSBALDISTON as PHŒBUS. CAPTAIN of the KINGs ARCHERS.
Nº 31.   *In Esmeralda*
Pub. by A.PARK & J.GOLDING, Nº 6. Old St. Road, City Road & 61.Oakley St. Lambeth
Jersey. Publd by G.SKELT, 24, Clearview St. Saint Helier.

was the tranquil rural landscape, domesticated by the cottage, the church spire, the harvesters in the field. These landscapes are pretty, gentle, inviting, bathed by the sun and washed by the limelight moon; they represent the domestic and romantic landscape at rest, before the turmoil of melodramatic action intrudes upon rural quiet and the harmony of nature. And it does so intrude: W. T. Moncrieff's *The Lear of Private Life* (1820) opens on a '*Rural Landscape, Corn-fields &c. &c. forming the estate of Fitzarden . . . Male and Female* REAPERS *discovered at work*,' said reapers singing a song in praise of their master. Some time later, however, Fitzarden's daughter, fleeing with her child from her seducer, encounters her now insane father in '*A Forest. Lights down and snow. Evening.*' The '*storm rises*'. Fitzarden gibbers wildly, the snow falls more heavily, and the demented father tries to strangle his distracted daughter—a long way, this scene, in terms of the physical and emotional landscape, from the opening harvest idyll, but only a short distance on the road of melodrama.

For these scenes are not just landscapes in the theatre, prettily painted back-cloths and ground rows and wings enclosing the dramatic action. They are also symbolic of the emotional states of melodrama, its violent alternation between repose and turbulent animation, the conflict between virtue and villainy—mental and spiritual landscapes expressive of melodramatic content and style as well as physical landscapes of the countryside. The forest, in particular, that ancient symbol of danger, the unknown, the mysterious, the sheer difficulty of man's journey through life, is employed in a visual way that encapsulates the whole genre.

The pictorial skills of the scene-painter were combined with those of the carpenter, the machinist, and the gasman to create the strong visual images that dominated the nineteenth-century theatre. Until the advent of the built-out three-dimensional set and the mechanized stage in the last quarter of the century, the theatre depended upon a combination of movable wings, shutters, drop-cloths, and ground rows to give dramatic settings their desired exterior locales. These canvas and wood surfaces were painted in perspective to convey a scenic environment. Some scenery was flown: that is, dropped and raised by a system of drums, ropes, and pulleys operated by stage crew in the fly galleries on each side of the stage; other scenery was raised by drums and winches from below the stage by operatives on what was called the mezzanine. In England and America the wings and shutters slid in grooves attached to the floor of the stage and to the underside of the lower fly galleries; Europe employed the chariot and pole system, whereby long poles from which scenery was hung went through lateral slits in the stage floor and were attached to trucks moving on rails in the mezzanine. Until late in the century all operations were manual, and the backstage technical staff for a big spectacle production could number in the hundreds.

This traditional scenic system served a theatre whose public knew perfectly well that they were watching a play and did not insist on the illusion of reality. As in

A Victorian scene-change. The drawing of a pantomime in progress, c.1860, is by the scene-painter William Telbin. The illustration, of an unidentified (and probably imagined) theatre, is richly revealing of stage technology: sliding flats, upper and lower grooves, gas battens, wing lights, and limelight are all in evidence.

the eighteenth century, though much more elaborately, scenery was a selective aesthetic representation of the world, not a replication of it. In the second half of the century such methods seemed to many inadequate and unrealistic; the idea of realism in art was growing stronger. Eventually the box-set, with walls and a ceiling for interiors—introduced around the turn of the century in Europe, later in England and America —and solid three-dimensional scenic structures for exteriors, became standard; scene-changes took longer and had to be concealed from the audience by the increasingly frequent use of a drop-curtain.

The eighteenth-century theatre had been lit by candles and oil-lamps, and lighting played no part in the creation of a stage environment that the audience was to take for reality. Except for romantic effects, light merely provided illumination so that audiences could see what was going on in the auditorium as well as upon the stage; it had no further purpose. In the nineteenth century all this changed with a changing technology. Gaslight, introduced to the stage and auditorium of major theatres by the 1820s, took some time to be generally adopted: not till 1857 at the Royal Theatre, Copenhagen, for instance. The English developed limelight for the stage in the late 1830s; this was followed by the electric carbon arc, first used in Paris, in the 1840s. Both these instruments were hand operated and could be focused by means of an attached lens, thus giving the theatre focused light for the first time. It was impossible to focus the main source of nineteenth-century stage lighting, gas, which blazed in hundreds of naked flames from footlights, wing lights, overhead gas battens, and various portable instruments. From the 1880s theatres were gradually electrified—cautiously, since back-up gas systems were installed in new theatres for years afterwards. (In 1881 the Savoy Theatre in London was the first to have a fully electrified lighting system.) The use of gas enabled scenery and actors to be lit much more brightly than before, thus encouraging more realistic scenic detail and a subtler, more realistic acting style. The use of limelight, which could be beautifully coloured by means of painted glass slides, stimulated romantic effects of all kinds: sunrises, sunsets, moonlight on water, reflections from jewelled armour and from spangles, glass, and foil sewn into rich costumes and painted scenery. Spectacle and sensational scenes in melodrama would have been almost inconceivable without limelight.

These, then, were the basic tools of the nineteenth-century theatre. They were essential in every kind of production, from one-act farce with a single interior set to spectacular pantomimes or historical dramas with huge casts and numerous heavy scenes. In the early melodrama and the romantic tragedy, the wild and forbidding landscape was a common accompaniment to the emotional agonies of the hero, heroine, and villain. If the scene were interior, darkness and mystery were stressed. '*Night. A gallery in a convent. A large Gothic window in the extremity, through which lightning is seen flashing. Two monks enter in terror*' is the fitting opening scenic direction for Charles Maturin's tragedy *Bertram* (1816). The new taste for the sublime, for the historical past, and for Gothic architecture powerfully affected the theatre.

The part of the brooding and fated Bertram in Maturin's play was undertaken with great success by the leading tragedian on the London stage, Edmund Kean. Kean, who had acted in melodrama for many years before his sensational London début as Shylock in 1814, has too often been regarded as a unique phenomenon, an actor springing, as it were, in full motley from the brow of some great theatrical Zeus. In fact, though blessed with a great individual talent, he was a product of his time, the Byron of the stage (Byron much admired him), an actor for romantic poets, romantic critics, and a romantic age. His physical intensity, his abrupt transitions of mood, his violent (though carefully controlled) expressions of emotion, and his famous point-making were drawn as much from melodrama as from current tragic theory and practice. He was the bodily equivalent of the thunder and lightning, the darkness and terror striven for in melodramatic stage effects.

Kean is best described by a sympathetic

Edmund Kean as Sir Giles Overreach. The sketch by George Clint is of Kean in one of his most famous parts, in Massinger's *A New Way to Pay Old Debts*, and shows something of his terrifying passion and intensity.

Rachel as Phèdre and Camille. Two of Rachel's leading tragic parts: in Racine's *Phèdre* (*left*) and Corneille's *Horace* (*right*). While the poses are static, something of Rachel's famous intensity can be seen in Phèdre's tight, slumped angularity and the disintegration of line and poise in Camille.

critic like William Hazlitt or a later student of acting, George Henry Lewes. Recalling Kean's Othello, Lewes wrote of the third act:

When shall we see again that lion-like power and lion-like grace—that dreadful culmination of wrath, alternating with bursts of agony—that Oriental and yet most natural gesture, which even in its *naturalness* preserved a grand ideal propriety . . . that exquisitely touching pathos, and that lurid flame of vengeance flashing from his eye?

In an interesting essay Lewes compared Kean with Félix Rachel, France's greatest tragic actress of the nineteenth century, who made her début with the Comédie Française in 1838 and, like Kean, died young—in his case from drink, that common affliction of the nineteenth-century actor, and in hers from overwork and tuberculosis. Rachel single-handedly brought back stature and passion to the leading female parts of Racine and Corneille, and revived the French classical

tragedy for Parisian audiences of the 1840s who had been thriving on Victor Hugo and Alexandre Dumas. Lewes called her 'the panther of the stage; with a panther's terrible beauty and undulating grace she moved and stood, glared and sprang'. Kean was the lion, and Rachel 'was very much as a woman, what he was as a man'. Her elocution, in her early days, was musically exquisite; her 'thin and nervous frame vibrated with emotion'; her face was capable of intense expression, and her gestures were fluent and graceful. Lewes's description of her Phèdre, her most famous part, is vivid: 'What a picture she was as she entered! You felt that she was wasting away under the fire within, that she was standing on the verge of the grave with pallid face, hot eyes, emaciated frame—an awful ghastly apparition.' It was an incandescent performance, one of the greatest in the history of acting, comparable with Sarah Siddons's Lady Macbeth and, Lewes felt, Kean's Othello; like Kean 'she had a power of concentrating into a single phrase a world of intense feeling; and even Kean himself could not have surpassed the terrific exclamation— "Misérable! et je vis!"'

Whether Rachel was a romantic artist is really irrelevant. One cannot, for example, classify Mrs Siddons's Lady Macbeth or categorize her as a 'classical' or 'romantic' actress. Like Kean and Siddons, Rachel excelled in tragedy, a performer of immense emotional power and intensity, and wonderful technique, acting in a century distinguished above all others by its tragedians.

The larger than life and (to us) powerfully extravagant strain of tragic acting that marked the late eighteenth and early nineteenth centuries, in melodrama as well as tragedy, was also found in America in the performance of Edwin Forrest, who is interesting not only for his acting but also for his repertory. Forrest was of that first generation of native-born (as opposed to British-born) American stars that also included Charlotte Cushman, James Hackett, and Edwin Booth. He first acted professionally in Philadelphia in 1820; his career lasted until 1872 and he was at the height of his powers between 1835 and 1855.

Edwin Forrest as Metamora. This engraving is from a painting by Frederick Agate. The actor, in a heroic pose, is set against a wild and romantic mountain setting, a suitable visual context for the heroic and romantic elements in the play and in Forrest's acting of the part.

Facing: Hugo's scene design for *Hernani*. Victor Hugo did his own watercolour designs and ground plans for *Hernani*; this is Act IV. Taken together, his designs for all five acts convey a strong sense of a brooding Gothic Romanticism appropriate to the play and the theatrical tastes of the avant-garde French dramatists of the time.

Forrest was a stocky barrel-chested actor with immense strength and a huge voice; though not tall he seemed a giant on the stage. At the première of his first great early success, John Augustus Stone's *Metamora* (1829), one critic described his voice as 'tremendous in its sustained crescendo swell and crashing force of utterance. His voice surged and roared like the angry sea; as it reached its boiling, seething climax, in which the serpent hiss of hate was heard, it was like the falls of Niagara, in its tremendous down-sweeping cadence; it was a whirlwind, a tornado, a cataract of illimitable rage.' When Forrest played Spartacus in Robert Montgomery Bird's *The Gladiator* (1831), his tense muscularity, burning eyes, and overwhelming energy astonished spectators. His eyes, in particular, rolled 'like the dilated and blazing eyes of a leopard, now white and set like the ferocious deathly eyes of a bull, while smothered passion seemed to threaten an actual explosion of the whole frame'.

The character of Metamora is that of a fearless Indian chief, savage yet noble, who tries to defend his tribe's territory from white encroachment and dies in the attempt. Significantly, James Fenimore Cooper's *The Last of the Mohicans* had been published in 1826. Forrest as Metamora (and Spartacus) spoke heroically and rhetorically of freedom and the cause of right against tyranny. In doing so he appealed to the deepest political instincts of his audiences in the new age of Jacksonian democracy: the populist Andrew Jackson was inaugurated as President in the same year as *Metamora* was first performed. The thunder of Forrest's voice boomed through such speeches as the following: 'I started to my feet and shouted the shrill battle cry of the Wampanoags. The high hills sent back the echo, and rock, hill, and ocean, earth and air, opened their giant throats and cried with me, "Red man, arouse! Freedom! Revenge or death!" [*Thunder and lightning. All quail but* METAMORA.]' In its concept of scenic environment as well as heroic individualism *Metamora* was profoundly romantic, and the scene directions reflect the practice of the contemporary Hudson River school of landscape painters with their mountain peaks, mighty forests, tumbling waterfalls, and sublime sunsets.

The last great explosion of Romanticism in the nineteenth-century theatre came in France in 1830 at the end of a decade in which English Gothic melodrama and melodramatic adaptations of the novels of Sir Walter Scott were at the height of their popularity. All through this decade in France there had been a struggle against the authority of the Comédie Française over the repertory. In England and America melodrama had invaded the principal theatres years before, but in Paris the Comédie still tried to maintain the purity of its classical repertory and encourage traditional modes of tragic writing in new playwrights. The ever-vigorous popular melodrama was still largely confined to boulevard theatres like the Ambigu, the Gaîté, the Vaudeville, the Variétés, and the Porte-Saint-Martin; and the lifting in 1815 of Napoleon's restrictions on the number of theatres permitted to perform a popular repertory led to an increase in the number of such theatres.

Hernani — Acte 4ème — Plantation

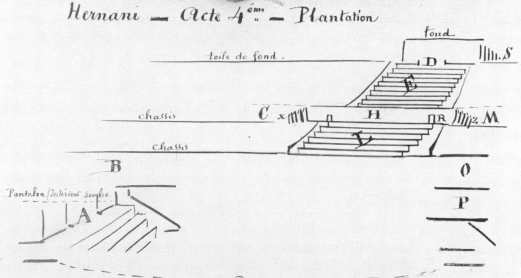

A. Porte du tombeau de Charlemagne. B. O. P. C. M. Coulisses. H. Palier.
S. X. Z. Escaliers pour monter de la coulisse. R. Petit socle ou pilier Sur lequel Ricardo pose
sa lanterne.

The riot at *Hernani*, 1830. The stage shows the end of the play, the auditorium the battle of the young radicals against the older classicists. The prompter's box down-stage centre is just visible; in England the prompter took his position off-stage left.

The death of Talma in 1826, the strict theatrical censorship instituted by Charles X, and the growing *embourgeoisement* of French culture were all factors contributing to a theatrical revolution which culminated in the performance of Hugo's *Hernani*, a play of love, honour, and revenge, at the Théâtre Français in 1830, not by coincidence the year of the revolution that forced Charles from the throne.

The famous first performance of *Hernani*, which was attended in force by both conservatives and radical writers and artists, and which resulted in a full-scale riot, was certainly a sensational event in the annals of theatre but has somewhat obscured the real issues. The conservatives bitterly opposed the liberties which Hugo took with long-established Comédie conventions of writing and staging, and the radicals vigorously supported them. Such liberties included violating the unities of time and place, showing violence and death on stage, mixing humour with tragic seriousness, and using words, phrases, and verse structures allegedly beneath the dignity of tragic vocabulary and style.

Nevertheless, although in many ways a definite break from the theatrical past, *Hernani* was a climax rather than a revolutionary beginning. The radical literary and dramatic opposition to establishment classicism had already applauded Dumas's *Henri III et sa cour* in 1829 at the Comédie, a romantic historical pageant influenced by Scott and Schiller, and Alfred de Vigny had adapted *Othello* as *Le More de Venise* in the same year with the same company. He would not have done so without the sudden fashion for Shakespeare, who enraptured the romantic radicals in the late 1820s. Although Shakespeare had previously been presented in translation and an English company had made an unsuccessful attempt to convince doubting and hostile Parisian audiences in 1822 of the merit of *Othello*, the real vogue for Shakespeare began with another visit by an English company in 1827, this time led by the eminent Charles Kemble. He offered *Hamlet, Romeo and Juliet*, and *Othello*. A few months later, in 1828, William Charles Macready arrived

Hamlet in Paris, 1827. Charles Kemble as Hamlet reclines at the feet of Harriet Smithson's Ophelia in the play scene. The tension of the scene is apparent in the contorted position of Hamlet and upon the faces of other characters, especially Gertrude and Claudius on the right of the picture.

from London with *Macbeth*; then Kean with *Richard III*, *The Merchant of Venice*, and *King Lear*.

This substantial exposure to Shakespeare on the stage overwhelmed many French spectators. A great number of them, understanding English imperfectly or not at all, focused on the acting and physical business, and were much struck by scenes of madness and violence actually placed upon the stage. Hector Berlioz was so overcome by Harriet Smithson—in England a relatively minor actress—as Ophelia and Juliet that he courted her passionately until their marriage in 1833. Indeed, la Smithson created a sensation. Dumas, who knew no English, exclaimed in his *Memoirs*, 'O Shakespeare, merci! Kemble et Smithson, merci! Merci à mon Dieu! merci à mes anges de poésies!' and claimed that when he went to *Hamlet* it was the first time he had seen men and women on stage animated by real passion. The revolutionaries perceived that productions of Shakespeare were not an end in themselves, but a means to the desired end of overturning the establishment theatre. The critic Sainte-Beuve said of the *Romeo and Juliet* that it was a weapon of war as well as a noble spectacle.

The new tragedies of the French romantics were close in feeling and plotting to the emotions and situations of melodrama, and they made admirable material for the composers and librettists of opera. One of Verdi's early works was *Ernani*, and he, too, was smitten with Shakespeare. *Rigoletto* is based on Hugo's *Le Roi s'amuse* and Donizetti's *Lucrezia Borgia* upon his *Lucrèce Borgia*. It is worth noting that Beethoven's Third and Fifth symphonies were given their first Parisian performances in 1828. Shakespeare and Smithson together inspired much of Berlioz's music, including a funeral march for *Hamlet*, a *Mort d'Ophélie*, and the opera *Roméo et Juliette*.

The performance of *Hernani* in 1830, with its attendant riot, is one of the focal points of nineteenth-century theatre history. Not only did it represent the rejection in France of the classical traditions of the past, but also the triumph of melodrama, now elevated to the status of literature and five-act tragedy and written by distinguished men of letters. It was also the peak of Romanticism in the theatre, with all that meant in terms of dramatic content, production method, and acting style. After *Bertram*, *Metamora*, and *Hernani*, after Kean, Forrest, and Harriet Smithson on the stage of the Odéon, after the wild mountains, desolate moors, dark forests, and Gothic chambers of the scene-painters, Romanticism had nothing new to offer the theatre.

This last statement needs some qualification. The growing middle-class hold over entertainment in England and America, and the new strength of the bourgeoisie in French society and culture, meant that eventually most forms of theatre were oriented towards middle-class audiences and their concerns. Melodrama, as popular with the middle class as with industrial and urban working-class audiences, now stressed domestic matters and offered audiences reproductions of the family parlour, the poor garret, the prosperous villa, the low tavern, the city street, the bank, and the shop counter. Thus was the material made manifest in scenic verisimilitude; exactly the same approaches to realism were being made in the novel. Yet there still existed a need for romantic fantasy, a need that strengthened as the century wore on, resulting in the hugely elaborate late Victorian pantomime, the Parisian *folie féerie*, and an American music spectacle melodrama like *The Black Crook* (1866). As bourgeois materialism and the art of daily life grew more prominent in the theatre, so did its opposite, the world of fairies and spirits. With the immense popularity in translation of Hans Christian Andersen from the 1840s (a decade earlier it was the brothers Grimm), the new fairy ballet (*Giselle* and *La Sylphide*), the fairy opera (Wagner's *Die Feen* and Lortzing's *Undine*), productions of *The Tempest* and *A Midsummer Night's Dream* that emphasized the fairy element, fairy poetry, and fairy paintings, it is not surprising that fairies thronged the stage of the most spectacular fantasy entertainments that England and France, especially, could offer. Was all this a variety of Romanticism? Suffice to say that at the very least the fairy cult had a lasting effect upon the content and production techniques of stage spectacles.

The 1840s appear to mark a turning-point, or a dividing line, between the old and the new in the theatre. In Paris the triumph of Romanticism was short-lived, and romantic tragedy gave way to the bourgeois prose dramas of Dumas *fils* and Émile Augier. In London melodrama transformed its Gothic and nautical stereotypes into a domestic drama dealing with the problems and struggles of contemporary life. In the German-speaking theatre new 'directors' emerged, such as Heinrich Laube at the Vienna Burgtheater and Franz von Dingelstedt in Munich and Weimar. In Italy front-rank tragedians appeared, on the road to stardom and international reputations: Adelaide Ristori, Ernesto Rossi, and Tommaso Salvini. In America two key events mark the middle of the century: the Astor Place riots in New York in 1849 and the George Aiken adaptation of Harriet Beecher Stowe's great anti-slavery novel *Uncle Tom's Cabin* in Troy, New York, in 1852. Both events had profound social and political implications, and both illuminate the course of American theatre.

The Astor Place riots pitted the supporters of Edwin Forrest, then playing a season at the Bowery Theatre, against the English actor Macready, performing at the Astor Place Opera House. There was already a considerable enmity between Forrest and Macready, since on a recent tour in Britain the former had publicly hissed the latter, from a box, during a performance of *Hamlet*. The Bowery audience and their sympathizers, very much working and lower middle class, combined with anti-British agitators in the city to attack the upper-class Opera House, its culturally effete, pro-British audience, and the middle-class colonially minded Macready. Or so they alleged. On the second night of the disturbances a violent assault was actually launched against the theatre, whose audience and

The Bowery Theatre, New York, 1878. The nineteenth-century theatre auditorium was much livelier than its modern counterpart. Here, a disturbance on stage and in the auditorium illustrates an intimate involvement of performers with audience, even in a large theatre.

Macready fled after the conclusion of *Macbeth*; the hastily summoned army opened fire. Thirty-one of the mob were killed and thirty-eight wounded.

The significance of the riots was twofold. First, they were an assertion of the rights of the Jacksonian 'common man' against the social and cultural domination of a middle-class power-structure. Forrest took no part in the riots but served, given the characters of his best parts, his powerful stage rhetoric, and his well-known democratic and patriotic tendencies, as a figurehead for the rioters. Secondly, the attack on the Opera House was a rejection of what has become more recently known as 'cultural cringe'. In 1849 only a small proportion of plays performed in major American theatres were by American authors. The most popular playwrights were all British, and plays were imported wholesale from Britain, as were actors and touring stars like Macready. The Astor Place rioters asserted their native Americanism in their own way. It took a long time after that, but at least their turbulent affirmation of American patriotism and the need for a separate American theatrical identity were a striking part of a process that ultimately bore rich and satisfying fruit in the twentieth century, with a uniquely American theatre.

Uncle Tom's Cabin on stage was also part of this process, which had really begun with a character like the idealized Metamora, a native American hero, at least in the play. The Christian and saintly Tom is not a hero whom Forrest could have played, but the persecution of the black man, Eliza's thrilling flight across the ice of the Ohio River, George Shelby's brave defence of his family, and the monstrous villainy of Simon Legree, all eminently theatrical in nature, made up the content of the most popular melodrama ever written. There were at least twelve dramatic versions of the novel, including Aiken's. Touring shows playing nothing but *Uncle Tom's Cabin*, usually in tents, were on the road by 1854; by the 1890s there were some 400 of them, and many incorporated minstrel show material into their productions. The popularity of the play and these tours lasted well into this century. Both novel and play were prime instruments in the campaign for the abolition of slavery, and together with the minstrel show and a play like Dion Boucicault's *The Octoroon* (1859), which also presents a slave auction on stage, established the racial images and stereotypes of the American South for a century to come.

Uncle Tom's Cabin also relied heavily upon the arts of the stage carpenter and the scene-painter to portray the heavenly reward awaiting Tom and Eva. The

Eliza flees in *Uncle Tom's Cabin*. In William Brady's spectacular 1901 production, Eliza crosses the ice of the Ohio River to freedom ahead of the pursuing slave dealers. The sky borders, the cut-out canvas branches, and the painted backcloth are clearly visible.

stage iconography carries the religious symbolism of the Assumption. The ending of the play, in whatever version, is pictorially notable. Here is the wordless finale for the Aiken version: '*Gorgeous clouds, tinted with sunlight. Eva, robed in white, is discovered on the back of a milk-white dove, with expanded wings, as if just soaring upwards. Her hands are extended in benediction over St Clare and Uncle Tom, who are kneeling and gazing up to her. Impressive music. Slow curtain.*' Such a scene was not confined to *Uncle Tom's Cabin* or American melodrama. The English stage knew it well in those infrequent melodramatic instances when suffering virtue perishes rather than receiving an earthly reward, the most famous example being the ending of some versions of *East Lynne*, where poor Isabel passes gloriously into heaven and is greeted by the outstretched arms of her dead son William.

The stage negro was, of course, not at all close to the real negro. He was used for colour, sentiment, and melodrama, with very little realism and no serious examination of racial problems or his actual way of life. There was nothing 'real', in this sense, about Metamora either. Much American melodrama of the second half of the nineteenth century propagated popular myths about American culture and American character. With the expansion of the country westward, this kind of melodrama was markedly different from its contemporary English equivalent, with its primary urban and domestic concerns—although the latter type was also widely available in America. By 1880 only 32 per cent of the American population lived in cities, as compared to nearly 80 per cent in Britain. Yet by mid-century 80 per cent of American theatres served urban audiences, and it was these audiences who thrived on melodramatic myths about the pioneer, the frontiersman, the miner, and the Indian. Plays incarnating such myths abounded. In Frank Murdoch's *Davy Crockett* (1872) the heroic backwoodsman saves the heroine's life by barring the door of a cabin with his arm, after the bolt has broken, against the ferocious wolves outside. The hero of Bartley Campbell's *My Partner* (1879) is a tough but gentle miner with great natural dignity. The myths of the West were especially powerful in the theatre, enduring in the plays of David Belasco, whose *The Girl of the Golden West* (1905), another mining-camp drama, ends with an especially gorgeous slow sunrise, a triumph of Belasco lighting technology. These myths passed effortlessly into the western film, as did the myths of the Civil War: William Gillette's *Held by the Enemy* (1886) and Bronson Howard's *Shenandoah* (1888) spawned a generation of Civil War plays, including Gillette's thrilling spy drama *Secret Service* (1895).

As is evident, the boundless western horizons of mid-century America left their thematic mark on the content of American drama and popular paratheatrical entertainment like Buffalo Bill's Wild West shows. They also shaped an extraordinarily vigorous and varied touring system. The American continent was crossed by rail in 1869, and that event, together with the earlier spread of railways in the east and south of the country, enabled touring stars and companies to travel with

The Park Theatre,
New York, 1822.
A typical early nine-
teenth-century theatre
auditorium:
proscenium doors still
in use, boxes gracefully
curved to join the
proscenium, pit
audience fairly close to
the actors, who are
playing downstage to
establish contact with
the audience and catch
as much light as
possible from the foot-
lights and auditorium
lights.

relative ease if a great deal of fatigue. This was also the case in the English
provinces, although of course the distances were far shorter. The basic structure
of the English theatre outside London from the late eighteenth century to the
1860s was the provincial Theatre Royal and the circuit company. The latter, often
based on a Theatre Royal in a provincial city like Bristol, York, or Norwich, might
comprise anywhere between two and twenty-two theatres. The manager moved
his company around his circuit, or divided it if he had the resources and if the
circuit were big enough, playing a short season in each town. This system was in

The stage western, 1893. Set-piece battles were popular on the melodramatic stage. This drawing shows the timely arrival of the US army to save the fort from Indian attack in Belasco's *The Girl I Left behind Me* at the Adelphi, London. The effect depends upon a skilful combination of mass and emphatic gesture.

decline after 1815, but it was the spread of railways, in both England and America, that ensured the rapid destruction of the circuits and the local stock companies, and the concomitant proliferation of the touring companies that replaced them. By the 1880s and the age of passenger steamships, international touring, especially to North and South America, became a significant source of revenue for major companies.

Not all actors travelled, or could afford to travel, by rail or by ship. Many English fit-up and strolling companies travelled on foot, with their scenery and properties in a cart, just as they had in Elizabethan times, playing in barns, assembly rooms, fairground booths, tents, and taverns. Such companies, sharing a frequently meagre box office, were commonly poor and sometimes close to starvation. 'The best judges of turnips is strolling professionals,' a travelling clown told the early Victorian journalist Henry Mayhew (alluding to hungry strollers hurling themselves on turnip fields), and a positively paradisiacal 'mummer's feast is potatoes and herrings'. In the United States, especially in the first half of the century when communications in many parts of the country were bad, actors travelled by horse, wagon, riverboat, and Shanks's pony. Often they traversed unsettled and sometimes dangerous territory and performed in primitive conditions. The pioneer touring manager Sol Smith recalled playing his way from Rochester to Cincinnati (then a small outpost on the western fringes of the country) in the 1820s, finally arriving by the Allegheny River, keeping a sharp look-out for river pirates, at a new settlement called Lewiston. His company acted in a log cabin with three rows of benches for the audience and home-made candles stuck into large potatoes for footlights. The Samuel Drake Company, travelling by wagon towards the Allegheny in New York state in 1815, lost one of its actresses who had wandered into the forest and did not find her again till the next day; she had spent the night in the crotch of a tree with wolves below her. Touring to the mining camps and 'opera-houses' of California, Colorado, and Nevada later in the century could be just as arduous and theatre venues just as rough and ready.

The audiences of these mining camps, bush towns, and river settlements were not, on the whole, members of the educated middle class. Neither were the audiences of the huge new theatres of the East End of London, such as the Britannia, the Pavilion, the Standard, and the Grecian, which catered almost entirely to a neighbourhood working and lower middle class. The Parisian melodrama audiences of the Boulevard du Temple were, in class terms, positioned well below the audiences of the Comédie Française. The taste of all these audiences tended to be broadly 'popular': they enjoyed the melodramatic clash between virtue and vice, physical sensations and physical skills, songs, stirring action, low comedy, fine rhetoric, and grand sentiments. One has to be very careful in generalizing about nineteenth-century theatre taste, for middle-class and working-class audiences enjoyed many of the same things. Shakespeare, for instance, was given successfully in the East End of London, and from October 1851 to December 1854 the new gold-rush city of San Francisco saw fourteen Shakespeare plays. Middle-class audiences, however, also liked middle-class comedy, satirical burlesque, comic opera, and a less vigorous acting style, more realism of setting, character, and language, and more creature comforts in the theatre. Certainly differences in class taste existed; nevertheless, before the advent of cinema and then television, the nineteenth century was the last time when the mainstream theatre catered to all social classes and all income levels, and the last time when theatre was a mass market entertainment. The presence of a huge working class (80 per cent of the population of London in the 1890s, for instance) in all industrially advanced countries meant the wholesale provision of theatre buildings and the rapid writing and rehearsing of plays in their thousands. Some have defined 'popular' in class terms, as a kind of theatre appealing to audiences low on the social scale. Others have taken 'popular' to imply the support of sheer numbers of people, and in this sense the nineteenth-century theatre was indeed popular. Still others have defined 'popular' according to characteristics of repertory, acting style, or types of entertainment either unavailable or uncommon in conventional theatre buildings. Such entertainment would include the cut-down Shakespeare, melodrama, and

PAVILION

THEATRE, WHITECHAPEL ROAD.

FALSE EARL

OR, THE ROVER OF THE NORTH SEA!

The Rescue of Emma Deane.

Julian's Discovery of his Mother.

NOW PUBLISHING In the HALFPENNY MISCELLANY.

Julian Overpowered by the Pirates.

TO CONCLUDE WITH THE MISER OF SHOREDITCH!

Mrs. Glendower imploring the Aid of Sir William Deane.

Captain Grunker commands Julian to Walk the Plank.

EVERY EVENING!

E. J. BATH, Printer, 42, Leman Street, Whitechapel.

pantomime offered in theatrical booths at fairs, melodramas (and Shakespeare) staged in circus rings with trained horses, street shows, spiritualists, mimes, magicians (who drew huge audiences with elaborate presentations), *tableaux vivants*, music hall, vaudeville, and Wild West shows.

In the larger British cities, especially London, music hall became a serious rival to theatre in its power to attract audiences. At first these audiences were predominantly working and lower middle class, with few women, since the halls had developed from the saloon entertainments and all-male drinking and music clubs of the 1830s and 1840s. In the 1850s and 1860s the first purpose-built music halls, as differentiated from adaptations of taverns or concert halls, were opened. At first such halls existed primarily to sell food and drink, with entertainment thrown in as an incentive; the profits came from the bar. Typically, these halls—like the Canterbury in Lambeth and the Middlesex in Drury Lane—had a simple stage at one end, like a concert platform, that employed little or no scenery; the floor of the rectangular unraked hall was filled with long tables and chairs for the drinking, smoking, and eating patrons (none of these activities being officially permitted in the theatre). A chairman presided at a sort of head table, armed with a gavel, to announce the artistes, keep order, and provide a jocular commentary on the proceedings. Toward the end of the century new music halls in the West End and

Facing: Pavilion Theatre poster, 1863. The Pavilion was a large theatre in the East End of London. The scenes depicted are typical of the content and performance style of melodrama staged for neighbourhood working- and lower middle-class audiences.

Below: an English music hall, 1874. The drawing by Alfred Concanen is an artistic fiction, but closely based on fact. Eating and drinking are just as important as listening to the artiste, a comic singer. The chairman can be seen with his gavel.

the suburbs eliminated tables, chairs, and chairman; the entertainment was now profitable in itself. Gentrification set in; audiences became middle class and even fashionable, and the acts were more tasteful and often more spectacular. Huge and elegant variety houses like the Coliseum (1904) and the London Palladium (1910) were the culmination of this process.

American vaudeville grew out of beer halls and did not develop as a separate organized form until the 1870s and 1880s. In the West it answered a desperate need for light entertainment amongst audiences of miners, lumbermen, cowboys, and gunfighters; and in the eastern cities it soon became big business, popular with respectable families and single men alike. Like British music hall, it began dying with the advent of the talking film in the 1920s. The pattern of a music hall or vaudeville evening was much the same: separate turns to orchestral accompaniment by comedians, serious and comic singers, jugglers, dancers, magicians, trick cyclists, etc., all structured in a 'bill'—which could be given twice nightly. The programme was carefully composed for balance, diversity, and audience appeal, but the turns did not relate to each other in style and content. They were completely presentational; that is, right out at the audience, an audience that was severely critical of poor work, smoked, drank, talked, and wandered in and out of the auditorium. In the later variety and vaudeville palaces, audience behaviour, like the acts themselves, was more refined. Both in the audience and on the stage the nineteenth-century music hall or vaudeville house had a unique atmosphere and immense character; it was truly 'popular' entertainment.

So was the minstrel show, American in origin but popular also in Britain and France. The negro impersonator T. D. Rice sang and danced his famous 'Jim Crow' number in Kentucky in 1828; after that it became the rage in American theatres everywhere. Burnt cork on his face, dressed in ragged, patched clothes and a white hat, holes in his shoes, the caricature of a plantation slave, Rice hopped and wheeled grotesquely about the stage as he sang his strange ditty. Rice was not a black-faced minstrel, but from Jim Crow to the minstrel show was only a short step. It was taken in the early 1840s: the Virginia Minstrels performed at the Bowery Theatre, and in 1846 the Ethiopian Serenaders caused a sensation when they appeared in London (Rice had toured Britain in 1836 and 1837) and performed before Queen Victoria. They were the first of many minstrel groups to tour abroad, the best known of these probably being the Christy Minstrels. Soon the English at least developed their own black-faced troupes. The appetite of English audiences for the minstrel show was prodigious, and it survived well into the twentieth century. Artistes crossed the Atlantic both ways, and minstrel songs and dances invaded the music hall, pantomime, and burlesque comedies. For example, in *The Enchanted Isle* of 1848, an English parody of *The Tempest*, Prospero plays the bones and accompanies Miranda in a minstrel song '*à la Ethiopian Serenaders*'. Seaside minstrels and the extraordinary longevity of the Black and

White Minstrels at the Victoria Palace in London and on BBC television in the 1960s and 1970s were the last gasp of this cultural phenomenon in Britain. In America growing racial sensitivity as well as changing taste had put a stop to minstrel shows years before. Minstrelsy was at its height there between 1850 and 1870; during the 1850s ten theatres in New York were devoted entirely to minstrels. Huge minstrel troupes became popular, but were short-lived for economic reasons.

Although in America there were eventually black minstrel shows, the vast majority of troupes were whites in black-face. The basic structure of the minstrel show was the rendition, by a company that varied considerably in size (originally four: violin, banjo, bones, and tambourine), of sentimental ballads and comic songs, jokes, stories, sketches, monologues, instrumental pieces, and eccentric individual or group dances by performers wearing a parody of negro fancy clothes and caricaturing negro life, humour, and character in every possible way. The closeness of this caricature to aspects of *Uncle Tom's Cabin* was soon remarked by blacks and white liberals, but their voices were too few to make a difference until the twentieth century.

In France the popular theatres of the Boulevard du Temple flourished in the 1820s, not only with the melodrama of Pixérécourt and his colleagues, but also in an amazing variety of fairground entertainment: acrobats, fire-eaters, tightrope-dancers, giants, dwarfs, performing fleas, snake-swallowers, sword-swallowers, human skeletons, marionettes—every conceivable kind of show. This activity took place along a street packed in the afternoons and evenings with crowds eating, drinking, strolling, and gawking (all splendidly recreated in the opening minutes of the classic French film *Les Enfants du paradis*). The great mime artist and Pierrot Jean-Gaspard Deburau was established at the Théâtre des Funambules, and Madame Saqui's troupe of acrobats and tightrope dancers at the Café d'Apollon. Here Paris thronged, and although the Boulevard changed character as the years went by—performing animals and masked balls in the theatres were especially popular in the 1830s—some of the atmosphere survived until all the theatres were swept away by the new street schemes of Baron Haussmann in 1864.

Out of the Boulevard du Temple came the greatest actor of nineteenth-century France. Frédérick Lemaître, 'the Talma of the boulevards', made his first entrance with a roar, on all fours, as a lion in *Pyrame et Thisbé* at the Variétés-Amusantes in 1815; he was 15, and had been born with the century. Barred all his life from acting with the Comédie Française, which refused to admit him, and therefore unable to play Molière, Racine, or Corneille, the legal monopoly of the Comédie, Lemaître was the quintessential boulevard actor of melodrama and the romantic dramas of Dumas, Hugo, and de Vigny, many of which were performed in the theatres of the Boulevard du Temple and many of which were written specially for Lemaître. Except for provincial tours and two successful visits to London in 1845

Amor and his Mother, 1853. Drawing by Honoré Daumier. The great French satirist was captivated by the theatre, where he found all the folly of life itself. Here the contrast between the wretched child off-stage and the part he is shortly to play in a pretty fairy extravaganza could not be sharper.

(when he greatly impressed Queen Victoria) and 1852, he spent his whole career on the Boulevard. To the characters of dramatists who wrote for him he added his own creation, Robert Macaire, in his play of that name (1834), a development of the character he originally burlesqued in a melodrama of serious intentions, *L'Auberge des Adrets* (1823). Macaire, a grotesque rogue, thief, and swindler of nightmarish stature, was a comic but bitter Aristophanic satire upon the greedy, selfish, and power-hungry bourgeois society of 1830s Paris; it was not a coincidence that the satiric cartoonist Honoré Daumier, an admirer of Lemaître, immediately developed a book of cartoons based on the character of Macaire.

Lemaître's great parts—Macaire, the gambler Georges de Germany in *Trente ans; ou, la vie d'un joueur* (1827), the wife-murderer Richard Darlington and the actor Edmund Kean in two of Dumas's collaborative efforts, *Richard Darlington* (1831) and *Kean* (1836), and the lackey-nobleman Ruy Blas in Hugo's eponymous drama (1838)—were predominantly tragic, but like Garrick (and unlike Kean) he was also a fine comedian, a talent most strongly exemplified in Macaire. Like Kean and Garrick, his effect on audiences could be overpowering. Dumas records that when he told his wife in *Richard Darlington* to say her prayers, 'a tremendous shudder went through the whole house, and a murmur of fear, escaping from every breast, turned into a terrified shriek'. Likewise, Charles Dickens, who saw

Trente ans in Paris in 1855 and wrote a long laudatory account of Lemaître's performance, noted that 'two or three times a cry of horror went round the house'. Finally, in his so-called 'decline', at the age of 63, one writer remembered his triumph in Édouard Plouvier's *Le Comte de Saulles* (1864):

He sprang about, filling the stage with his ample, supple gestures and his giant strides. They were real tears that he shed, and the flame of passion burned in his eyes. His face, flushed with genuine anger, went pale with real terror, softened with sincere pity. His voice, so faint to begin with, burst forth in cries, groans, sobs. This was truth itself, since it was life, but truth as it should be revealed to the people, that is to say magnified by art, poetic, poignant, and grandiose!

Lemaître and his frequent acting partner, Marie Dorval, declared Théophile Gautier, 'were the actors of an age that was passionate to the point of delirium, and they experienced all the turmoil, all the fever, all the fire of those times'. Thus

Audience at a melodrama, 1864. Painting by Honoré Daumier. The total and intense absorption of the audience in, even identification with, what is happening on stage is striking.

does a great actor always express his time, and it must be remembered that Lemaître's own brand of electrifying Romanticism, his burning intensity, his vocal and physical power, his satanic comedy, his ability in a moment to paralyse an audience with horror or convulse it with mirth, were all matured in the popular theatres of the Boulevard.

Colourful, lively, eccentric, extravagant, eminently democratic, and on many occasions commanding the heights of acting and playwriting, the French popular theatre, like all popular theatre of the nineteenth century, related to and yet was distinct from the more conventional theatre of the literary drama and the passive, well-dressed, middle-class audience. In France as in England the heritage of mid-century theatre included the lusty melodrama, the extravagant acting styles, and the physical sensations of the dumb show and fairground theatres of the late eighteenth and early nineteenth centuries. The popular theatre in its many manifestations remained an identifiable and important part of nineteenth-century theatre as a whole, its techniques and content sometimes appropriated by the mainstream, at other times sharply divergent from it.

In the mainstream, the dominant trend by mid-century was not the romantic, rhetorical, sensational, and presentational, but a content and a matching acting and production style that stressed domestic verisimilitude, a half-way house to the naturalism of the 1880s and 1890s.

This was so in Russia as well as in France, England, and Germany. The Russian theatre emerged from its own classical period into a short-lived Romanticism in which Shakespeare, as in France, was a leading figure, but in this case a Shakespeare performed in Russian. Romanticism, best represented in the native drama by Alexander Pushkin's tragedy *Boris Godunov* (1825, performed 1870), was largely suppressed by a rigorous state censorship even more severe than Prussia's. Well after the accession of the liberal Alexander III in 1855, a police state mentality effectively crippled the new drama, delaying the first performances of important new plays by many years. It was only the surprising personal support of Nicholas I that ensured the production of Gogol's comedy *The Government Inspector* (1836). The uproar over the production was as significant in St Petersburg as the furore created by *Hernani* in Paris, and Gogol felt compelled to leave the country, not returning until 1848. With this play and Alexander Griboyedev's *Woe from Wit* (1823, performed 1831), the Russian theatre developed a vein of savage, sometimes grotesque, and essentially political satire (despite the censorship) that far exceeded in strength and comic venom anything written elsewhere in Europe. The new domesticism was well represented in the 1850s by Ivan Turgenev's *A Month in the Country* (1850, performed 1872), a play whose concern with the theme of unrequited love and the inner life of an apparently quiescent rural gentry is a precursor of Chekhov. This particular concern, as well as a deeply political and satirically expressed hostility to the aristocracy and the virtually feudal landowning class, also marked the plays of Alexander Ostrovsky, Russia's first fully professional dramatist, whose best-known play in the West is *The Storm* (1859), a tragedy about intolerance.

The story of naturalistic theatre properly belongs in the next chapter, but it is worth noting two Russian plays in this genre: Pisemsky's striking precursor of fully blown naturalism, the rural tragedy *A Bitter Fate* (1859, performed 1863), and Tolstoy's grimly environmental tragedy of greed, murder, and betrayal in a peasant community, *The Power of Darkness* (1866, performed 1895—and by Antoine in Paris, 1888). In Russia as in France, but not in England, great novelists wrote major plays. In Russia also, as in France, the theatre was heavily centralized; the principal theatres in Moscow and St Petersburg were controlled by the court and the government. The provinces had to manage with sometimes worse than mediocre touring companies. It was, in fact, the star-oriented Alexandrinsky Theatre in St Petersburg that nearly finished off Chekhov's playwriting career with the disastrous première of *The Seagull* in 1896. All over Russia, however, even in the imperial theatres, the art of production was not nearly so advanced as in the rest of Europe; in particular theatres relied on a very limited set of stock scenery for the production of the entire repertory.

In case the preceding account conveys the impression that Russian theatre in the nineteenth century dealt only in intellectually demanding plays by serious dramatists, it must be remembered that Russian audiences, wherever they were to be found, were just as devoted to light comedy, musical pieces, and sensational melodrama as audiences anywhere else in Europe, and these kinds of drama constituted the vast majority of the repertory.

By 1851 and *La Dame aux camélias* by Dumas *fils*, the old melodrama had been absorbed in France by the bourgeois prose tragedy. Only a few years later both Dumas and Augier were writing the highly moralistic *drame à thèse*, or thesis play, as in the former's *Le Demi-monde* (1855) and the latter's *Le Mariage d'Olympe* in the same year. In *Le Demi-monde* the hero Oliver tells the duplicitous Susan that 'reason, justice, and the laws of society forbid an honourable man from marrying any but a virtuous woman'. Plays dealing with courtesans and the life of the *demi-monde* were unacceptable on the English and American stage. *La Dame aux camélias* was forbidden a licence by the Lord Chamberlain, although its operatic version, *La Traviata*, was allowed, not without considerable public controversy. Such a morality was another legacy of melodrama, in which sin is ugly, as well as a manifestation of contemporary social attitudes. Even French farce had to be sanitized on the English stage. Its flowering in the 1850s and 1860s with the plays of Eugène Labiche, especially *Un chapeau de paille d'Italie* (1851) and *Le Voyage de M. Perrichon* (1860), provides delightful counterpoint to the high seriousness of Dumas and Augier.

Contemporary attitudes were now embodied in plays that dealt with modern social, business, and family life, not with the life of banditti in mountains and forests, or heroic sailors on their ships. This sort of life was presented in recognizably domestic and material forms, no matter how extravagant the dramatic content. The staging of Tom Taylor's *Still Waters Run Deep* (1855) at the Olympic Theatre in the West End exemplifies these new directions. Its hero is John Mildmay, a quiet, gentle man who is viciously henpecked by his flighty wife and her domineering aunt. In the course of the action, however, he exposes as a business fraud, blackmailer, and sexual predator the blustering Captain Hawksley, and in doing so wins the humbled and adoring respect of wife and aunt. The play opens in Mildmay's villa at Brompton; the drawing-room leads into a conservatory with plants. We see the characters in a pleasant, easy, and utterly domestic setting: 'MR POTTER, *in easy chair by fire;* MRS MILDMAY *on ottoman;* MRS STERNHOLD *seated by table,* R.C. *and* MILDMAY *seated, looking at book, by writing table.*' The only other settings are the breakfast room at the villa, in which the family quietly eats breakfast, and Hawksley's fully furnished apartments. Mildmay was played by Alfred Wigan, then manager of the Olympic, a restrained and polished actor with an intimate knowledge of French theatre. It was Wigan who took the part of the elegant villain Château-Renaud in Charles Kean's production of *The Corsican Brothers* (1852); this and other productions by Kean in the 1850s were described by the term 'gentlemanly melodrama', which comprehended acting as well as tone and characterization. Wigan had a classical education and had been a school-

Nora dances the tarantella. The first performance of Ibsen's *A Doll's House*, at the Royal Theatre, Copenhagen, in 1879. The formal dark gravity of the setting and costumes is subverted by the dramatic gestures of the actress, thus setting up a tension between Nora and her environment.

teacher; he was described as an actor 'of extreme refinement, delicate perception, and truth to nature, combined with deep, though quiet feeling'. He was perfectly suited to playing English as well as French gentlemen, especially in a play like *Still Waters Run Deep*, in which, one critic said, 'common-place exaggeration is shunned and the language is made to approximate as much as possible to that of real life'.

Hawksley, it can be seen, gets nowhere near seducing Mrs Mildmay. A sexual problem like this, until at least the 1890s, was sanitized in the English theatre, unlike the French and German, which from Georg Büchner's unperformed *Woyzeck* (1837) and Friedrich Hebbel's *Maria Magdalena* (1844) to Gerhart Hauptmann's *Before Sunrise* (1889) had often dealt unflinchingly with sexual and social problems and human suffering in a realistic environment. Furthermore, English and American theatre, unlike the French and German of the same period, did not care for unhappy endings—except in the occasional melodrama—and did not begin to come to terms with them until the 1890s. Yet the theme of the family under stress, a family frequently introduced in Act 1 as seemingly comfortable and happy, is a marked feature of early Victorian domestic melodrama, as is the theme of the inescapable consequences of past choices and actions. This is Ibsen country, and it is worth noting that in his 'social' plays of the 1870s, when he first began to work with these themes, Ibsen, like the melodramatists, developed his action in carefully detailed domestic settings, such as the garden room of *The Pillars of Society* (1877) and the living-room of *A Doll's House* (1879). In these rooms the family, under external and internal pressure, begins to disintegrate.

Here indeed we have 'the illusion that everything is real' and that we as spectators are watching 'something that is actually taking place in real life'. But that is also true of *Still Waters Run Deep* and many of its English and American contemporaries; it is only when we get to human behaviour and motivation and the decisions people make about their lives that Ibsen is clearly more real, more complex, more truthful. The theatrical *technique*, however (as distinguished from the dramatic *content*), is the same as in the best current European practice. Ibsen's verse-dramas *Brand* (1865) and *Peer Gynt* (1867)—a great, sprawling, epic play— though powerfully dramatizing man's inner struggles, the force of his imagination, and the tensions between society and the individual will, were not intended for the stage, and stand apart in method from the later work. The staging of Ibsen's early social plays followed standard methods, and even the dramatic form is that of the well-made play of Eugène Scribe, with its climaxes and strong curtains, its economy of means, its careful planting of clues and significant properties. In terms of his place in the repertory of every European theatre and the total number of performances given, Scribe, who flourished as a prolific dramatist in all genres for nearly fifty years before his death in 1861, was undoubtedly the most popular European playwright of the nineteenth century. French plays were favoured in Norway, and in his theatre jobs at Bergen and Christiania in the 1850s and 1860s Ibsen had plenty of opportunity to study French dramatic construction and technique.

It is extraordinary that Norway produced such a giant of the theatre as Ibsen even before it became independent in 1906, and almost at the beginning of the history of professional theatre in the country. In the first thirty or forty years of the nineteenth century, Norwegian theatre depended upon amateur societies and touring foreign companies. The first professional theatre opened in Christiania in 1827, but it was dominated by a Danish repertory—plays were performed in Danish, not Norwegian—and Danish actors (Norway had belonged to Denmark until 1814, and then passed under Swedish rule). The first professional Norwegian-speaking theatre was founded by the violinist Ole Bull in Bergen in 1850, and another started in Christiania in 1852. It was to Bergen that Ibsen went as Bull's assistant in 1851. Det Norske Teater went bankrupt in 1863, reopening in 1876. In the mean time Ibsen had gone to the Christiania Norske Teater in 1857; this closed in 1862. The Norwegian repertory was absorbed by the rival Christiania Theatre, under the direction of Bjørn Bjørnson from 1865 to 1867; he also offered Danish and French plays and Shakespeare.

Until Ibsen and Bjørnson begin to write significant plays, there is little to distinguish Scandinavian theatre from the rest of Europe. Particularly susceptible to French and German influences, the Copenhagen theatre nevertheless retained romantic and idealized styles of acting and production for a long time, notably in the plays of Hans Christian Andersen in the 1840s and the acting of Johanne Luise Heiberg. The Royal Theatre, or Dramaten, in Stockholm offered *Hamlet* in 1819, but its next really distinguished dramatist was Strindberg, some of whose early history plays were produced there and at Ludwig Josephson's New Theatre in the 1870s and 1880s, together with Ibsen and Bjørnson.

When Ibsen took up his appointment in Bergen, he was specifically hired to organize the scenic arrangements, including costumes, to arrange exits, entrances, and appropriate groupings, to ensure that an actor's physical expression was suited to the text and the character, and to show individual actors what part they played in the action as a whole. He was also responsible for co-ordinating the whole production. His colleague, Herman Laading, formerly a schoolmaster, was responsible for overseeing diction, vocabulary, and language generally, for giving actors any necessary historical context, for making sure that roles were interpreted correctly and parts memorized well enough so that actors could appear at the first stage rehearsal without books. This division of duties was common enough in European theatres and had been advocated by a leading German director, Heinrich Laube. Ibsen actually did more than his share: he also designed sets and costumes, stage-managed, wrote prologues, checked the accounts, handled public relations, and wrote his own plays for production.

Clearly neither Ibsen nor Laading served as a 'director' in our modern sense of the word. The knowledge of how plays reached the stage in the nineteenth century is vital to an understanding of the theatre practice of the time. Were plays 'directed', and, if so, when did this happen? Certainly, calls for control of the acting ensemble and all aspects of production by one person were heard years before Edward Gordon Craig, and even before Wagner enunciated the principle in *The Art Work of the Future* (1849). Certainly, elements of modern directorial practice can be found all through the century. Karl Immermann at Düsseldorf in the 1830s declared that the reproduction of the poet's work on the stage must come from a single mind. Therefore, after reading the play to the actors, he prescribed a special reading rehearsal with each actor, followed by a general reading rehearsal. Disparities were corrected until the vocal interpretation of the whole play was acceptable. Before this, the Hoftheater in Kassel posted regulations governing reading rehearsals and included the injunction that the *Totaleffekt* of a play can be achieved only by harmony; in these rehearsals, then, every actor must learn how his part is related to the whole. (The German theatre, especially early in the century, was fond of drafting an excessive number of rules for its actors.) In the arrangement of visual elements, the movement of actors, the organization of rehearsals, and the stress on unity, German 'directors' like Laube, who worked at the Vienna Burgtheater between 1849 and 1867 and then at the Stadtstheater in the same city, and Dingelstedt, who had artistic control of three major theatres between 1851 and 1881—Munich, Weimar, and the Burgtheater—undoubtedly accomplished much of what modern directors do. So did Lemoine Montigny at the Gymnase in Paris in the 1850s, with his careful superintendence of a rehearsal period longer than usual on the French stage and his work on the socially realistic plays of Dumas *fils* and Augier. In England both Macready at Covent Garden and Drury Lane in the late 1830s and early 1840s and Charles Kean at the Princess's in the 1850s were well known for their careful rehearsals and their authoritative supervision of every aspect of production.

Yet the larger question remains. Was this 'directing' as we understand it today, and were these men 'directors' (many of them were *called* directors in their own time)? The short answer is 'no', and this 'no' can be explained on several grounds. The nineteenth-century actor was his or her own master. He or she, if an experienced professional, had played a great many parts in a relatively short time and was responsible for the interpretation of all of them. This was possible for three reasons.

First, classical acting styles, fully developed in theory and practice in the eighteenth century and bequeathed to the nineteenth century, were based on traditional ways of acting the passions, such as Grief, Remorse, Anger, or Jealousy, that constituted a pictorial code understood by performers and audience alike. Gesture, facial expression, and bodily attitude conveyed emotion in a pictorial manner appropriate to the text spoken. Or not spoken: actors in this tradition were expected to convey a full range of feeling through facial expression alone, without speaking a word.

The actor's emphasis in the classical style was upon the universal, upon passion as it was felt by all men and all women, upon character as a unity governed by a general overriding passion, not—unless it were low comedy—particular idiosyncrasies of character arising out of individual oddities and psychological quirks. All this was physically externalized and pictorialized, always with grace and beauty of form, no matter how ugly the passion or the deed. Thus a long-established set of acting principles defined the expression of tragic and comic passions and served in total as a guide to every emotional situation encountered on the stage. A director here would be entirely superfluous.

Secondly, unless the actor or actress were great enough to rise above all distinctions of type, he or she specialized within general character stereotypes. In England and America in the nineteenth century these were known as 'lines of business' and are exemplified by melodrama's character-types: hero, villain, heroine, good old man, comic man, comic woman, etc. The same general pattern held

The New Burgtheater, Vienna, 1888. The late nineteenth-century theatre auditorium became increasingly grand and opulent in architecture and decoration. Of interest also is the painted backcloth of a country scene visible on stage, as well as the great swagged front-curtain, typical of the period.

in Europe. An actor who specialized in a particular line of business would have an entire set of characteristics of speech, facial expression, costuming, deportment, and general behaviour at his command in order to express his character expertly, individualizing, of course, within the general stereotype, since all heroes or old men are not the same, but nevertheless drawing on his experience of the whole type in order to play any individual within that type. No director instructed such an actor how to play a line of business; he learnt by doing it.

Thirdly, the relationship of the repertory to the rehearsal process was a crucial factor in the necessity, or lack of it, for a director. Until the second half of the nineteenth century, the population of cities had not reached a sufficient size to justify long runs, and until this happened and the spread of railways enabled whole companies to tour with one play, or a handful of plays, the theatre depended upon frequent changes of repertory to attract the relatively small potential audience of a town or city. The bill might change two or three times a week, or even nightly. Under these conditions it was impossible to have more than six or seven, perhaps eight or nine rehearsals in a major theatre for a new play. Even these often consisted only of muttering lines or just cues, arranging exits and entrances and relative positions; in no sense were they 'rehearsals' as we understand them. A play already in the repertory but not performed for a year or two might receive one or two rehearsals if a new actor were playing a principal part, or none at all. Under this system an actor played a great many parts. Learning his business in the English provinces in the 1850s, Henry Irving played 428 different parts in his first two and a half years on the stage; it is unlikely that he had much rehearsal time for any of them. In any case, as we have seen, an actor was in command of his own parts; he might receive, or ask, advice from his fellows or his manager, but his interpretation of his part was his own business, although in Europe he might be subject to correction if he were not a star. To provide a longer rehearsal period for a new play was not only logistically out of the question, but also unnecessary.

Thus, until about the 1880s, a play reached the stage without the services of a director. In England and America a play in rehearsal was supervised by the manager, or the leading actor (frequently the same person), or the deputy manager, or the stage manager, or any combination of these. In Europe the system was more formal: reading rehearsals were common; the directorial staff in major theatres was larger, and managers or directors paid a great deal of attention to visual arrangements—not surprising in a primarily pictorial theatre—exits and entrances, blocking, the disposition of furniture on the stage, lighting, and sometimes clarity of speech. The position of *Dramaturg* in the German theatre, unknown in Britain and America, enabled more attention to be paid to literary matters and historical context. All the business of *production* was in the best theatres organized to a very high standard and frequently under the control of a single powerful figure. What was missing was a director working with actors to develop character, the interpretation of a role, and the interpretation of the play itself.

The old system had to change under the impact of changing conceptions of character and dramatic writing and new ideas of psychology, where motivation was hidden or obscure, where abnormal psychology and inner struggle were significant determinants of personality, and where text was complex and enigmatic. There were no traditional ways of doing Ibsen and Strindberg, and no way at all in which five or six rehearsals might be adequate for their plays. Actors now needed help, and rehearsal systems were eventually evolved to cope with these new complexities. There is all the difference in the world between the treatment of *A Doll's House* by the Royal Theatre, Stockholm, in 1879 (two blocking rehearsals, eight general rehearsals, one dress rehearsal) and William Bloch's production of *An Enemy of the People* at the same theatre in 1883 (thirty-two rehearsals altogether, twelve for the crowd scene in Act IV alone). Bloch was a full-blown naturalist, and the increasing intricacy of the means of production, especially the new possibilities of electric lighting, also added weight to the argument for centralized artistic control and the need for the new positions of lighting designer, set designer, and costume designer. The director's interpretation of character and the play soon became influential and indeed paramount.

The movement towards the absolute authority of a director was strongest in Germany and Austria, as was the strict codification of the rehearsal process. The development of the German theatre had been hampered by political fragmentation into hundreds of states and the absence of a capital city. Before 1848 Berlin had only three theatres, and theatrical life was dominated, even monopolized, by the theatres of princes. In Prussia the police exercised a rigid control over the theatre, and censorship of the drama was restrictive. After the war within the German Federation in 1866 and the Franco-Prussian War of 1870, Prussia emerged as the most powerful of all German states and engineered a union with the south German states. The Second Reich was proclaimed in 1871 with Wilhelm, the King of Prussia, as Kaiser Wilhelm I. After this, with the rapid expansion of cities and Germany's belated industrial revolution, hundreds of new theatres were built everywhere and the court theatres, by law, lost their privileged position. Vienna, Berlin, and Munich emerged as the dominant theatre centres.

None the less, a striking illustration of the importance of the court theatre and the total authority of the director is to be found in the small ducal state of Saxe-Meiningen, ruled from 1866 by Georg II. The Duke was a passionate devotee of the theatre and soon transformed his court theatre at Meiningen, a town of only a few thousand, into one of Germany's leading theatres. The Meininger Company, completely financed from the Duke's personal revenues, came to national prominence in 1874, when it toured to Berlin with a repertory including a highly acclaimed *Julius Caesar*; later the company toured extensively within Europe (and to London in 1881): thirty-eight cities altogether, including Paris, and Moscow twice (1885, 1890), where Stanislavski saw it. Altogether the Meininger on tour gave nearly 3,000 performances of forty-one plays. Schiller was the mainstay of the repertory, which ranged widely in the German classics, and six of the forty-one were by Shakespeare. During the seventeen years of touring, *Julius Caesar* was performed 330 times. At home in Meiningen a total of sixteen Shakespeare plays were presented.

The Germans had always been keener on Shakespeare than the French. Despite the romantic enthusiasm for him among the Parisian revolutionaries of the 1820s and 1830s, Shakespeare could make French spectators most uncomfortable. It was one thing to admire Shakespeare performed in English and declare him a great dramatist, but quite another to do him successfully in French. Translations and adaptations often distorted the original; unclassical ghosts and violence on stage evoked laughter and ridicule; a mixture of comedy and tragedy was still not acceptable, and the great number of scenes set at different times and in different places gave production staff headaches—as they also did in Victorian England. It was only toward the end of the century that complete performances of Shakespeare's plays were tentatively offered in Paris. There is no great nineteenth-century French Shakespearian actor, although the romantically overpowering Sarah Bernhardt played Hamlet, Cordelia, Desdemona, and Ophelia. In contrast Ristori, Rossi, and Salvini in Italy made international reputations playing Shakespeare; even the psychologically realistic Ibsen actress Eleonora Duse attempted Cleopatra.

Germany was quite different from France. Shakespeare was well served by the uncut translations of August Wilhelm von Schlegel, many of which had appeared by 1803. The romantic actor Ludwig Devrient, often compared to Kean, and like Kean an alcoholic—they died a year apart, Devrient in 1832, Kean in 1833—was a noted Lear, Shylock, and Falstaff; he also played Hamlet, Mercutio, and Parolles. The eminent critic, dramatist, editor, and translator Ludwig Tieck encouraged the production of Shakespeare at Dresden in the 1820s, and was himself responsible for a scenically minimalist *Romeo and Juliet* at the court theatre there in 1823, a permanent-setting *Macbeth* in 1836, and a multi-level open stage *Midsummer Night's Dream* at Potsdam in 1843, a radical production for its time considering the consistent German emphasis on spectacle in Shakespeare. At Weimar in 1864 Dingelstedt produced a cycle of the history plays, and at the Burgtheater in Vienna in 1875 another such cycle, at huge expense. Among the audience for the Weimar cycle, as he had been for Charles Kean's Shakespeare in London in 1857, was the Duke of Saxe-Meiningen.

A brief examination of the 1874 *Julius Caesar* will not only reveal the Duke's methods but also illuminate the nineteenth-century treatment of Shakespeare.

Georg II was, for his age, a very complete director. He put up the money for the theatre, did all the casting, designed sets, costumes, and properties, and controlled all aspects of production through his stage manager Ludwig Chronegk, and his extensive correspondence with his scene-painters in Coburg. He insisted that actors rehearse with full sets, costumes, and properties as early as possible. In performance actors should not look as if they had just put on 'costumes', but instead appear to be wearing clothes with which they were thoroughly familiar. He stressed the importance of the visual relationship of actors to each other— symmetry was to be discouraged—and to the scenery, and was much concerned, like all directors of his time, with perfecting the visual arrangements of the stage, especially the appearance and lighting of the scenery. In his many published guidelines for production, however, he never mentions working on the interpretation of character with (or without) actors, let alone attaching any significance to bringing out the meaning of a play. Neither did he actually direct rehearsals, although from the balcony he would send instructions to the efficient Chronegk, who sat at the front of the orchestra armed with a dinner bell. The direction of rehearsals by the stage manager was standard nineteenth-century procedure.

It was particularly appropriate to the *Julius Caesar* that the Duke had an interest in Roman history and had studied archaeology at the University of Bonn in the 1840s. *Julius Caesar* was conceived as a large-scale illusionist spectacle with archaeologically correct scenery, properties, and costumes. The importance of archaeology in the nineteenth-century theatre and the influence of archaeological discoveries upon production methods must not be underestimated: Shakespeare and plays both classical and modern set in the historical past were the beneficiaries of this influence. At the very time the Meininger Company was preparing *Julius Caesar*, Schliemann was about to start digging in Mycenae, after excavating at Troy for years. German archaeologists were everywhere on ancient sites. All this activity was strengthened by scholarly publication and popular illustrated magazines; the public's appetite for archaeology seemed inexhaustible. This enthusiasm and commitment to archaeology also marked the English theatre, although it was clear that as in Germany spectacle and archaeology walked hand in hand. The

The Meininger *Julius Caesar*, 1874. The Duke of Saxe-Meiningen's sketch for Act III, scene ii: Antony addresses the mob. This was a spectacle scene, and the impact of mass and numbers is apparent.

richly adorned palace, the luxurious banquet, the colourful market-place, the splendid procession—these were the habitats of stage archaeology. Archaeological zeal in the theatre did not extend to the mean street or the wretched hovel. Archaeology also satisfied the realistic imperative in mid-century theatre, in this case historical realism. To present an archaeologically authentic Rome on stage was as important as accurately reproducing Union Square in the snow (Boucicault's *The Poor of New York*, 1857); the impulse for both was the same. Faithful replications of historical settings and costumes also had an educational value, or so claimed managers and reviewers. Where else could the public see living pictures of past ages? Shakespeare loomed large in this argument, and the Victorians valued him as a historian and educator as well as a dramatist. The Duke even corrected Shakespeare by moving the assassination of Caesar to the Curia of Pompey, rather than the Capitol, where Shakespeare places it. Whether this mattered is a question that would not have occurred to him, nor to any serious theatrical archaeologist. Indeed, Georg II undertook a great deal of archaeological research for *Julius Caesar*, checking his findings with the most eminent authorities for fidelity to the past; before the production went to Berlin it was viewed in Meiningen by a specialist in Roman history from the University of Heidelberg. Likewise, Charles Kean in his programme notes cited external authorities at length as guarantors of the authenticity of his archaeologically detailed (and lavishly spectacular) Shakespeare in the 1850s.

The principal attractions of the Meininger *Julius Caesar* in performance were the elaborate historicity of the sets, the spectacular nature of Acts I and II, and the disciplined use of crowds. Of the two scenes in Act III depicting the murder of Caesar and the addresses of Brutus and Antony to the Roman mob, the Duke believed that the scenic effects alone were a greater attraction than all five acts of Shakespeare's play. The crowd of Act III was a rampaging mob, easily inflamed and very dangerous. Its ugly behaviour when it attacked the poet Cinna was strongly emphasized. The year in which the production opened, 1874, was only three years after the Paris Commune, whose initial triumph but speedy destruction reverberated throughout a Europe which had not forgotten the revolutions of 1848 or, closer to home, the civil war within the German Federation in 1866. The connection between the Duke's conservative politics and the actions of the mob in his *Julius Caesar* was a meaningful one, and although the play was not given a general political interpretation it had at least a political context.

The mob's stage business had been carefully programmed by Chronegk: it was broken down into small groups of well-drilled supers, each with specially written responses and guided by a regular actor in the company. Trained in this way the supers became totally involved in the action and behaved convincingly as individuals as well as forming distinct units in a cohesive mass. The result, as *The Times* of London put it in 1881, was 'a total absence of that lumping of masses, that rigidity of form and feature, which chills the spectator at ordinary performances'. Georg laid great stress on ensemble; he was opposed to the star system and required leading actors in one production to take minor roles and even walk on in another, if necessary. Lengthy rehearsal periods also ensured the perfection of the crowd scenes, which much impressed European reviewers and theatre people.

The use of large numbers of supers on stage was a distinctive feature of nine-teenth-century theatre, especially in Shakespeare, the French *folie féerie*, or fairy extravaganza, and the English Christmas pantomime. It did not originate in Meiningen. Macready had used between 100 and 200 senators and soldiers for *Coriolanus* in 1842. His approach to stage crowds was very similar to the Duke's. He noted, 'I thought for and acted to myself every character and every supernu-merary figure, and taught them to act as I would have done had I been cast in their place. Thus there was the mind of a first actor moving and harmonising the action of the mass.' The Duke had seen Kean's *Richard II* in 1857 with its spectacular interpolated entry for Richard and Bolingbroke on horseback into London and its 300 supers. In 1885 a *Comedy of Errors* opened at the Star Theatre in New York. A critic for the *Theatre*, an English magazine, was enraptured by the beauty of an interpolated scene in which the Courtesan entertains Antipholus of Ephesus and Angelo the Jeweller. The entertainment included a song by a beautiful blonde and a dance of the Bacchantes. But it was the sheer sensuous pictorialism of the scene that most impressed the reviewer. 'Tapestries and skins of the richest hues are scat-tered over the mosaic floor. Columns of tinted marble support canopies of gaily embroidered and bespangled lace. Palms and brilliant flowers drink in the fresh-ness of fountains of sparkling waters. Genuine negroes . . . draped in scarlet tunics perform the part of slaves-in-waiting.' Nowhere does the review mention that this scene is not in the text. To make room for such interpolations Shakespeare's text was commonly reduced by about one-third, a reduction that also allowed for leisurely scene-changes necessitated by the use of heavy sets.

In 1882 the manager of Drury Lane, Augustus Harris, put 650 people (wearing costumes made from 300 separate designs) into the pantomime *Sindbad the Sailor* for a monster procession of the Kings and Queens of England and their retinues. This kind of spectacle was to be anticipated in pantomime and the *folie féerie*, but in more classical drama critics felt—and the criticism was made of the Meininger *Julius Caesar*—that the stress on crowds reduced the principal actors to subordi-nate roles. This sort of complaint was familiar in the last half of the century: the actor and his words, it was alleged, were overwhelmed by the large-scale effects of mass, colour, scenery, light, and costume. The fact that it was the complaint of a minority of critics rather than a majority of the audience, who greeted this style enthusiastically, ensured the continuance of such production methods into the twentieth century.

The staging of *The Comedy of Errors* in New York is an excellent example of the

way in which theatre approached as closely as possible to painting (in this case a painting by Alma-Tadema), to replicating a picture in a gilt frame, albeit a picture that moved. It is no wonder that managers and directors fussed so much over their scenic and lighting arrangements and the visual effect of their work. In large part this is what theatre was for: to convey a visual image of the historical or contemporary world, to provide a complete visual environment for the actor and the text, to *illustrate* the drama as if by pictures in a book. 'Each scene is like his picture to a painter,' said Irving of his production of *Faust* in 1885, 'You have to combine colours, group figures, and arrange the mountings.' By the 1880s the theatre-going public was well trained in looking at the stage as if it were a picture. Since early in the century the theatre had 'realized' well-known paintings for its audiences; that is, positioning actors, properties, and scenery in a frozen tableau at the end of an act that would exactly reproduce a popular painting.

By the 1880s also, theatre architecture had evolved to its final nineteenth-

The fly gallery at Drury Lane, 1883. The pantomime on stage is *Cinderella*. Gas battens and concealing borders are visible at the top of the picture, as are the wooden pipe down which rattled cannonballs to produce thunder, and the ropes attached to the flown scenery.

TWO CONTEMPORARY DESCRIPTIONS OF SCENES FROM HENRY IRVING'S PRODUCTION OF *FAUST* AT THE LYCEUM THEATRE, LONDON, IN 1885

The first, Martha's Garden, illustrates the pictorialism of stage setting; the second, Mephistopheles' revels on the Brocken with 250 other evil beings, the character of spectacle.

(*a*) Quiet red-brick garden walls with climbing roses; an old moss-grown apple-tree with an ancient bench; a rustic cottage porch; a background of city towers; the whole a characteristic town garden. . . . Margaret is in a dress of pale yellow and white brightened by a slight touch of black; then Faust in a costume of rich brown, a connecting link of colour, leading up to Mephistopheles, the antithesis of Margaret; forming as it were a second focus of colour in the picture, the splendour of it relieved by the sombre tones of the gateway against which it is massed. Over all there is a rich glow of summer evening colour deepening towards twilight.

(*b*) There is nothing more powerful than this single scene,—one minute a wild shrieking, singing crowd of misty shapes, moving hither and thither, clambering over the rocks and up the trees, dancing and turning; the next, after one last shriek, wilder, shriller than the rest, a silent, storm-beaten mountain top deserted but for one flaming form. Then, summoning them once more, he [Mephistopheles] plunges into the midst of the reveling. Now the dreary light, that has been strangely glimmering, here glows through film and haze, there sweeps in a rolling vapour; now creeps like a thread, now leaps and plays, lighting up the great mountain and all the rugged shapes, and finally gushes forth, a shower of fiery rain, over the wild and howling crowd of witches, while the rocky ramparts on all their heights are set ablaze.

century form, which can still be seen today in those theatres that survive from this period. A semicircular or horseshoe-shaped auditorium converged on either side upon a richly framed proscenium; class and price divisions were precisely observed in a rigid separation between boxes, stalls, and galleries, with different refreshment areas and exits and entrances provided for the different parts of the theatre, and comfort levels appropriate to the seat price. This proscenium usually displayed an actual picture-frame, moulded and gilded, flush with the front of the stage. What happened behind it, thanks to the skills of the scene-painter, the gasman, the carpenter, and the limelight operator, was thoroughly painterly; the framed proscenium heightened the pictorial nature of acting and production.

Irving's spectacular *Faust* at the Lyceum, adapted from Goethe's Part 1 in a workmanlike but highly melodramatic fashion by W. G. Wills, tells us all we need to know about the function of pictorialism in the nineteenth-century theatre. It also shows that the late romantic theatre flourished alongside the new drama and the new ways of doing theatre. Despite Ibsen, Strindberg, Chekhov, Antoine, Craig, and other reformers, the romantic and thoroughly anti-realistic theatre, bathed in limelight and wallowing in spectacle, lived a veritable golden age that lasted in Britain, Europe, and America until at least 1914.

Irving had done the standard archaeological research for his new production, spending two weeks in Nuremberg and Rothenburg with his chief scene-painter and his leading actress, Ellen Terry. Here the party not only made sketches and absorbed the atmosphere, but also bought antique properties and costume fabrics and shipped them back to London. Irving clearly intended *Faust* to be a pictorial recreation of medieval Germany as well as a starring vehicle for his own Mephistopheles. Critics of the production stressed its artistic qualities, comparing it variously to Dürer, Rembrandt, John Martin, and Gustave Doré. Of the first scene in Faust's study, a musty Gothic chamber entirely medieval in character and furnishings, the artists Joseph and Elizabeth Pennell wrote that Irving saw Faust and Mephistopheles as 'the principal figures in a picture rich in colour, vigorous in composition. Their every pose is a subject for a painter.'

At the Lyceum Irving was his own lighting 'designer', and he used gas and lime-light with great subtlety. The stunning impact of the Brocken scene would have been impossible without the by now sophisticated pre-electric lighting technology of the nineteenth-century theatre. Whether the kind of theatre to be seen was romantic or realistic, spectacular or intimate, it was lighting technology that now permitted the withdrawal of the actor behind the proscenium arch and his integration with a scenic environment as realistic or romantic as stage art wished to make it. The taste of nineteenth-century audiences for the pictorial, the material, and the archaeological meant that before them unfolded pictures of the 'real' world—modern or ancient, exotic or domestic. A pictorial reality was not possible when the actor was in front of the proscenium; he became part of a living picture

---

SHERIDAN KNOWLES'S *THE HUNCHBACK*, 1832

A description of a performance in 1847 by the English actress Fanny Kemble of a speech in which she pleads with her guardian to break off her coming marriage with a man she does not love. The description illustrates not only current acting style, but the ability of a powerful actress to bring to vivid life and colour a passage of text that to us, on the page, seems quite uninspired.

Tortured, despairing, maddened, she sprang to her feet erect and terrible. With fiery eyes and dilated form she turned at bay, even as a wounded hind might turn upon the hunter's spear, then with quivering lips she commenced the famous speech, extending over some thirty lines. As it proceeded her voice gained strength, changing from the flute to the bell—from the bell to the clarion. Then upon a rising *sostenuto* of concentrated agony and defiance, she smote and stabbed Walter with that awful 'Do it! Nor leave the task to me!' Even as the last word left her lips, she strode down to the right hand corner, returned to the centre, and then came to anchor, her right hand clutched on the back of the great oaken chair, her left thrown out towards Walter, her blazing eyes fixed on him in an attitude of denunciation and defiance. Then it was, and not till then, that the breathless and enthralled auditors rose in such an outburst of wild enthusiasm as I have never heard equalled before or since.

---

only when he moved behind it. The ideal pictorial beauty so treasured in the late nineteenth-century theatre was one that the twentieth century, and those earlier reformers who cohabited unhappily with this ideal, deliberately set out to destroy.

10

MODERN THEATRE 1890–1920

MARTIN ESSLIN

As the century entered its last decade, the theatres of the great metropolitan centres continued to perform their accustomed repertoire: some classics, above all Shakespeare as vehicle for virtuso star performers; an abundance of melodramas and sentimental comedies; farces, operettas, vaudevilles. Touring companies carried these to the provinces. The great national theatres like the Comédie-Française or the Vienna Burgtheater gave the national classics in traditional histrionic style. In London pioneers like Irving attempted Shakespeare and other classics in somewhat more ambitious productions, striving for historical authenticity in costume and design.

Yet, as so often in the history of the arts, new impulses, as yet barely noticed by the public, began to stir on the fringes of the theatrical world. To the contemporaries who witnessed them they may, at first, have appeared as no more than marginal aberrations, but, in fact, they contained the seeds of developments that would in due course bring about great changes in theatrical practice as well as a revaluation of the role of theatre in society.

For the nineteenth-century middle classes the theatre had been a source of entertainment, a provider of laughs, glitter, and maudlin sentimentality; it had lost the more lofty functions it had served in some earlier epochs: poetic, religious, political. But voices began to make themselves heard that the time had come to put the theatre—and drama—back to its formerly hallowed position in the culture as a place of serious reflection on the state of society and, indeed, a source of sublime emotion and almost religious uplift.

Foremost among those voices were that of Richard Wagner, who wanted to make drama a 'total work of art' combining poetry, music, architecture, design, to create sublime emotional experiences and profound insights in audiences attending what came near to being religious rites, and that of his erstwhile friend and later antagonist Friedrich Nietzsche, intent on recreating great tragedy from the 'spirit of music', and advocating the 'revaluation of all values'.

By the last quarter of the century, the time seemed ripe for such a revaluation. The nineteenth century had seen cataclysmic ideological, social, and technological change—Darwin, Marx, the positivist sociology of Auguste Comte, the rise of the natural sciences and their methods, the explosive use of technology and its inventions, machine-driven industries, railways, the telegraph, had undermined most of the long-held beliefs of the Western middle classes. In the arts new revolutionary movements had transformed the traditional assumptions about what was beautiful and moral, what could be the subject of representation: impressionism in painting, the ruthless realism of writers like Zola, the moral ambivalence of poets like Baudelaire had already shocked the complacent public. The theatre, being a rather slow-moving and complex institution, serving a large and unsophisticated public, was among the last to feel the impact of these changes.

Yet in the last decades of the century these new tendencies gradually made themselves felt: in France Henri Becque had shocked Parisians with plays boldly dealing with modern moral dilemmas, characterized as *comédies rosses* (nasty plays); and Henrik Ibsen, the Norwegian poet who had achieved great success with monumental verse plays, turned from 1877 onwards to contemporary subjects in prose, dramas like *Pillars of Society* (1877) or *A Doll's House* (1879) which not only criticized prevailing moral dogmas but advocated radical new solutions.

In this Ibsen reflected the new ideas about the function of the arts, a tendency that one of its pioneers, the French novelist Émile Zola, had called 'naturalism': no longer was the drama merely to make people feel good, it should become a branch of scientific inquiry, an experimental laboratory to explore human relations presenting images of the world and society as it really was, with all its ugliness, all its blemishes, all the aspects of human existence which hitherto had been regarded as unmentionable in public. 'Drama is dying of its extravagances, its lies and its platitudes,' Zola wrote as early as 1873:

If comedy still keeps on its feet in the collapse of our stage, that is because it contains more of real life, because it is often true . . . The time has come to create works of truth. . . . There must be no more schools, no more formulae, no more literary panjandrums of any kind. There is just life itself, an immense field where everybody can explore and create to his heart's content.

Increasingly, young intellectuals, natural rebels, and dissidents rallied round the banners of Ibsen, Nietzsche, Wagner, and Zola, to create a platform for this new kind of drama which frequently met censorship and interdiction on the part of the authorities, as being indecent and subversive. When Ibsen was occasionally performed in established theatres the reviews were usually devastatingly harsh. Yet there were also increasingly vocal advocates of the new drama—and the new man and woman—like the London critic William Archer, who began to translate Ibsen into English, or Archer's friend George Bernard Shaw, a dedicated champion of Ibsen, an equally enthusiastic Wagnerite and socialist; or the Berlin critic Otto Brahm.

These pioneers pressed for the creation of private clubs or societies, where the new drama could be performed free of censorship for a select audience of like-minded adherents of these revolutionary ideas: in Paris the Théâtre Libre started in 1887; in Berlin Otto Brahm's Freie Bühne in 1889, J. T. Grein's Independent Theatre in London in 1891, and the Moscow Society for Art and Literature, which later developed into the Moscow Art Theatre, in 1888.

The first of these pioneer organizations, the Théâtre Libre, was the creation of a young man of 29, André Antoine, then still an employee of the Paris gasworks, self-educated and an enthusiastic amateur actor. Dissatisfied with the usual repertoire of his amateur group, the Cercle Gaulois, Antoine let it be known that he was looking for plays by hitherto unknown playwrights. Having assembled a number of one-acts of the progressive type he wanted, but failing to persuade the club to perform them, he decided to put them on at his own expense. The Cercle Gaulois rented him their theatre near the Place Pigalle and he assembled a group of actors from his own and some other amateur societies. One of the plays in the programme was an adaptation by Léon Hennique of a novella by Zola, *Jacques Damour*. True to the principles of naturalism Antoine wanted his stage to be as real as possible without the usual painted scenery but with solid three-dimensional furniture. The story goes that he borrowed pieces from his mother's living-room and transported them to the theatre himself, pushing a handcart. Zola came to a rehearsal, and was deeply impressed.

On 30 March 1887 the production opened in the dingy little theatre, holding fewer than 400 spectators, in the Passage des Élysées-des-beaux-arts on Montmartre. The main attraction—there were four short plays on the programme—was Zola's tale about a Communard banished to New Caledonia, who returns home after ten years of exile to find that his wife has remarried. Antoine himself gave a memorable performance as the haggard and bedraggled Jacques Damour. The realism of the scene and the natural undeclamatory acting style made an immense impact. The *République française* critic wrote: 'If the naturalistic theatre has more such plays, it need not worry about its future.' The Théâtre Libre was born.

The group gradually developed into a professional organization, although

André Antoine as Père
Rousset in Brieux's
comedy *Blanchette*,
Théâtre Libre, 1892.
Antoine was
outstanding in
character parts.

many of the actors still appeared without
pay, and subsequent productions consolidated and perfected the new approach to
acting and direction, discarding many practices hitherto obligatory in theatre. Antoine
decisively broke with the dogma that actors
should never turn their backs to the audience. Only exceptionally had great stars been
able to break with this convention, but the
need to present as natural a picture of life as
possible clearly could not tolerate adherence
to it. Whereas actors hitherto had always
been conscious of the presence of the
audience, confiding the characters' secret
thoughts to them in 'asides' and monologues, the new approach demanded complete separation between the stage and the
auditorium, which had still been fairly well
lit but now had to become as dark as possible. Nor could painted scenery be compat-
ible with naturalness: the box-set of a room
with seemingly solid walls, and later even a
ceiling, and real furniture became hallmarks
of the naturalistic theatre. The audience was
eavesdropping on people living their lives as
though looking through a transparent fourth wall. Sometimes the stage presented
two walls of a room at an angle, giving the spectators the illusion of actually
watching the action from one of its corners.

While classical drama had concentrated on the lives of kings and heroes, and
boulevard theatre on the drawing-rooms of the upper and middle classes, the
subject-matter of naturalistic drama extended the scope of its offerings to the lives
and tragedies of the lower classes. Not all Antoine's productions were of naturalistic drama. In the second soirée of the new organization one of the plays, Berg-
erat's *La Nuit bergamasque*, was, in fact, a romantic piece in the style of the
commedia dell'arte. But here too scenery was three-dimensional and the actors
seemed unaware of the presence of an audience—even a verse play could thus aim
at the maximum of naturalness, of reality within its own convention.

Success enabled Antoine to transfer to a better theatre on Montparnasse; later
each production was extended to three performances and a subscription system
introduced to secure a degree of financial stability. Gradually substantial full-
length plays were brought into the repertoire: Tolstoy's *The Power of Darkness* in

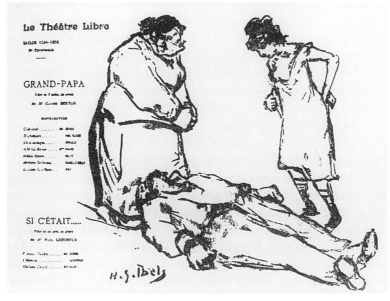

Above: *La Terre* by
Zola at the Théâtre
Antoine, 1900. Antoine
is the third from left.
Naturalism brought the
environment of the
poorest members of
society into the theatre.

Left: Programme of
two plays at the
Théâtre Libre, 1894/5.
The costumes—in
contrast to the elegance
displayed by
conventional drawing-
room drama of the
period—show poverty
and do not even shy
away from under-
clothes.

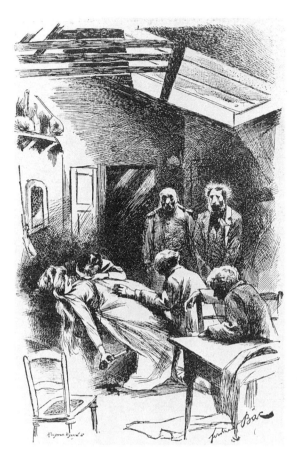

The Wild Duck at the Théâtre Libre, 1891. Naturalism called for the 'box-set'—walls and ceiling of a room.

February 1888, Ibsen's *Ghosts* in May 1890, and his *Wild Duck* in April 1891. And from the very beginning the Théâtre Libre spread its gospel by touring: Brussels and Berlin in January 1888, London in 1889. The readiness with which pioneers like the Théâtre Libre were invited and appreciated abroad is an indication of how truly international was the revolutionary artistic and ideological ferment of the period.

As news of the Théâtre Libre reached him, August Strindberg, Ibsen's great Scandinavian rival, was so impressed with Antoine's ideas that he sent him a French translation of his hitherto most naturalistic play, *The Father*. A correspondence developed between them, which, though *The Father* was not taken up by Antoine, led to Strindberg setting up a Scandinavian Experimental Theatre in Copenhagen in 1888. It opened in March 1889 with a triple bill of short plays by him. Originally it was to contain a fourth piece, but the most daring of these plays, *Miss Julie*, had been banned by the police. It could only be shown to a severely restricted audience at the University Students' Union in a single performance on 14 March 1889.

Miss Julie, the story of the aristocratic girl seducing her father's valet, was Strindberg's most radical implementation of naturalistic ideas, which he outlined in his preface to the play. It took sixteen years, till 1905, before *Miss Julie* could be publicly performed in Scandinavia. But Antoine produced it in January 1893. Even earlier, on 3 April 1892, it was shown in Berlin by an offspring of Antoine's Théâtre Libre, the Freie Bühne.

Germany had long been one of the focal points of modernist strivings in the theatre: the country was rich in impressive theatre buildings, derived from the playhouses that had, since the eighteenth century, served to bolster the prestige of the kingdoms, dukedoms, and principalities that had been unified by Bismarck in 1871. And there was a strong tendency in the country to embrace new ideas. The autocratic Prussian regime of the new Reich was vigorously opposed by a strong socialist movement. Ibsen had settled in Germany in 1868; he lived first in Dresden and from 1875 in Munich. His plays, which had regularly first appeared in book form, were immediately translated into German and had been frequently produced in spite of violent attacks and censorship bans. In 1880, for example, *A*

Doll's House in Berlin had to be given a good ending with Nora ruefully returning to her husband. *Ghosts* had been banned in Berlin by the police, but did reach a matinée performance there in 1887.

Among the leading champions of Ibsen and all he stood for was Otto Brahm, second critic of the *Vossische Zeitung* until he was dismissed for being too virulent in his condemnation of old-fashioned theatre. When a group of radical intellectuals decided to emulate Antoine by founding a society on the model of the Théâtre Libre, to present matinées of radical plays in performances closed to the general public, they invited Brahm, early in 1889, to take over its direction and to become the editor of a periodical, *Freie Bühne* (*Free Stage*), which would act as a focus for the activities of the enterprise. His manifesto published in its first issue summed up the naturalist creed:

The device on the banner of the new art is one word: Truth. And truth, truth in every

NATURALISM IN THE THEATRE

The basic principles of naturalism as outlined by Strindberg in the preface to *Miss Julie*, 1888.

. . . judgement of authors—this man is stupid, that one brutal, this jealous, that stingy and so forth—should be challenged by the naturalists who know the richness of the soul-complex and realise that vice has a reverse side very much like virtue. Because they are modern characters, living in a period of transition more feverish than its predecessor at least, I have drawn my figures vacillating, disintegrated, a blend of old and new . . .

My souls (characters) are conglomerations of past and present stages of civilisation, bits from books and newspapers, scraps of humanity, rags and tatters of fine clothing, patched together as is the human soul . . .

As far as the technical side of the work is concerned I have made the experiment of abolishing the division into acts. This is because I have come to the conclusion that our capacity for illusion is disturbed by the intervals . . .

As regards the scenery I have borrowed from impressionist paintings its asymmetry and its economy; thus, I think, strengthening the illusion. For the fact that one does not see the whole room and all the furniture leaves room for conjecture . . .

Another much needed innovation is the abolition of footlights. This lighting from below is said to have the purpose of making the actors' faces fatter. But why, I ask, should all actors have fat faces?

aspect of life is what we too aim at and demand. . . . the truth of the independent spirit that does not need euphemisms, that does not want to conceal anything. And which therefore knows only one adversary, its enemy to the death: the lie in every form.

On 29 September 1889 the Freie Bühne opened with Ibsen's *Ghosts*. So great was the impact of this performance that within a short time the number of members had reached 900. Even more controversial was the second presentation, the début of Germany's own pioneer of naturalistic drama, Gerhart Hauptmann, with *Before Sunrise*, a sombre and violent tragedy about the ravages of hereditary alcoholism. It caused an uproar.

So great was the impact of the Freie Bühne that members of the Social Democratic Party initiated an organization aimed at bringing this new type of drama to the working classes: the Freie Volksbühne (Free People's Theatre), founded in 1890, organized special performances for working-class audiences who paid only 50 pfennigs for their seats. Here the ideas about the educational and spiritual function of the new theatre seemed to come to fruition. Fritz Mauthner, the critic and philosopher, described in 1891 how moved he was observing the hushed concentration of such a working-class audience, how 'a congregation of believers listened with rapt attention to what gave them a new symbol for their beliefs; who could fail to feel that, after two thousand years, drama, secular drama, could once again have the effect of a sacrificial rite'.

Unlike Antoine, who had been an enthusiastic and experienced amateur actor, Brahm was a scholar and critic. He relied on professional stage managers to direct, but supervised the productions and discussed and explained his concepts to the actors, and so acquired an immense influence on the style of production. Ibsen, Strindberg, Hauptmann, and another German realist, Hermann Sudermann, formed the core of his repertoire. His most outstanding success was Hauptmann's *The Weavers*, which had appeared in print at the beginning of 1892, but only reached the stage in a 'closed' matinée on 26 February 1893 (being still under interdiction by the police). In some ways this play represented the culmination of naturalistic drama; it dispensed with a hero or, indeed, leading characters. Its hero was an entire social class; the play depicted the uprising of the Silesian weavers of the 1840s against the introduction of mechanical looms, each act showing a different stage of their plight and introducing new characters. Hauptmann had, in the interests of complete 'naturalness', written the play in strong Silesian dialect. For the Berlin performance it had to be translated into a more comprehensible idiom. As early as May of the same year Antoine directed *Les Tisserands* in Paris.

So fundamental was the impact of the Théâtre Libre and the Freie Bühne that as early as 1894 Brahm was offered the directorship of Berlin's most prestigious private theatre, the Deutsches Theater, while Antoine was made co-director of France's second national theatre, the Odéon, in 1896 (albeit for a brief period

only, as he resigned in a huff at the end of the year); he had received official recognition, the Legion of Honour, as early as 1900.

In England, as elsewhere, the cause of the new drama was linked with Ibsen's socially aware prose plays, the 'new woman', and the need to expose the hypocrisies of Victorian society. Chief advocate of the new playwright was the critic William Archer, who at the age of 24, in 1880, met his chief ally in this crusade, George Bernard Shaw. They soon recognized each others' affinities: Archer introduced Shaw to Ibsen and they even planned to write a play together. Archer, whose family had business connections in Norway, was able to translate the works of his idol. An adaptation of his version of *Pillars of Society* was performed in December 1880 at the Gaiety Theatre in the Strand, very radically toned down, with the new title *Quicksands*. It made no impact.

Four years later the time was not yet ripe for undiluted Ibsen: *A Doll's House* appeared in 1884 as *Breaking the Butterfly* and the unadulterated text could merely be given a private reading in January 1886 with Marx's daughter Eleanor in the part of Nora and Shaw himself as Hellmer. The breakthrough came in June 1889, when the actor-manager Charles Charrington presented *A Doll's House* at the Novelty Theatre in Kingsway with his wife Janet Achurch, a brilliant exponent of Ibsen's female roles adored by Shaw, as Nora. In spite of still vitriolic reviews this production did well—and Charrington and Achurch took it to Australia.

Very soon now London did acquire an equivalent of the Théâtre Libre. J. T. (Jacob Thomas) Grein had already been a theatre critic in his native Holland when circumstances forced the family to emigrate to England, where he earned his living working for a tea merchant in Mincing Lane. But he travelled, and was present at the first performance of the Théâtre Libre in Paris. He also produced English plays in Holland, and it was with £50 he had earned from this, augmented by £30 from supporters of the cause, that he decided to start a 'British Théâtre Libre' (later renamed the Independent Theatre) in London. He won support from prominent literary figures, among them Thomas Hardy, Meredith, George Moore, Henry James, and, of course, Bernard Shaw—who became founding members of the society. The opening performance on 13 March 1891 was Ibsen's controversial and frequently banned *Ghosts*. After the first night, at the Royalty Theatre in Dean Street, Soho, Shaw could report to Charrington in Australia that the play had been a success.

The Independent Theatre had made its mark, but was always in financial difficulties. It struggled on for seven years, producing a succession of interesting work by Ibsen, George Moore, Zola, and other contemporary pioneers, or rediscovering Webster's *Duchess of Malfi* and staging it as a 'realistic classic'. Grein can also claim to have set Shaw off on his career as a playwright. When he complained that he was lacking new plays by British authors, Shaw remembered the unfinished draft of the play he had started with Archer years before, sat down, and

IBSEN'S *GHOSTS* IN LONDON

George Bernard Shaw's account of the opening of the Independent Theatre on 13 March 1891, in a letter to Charles Charrington.

After the first act the applause was immense. After the second, a third of the applauders were startled into silence. After the third four fifths of them were awestruck. When Grein came out, very nervous, to make his speech, a lady in the stalls said naively: 'Oh, is that Ibsen?' . . . Then a man in the gallery cried out to Grein: 'It's *too* horrible' and was instantly met with a shout of 'Why dont you go to the Adelphi?' . . . Next day the devil was to pay in the papers. Scott [the critic of the *Daily Telegraph*] . . . went stark raving mad, and produced not only a column of criticism, but a leading article . . . in which he compared an Ibsen play to 'a dirty act done publicly', 'an open drain' and so on, demanding that the Independent Theatre should be prosecuted, suppressed, fined and the deuce knows what not.

finished it. *Widowers' Houses* had its first performance by the Independent Theatre at the Royalty in November 1892. But Grein was not, like Antoine or Brahm, an inspired director, nor was he a very good administrator, and he gradually lost his predominant position in the management of the Independent Theatre. But he had helped to establish the concept of a theatre aiming at more than shallow entertainment, that would contribute to the airing and discussion of the major issues of the time in Britain.

The cause was taken up by a rich and eccentric lady, Miss Annie Horniman, heiress of a vast tea fortune. She enabled one of Shaw's favourite actresses, Florence Farr, to start a season of advanced plays at the Avenue Theatre in London in 1893. Here Shaw's first play to be performed outside a closed society, *Arms and the Man*, received its first production. Miss Horniman was also a friend of W. B. Yeats and supported his efforts to create a theatre that would contribute to the movement for the cultural and political autonomy of Ireland. Yeats and Edward Martyn had started the Irish Literary Theatre in 1899, but it was through Miss Horniman's financial assistance that the Abbey Theatre in Dublin could open its doors in 1904—to become the centre of a renaissance of Irish drama (Synge, O'Casey) and the breeding ground of a tradition of fine realistic acting.

When the quarrels and contentiousness of the Abbey became too much for her, Miss Horniman transferred her attention to England: in 1908 she acquired the Gaiety Theatre in Manchester and opened it as a 'repertory theatre', that is, a locally established house producing serious drama. From there the repertory movement spread to Glasgow (1910), Liverpool (1911), Birmingham (1913), and in subsequent years throughout Britain and the English-speaking world.

London had acquired the equivalent of such a repertory house as a direct successor to the Independent Theatre in the Stage Society, led by Shaw's follower and friend, the actor, director, and important naturalistic playwright Harley Granville-Barker. Together with the impresario J. E. Vedrenne, Barker presented a succession of productions, including eleven plays by Shaw, at the Court (now Royal Court) Theatre in Sloane Square from 1904 to 1907.

In the United States the movement towards a 'new theatre', socially conscious and of high artistic ambition, got into its stride with the formation of Maurice Browne's Little Theatre in Chicago in 1912. Among the supporters of this venture were the writer and producer George Cram Cook and his wife, the playwright Susan Glaspell. They moved to New York in 1913 and started the Washington Square Players in Greenwich Village (1914) which, during their summers spent on Cape Cod, became the Provincetown Players, where Eugene O'Neill's first short plays were performed in 1914 (*Thirst*) and 1916 (*Bound East for Cardiff*). With the work of O'Neill and Glaspell the modern American theatre broke on to the world scene, while the Broadway theatre continued with its traditional fare—very like that of the London West End—of musicals and melodrama; in the first decades of the century the outstanding personality here was David Belasco, adaptor, playwright, and, above all, producer of performances rich in spectacular scenic effects of great realism.

It was one of the features of the modern movement in literature that many of its new impulses had come from the fringes of Europe: Ibsen and Strindberg from Scandinavia; and Gogol, Tolstoy, Turgenev, and Dostoevsky from Russia, where the opening of the country towards the west had released immense new creative economic and artistic resources. By the end of the century Russia had already produced a major realistic playwright, Alexander Ostrovsky, and brilliant plays by Tolstoy, Gogol, and Turgenev, but the style of acting and production had remained largely declamatory and old-fashioned. Here too the impulse for a radical renewal came from a young man who was a fanatically devoted amateur actor.

Konstantin Alexeyev, who took the stage name Stanislavski to spare his relatives embarrassment when he appeared in improper French farces, came from a wealthy family of merchants and industrialists. While working in the family firm, he devoted his spare time—and a good deal of his money—to appearing with his own and various other amateur companies. Although an admirer of some of the great old actors of the Moscow Imperial Theatre, the Maly, or 'little', Theatre (so called merely to distinguish it from the other imperial house devoted to ballet and opera, the 'large', or Bolshoi, Theatre), Stanislavski grew more and more dissatisfied with his own work and the prevailing style of acting and production in general. When a group of theatre artists and intellectuals founded the Society of Art and Literature in 1888 he enthusiastically joined; and, thanks to his financial contributions, he gradually assumed a leading position in its section devoted to dramatic performance and became responsible for directing its still largely amateur productions. Visits by the troupe of the Duke of Meiningen in 1885, and particularly in 1890, made an overwhelming impression on Stanislavski through the historical accuracy of their costume and décor, the brilliant use of space, the subtle movement and grouping of crowd scenes, and the freedom from pre-set clichés in the acting. He realized that to co-ordinate all these elements a new artistic function was needed in the theatre: the director.

Stanislavski's productions for the Society of Art and Literature, in which he continued to play leading parts (Othello, Benedict in *Much Ado*, Heinrich in Hauptmann's *Sunken Bell*), attracted the attention of another champion of a new approach to acting and production, Vladimir Nemirovich-Danchenko, playwright, literary manager at the Maly Theatre, and teacher at the theatre school of the Moscow Philharmonic Society. The class of 1897/8 at this conservatory contained so many young actors of outstanding promise that Nemirovich-Danchenko was looking for ways to keep them together as a company. It was in June 1897 that he and Stanislavski met at a café, the Slavic Bazaar, and in a historic

APHORISMS FOR A NEW THEATRE

Concepts governing the policy of the Moscow Art Theatre as jotted down by Stanislavski and Nemirovich-Danchenko at their historic meeting at the Slavic Bazaar in June 1897.

1. There are no small parts, there are only small actors;

2. Today Hamlet, tomorrow an extra, but even as an extra the actor must be an artist;

3. The playwright, the actor, the scene-designer, the dresser and the stage-hand all serve one purpose, namely, to express the playwright's main idea in writing his play;

4. The theatre begins with the cloakroom;

5. Every violation of the creative life of the theatre is a crime;

6. Arriving late at the theatre, laziness, capriciousness, hysteria, ignorance of parts, the necessity of repeating the same thing twice, are all equally harmful and must be rooted out.

The auditorium of the Moscow Art Theatre after reconstruction, 1902. Stanislavski's curtain did not rise, but parted.

conversation, which lasted from 10 in the morning till 3 a.m. the following day, hammered out the principles of a new professional theatre combining the best talents of Nemirovich's pupils and Stanislavski's semi-amateurs, which they planned to finance through acquiring shareholders: a new type of theatre which would cleanse production and acting of all the hoary stereotypes of the past. The gist of their creed was enshrined in a series of 'aphorisms'. Stanislavski and Nemirovich-Danchenko would be joint directors of the new enterprise, with Nemirovich responsible for the administrative and literary, Stanislavski for the production side. Throughout the summer rehearsals were conducted at a mansion outside Moscow lent to the company by one of its supporters. Stanislavski and the designer Simov travelled widely to collect authentic props and visual impressions for the opening play, Alexei Tolstoy's *Tsar Fyodor*, while the Hermitage, a dingy theatre in Moscow, was being readied for the new organization, at first called the Moscow Art and Popular Theatre, later shortened to Moscow Art Theatre. On 14 October 1898 the curtain parted (unlike all other theatres of the time Stanislavski's curtain did not rise, but opened to the sides) on his production of *Tsar Fyodor*. It proved an immense success. But financially the new enterprise remained on very shaky ground. Everything depended on the fourth production (after a *Merchant of Venice* and a Goldoni), a play which had already been performed in St Petersburg and failed—Anton Chekhov's *The Seagull.*

It was Nemirovich-Danchenko who had suggested the play to Stanislavski as particularly suitable: it contained (like *Hamlet*) theatre-within-the-theatre (a performance of an experimental, 'modernist' play); a shallow provincial actress who tried to thwart the talents of her much more gifted son, Treplev, a writer of great promise; a heroine seduced by the actress's lover, the equally superficial writer Trigorin; it was funny and satirical in places, but ended tragicomically with Treplev's suicide. Stanislavski at first failed to respond to this seemingly diffuse and understated text. But, gradually, when working on it, the beauties of this new type of drama become apparent to him: a play in which the surface dialogue concealed a much more important stream of subterranean action and emotion, where, as in real life, people hardly ever spoke openly about their feelings but covered them in trivial small-talk, a drama of atmosphere and subtle ironies.

The author, Anton Pavlovich Chekhov, a doctor who had made his reputation as a writer of short stories, had been so disappointed by the failure of his first major dramatic work that he was reluctant to let the Art Theatre do it. Only after he had witnessed some rehearsals did he relent. Chekhov's work required a new style of acting, without any flamboyant theatricality. *The Seagull* had failed because the old-fashioned approach of the St Petersburg production had been incapable of handling its dialogue. It was Stanislavski's conviction, which he had

Stanislavski as Satin in Gorky's play *Lower Depths* (playing the card, on the right), Moscow Art Theatre, 1902. Gorky was the most extreme representative of naturalism in Russia. *The Lower Depths* takes place in a doss-house inhabited by a selection of failed and criminal individuals.

reached after many failures in his own development as an actor, that only total emotional truth, the complete immersion of the actor in the character's feelings, could remove the cobwebs of the old stereotypes, and would enable him to master the demands of Chekhov's work. The opening night, on 17 December 1898, was a test of nerves. Stanislavski, who appeared as Trigorin, was terrified. But the immense success of *The Seagull* saved the Art Theatre. Momentous performances of Chekhov's other major plays followed: *Uncle Vanya* (October 1899), *Three Sisters* (January 1901), and *The Cherry Orchard* (January 1904). Chekhov died in July of that year, mourned by the actors who owed him so much. *The Seagull* became the Moscow Art Theatre's emblem, and still adorns its curtain in the Kamergersky Theatre, to which the company moved in the autumn of 1902. The conversion of this house, with the latest technical and lighting innovations, including an elaborate revolving stage, was financed by the railway millionaire Morozov, who had become a fervent supporter of the Art Theatre.

Ever since his beginnings as an amateur actor Stanislavski had been striving to get an insight into the secret of the actor's art: what it was that might lift it above mere exhibitionistic posturing and empty declamation into the realm of deep human truth. He had experienced such moments in the work of great star actors like Tommaso Salvini, the result of divine intuition, instinctive genius. But how could one find ways of achieving such results through systematic work, thought, and training? Gradually Stanislavski elaborated his own system, designed to enable the actor to achieve inner emotional truth: in order to live the part, he or she must get inside the mind of the character, must develop a thorough knowledge of the character's background and history. Even if appearing for only a brief moment, the performer should show the outward signs of the character's past history. In order to relive the character's emotions, the actor must call upon his own emotional memory and use the recollection of feelings of a similar nature he had experienced in analogous situations. Once that emotional truth was achieved, the actor's technical knowledge having become instinctive, these states of mind would find their natural expression on the stage.

The intense emphasis on truth in Stanislavski's system coincides with the basic impulse behind naturalism. Yet Stanislavski has stressed that his aims went beyond that: 'Those who thought that we strove for naturalism on the stage are mistaken. . . . Always we sought for inner truth, for the truth of feeling and experience, but as spiritual technique was only in its embryo stage among the actors of our company, because of necessity and helplessness . . . we fell now and again into an outward and coarse naturalism.' And, indeed, the Moscow Art Theatre soon went beyond mere naturalism; one of its greatest successes was a work of the leading playwright of symbolism, Maurice Maeterlinck's *The Blue Bird*; and two of the most experimental directors of the period, Yevgeny Vakhtangov and Vsevolod Meyerhold, started with Stanislavski.

Naturalism carried the seeds of these further developments within it and opened the way to 'symbolism', which explored deeper aspects of the human condition, by delving into dreams, strange states of mind, subtle feelings, atmosphere. Indeed, Chekhov's work in reflecting the haphazard casualness and aimlessness of outward existence already represents that step beyond a mere photographic realism. In his plays, while they give meticulously accurate pictures of reality, emphasis is on feelings underneath the surface, the subtlest of atmospheric nuances; and real objects like the cherry orchard become symbols for far more complex social and psychological aspects of life.

As had very soon become apparent, the same applied to the other great masters of 'modern' drama: Ibsen's realism had at first been perceived mainly as social and moral criticism, but his plays too depicted complex emotional conflicts underneath the surface of the dialogue and were rich in symbols like the wild duck, the master builder's desire to scale his tower, the white horses of the legend surrounding Rosmersholm. And Strindberg's determination to give a truthful account of his own states of mind soon led him into writing a 'dream play', a 'ghost sonata', and other plays depicting his own internal reality, rather than an 'objective' external reality.

Yet the playwright who became the embodiment of this new trend was the Belgian poet Maurice Maeterlinck, around whose first play, *La Princesse Maleine*, first published in 1889, the Parisian avant-garde rallied, under the banner of 'symbolism', in contrast to Antoine's naturalism. Maeterlinck called this dramatic fairy-tale, based on the brothers Grimm and set in a mythical past, 'a play in the spirit of Shakespeare for a puppet theatre'. It is the story of an unhappy princess whose love is thwarted in a series of reversals of fate; its theme is the mystery of life and death. Although Antoine also wanted to obtain the rights for the play, the young poet Paul Fort succeeded in securing them for his experimental group, the Théâtre Mixte (that is to say, a theatre open to all kinds of tendencies) which he later rechristened Théâtre d'Art. *La Princesse Maleine* did not at that time receive a performance, but it was Paul Fort's theatre that staged, in 1891, seminal performances of two of Maeterlinck's short symbolist plays, *L'Intruse* (*The Intruder*) and *Les Aveugles* (*The Blind*), typical of this new trend of plotless, atmospheric images of life.

L'Intruse is a scene showing a family (grandfather, father, uncle, daughters) in a room next to which the mother is lying gravely ill. The atmosphere is one of fear and waiting for something undefinable. The old, half-blind grandfather thinks he has heard someone enter the house. Soon after, the nurse appears to tell them that the mother has died. The intruder was Death. In *Les Aveugles* a group of blind people who have been taken on an outing by a priest are awaiting his return while resting in a clearing in the woods. But the audience can see that the old man is lying there, dead, having succumbed to a stroke. Thus, the play shows, we all pass our lives blind to what is happening to us in reality. The actor who played one of the blind men in this performance was a young man who soon after emerged as the leader of the symbolist movement in the theatre—Aurélien Lugné-Poë. Unlike Antoine and Stanislavski, Lugné-Poë was a fully trained professional actor. In 1888 he had entered the *conservatoire*, but, at the same time, under an assumed name (as the *conservatoire* did not allow its pupils to take professional engagements), he appeared as an actor in Antoine's Théâtre Libre, excelling, in spite of his youth, in old men's parts. He achieved his breakthrough as a director in 1892, with a production of Ibsen's most 'symbolist' play, *The Lady from the Sea*, with such success that Ibsen thereafter granted him sole rights to all his plays for France.

Maeterlinck entrusted Lugné-Poë with the world première of his new play, *Pelléas et Mélisande*, taking an active part in the rehearsal process and even suggesting the colour scheme for the costumes. The play was scheduled for a single performance at the Bouffes-du-Nord in Paris on 17 May 1893. This haunting play tells the story of the mysterious maiden Mélisande found wandering in the woods; Prince Golaud (played by Lugné-Poë himself) finds and marries her only to see her fall in love with his younger brother (it has survived in the world repertoire as the text of Debussy's opera). The performance made an immense impression: the characters moved in beautiful stylized scenery with the stately slowness of figures in a dream. Its resounding success enabled Lugné-Poë to launch his own company, the Théâtre de l'Œuvre, which, he proclaimed in its opening manifesto, was destined to be 'le temple du drame symboliste'. It opened with a performance of Ibsen's *Rosmersholm* on 5 October 1893.

Lugné-Poë's production of *Measure for Measure* at the Cirque d'Hiver, Paris, 1898. The influence of William Poël's ideas on authentic Shakespearian production is clearly visible (see p. 361).

The Théâtre de l'Œuvre did not acquire a permanent home till after World War I, but used a number of venues as they became available. Lugné-Poë also did a great deal of touring abroad, to Belgium, England, and Scandinavia, where he was acclaimed as a superb interpreter of Ibsen and Strindberg. The repertory was extremely wide-ranging—from Ibsen, seen as a master of subtle moods and mysterious symbols, to international classics, ranging from ancient Indian masterpieces like *Sakuntala* and *The Little Clay Cart*, to Maeterlinck's adaptation of John Ford's *'Tis Pity she's a Whore*, and Shakespeare's *Measure for Measure*. The symbolist creed, moreover, not only embraced the delicately lyrical and poetic, it also extended to the grotesque, caricature and distortion being another aspect of the subjective experience of reality.

One of the most momentous performances of the Théâtre de l'Œuvre was that of Jarry's *Ubu roi* on 10 December 1896. Alfred Jarry was a bizarre Bohemian genius, small of stature—almost a dwarf—an inveterate practical joker, absinthe drinker, and Rabelaisian philosophizer. His play *Ubu roi* had started as a schoolboy joke—a lampoon of an unpopular teacher, Hébert, nicknamed 'père Hébé', which became 'Ubu'. It is a savage satire on the unlimited greed and unscrupulous thirst for power of the French bourgeois; and a parody of *Macbeth* as well. Ubu and his wife, mère Ubu, make themselves rulers of Poland by ruthlessly killing anyone opposing them. Ubu even invents a 'debraining machine' for the purpose. Firmin Gémier played the part at the first performance. The first word he uttered was 'merdre!' (which might be rendered as 'sh-r-it!'), probably the first time the taboo word had been heard on the French stage. It created pandemonium in the audience: only after fifteen minutes could silence be re-established among the excited protesters and defenders of the outrage. Among the audience who witnessed this breaking of all taboos was not only the great symbolist poet Stéphane Mallarmé, but also a number of distinguished British visitors. Arthur Symonds, one of the main advocates of symbolism in Britain, has described the stage as the curtain opened: 'The scenery was painted to represent, by a child's convention, indoors and out of doors, and even the torrid, temperate and arctic zones at once. . . . On the left was painted a bed, and at the foot of the bed a bare tree and snow falling. On the right there were palm trees.' W. B. Yeats was also present. He later recalled that the players were 'supposed to be dolls, toys, marionettes' and that 'the chief personage seemed to be some kind of King [carrying] for a sceptre the kind of brush that we use to clean a closet'. And Yeats sadly reflected that, after all the strivings of the symbolists, 'after Stéphane Mallarmé, after Paul Verlaine, after Gustave Moreau . . . after our own verse, after all our subtle colour and nervous rhythm . . . what

Père Ubu. Costume design by Alfred Jarry for the first performance of *Ubu roi* at the Théâtre de l'Œuvre, Paris, 1896. The emphasis is on the grotesque marionette-like nature of the character, its deliberate evocation of childish cruelty.

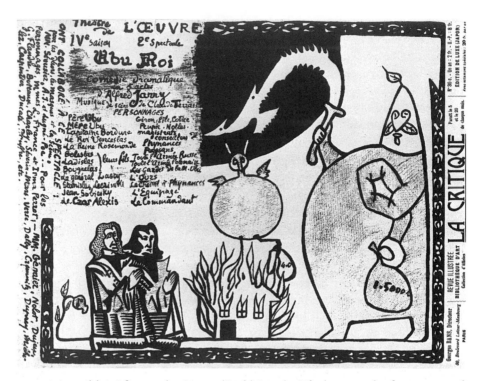

Ubu roi. Illustration by Jarry in the programme of the first performance of *Ubu roi*, 1896.

more is possible? After us the Savage God.' Jarry's *Ubu* became the forerunner of many influential forms of subversive art—Dada, surrealism, the theatre of the absurd—but its immediate impact, beyond the sensation of its scandal, was relatively slight.

In England the reaction of the theatrical avant-garde against naturalism was less pronounced. The aestheticism of the upholders of an art free of social, political, or moral reforming tendencies, the champions of 'l'art pour l'art', found a powerful spokesman in Oscar Wilde, who wrote his most characteristically symbolist play *Salomé* (1893) in French. Yet most of his other very successful plays—seemingly repeating the pattern of the prevailing drawing-room comedy—contain in their witticism and parody of upper-class manners and morals clear, if gentle, parallels to Jarry's anti-bourgeois iconoclasm. *The Importance of Being Earnest* (1895) is the prime example of this tendency and one of the finest comedies in the English language.

Another aspect of the reaction against naturalism in the English theatre, as well as a decisive rejection of the prevailing style of presenting the classics, was an attempt to return to the authentic performance of Shakespeare. The pioneer of this movement, William Poël, came from student theatre: he had directed a performance of the text of the first quarto of *Hamlet* at St George's Hall as early as 1881. In 1895 he founded the Elizabethan Stage Society. For the next ten years he continuously experimented with ways and means to re-create Shakespeare's own approach to staging, without painted sets and with uncut and unmanipulated texts, letting the words set the scene and allowing the flow of the action to proceed uninterrupted by long breaks for elaborate scene-changes. Poël was in personal touch with Lugné-Poë and they influenced each other. Poël sometimes had to use conventional theatres, but he preferred to put up his simple platform, surrounded by the audience on three sides, in the halls of livery companies or other authentic Elizabethan buildings: the hall of Gray's Inn, Merchant Taylors Hall, or the London Charterhouse. Lugné-Poë, who had seen some of these performances, described them in an influential article 'Shakespeare sans décor' in 1897, and a year later staged *Measure for Measure* in a circus, the Cirque d'Été, in Paris.

Poël's productions may at times have been over-pedantic and eccentric, yet his influence on the subsequent development of Shakespeare production has been immense. Above all, he showed that the seemingly haphazard structure of Shakespeare's plays, which the nineteenth century had tried to remedy by reordering the sequence of scenes and by radical cuts, was in fact a most efficient and rapid method of dramatic story-telling, a truly 'epic' theatre; and that historical authen-

William Poël's production of *Measure for Measure*, Royalty Theatre, London, 1893. Poël not only returned to a reconstructed Elizabethan stage, he also showed that the structure of Shakespeare's plays achieved an epic flow of the action superior to the Victorian attempts at pressing them into a five-act form.

ticity in costumes and décor was far from essential. Many actors who later achieved fame started with Poël, among them Edith Evans, Lewis Casson, and Robert Atkins. The subsequent development of the Shakespeare Memorial Theatre in Stratford-upon-Avon and of the Old Vic, after Lilian Baylis made it the London home of Shakespeare in 1912, owed a great deal to Poël's pioneering efforts.

But the most radical exponent of the reaction against naturalistic scenery and acting was an artist who exercised great influence on the future development of theatre design and directing (although he himself did not carry many of his own directing projects to fruition): Edward Gordon Craig. Son of the actress Ellen Terry and the architect E. W. Godwin (she was still married to the painter G. F. Watts when her son was born), Craig legally assumed the name Edward Gordon Craig in 1889. Ellen Terry later lived with Henry Irving, so Craig grew up in the milieu of the Lyceum Theatre and soon played parts in Irving's productions, and in the provinces. He even appeared as Hamlet, in Ben Greet's company at the Olympic Theatre in London in 1897 at the age of 17. Outstandingly handsome— and a great ladies' man all his life—his mother regarded him as an actor of exceptional talent. Yet he became more and more interested in the graphic arts and theatre design and directing. Through his friendship with the musician Martin Shaw he joined the Purcell Operatic Society at Hampstead; for them he directed, among other minor projects, Purcell's *Dido and Aeneas* (1900) and Handel's *Acis and Galatea* (1902). His designs and simple but highly effective direction of the amateur chorus were generally acclaimed by the critics. When, in 1903, Ellen Terry for a short time managed the Imperial Theatre, Westminster, she entrusted Craig with productions of Ibsen's *The Vikings* and *Much Ado about Nothing*. Although here he had to contend with problems of time and the resistance of established actors, he again achieved remarkable effects, largely through brilliant use of light. His fame as a designer had spread to Germany and in 1904 Craig was invited by Otto Brahm to design a production of an adaptation of Otway's *Venice Preserved* at the Lessing Theater in Berlin. Disagreements with Brahm aborted that project, but henceforth Craig lived outside England—chiefly in Italy and France—for the rest of his life. Increasingly disillusioned with the theatre as he had found it, he embarked on plans for a new kind of theatre school, and between 1904 and 1911 published a series of essays and Socratic dialogues, later issued as a book, *On the Art of the Theatre*.

Craig's ideas were extravagantly novel and utopian, yet he had no illusion that they could be put into practice immediately: 'I believe in the time when we shall be able to create works in the Theatre without the use of the written play, without the use of actors; but I believe also in the necessity of daily work under the conditions which are today offered us.' Rejecting any attempts at a photographic reproduction of reality, Craig called for simple designs, using large three-dimen-

sional structures on a variety of levels to create a rhythmically harmonious space.
He felt that light falling diagonally across the stage and creating strongly contrasting areas of illumination and shadow would create a new poetry of movement: 'it
is the large and sweeping impression produced by means of scene and the movement of the figures which is undoubtedly the most valuable means at [the director's] disposal' and 'by means of suggestion in movement you may translate all the
passions and the thoughts of vast numbers of people, or by means of the same you
can assist your actor to convey the thoughts and the emotions of the particular
character he impersonates. Actuality, accuracy of detail, is useless upon the stage.'
Movement and light: in 1930, listing some of the artists who had most been in
harmony with his ideas, Craig put in the very first place Isadora Duncan, the

American who had laid the foundations of modern dance (and with whom he had had a torrid affair), and two pioneers of lighting design—the Italian Mariano Fortuny, the inventor of a technique using reflected light, and Adolf Linnebach of the Dresden Court Theatre, who had developed 'shadowgraphs', magical lighting effects produced by the use of cut-outs placed in the path of direct-beam projectors.

But Craig went even further: his vision was of the director as a wholly autonomous creative artist:

When he interprets the plays of the dramatist by means of his actors, his scene-painters and his other craftsmen, then he is a craftsman; when he will have mastered the uses of actions, words, line, colour and rhythm, then he may become an artist. Then we shall no longer need the assistance of the playwright—for our art will then be self-reliant.

A theatre without playwrights, but also—this was Craig's most revolutionary, and often misunderstood, notion—without the actor as he then existed; Craig called for *Über-marionettes* instead of actors who relied on their personal appearance and charisma: '[The actors] must create for themselves a new form of acting, consisting for the main part of symbolical gesture. Today they *impersonate* and interpret; tomorrow they must *represent* and interpret; and the third day they must create. By this means style may return.' In fact, the *Über-marionette* would be 'an actor who reached such a state of mechanical perfection that his body was *absolutely* the slave of his mind'. At the time Craig's ideas seemed wholly utopian. Yet they foreshadow the late twentieth-century phenomenon of 'performance artists' like Robert Wilson, who create spectacles with a minimum of dialogue and rely mainly on brilliant design, lighting, sound, and actors moving in highly stylized attitudes.

While elaborating these ideas Craig undertook various projects, few of which came to be fully realized. He created sets for *Rosmersholm* for the great Italian actress Eleonora Duse. His most important commission was an invitation he received in 1908 to design and co-direct *Hamlet* with Stanislavski at the Moscow Art Theatre. Craig developed a concept based on a number of large panels that could be moved, in full view of the audience, into different configurations to suggest—without actually representing—the backgrounds to the action. But it was found impossible to move the panels with the flow of the action; and, to Craig's chagrin, the curtain had to be closed while the scenery was changed. The production opened at the beginning of 1912 and became an immense success, remaining in the repertoire of the Moscow Art Theatre for more than 400 performances. Yet it was not as Craig had conceived it and the experience finally convinced Craig that the theatre as it existed then was not for him. Many of Craig's ideas found a parallel in the work of an artist who later became his firm friend—Adolphe Appia, a French-speaking Swiss from Geneva, trained as a musician. Having seen Richard Wagner's last production, *Parsifal* in Bayreuth in 1882, Appia became obsessed with Wagner and his ideas for the renewal of drama and theatre as festive events of deep spiritual value, on the model of ancient Greek tragedy. But he felt equally strongly that there was an immense gap between the music and poetry of these works and the style in which they were being staged, even under Wagner's own auspices. To develop a new approach to their production Appia took lessons in drawing and acquired a thorough knowledge of stagecraft by working in a humble capacity backstage at theatres like the Vienna Burgtheater and the Dresden Hoftheater. In 1892 he began working on detailed plans for designs for Wagner's *Ring* and in 1895 published the first of a series of books elaborating his ideas in detail—*La Mise-en-scène du drame Wagnérien.*

Appia rejected the tradition of painted scenery. He felt that the two-dimensional pictures that surrounded the actors in the traditional staging of plays and operas contradicted the very essence of the theatrical action, which was the three-dimensional human figure moving in a three-dimensional space. The actor had to be seen against solid forms and move through a structured space on varied levels. Being essentially committed to music drama Appia insisted on harmonizing this movement in space with that of the music itself. The third essential element, producing its own poetry, was the light which would mould the space and bring out the sculptural and rhythmical aspects of the visual and musical experience. Appia's ideas remained theories for a considerable time—he was a withdrawn person and stuttered—although they were soon taken up by more astute practitioners of design. He had an opportunity to show what he could do when, in March 1903, he was offered a chance to direct—in the private theatre of the comtesse de Béarn in Paris—scenes from Byron's *Manfred* (with Schumann's music) and Bizet's *Carmen*. The daring lighting effects and beautifully stylized décor made a stir, but Appia received no further commissions. His career took a decisive new turn, however, when, in 1906, he met his fellow Swiss, the music-pedagogue Émile Jacques-Dalcroze, who had elaborated a system to teach music and rhythm to children through movement (eurhythmics) which in due course developed into one of the foundations of modern dance. Appia enthusiastically endorsed these ideas and became a close collaborator with Dalcroze. When Dalcroze was invited (in 1910) by the founders of an advanced garden-city-type utopian settlement at Hellerau near Dresden to establish a Eurythmics Institute there, Appia worked with him on the design of the building that would house the school and its theatre-space.

Appia opted for a space in which there was no difference in level between the auditorium and the area at the back which might be used as the stage—it was in fact one of the first entirely flexible theatre-spaces. In the summer of 1912 the new building was the scene of the first Eurythmics festival, for which Appia designed and directed scenes from Gluck's *Orfeo ed Euridice*. It was above all a triumph of

Appia's ideas about lighting: he used a background of light diffused through gauze, against which beams of concentrated light could mould the moving figures of the singers and chorus and produce fascinating contrasts of illumination and shadow. In the following year the whole of *Orfeo* was presented by Appia at the Hellerau Festival; he also designed Claudel's *L'Annonce faite à Marie* (which the author himself directed). Appia's ideas had a very great influence not only on the production of music drama, but on theatrical practice in general. He was one of the first to elaborate in detail the concept of the director as the principal creative force in modern theatre and to develop, in both theory and practice, new forms of design and, above all, lighting.

Theatre design by Appia. For unspecified Wagnerian music dramas, his use of simple forms and subtle lighting effects produce three-dimensional space (1909).

The master director who combined Craig's and Appia's insights with those of Antoine, Lugné-Poë, and the other pioneers of the new theatre, while also adding a strong component of original ideas and practices, was Max Reinhardt, perhaps the most complete embodiment of that new phenomenon in the theatre: the director as an independent creative force. Reinhardt (originally Goldmann) came from a family of Viennese Jewish merchants. In later years he used to stress the influence of the Viennese theatrical tradition on his outlook: a tradition of extravagant baroque theatricality, mixing sublime poetry with spectacular visual effects, broad humour, and a dose of vulgarity—Mozart's *Magic Flute* is the archetypal Viennese folk-play. From his earliest youth Reinhardt became addicted to theatre; as an actor he excelled in playing old men even before he was 20. Otto Brahm saw him and eventually gave him an engagement at the Berlin Deutsches Theater, the temple of naturalism, from the season of 1895.

For seven years he acted in Brahm's repertoire of naturalistic drama—mostly about the lives of poor people in mean rooms. He became a pillar of the company, but more and more bored and disillusioned with the drabness of the work he was offered. So, to diversify their material, Reinhardt and a group of his colleagues started a cabaret, *Schall und Rauch* (*Sound and Smoke*), in which they performed parodies of classical drama, sang satirical ditties, and enjoyed themselves. In 1902 this group acquired its own premises, the Kleines Theater (Little Theatre), in which they also began to present full-length plays. Reinhardt, who had become the leader of the group, formally severed his connection with Brahm on 1 January 1903. The new venture was so successful that soon they could take over another house, the Neues Theater am Schiffbauerdamm ('New Theatre on the Shipwrights' Quay'), which fifty years later became the home of Brecht's Berliner Ensemble.

In these two theatres the new company, with Reinhardt soon the unchallenged leader, cultivated a repertoire representing the reaction against naturalism as practised by Brahm: the later, dreamlike Strindberg; the extravagant advocate of sexual emancipation, Frank Wedekind; Chekhov; the Austrian neo-romantic poet Hugo von Hofmannsthal; the Viennese portrayer of subtle eroticism, Arthur Schnitzler;

the Oscar Wilde of *Salomé*; these were the mainstay of their modern repertoire, together with the established German classics and, above all, Shakespeare. It was with a production of *A Midsummer Night's Dream* (31 January 1905) that Reinhardt achieved his breakthrough: the enchanted wood was both real, with three-dimensional trees and plants, and poetical. The action flowed rapidly on a revolving stage, and the contrasts between elemental spirits, courtiers, and mechanicals were brilliantly highlighted in an acting style which combined the naturalness of Brahm with the melody of classical verse-speaking, and could also produce gales of laughter. *A Midsummer Night's Dream* was for Reinhardt the complete embodiment of all aspects of the theatre; he returned to it over and over again in the course of his long career.

SOMMERNACHTSTRAUM

The revolving stage. Model for Reinhardt's production of *A Midsummer Night's Dream*, Berlin, 1905. By turning this three-dimensional design in various directions different scenes could swiftly be established.

When Brahm gave up the directorship of the Deutsches Theater, Reinhardt was appointed his successor. On 31 August 1905 he entered the theatre where he had earned his laurels as an actor as its artistic director. He was not yet quite 32 years old.

Reinhardt was a highly intelligent and well-read man, but he was also a consummate showman of truly catholic range—from sublime cerebral drama to operetta and pantomime. His ambition was to leave no aspect of the theatre's potential unexploited. The Deutsches Theater was a large house, suitable for the classics. But Reinhardt agreed with Strindberg's ideas that plays exploring psychological subtleties required a more intimate environment. One of his first objectives on taking over the Deutsches Theater was the construction of a second, more intimate venue next to the old building—the Kammerspiele (Chamber Theatre). Here the rigid separation of stage and audience was broken down by letting the architectural treatment of the auditorium carry across on to the stage, and by keeping the stage only a small step above its floor level. The spectators—no more than 400—could thus get the illusion of being almost in the same room with the characters of the play. The Kammerspiele was opened on 8 November 1906 with a performance of Ibsen's *Ghosts*, with sets designed by Edvard Munch—this moved the old masterpiece of naturalism on to a new, almost expressionist level, the colour scheme of the Alvings' home exuding evil.

In his larger auditorium Reinhardt dazzled his contemporaries with a succession of productions of the classics which dispensed with flat, painted scenery and assured a smooth progress of the action by an ingenious use of solid elements like towers or arches that could be moved into different positions (analogous to ideas expressed by Craig, whom Reinhardt repeatedly tried to get to design for him, but always failed because of Craig's insistence on complete control), and, above all, by the revolving stage: a single construction containing, for example, the outside and inside of a house, or a landscape with a river between two banks, which could, by being turned into different positions, present a number of distinct environments, yet retain a stylistic and organic unity.

Reinhardt, aware that Germany's electrical industry was rapidly developing new technologies, actively encouraged their researchers to provide him with ever more ingenious lighting devices. He spoke of painting the stage with light. The more he saw the newly emerging function of the director as that of an independent creative artist, the more Reinhardt was drawn to spectacle in which the visual elements and the expressive movement of the actors would predominate over the 'script'. This led to his producing a number of *Pantomimen.* They were not 'pantomimes' in the English sense, but mime plays in which there was a minimum of text and the director merely filled a brief scenario with visual delights: movement, dance, colour, gesture, and dazzling light effects. *Sumurun*, an oriental tale from the Arabian Nights, was the first of these; it opened in the Kammerspiele in April 1910 and caused such a stir that Reinhardt was invited to take it to London (Coliseum, February 1911), New York (January 1912), and Paris (May 1912).

The most ambitious of these ventures, *The Miracle* (subtitled 'a wordless mystery spectacle'), was first staged by Reinhardt in London at the Olympia exhibition hall at Christmas 1911. Here Reinhardt had freed himself from the constraints of the traditional theatre building. The huge hall was transformed into the interior of a cathedral with spectacular stained glass windows; the spectators were placed at the sides, while the action took place in the centre. The story, devised by the neo-romantic German poet Karl Vollmöller, was that of the little nun enticed into

The Miracle. Sketch by Ernst Stern for Reinhardt's production of the spectacle at Olympia, London, Christmas, 1911. The vast hall was transformed into a Gothic cathedral, with the audience sitting at the sides, the action taking place in the centre.

the world by a seducer, who, after many misfortunes and tribulations, returns to her convent to find that the statue of the Madonna over the altar has taken her place and fulfilled her humble duties during her absence. After the nun had left the cathedral the stained-glass windows faded away and the acting area became the scene of her adventures in the world. Then, when she returned, the cathedral magically reappeared. With this spectacle Reinhardt established himself as master showman to the world. *The Miracle* was restaged in Vienna (September 1912), Berlin (at the Zirkus Busch, April 1914), and after the war in New York (1924), touring major American cities until 1929.

So universal and all-embracing (his detractors would have said eclectic) was Reinhardt's range that he sought ever wider venues for his experiments. He who had created the Kammerspiele also craved for vast arenas in which to deploy large masses of humanity. As early as September 1910 he had staged an austere *Oedipus* in a concert hall in Munich; this production later transferred to the Zirkus Schumann in Berlin; here crowds of extras, the people of Thebes, formed a substitute for scenery. Reinhardt liked the Zirkus Schumann so much that in December 1911 he used it again for an equally austere production of the English medieval mystery play *Everyman* (which William Poël had earlier resurrected). Reinhardt's friend Hugo von Hofmannsthal adapted this text into a sinuous archaic German. The circus building suited Reinhardt so well that he acquired it and decided to transform it into a 'theatre for five thousand'. The architect Hans Poelzig was entrusted with the task of creating this giant house: the plan was that, from a relatively low-ceilinged foyer, the audience was to be led, with a sudden shock, into the vast arena that opened up like a giant fairytale castle bathed in light. The open arena stage allowed the action to be taken right into the centre of the auditorium. World War I delayed its completion, but Reinhardt opened it, renamed the Grosses Schauspielhaus, in November 1919, with the *Oresteia* of Aeschylus.

So now Reinhardt had at his disposal in Berlin a chamber theatre, a traditional proscenium house, and a vast open stage. Yet, as the experiment with *The Miracle* had shown, he wanted to widen his scope even further and to take drama out of traditional theatre spaces altogether, into settings naturally fitted for a particular spectacle. Having started his acting career at the municipal theatre of Salzburg, Mozart's birthplace and one of the gems of baroque architecture, Reinhardt conceived the idea of a performance that would exploit the setting and literally involve the whole city. The cathedral square in Salzburg is completely enclosed, and one enters it through arcaded archways. The façade of the cathedral fills one side, its entrance adorned by large statues of saints. Reinhardt decided to perform Hofmannsthal's adaptation of *Everyman* against this background. A simple podium in front of this façade was the stage; but the whole town, nature itself, played its part in the performance. When Death called Everyman, voices from all the church-towers in the vicinity echoed Everyman's name. As Faith and Good

Works in the end accompanied the dying sinner to the Last Judgement, the al-legorical female figures appeared as though the baroque statues in front of the church had come to life; and as they entered the brightly lit interior of the cathed-ral, the sun was just setting and darkness fell over the square. The start of the performance each afternoon was synchronized with the time of the sunset. And all the bells of Salzburg pealed as the repentant rich man's soul entered eternity. This performance, first presented on 22 August 1920, became the centrepiece of the Salzburg Festival (the first of a growing number of similar theatre and music festivals throughout Europe), of which Reinhardt remained the artistic director until the Nazi occupation of Austria in 1938. And each year *Everyman* was performed in Reinhardt's production.

Inside his theatres Reinhardt relied for his visual magic on a series of outstand-ing designers, Ernst Stern, Karl Walser, Alfred Roller among them; he was, above all, a supreme director of actors. Being himself an outstanding character actor he had the uncanny knack of showing his actors how to use their own essential personalities, not how he, Reinhardt, would have done the scene. Thus Rein-hardt's actors all developed their own individuality to the full, without any unifor-mity of approach or style. A whole generation of great actors and actresses, many of whom attained world fame through the cinema, was thus formed by Reinhardt: Alexander Moissi, Emil Jannings, Paul Wegener, Werner Krauss, Lucie Höflich, Getrud Eysoldt, Tilla Durieux, Elisabeth Bergner, among many others.

Reinhardt's world-wide success before the Great War clearly established the function of the creative director as a new artistic force in the theatre. His theatres in Berlin and his touring ventures were commercial enterprises, without state subsidies of any kind, but the Prussian state theatres and the other court and municipal theatres of Germany and Austria were gradually influenced by Rein-hardt's methods and ideas and continued to function side by side with the boule-vard theatre in Berlin and other larger cities.

In Britain it was the repertory movement which attempted to establish some-thing like the Central and East European type of non-commercial permanent home for the classics and innovative contemporary drama; and in the first decade of the century it had achieved impressive results—not only in the revival of Irish drama at the Abbey Theatre in Dublin, but also with the emergence of the vigor-ous Manchester school of north-country comedy (Stanley Houghton, Harold Brighouse) at the Gaiety in Manchester. But in London the scene was still domi-nated by a breed of powerful actor-managers who continued the traditional approach to production, an antithesis to the new concepts advocated by Craig or Appia, and practised by Stanislavski or Reinhardt. While new ideas of scenery and lighting made themselves felt in this actor-manager-dominated theatre, the emphasis still remained on the leading player and there was much less of a unitary concept behind productions. Shaw, aware of this feature of the London stage, had aimed at star actors, in order to reach the largest possible audience with his political and social message. He wrote *The Man of Destiny* for Irving—but Irving rejected it; his Caesar in *Caesar and Cleopatra* was written for one of the finest actors among the actor-managers, Johnston Forbes-Robertson, the outstanding Hamlet of his generation. Having opened in America, Forbes-Robertson appeared in Shaw's play at the Savoy in November 1907.

Herbert Beerbohm Tree, also a fine actor, had built himself His Majesty's Theatre in the Haymarket in 1897 and continued Irving's tradition of sumptuously spectacular productions of Shakespeare; he ventured on Shaw in 1914 with *Pygmalion*. Tree appeared as Higgins, Mrs Patrick Campbell, much beloved by Shaw, as Eliza Doolittle. The rehearsals, conducted by Shaw himself, as Tree was mainly concerned with his own performance, were stormy—and interrupted by Mrs Patrick Campbell's elopement with a new bridegroom. The rumour had spread in advance of the first night that one of the great taboos of stage decorum would be broken. Basil Dean, later an important London producer who acted as factotum to Tree, reports in his memoirs that

on the first night everybody was on tiptoe with expectation. Would Mrs. Patrick Campbell really say 'bloody' on the stage? Rumours of a public protest at the threatened depravity had been circulating. When the moment came I was standing in the prompt-corner . . . We heard a vast intake of breath, and then a vast sigh of relief. The trumpets of Shaw had at last breached the walls of the Lord Chamberlain's Jericho.

For twenty-seven years Sir George Alexander, the most suave and elegant of drawing-room bon viveurs, ran the St James's Theatre. It was he who, in 1895, had staged Oscar Wilde's greatest triumph before his fall, *The Importance of Being Earnest*, and numerous plays by the dramatists who, having also excelled in farce and melodrama, had made the socially conscious problem play à la Ibsen into a commercial commodity, Arthur Wing Pinero and Henry Arthur Jones.

At Wyndham's theatre Gerald du Maurier, who had created Captain Hook in Barrie's *Peter Pan* (1904) and the amiable rogue Raffles (1906), excelled in light comedy. Lewis Waller, the handsomest of men, specialized in romantic melodrama from d'Artagnan in *The Three Musketeers* to another French charmer, *Monsieur Beaucaire*, while Sir John Martin-Harvey broke every heart as Sidney Carton in the *Tale of Two Cities*.

The outstanding pioneer of new drama and the modern approach to production, Harley Granville-Barker, had also, after the end of his repertory season at the Court Theatre, gone into the commercial theatre. In an attempt to start a repertoire in the West End, at the Duke of York's Theatre, financed by the American impresario Charles Frohmann, Granville-Barker directed his own play *The Madras House* and *Justice* by Galsworthy, as well as Shaw's *Misalliance*. When this failed, he and his wife Lillah McCarthy went into management themselves, stag-

Much Ado About a Kiss
The opening scene of Act II. of *The Madras House*
SKETCHED AT THE DUKE OF YORK'S THEATRE BY H. M. BATEMAN

ing, among many new plays by Shaw, an ambitious production of Thomas Hardy's vast epic *The Dynasts* in 1914.

Granville-Barker's marriage came to an end; and after serving in the war he married a rich American lady and virtually withdrew from active work in the theatre. He could have become the English equivalent to Reinhardt; as his brilliant *Prefaces to Shakespeare* show, he had a similar approach in analysing a text and, as has only decades later become apparent, he was perhaps, next to his friend and ally Shaw, the outstanding playwright of the period in England. His *The Voysey Inheritance* (1905), *Waste* (banned by the censor, 1907), and *The Madras House* (1910) are the finest fruits of naturalism in English. Granville-Barker was also one of the movers of a renewed effort to persuade the powers that be to create a national theatre in Britain; together with William Archer he published a detailed blueprint of such a project as early as 1907, but it took more than half a century till it was finally realized.

Another actor-manager who helped to lay the foundations of the present structure of Britain's national theatres was Frank Benson. He had started his own touring company mainly devoted to Shakespeare as early as 1883. Himself a notable sportsman, he was reputed to recruit his company as much for their cricketing as for their acting skills. This company provided the annual Shakespeare season at Stratford-upon-Avon each summer for almost thirty years (1891–1919) while touring with their productions for the rest of the year. From these beginnings the Royal Shakespeare Company evolved.

A 'harem of industry' on stage. The scene from Granville-Barker's *Madras House* (1910) shows Miss Yates, an unmarried and pregnant employee of a fashion house, being investigated having been seen kissing Mr Brigstock, another employee.

Shakespeare was also the mainstay of the Old Vic, another institution which gradually evolved into a non-commercial theatre. One of the great Victorian social reformers, Emma Cons, had taken over the theatre south of the river, then named the New Victoria Palace and operated as a music hall known as the haunt of drunks and prostitutes. She remodelled it, to provide clean, non-alcoholic entertainment for the labouring masses, as the Royal Victoria Hall and Coffee Tavern. William Poël, the reforming producer of Shakespeare, was named its first manager, from 1880 to 1883. At first scenes from Shakespeare together with musical numbers were offered at low prices to working-class audiences. It was only after Emma Cons's niece Lilian Baylis was entrusted with running the Old Vic in 1912 that it became the home of a continuing Shakespeare cycle and developed into something like an unofficial national theatre run as a charitable trust.

On the other side of the channel the Comédie-Française and the other state-assisted theatres in Paris continued on their centuries-old course of presenting the classics and were more and more influenced by the innovations brought about by pioneers like Antoine and Lugné-Poë. These state theatres were the nurseries of great acting talent; but the star actors they produced often tended to exploit their fame by starting their own companies and dominating the Boulevard as actor-managers as powerful as their London counterparts. The Paris boulevard theatre continued to produce large numbers of comedies, farces (like those of the brilliant Georges Feydeau), melodramas, and operettas. But here too the influence of Ibsen, and of the new ideas about scenery and lighting began to be felt.

The most successful playwright who succeded in blending the anti-naturalistic strivings of the symbolists with solid melodramatic effects using witty and free-flowing verse was Edmond Rostand. Having achieved his first success in 1894 at the Comédie-Française with *Les Romanesques* (which became a record hit in its English adaptation, *The Fantastics*), his most famous work was produced on the Boulevard, at the Porte-Saint-Martin in 1897 by one of France's most celebrated actor-managers of the period, Constant-Benoît Coquelin. A brilliant character-comedian and scion of a whole family of actors, Coquelin created the title-role in *Cyrano de Bergerac*, while Sarah Bernhardt, acknowledged as one of the world's greatest actresses, appeared as the female lead, Roxane. For her Rostand wrote his next play *L'Aiglon*, the story of Napoleon's ill-fated son by the Austrian Emperor's daughter. For Coquelin Rostand wrote another play, *Chantecler*, set in a barnyard, with the rooster as the hero. But before the opening—in February 1910—Coquelin died; his place was taken by another leading actor and actor-manager of the Parisian scene, Lucien Guitry, whose son Sacha Guitry continued the tradition both as author of light comedies and as accomplished bon vivant.

Sarah Bernhardt not only managed her own theatres in Paris, she was one of a breed of world star actors who toured Europe and America with their own companies in this epoch; her golden voice and svelte figure and the intensity of her emotions turned heads everywhere. She even attempted Hamlet. Her rival for the title of the world's greatest actress at the time was Eleonora Duse, who dazzled her public with her serene beauty and stillness at moments of high passion, and at a very early stage ventured into the new drama—Zola, Ibsen, the Italian naturalists (Verga's *Cavalleria rusticana*), and, above all, the work of her lover, the Italian poet Gabriele d'Annunzio. Touring the world with her own company she captivated the imagination of poets and artists everywhere.

Another touring company that burst upon Paris in 1909—and later London—with an impact that affected stage design and practice in multifarious ways was that of the Ballets Russes, managed by the flamboyant impresario Sergei Diaghilev. The Russian ballet had always been pre-eminent, but under the influence of modern dancers like Isadora Duncan the rigid formalism of the classical tradition had been modified by a choreographer of genius, Michel Fokine, who introduced a far greater element of mime and expressive acting and costume, while retaining the essence of the classical steps. The upsurge of modernism in Russian painting at that period enabled Diaghilev, moreover, to enlist a number of outstanding painters to create scenery and costumes of overwhelming vitality with daring combinations of the most vivid colours: Alexandre Benois, Leon Bakst, and Natalia Goncharova. Outstanding dancers like Vaslav Nijinsky, Ida Rubinstein, Tamara Karsavina, and Anna Pavlova assured the triumph of the first season, which was followed by others, until finally, in 1913, Diaghilev severed his connection with the Russian state theatres and settled his company in Paris.

Enlisting the collaboration of leading French as well as Russian composers and painters Diaghilev created a series of dazzling spectacles—*The Firebird* (1910, music by Stravinsky; scenery by Bakst and Golovine); *Daphnis and Chloe* (1911, music by Ravel; scenery by Bakst); *L'Après-midi d'un faune* (1912, music by Debussy; choreography by Nijinsky; design by Bakst); *The Rites of Spring* (1913, music by Stravinsky; choreography by Nijinsky; design by Roerich); *Parade* (1917, libretto by Cocteau; music by Satie; design by Picasso), to name but the most momentous. The exuberance of colour and the wild emotional intensity of these spectacles greatly influenced the use of colour in the theatre: colour became an expressive element outside any merely descriptive or representational function. The Ballets Russes pointed towards a theatre of temperament, overflowing vitality, and intoxicating spectacle. There were even attempts to create ballets without dancers, relying solely on shifting forms and colours. The Italian futurist painter Giacomo Balla designed such a spectacle for Diaghilev: *Feu d'artifice* (music by Stravinsky), which was staged in Rome in 1917.

Yet at the same time a reaction against spectacularly visual theatre made itself felt in France, a call for a return to simplicity, intellectual clarity, and the pure impact of the poetic word. In November 1910 the critic and writer Jacques Rouché revived the Théâtre des Arts (from which Paul Fort and Lugné-Poë had emerged some twenty years earlier). Rouché had just returned from a journey to Germany and Russia and had published his impressions in *L'Art théâtral moderne*, which for the first time made the ideas of the pioneers of modern production known in France. Rouché reported on the work of Reinhardt and on the experiments of the Munich Arts Theatre of Georg Fuchs, who had created what he called a 'bas-relief' stage, a shallow acting area where the scenery was restricted to a backdrop and a few properties, and had laid the main emphasis on dialogue. He had also been deeply impressed with the experiments of the studio of the Moscow Art Theatre, then directed by Vsevolod Meyerhold in his symbolist phase (which preceded his more famous revolutionary constructivism of the 1920s); and he gave an account of the ideas of Craig and Appia.

These ideas had an important impact on a young critic, Jacques Copeau, a contributor to the *Grande Revue*, edited by Rouché. Copeau was a member of the circle of intellectuals around the great man of letters André Gide and became, with Gide, one of the co-founders of France's most influential literary periodical *La Nouvelle Revue française*, which first appeared in February 1909. Having discovered Craig and Appia through Rouché's book, Copeau became a fervent advocate of the creative director as the true artist of the theatre. As he put it in an article in the September 1913 issue of *NRF*: 'It is the totality of the movements, gestures, and attitudes, the congruence of facial expressions, voices, and silences, it is the whole of the theatrical performance, which is the product of a single thought that conceives, regulates, and harmonizes it.' Copeau opened his own theatre, the Vieux-Colombier (Old Dovecote) on 22 October 1913 with a performance of an Elizabethan play, *Une femme tuée par la douceur* (Heywood's *A Woman Killed with Kindness*), followed by a short play by Molière, *L'Amour médecin*. The choice of an Elizabethan text was significant in that Copeau too believed in using a minimum of scenery, relying on the words to paint the environment. The greatest success of the first season was *La Nuit des rois* (Shakespeare's *Twelfth Night*); the repertoire of modern plays was based around the authors of the *NRF*, above all Claudel, Henri Ghéon, and Jules Renard.

Wanting to free the stage from all clutter, all inessential ostentation, all ranting and flamboyant declamation, Copeau demanded a 'tréteau nu' (a naked stage).

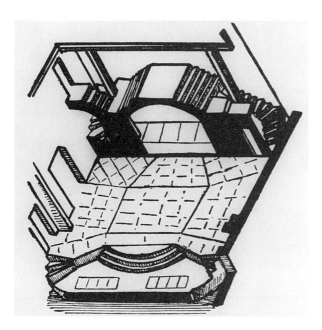

The Vieux-Colombier. Design for the reconstruction of the stage by Jacques Copeau and Louis Jouvet, 1919–24. Steps leading to a platform at the back create an elevated section on the lines of the Elizabethan playhouse, the venue of the action often being indicated by a backdrop below the platform.

While he believed in the 'lustre et grandeur' of the theatre, its high intellectual and moral mission, he also saw it as essentially free play, playfulness, hence his fascination with mime, improvisation, *commedia dell'arte* (in the same spirit as Reinhardt in Germany, Meyerhold in Russia).

If Copeau reacted against spectacle with a demand for simplicity, clarity, and stylization, the opposite tendency dominated the newest and most radical avant-garde in France, Russia, Germany, and Italy. As so often it was painting which blazed the trail: the impressionists' depiction of light and atmosphere engendered a violent response in the ideas of cubists, 'Fauves' (wild ones), and expressionists. In Italy Filippo Tommaso Marinetti proclaimed the creed of 'futurism'—an art capable of depicting the speed, mechanization, and machine-like rhythm of modern life by explosive images of simultaneity of movement and abbreviated expression. In Russia the term 'futurism' was taken up, though in a somewhat different sense, by poets like the young Vladimir Mayakovsky and painters like Kasimir Malevitch, one of the first to introduce the ideas of abstract painting.

In Germany similar iconoclastic tendencies emerged as a vigorous 'expressionist' movement which produced some great poetry and much very interesting drama. Its roots go back to the Strindberg of works like the *Dream Play* or *To Damascus*—in which the dramatist projects his dreams and neuroses on to the stage, turning the drama into an expression of an individual's highly subjective, and necessarily distorted, view of the world—and to the grotesque images of Wedekind's plays like *Spring Awakening* or *Earth Spirit*. Depicting an entirely subjective view (the protagonist's, who tends to be a double of the author), expressionist drama puts the leading character centre-stage—usually in the cone of a spotlight focused on him—while all other characters become secondary, and often mere abstractions (The Woman, the Boss, the Father). If impressionism focused on the loving depiction of detail and atmosphere (as in Maeterlinck and Chekhov), expressionism burst forth in direct address to the audience; it shunned the slowness of building a picture by piling detail on detail, and used an abbreviated, impatient approach. Its language (like that of Marinetti's futurism) became a kind of urgent telegraphese or, in contrast at climactic points, long, elaborate, self-defining tirades thrown by the protagonist at the audience in explosive bursts.

The themes of German expressionist drama circled around the need to create a 'new man' and the rejection of the stuffy respectability of the nineteenth century, the world the parents of these writers had created. Many of their plays depicted deadly conflicts between generations. *The Beggar* by Reinhard Johannes Sorge, usually regarded as the first German expressionist drama, shows a young poet who kills his parents because he cannot endure the sight of the miserable bourgeois life they are leading. Written as early as 1911–12, it was staged by Reinhardt in December 1917, a year after its author had been killed on the Western Front. *The Son* by Walter Hasenclever, first performed in Prague in 1916, also deals with parricide;

EXPRESSIONISM IN THE THEATRE

A definition of the essence of Expressionism by the German poet
Theodor Däubler, 1916.

It is a popular belief: when someone is being hanged, in his last moment he lives through his whole life. That must be Expressionism!

Speed, simultaneity, the highest tension about the interrelatedness of one's vision—these are the preconditions for the style. The style is Expression of the Idea.

A vision wants to express itself in its ultimate concision in the sphere of extremest simplification. That is expressionism in any style.

Everything that is, experiences, culminates in something spiritual. Every event becomes typical of itself.

The centre of the world is within each ego. . . . The frame of the picture becomes the limit of its comprehensibility. . . . When we are creating, it is we who give the thing its spiritual focus, which at the same time must be the centre of its being as an object. Through it, it has a core. From that core streams our own passion. We are astonished that we could thus sum it up, thus intensify it.

the protagonist, held in the spotlight, monologizes about his conflict with his father.

Carl Sternheim excelled in bitter satirical comedies about the life of the stuffy bourgeosie. Georg Kaiser, the most prolific among the expressionist dramatists, constructed intellectual conundrums with immense ingenuity; his play *From Morning to Midnight* (first performed in Munich in 1917) is the most often cited example of the genre: it shows the adventures of a little clerk who embezzles a large sum of money and tries to achieve happiness with it, only to end in martyrdom. The structure of this play, as of many others of its kind, is that of *Stationendrama* in which the protagonist passes through a number of distinct encounters analogous to the stations of the cross in Catholic churches. It took some time to find the right style of performance for these expressionist works. Even Reinhardt and his directors, who staged many of them in a cycle entitled *Das junge Deutschland* (*Young Germany*), tended to keep them within a more realistic convention.

In France, where Jarry's *Ubu roi* had anticipated this radical avant-garde by almost twenty years and where cubism in painting had also dissolved the representation of nature into broken images of simultaneity, the champion of cubism among poets, Guillaume Apollinaire, tried to embody its principles in a satirical play *Les Mamelles de Tirésias* (*The Breasts of Tiresias*, 1917), which he subtitled 'drame surréaliste'—the first use of that term. Here too all external realism is abandoned in favour of a 'true' super-realism, which depicts the essence rather than outward appearance; a woman Thérèse becomes a man, Tirésias, and thousands of children are engendered to replace the losses of the war.

Even more radically disruptive of the conventions were the efforts of the group of artists from both sides in the conflict who had escaped it by living in neutral Switzerland. In the Cabaret Voltaire in Zurich they produced nonsense spectacles under the banner of anti-art, Dada. The Austrian painter Oskar Kokoschka, the Romanian poet Tristan Tzara, the German painter-poets Hugo Ball and Hans Arp presented spectacles designed to subvert the very concept of 'art': several poems recited simultaneously, nonsense playlets, strings of inarticulate sounds, anything to shock the complacency of the placid citizens and to subvert established notions of beauty and seemliness. It was from these roots, together with the pioneering work of Jarry and Apollinaire, that after the war 'surrealism' developed in Paris and a radical left-wing movement in Berlin—to which painters like George Grosz and playwrights like the early Brecht adhered.

In the warring nations themselves theatre tended to decline into a means of distraction for the soldiers on leave and the populations, suffering varied hardships, who had remained at home. The war also saw the emergence of a new rival to the theatre, the cinema, which established itself as a preferred means of mass entertainment, with immense consequences for the subsequent course of development of the theatre.

It was in the aftermath of the tremendous upheavals, material as well as ideological and spiritual, of the First World War that new and radical impulses spread through the arts; there was an influx of refugees from Russia, including theatre artists like the Pitoëffs in France, Fedor Komisarjevsky in Britain, and Mikhael Chekhov, who spread the gospel of Stanislavski, in the United States. They contributed much to the evolution of a 'director's theatre' in the post-war world. By then the number of elements to be kept in play in staging drama had increased so hugely that the need for a controlling mind, a controlling concept, a conductor to command the rhythms and the harmonic interplay of all the ingredients that created the total effect, became overwhelming.

II

THEATRE AFTER TWO WORLD WARS

CHRISTOPHER INNES

IN no period is the connection between theatre and society clearer than in the aftermath of the two World Wars, unprecedented in scale and intensity, that have been the major historical events of the twentieth century. Along with ideological fall-out, the material and psychological disruption found immediate reflection on the stage. Cultural traditions had largely survived intact through previous conflicts, which were relatively limited in scope. Even the French Revolution or the following Napoleonic Wars hardly marked a break in the form of theatre or type of drama presented in it. By contrast the hostilities that opened in 1914 engulfed the whole of Europe and its colonies, together with the Turkish and Russian Empires, eventually drawing in the United States as well. By the time peace descended on the exhausted combatants at the close of 1918, almost a complete European generation had been wiped out. Revolution had radically changed the nature of the largest empire, Russia, and two others had been dismembered, while the old political and moral authorities were as bankrupt as many national economies.

At the same time, the very extent of the devastation in that 'war to end all wars', compounded by the wide discrediting of previously accepted beliefs and social hierarchies, created the hope for a fresh start: the idealistically anticipated birth of a new civilization, which fuelled not only messianic communism, but also its fascist mirror-image. All this became the inspiration for radically novel art-forms, which were given their most extreme expression in theatre, while the newly independent countries in Eastern Europe also found their national voices, particularly on the stage. The advances in mechanization that produced new weapons of war—the tank, aeroplane, and submarine—also introduced new theatrical machinery, heightening the dominance of the director during the 1920s (as the new technology of computers and lasers would again in the 1980s). Yet, however representative of modern existence, this proved an extension of nineteenth-century modes. New directions came from individual artists, painters, and poets performing in cafés and on the streets without technical resources; and by the

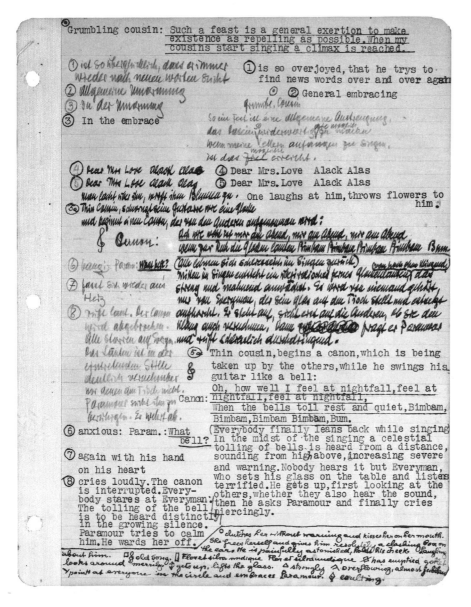

Symbol of the director's dominance. Reinhardt's prompt-book for a 1940 English production of *Everyman* in Hollywood: barely fifty words of speech (underlined type-script), a quarter of which are repetitions, with every detail of movement, gesture, sound, and lighting effect highlighted in different coloured inks.

1930s the age of the director had passed. Although there were some exceptions—notably Artaud in France, or Kazan and the Actors' Studio in America—over the next half-century dramatists took the lead in expanding the theatrical boundaries, just as it was the playwrights who articulated the aspirations of newly (and briefly) independent countries in Eastern Europe. Only under the communist dictatorships, where freedom of expression was drastically curbed, did the director remain a primary source of theatrical excitement, presentation being harder to censor than dialogue.

In hindsight, the peace that had been hailed in 1919 was an illusion, and the way it was imposed sowed the seeds of later hostilities. Indeed, the fighting continued in Russia until 1921, along with foreign intervention; and the secret treaty of Rapallo that heralded rearmament was already signed in 1922. The severity of the Great 1930s Depression, exacerbated in Germany by the crushing weight of financial reparations exacted by the victors, created the conditions for the rise of Hitler. Barely a year after Allied occupation forces were withdrawn from the Rhineland in 1930, Japan invaded Manchuria. The social destabilization of 1914–19 was directly responsible for Mussolini's take-over of Italy in 1922, and (even though the Spanish had earlier preserved their neutrality) for the 1931 left-wing revolution in Spain, followed five years later by civil war that only ended immediately before the outbreak of the Second World War. The twenty years of unstable coexistence between the armistice of 1918 and the Nazi invasion of Poland in 1939 can be seen as merely a strategic retreat for regrouping and resupply between one extended battle and the next. And such events shaped the theatre and were directly reflected on the stage. The reciprocal relationship is as evident in the use of the stage for propaganda, as it is in the flight of German dramatists and directors into exile after 1993 or the untimely murder of a Spanish playwright such as Lorca in 1936, which brought an abrupt end to theatrical developments of great promise.

In a sense drama became a continuation of war by other means during the inter-war years, with the slogan 'Theatre is a Weapon' being coined in 1920. And this essential continuity, right up to the declaration of peace in 1945, perhaps helps to explain why there was no artistic ferment and radical innovation in the theatre—such as characterizes the 1920s—in response to the Second World War. Instead, particularly in England and the USA (as well as in Germany, where almost all the major theatre centres had been reduced to rubble by Allied bombing), energy went into bricks and mortar: governments subsidized new theatre buildings, leading to new actor/audience relationships in 'environmental' spaces, thrust stages, 'theatres-in-the-round'; and theatre became part of the educational establishment. Drama departments blossomed in universities; theatrical training became a tool for developing children's expressiveness in schools; troupes were employed—by both governments and trade unions—to perform socially relevant pieces in community centres. Despite the far greater extent to which civilians had become victims, and cities along with the whole infrastructure of society had been destroyed, drama in the decade from 1945 to 1955 largely picked up on what had gone before.

The unusually close integration of theatre with its social context after 1919 can be seen in the different kinds of drama on the stages of defeated countries, and on those of the victors. There seems to be an equation between the degree of war damage a country had suffered, and theatrical experiment. In Germany, where the theatre had long held the reputation of a 'moral tribunal', Expressionism had already burst on to the scene in 1916—the year of Verdun, a slaughter that marked the turning-point in German fortunes—and the violence of communist revolts and right-wing *putsches* that followed defeat went along with the introduction of radically new forms of drama in the short-lived Weimar Republic. Russia had been even more soundly defeated, although initially on the victorious Allied side, and already by 1920 Meyerhold was raising the banner: 'Put the October Revolution into the Theatre!'

By contrast in England, where national pride and social complacency were (superficially at least) confirmed by the outcome of the war, drama retained its traditional shape. The major voices were still those of Shaw and Somerset Maugham, while the new tone of Noel Coward remained conventional in form despite morally shocking themes. Perhaps characteristically, whereas expressionistic transcendence or Meyerholdian revolution swept the stage in Germany, Austria, and Russia, in London audiences were flocking to what must count as the first of the block-buster musicals, *Chu Chin Chow*. A fantastical extravaganza combining the traditional romance of Ali Baba and the Forty Thieves with oriental exoticism, this broke box-office records in its continuous run from 1916 to 1921.

France and Italy, where much of the fighting had taken place though they were counted among the victors, stand between these extremes. Only recently united as a nation, Italy was the more destabilized, and this was reflected in the aggressive iconoclasm of the futurists, as well as the conceptual experiments of Pirandello. In France there were the surrealists, though their presence on the stage was marginal, and even Artaud had relatively little influence until the 1960s. But in both countries these challenges to traditional or naturalistic theatre lacked the ideological element that characterized Russian and German innovations.

Perhaps the least affected of all the combatants was the United States, where war had barely impinged on national life. In place of clinging to the past or completely rejecting it, continuity balanced change, merging naturalism with expressionism in the work of Elmer Rice and Eugene O'Neill. Though still very much present in the 1920s, the waning naturalistic influence of Belasco was carried forward by the Theatre Guild (founded in 1919), which developed into the Group Theatre, making this stylistic amalgam the defining mark of American mainstream drama right through to the late 1960s.

Of course, however exciting and ultimately influential, to the general public of the time such avant-garde experiments seemed marginal. In all these countries the classics were still widely performed; the nineteenth-century architecture of the stage, and the audience relationship that imposed, remained standard; while a large section of the theatre continued to be part of the commercial entertainment industry. Shakespeare was almost as much a presence on the stages of the Weimar Republic—along with Goethe, Schiller, and Wagner—as in English-speaking countries, where his plays were given still greater prominence by the rebuilding of the Stratford Memorial Theatre in 1932, and later the establishing of Shakespeare festivals in North America in the 1950s. On a more popular level, in New York the annual 'Ziegfeld Follies' (started in 1907) continued to appear with ever more scantily clad beauties and more elaborate chorus-lines until 1931, while the Broadway musical developed its characteristic form during the 1920s. In France, light boulevard plays continued to be the main theatrical fare right up to the 1960s. The symbolist Paul Claudel—whose first play had been written in 1889—remained the leading modern French playwright, setting the tone for serious drama with his highly poetic and religious plays as late as the 1940s, while even a major reformer like Jacques Copeau based his repertoire on Shakespeare and Molière. Spanish theatre during the 1920s continued to be dominated by conservative playwrights like Jacinto Benavente, who became director of the Teatro Español in 1920; and in 1926 Berlin no less than nine theatres were given up to revues with self-explanatory titles like *Donnerwetter—1000 Women*.

In terms of sheer audience numbers, such classical and commercial theatre far outweighed any avant-garde experiments. Its predominance is as clear at the end of the 1960s as in the 1920s, even if Charles Laughton's American tour of the Hell scene from Shaw's *Man and Superman* drew more spectators than the hit musical *South Pacific* in the 1950s. This variety is a measure of the theatrical vitality of the period. Yet it is the unconventional, rejectionist forms of alternative or minority drama that embody the creative energies of the time. Though short-lived in their intensity, and generating opposition rather than imitation because of their extremism, it is the expressionist, dadaist, futurist, and surrealist movements, the theatres of the grotesque or cruelty, cabaret and agitprop, that become representative of 'modern' theatre.

Indeed, all this—together with the exceptionally close linkage of theatre and politics—makes the period following 1919 very different from anything that had gone before. Each of the earlier theatrical eras displays an essential continuity that can be summed up in key representative moments. By contrast, one of the distinguishing features of 'modernism' is its multi-faceted complexity, the repeated breaks with previous generations and diversity of competing voices: all of which is reflected by the way this part of the story has been told.

Facing: tableau from the Ziegfeld Follies; Hal Prince's *Cabaret*.

Inset: the symbol of commercial theatre: glamour and spectacle displayed in the female body. The archetypal revue in 1919; and a parody of its down-market night-club equivalent as dark political commentary in 1966.

The roots of expressionism may lie in Strindberg and Wedekind, but its development was (literally) revolutionary. Abolition of the monarchy and the end of censorship brought an exhilarating release, together with a sense of political urgency from the chaos of defeat and the abortive communist uprising in Berlin, which placed the public voice of the stage in the forefront of social debate. As Herbert Jhering, who became the most influential of the Weimar critics, put it: 'The theatre only needed to be relieved of the pressures of war, and plays and productions came sizzling out of it . . . the dynamism that the revolution lacked was to be found in the theatre.'

Up to then expressionism had remained script-bound and thematic. Now directors such as Max Reinhardt and Leopold Jessner evolved new forms of stagecraft to present the expressionist vision. The immediate post-war plays—Ernst Toller's utopian *Transfiguration* or Walter Hasenclever's *Humanity*—are montages of poetically charged and universalized images, either idealized and abstract, or grotesque and distorted. With speeches concentrated into single emotive words or declamatory monologues, the texts are little more than cues for stage action. Wounded in the trenches and haunted by the skeletons left hanging on the barbed wire of the battlefield, the pilgrimage of Toller's highly autobiographical hero makes him the Christ-like instrument for a spiritual transformation of the enslaved population into ecstatic 'new men'. The production was intended to have exactly the same quasi-religious effect on the German public. And indeed the play directly overlapped reality, with its 25-year-old author imprisoned for having attempted what his hero achieves on the stage. Elected President of the short-lived Bavarian Soviet Republic—the only dramatist before Vaclav Havel to hold such a position—he had unsuccessfully tried to stop civil war by placing himself between the battle-lines, exhorting the civilian militia to passive resistance against advancing right-wing troops, who crushed the fledgeling communist state with extreme ferocity. (His experience was itself dramatized half a century later in Tankred Dorst's *Toller*, written in response to the student uprising of 1968: a signal instance of the reciprocal relationship of politics and theatre.)

Performed on the tiny Tribune stage in 1919, the pacifist and revolutionary topicality of *Transfiguration*, its utopian appeal and macabre distortion, set the pattern for German expressionist drama. It gave the 25-year-old Toller the status of a cult figure: Artist as Messiah. It also made Fritz Kortner, who played Toller's author/hero, the leading actor in Berlin. The aim of expressionist acting was to achieve emotional transcendence, communicated physically through dynamic and stylized movement, rigid muscular tension, and harshly staccato vocalization. The impassioned, ecstatic actor was 'a man possessed'; and the style was also applied to the classics, as in Leopold Jessner's staging of *Wilhelm Tell* or *Richard III* in 1919/20.

For these productions, Jessner conceived a setting of platforms and broad staircases for three-dimensional movement (with Kortner tumbling from the top all the way down to the stage floor in Richard's death agonies). The visual economy focused on the human figures. Their passions were reinforced by symbolic lighting: a lurid red for Richard's bloodthirsty anger; a sunlit glow for the Tell family. And the emotions of the audience certainly ran high at the opening of *Wilhelm Tell*, the inaugural production of the new Staatstheater, formerly the court theatre. Schiller's revolutionary theme was intensified by the change in the theatre organization from an establishment supported by the dethroned Kaiser to one financed by a socialist government, and by the recent bloodshed in the streets of Berlin. Jessner's identification of the villain with the old regime, coupled with anti-Semitic feelings (as a conspicuous Jewish socialist, Jessner was seen as personifying the Republican take-over), led to a shouting-match between the gallery and the actors, which engulfed the whole audience including the first-night critics.

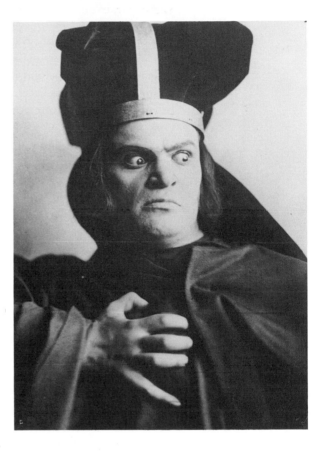

Such antagonism between actor and public was hardly the spiritual unity that visionaries like Toller believed would transfigure society. However, it was exploited by the aggressive young Bert Brecht in his early expressionist plays. Thus *Drums in the Night* was a deliberate insult to practically every political faction in the 1922 audience. The soldier returning from the trenches is no hero, but a physically decayed and repulsive 'swine' (denigrating all who had served in the war). His fiancée is unfaithful and throws him over (a defamation of German womanhood). The society he has suffered for, exposed as crassly materialistic, despises him; and the Spartakist revolution is rejected as irrelevant (offending both the bourgeoisie and the left wing). Self-interest is all that counts. Even though the distortions of set and acting were typically expressionistic, expressionist pretensions were equally deflated by Brecht's use of crude placards, like one admonishing the audience:

Fritz Kortner as Richard III in 1920. The image of the impassioned, ecstatic actor of German expressionist drama, face transformed into a mask and hands into talons, with glaring eyes.

DON'T GOGGLE SO ROMANTICALLY!

Expressionism was exported as far as America, where its influence can still be seen over a quarter of a century later in memory plays like Tennessee Williams's

The Glass Menagerie or Arthur Miller's autobiographical *After the Fall* in the mid-1960s. However, already by 1923 the expressionist wave was receding in its German heartland, undermined as much by its own emotional inflation as by opposing artistic and ideological tendencies.

In a sense, expressionism was still too tied to conventional theatre. It relied on empathy and (for all its messianic attempts at direct emotional contact with the audience) remained behind the proscenium arch. Satirical cabaret had far more political immediacy, as Brecht was quick to realize in becoming a singer of protest ballads in workers' clubs. And the dadaists had evolved a radically iconoclastic type of performance at the Cabaret Voltaire in Zurich during the First World War. Reinhardt's pre-war cabaret, revived in the basement of his Grosses Schauspielhaus, was an entirely appropriate setting for the dadaists, whose aim was to undermine the foundations of culture. They dominated its programme, and spread to rival cabarets with names like 'Megalomania' or 'Wild Stage'.

In post-war Berlin Dada also took to the streets, with politically destabilizing 'Happenings', such as the orchestrated disruption of a parliamentary debate, or with events mounted to demolish cherished cultural values, like a 1920 exhibition in a public lavatory. There spectators were offered an axe to hack up artworks while a prepubescent girl in white communion dress recited obscene poems. However, behind the anarchistic rejectionism lay positive artistic principles. Active involvement, immediacy, and authenticity were the criteria of dadaist 'anti-art' forms. 'Bruitism' used sound to force an audience into a fresh perception of their environment, as with Marinetti's evocation of 'the awakening city' through a concatenation of motor-horns, typewriters, saucepan-lids, and babies' rattles. 'Collage', and 'Montage' (its photographic equivalent), formed pictures from juxtaposed fragments of reality: pieces of metal, bits of newspaper, cloth, 'found' objects—an urinal or a stencilled board from a packing-case, as in Kurt Schwitters's 'Merz' art (which derived its name from the central syllable of Kommerzial in one of his collages). This became the basis for scenery and costumes; its verbal equivalent was 'Simultaneity' and 'Sound Poetry', designed to encompass all the different aspects of existence, and to liberate the listeners' imagination.

In Paris, Dada evolved into surrealism: a development marked by the 1923 performance of Tristan Tzara's *Gas Heart*, which provoked one of the most intense of the theatre riots that characterize the period. The stage became a battlefield with one hostile artist being knocked into the footlights, broken arms, and shredded costumes, before the police arrived and closed the theatre. Wrecking the theatre was a logical extension of Tzara's satiric dismemberment of human anatomy and attack on rationalism. Yet it also points to the surrealists' rejection of any compromise between art and material existence. Their aim was to transcribe the subconscious, presenting dreams as a superior reality, which led to purely psychological theatre: the product (in ideal terms) of automatic writing. This is embodied in plays from the early 1920s like *Free Entry* or *Poison*, 'a drama without words', by Roger Vitrac, one of the few successful playwrights associated with the movement. Time-frames and physical scale are distorted in montages of violent and arbitrary images. Syntax is dislocated and characters become interchangeable. But the more it aspired to stream of consciousness, the less theatrical surrealism became; and its full expression was reached outside the material constraints of stage presentation, in films such as the Dali/Buñuel *Un chien andalou*.

By contrast, in Italy the dadaist principles of spontaneity and simultaneity were adopted as the basis for futurism (first proclaimed in 1909): a response to scientific modernism, which exalted technology and embodied the machine age in an 'Aeroplane Sonata' or 'electric poetry'. Manifestos were developed into a new literary form by Filippo Marinetti, who called for 'The Anti-psychological Abstract Theatre of Pure Forms and Tactilism' in 1924, or 'Total Theatre for the Masses' in 1933. These were accompanied by plays like *Rapid Light* or *Simultanina*, and by 'syntheses': short, highly performance-oriented pieces, in which the complete physical involvement of the spectators was required to complete the 'synthesis'. To this end disputes or even fights were deliberately provoked among spectators, and images were designed to shock or outrage the public. Futurism had its imitators—Enrico Prampolini's 'mechanistic theatre'—and its equivalents, such as the German Bauhaus. This architectural school also housed a stage work-

A 1927 futurist production by Prampolini. 'Alogical' action and 'concrete' presentation reflecting a mechanistic vision of existence that reduced actors to moving geometrical shapes in a polydimensional space, intended to transform the stage into 'the temple of spiritual abstraction'.

shop, which under the designer Oskar Schlemmer focused on the mathematical patterning of movement, in which the human body was reduced to abstract designs through geometrical costumes or long sticks attached to the arms. The title of Moholy-Nagy's *Score Sketch for Mechanized Eccentrics* is representative of Bauhaus aims; and the most theatrical realization of their experiments, Schlemmer's *Triadic Ballet* in 1922, was a non-representational and 'clinically isolated concentration of action on the stage', explicitly 'without purpose' or even story-line, designed to investigate the phenomena of form and space, as well as the process of human perception.

The nature of perception was also Pirandello's major theme, when he turned full-time to the stage in 1916 after a long career as a novelist. There were two types of reaction against the neo-romantic lyricism of D'Annunzio and the naturalistic problem plays that had filled the pre-war Italian stage. The first was a peasant dialect drama, based on the surviving tradition of regional travelling troupes, to which Pirandello contributed several one-act pieces. The second was a form of parody, developed towards the end of the First World War, which explored the absurd or bizarre possibilities of social existence while ridiculing the illusionistic conventions of standard drama. One of the major plays in this 'Theatre of the Grotesque' (as the short-lived group came to be labelled) was Luigi Chiarelli's *The Mask and the Face*: a title consciously echoed by Pirandello, whose 'theatre-in-the-theatre' trilogy was published as *Naked Masks*. The discrepancy between the public persona imposed on individuals by society and their private personality provides both the method of characterization and the turning-point of the action in most of the plays Pirandello wrote during the 1920s. This is paralleled by the 'tragic conflict' between the flux of life and the immutable forms of material existence—all of which is epitomized in drama, since a play fixes supposedly living characters in the script, as well as exemplifying a similar split between the actor and the part he impersonates. In stage terms this is clearest in the play that established Pirandello's international reputation, *Six Characters in Search of an Author*.

Actors rehearsing a Pirandello play are interrupted by a family of fictional figures, who demand that the drama of their own lives be performed and thus given a reality denied them as mere figments of their author's imagination. They act out an exaggeratedly melodramatic series of scenes for the actors to copy. But each of the characters has a different, self-justifying view of their adulterous or incestuous actions; and the basic question—one that continues to echo throughout twentieth-century drama right up to Albee's *Who's Afraid of Virginia Woolf* in 1962—is: 'illusion or reality', which is which? During this process, it becomes clear that the real people are socially conditioned puppets, while the imaginary figures come to be accepted as human beings. In Pirandello's own production of 1925, the contrast between art and life was emphasized by costuming: the actors' clothes being lightweight, flowing materials that fluttered as they moved, while the six characters were dressed in heavy, dark fabric with rigid pleats to give 'a statue-like appearance', and their grotesquely blackened faces reminded reviewers of 'antique masks of tragedy'. Eternally trapped in their unfinished story, they were the opposite of ghosts (as, for example, Reinhardt had expressionistically presented them). And in contrast to their fixed solidity, Pirandello's direction stressed the ephemerality of all the elements of the staging. The scenery was makeshift, the actors continually in motion, lights hastily adjusted to meet the requirements of the 'comedy in the making'. As Pirandello said, 'everything is in process . . . experimental, improvised'. At the same time, conventional theatre is parodied. The Stage Manager is unable to comprehend the Pirandello play he has chosen to direct, while the actors reduce the characters' story to farce.

In 1921, when *Six Characters* opened in Rome, it caused yet another of the era's notorious theatre riots. The audience—already protesting before the performance started, on seeing a bare stage with absolutely no scenery—booed the first-scene 'rehearsal', greeted the entry of the characters with shouts of disapproval, and came to blows among themselves. A poet who supported Pirandello hurled himself from the stalls into a box where some of the most vociferous whistlers were gathered; and the fighting continued on the streets after the curtain fell. Yet by 1925, when Pirandello formed his own company, the Teatro d'Arte (funded by Mussolini's fascist government), the play had already become a contemporary classic, performed widely throughout Europe. Pirandello's work was particularly popular in Germany, where its stripping of the stage and anti-illusionistic theatricality paralleled (and perhaps influenced) Brecht's development of Epic Theatre. Due largely to *Six Characters*, 'theatre-in-the-theatre' became one of the defining features of modern drama, although Pirandello was alone in using it to express such highly philosophical issues: indeed, underlining its intellectual nature, when the Teatro d'Arte played *Six Characters* in London, Pirandello followed each performance with a lecture to explain its meaning.

The major catalyst for all this avant-garde experimentation came from Russia. At the time, the triumph of the 1917 Soviet Revolution seemed to mark a new age, and had an immediate effect in theatrical terms. In Russia too the stage initially seemed central to the political process. Lunacharsky, Lenin's Commissar for Art, was also a playwright; Trotsky, the leader of the Red Army, wrote a treatise on *Literature and Revolution*; the Proletcult movement made theatre the mouthpiece of the state. So it was hardly surprising that the apparent achievement of socialism was the catalyst for a whole range of avant-garde artists: Tairov's 'abstract formalist' theatre, Prokofiev's operas, Eisenstein's films, Mayakovsky's poetry, and above all Meyerhold's constructivist innovations in performance. Vsevolod Meyerhold, who began as an actor in the Moscow Art Theatre and had been appointed director of its experimental Studio Theatre by Stanislavski in 1905, was an enthusiastic supporter of Bolshevik ideals. Briefly enrolled in the Red Army, and a friend of Lunacharsky, he picked up on Trotsky's claim that one of the ultimate aims of revolution was to do away with any separation between physical labour and intellectual work (including art).

Based on the analysis of production-line workers by Frederick Taylor, the inventor of time-and-motion studies, and on Pavlov's behaviourist psychology, Meyerhold's novel system of Biomechanics was intended to create not only a new type of actor, but the ideal 'new man' of revolutionary theory. Precision and economy of movement were to be achieved by extensive training in acrobatic exercises, and through breaking down stage action into 'acting cycles' (by analogy to Taylorist 'working cycles') of intention, realization, and reaction. This was combined with the Pavlovian notion of emotional reflex, in which inner feeling and physical posture are mutually reinforcing, each heightening the other. Initially a method of actor-training, Biomechanics became a highly theatricalized style of performance, and went along with Lunacharsky's notion of Socio-mechanics, in which political consciousness or social position was used as a basis for characterization, not individual psychology. This approach derived partly from Meyerhold's interest in *commedia dell'arte*, which he shared with other directors of the time, who were also searching for alternatives to naturalism. On Meyerhold's stage figures became types that tended towards grotesques; and the actor's role was turned into a construct of movements, while the emphasis was always on the group. Many of the gymnastic exercises Meyerhold developed—the 'Slap in the Face', the 'Leap on the Chest', 'Horse and Rider', the 'Pyramid'—require the close co-operation of at least two or three actors, so that his productions raised ensemble playing to a new level, explicitly embodying the ideal of collectivism.

Transferred directly into performance, these exercises were extended into sequences of vigorous gymnastic action, and became the basis for what Meyerhold called 'Attractions': moments of virtuoso display, symbolically incorporating political attitudes or thematic points. Eisenstein—who began his career as a stage director, where he formulated the techniques that were to revolutionize the cinema—described his films as 'a montage of attractions', in which independent images were given meaning by juxtaposition and sequence alone. Meyerhold's practice paralleled Eisenstein in isolating each beat in a production, every 'episode' being played as an independent 'turn' that was integrated in an overarching rhythm. This intensified the theatricality of the presentation, 'baring the device': a term that also applied to his 'constructivist' settings. Redefining stage performance, Meyerhold abolished the framework of illusion—proscenium, wings, representational scenery—to make theatre an 'action space'. Stripping the stage to its bare brick walls, he provided multi-level platforms, ramps, ladders, and three-dimensional 'constructs' to create a 'machine for acting' that also carried a

БОКС
БИОМЕХАНИКА
КОНСТРУКТИВИЗМ

Meyerhold's constructivist stage. Wooden, mechanical shapes—derived from industrial design and architecture—as a basis for three-dimensional and interactive movement used for the 1922 production of *The Death of Tarlekin*.

latent ideological meaning. The aim of constructivist art was 'to realize the Communist expression of material structures'; and this utilitarianism also carried over into uniform-like costumes, and even geometrically patterned work-clothes for his actors.

As well as radically reinterpreting the classics, Meyerhold worked with new playwrights like Tretyakov and particularly Mayakovsky, whose poetic paean to revolution, *Mystery Bouffe*, he produced in 1918 and again in 1921. As the Prologue announced, 'We too will show you a life that's real / But life transformed by theatre into a spectacle most extraordinary.' Meyerhold's staging projected the action into the auditorium, and accentuated the circus qualities of the script. Combining satiric farce and visionary symbolism, *Mystery Bouffe* was proclaimed the first distinctively Soviet play: precursor of a completely new form of drama. However, if anything qualified as such it was the mass spectacle of the *Storming of the Winter Palace*, mounted in 1920.

The Proletcult ('Proletarian Culture') movement aimed to abolish the professional theatre. It would be replaced with public celebrations of working-class struggles, in which 'the boundaries between labour and festival, between life and art, between spectator and artist' would be dissolved. This was best realized in the vast pageants of the revolution, involving casts of thousands and staged in Petrograd during the Civil War, which culminated in the *Storming of the Winter Palace*. Re-enacting the turning-point of the Bolshevik Revolution in its actual location and on the third anniversary, it had an electrifying effect. Events leading up to the fighting were depicted on two stepped platforms either side of the square—with the White leaders portrayed as caricatures by professional actors in contrast to the unified groups of Reds—and the confrontation between the forces of good and evil took place in the Palace between. Silhouettes of hand-to-hand combat filled all the lighted windows behind white blinds, and spilled out over the square. Searchlights swept over the massed figures (2,865 performers were required for the White platform alone, and three to four times that number for the Revolutionaries, with Red Army infantry and motorized units taking part) and gunfire came from warships moored in the river.

A gigantic inflation of the medieval mystery play—on which such revolutionary pageants were explicitly modelled, even if a closer parallel was the romantic monumentalism of Reinhardt's *The Miracle* at the London Olympia exhibition hall in 1912—mass spectacles on this scale were judged to be redundant once the communist state was firmly established. The Proletcult movement was diverted into agitprop activities: mainly the staging of mock trials of contemporary hate-

Political theatre at its most basic. An Agitprop troupe, performing on a bare platform at a workers' sports meeting in the Brecht–Eisler–Dudow film *Kuehle Wampe* (1932).

figures, or Living Newspaper performances by travelling troupes (the most famous of which, the Blue Blouses, was run by the Moscow Institute of Journalists). This was specifically propaganda theatre; and both dramatic forms were exported to the West. Agitprop, with its minimal staging, crude caricature, and topical immediacy, remained a primary vehicle for left-wing performance: adopted by Piscator and Brecht in Germany in the 1920s, by Joan Littlewood's Theatre of Action in England during the 1930s, and in America by Clifford Odets, whose vignettes of working-class exploitation made *Waiting for Lefty* the one lasting agitprop classic. But no theatrical form has an intrinsic ideological content; and the mass spectacles of Communist Revolution prefigured the Laban dance display celebrating the spirit of National Socialism at the Berlin Olympics of 1936, and the monster Nazi rituals in Germany. However, the immediate inheritor of the revolutionary pageant was Erwin Piscator.

After an agitprop apprenticeship, Piscator began his career by mounting Red Revues, commissioned by the Communist Party for the German elections of 1924 and for the Party Congress of 1925, and performed in Reinhardt's Theatre of the 5,000. Lacking the massed performers of the Petrograd spectacles, Piscator experimented with combining film-sequences and staged action. The new standards of actuality and objectivity this provided were extended by projection screens, on which facts and figures could be displayed. And this multi-media presentation allowed actions to be shown from multiple viewpoints. Film could also open up the focus of the stage to encompass major world events in a direct way; and it could present the masses (the prime movers of history, according to Marxist theory) in place of great individuals. As his use of film developed, Piscator began to explore the total mechanization of the stage—and this had a symbolic function too, transforming the conventional theatre into the image of modern technological science. Very much a sign of the times, the plans for a 'total theatre' by the Bauhaus architect Walter Gropius, commissioned by Piscator in 1927, were exactly duplicated by plans for a projected Meyerhold Theatre drawn up in 1931.

Stanislavski, Antoine, and Otto Brahm had already established the director as the guiding and unifying creative force in a production, which proved the most important single factor in promoting the theatrical innovation so characteristic of twentieth-century staging. But the degree of control required by Piscator's inventive deployment of all the contemporary technological resources made him the epitome of the modern director. In addition, he was the only theatre-artist to come to grips with the challenge of film, and explore ways of adapting it to the stage. The way film was already conditioning audience expectations is clear from the 1926 design of the Ziegfeld Theatre in New York, which imitated the cinema in the narrow sight-lines of its long auditorium without side-boxes and the plaster cyclorama that enclosed the back of the stage. When 'talkies' arrived, film

Technological theatre: *Rasputin*, 1927. The diagram shows twelve interior acting spaces, indicating how the huge completely mechanized half-globe constructed by Piscator could be rotated or opened up, and the integration of back-projected film. The effect is illustrated by the scene in which the Czarina ridiculed reports of popular uprisings, while film of the coming massacre of the Russian royal family appeared shockingly above the actors' heads.

became a major threat, drawing away not only spectators but creative talent from the live theatre—a trend that intensified in the 1950s with the advent of television. The way the stage began to exploit theatricality in the 1920s, the growing search for ways of achieving audience participation, the use of improvisation, and the emphasis on theatre as 'process' rather than a 'product' can all be seen as a reaction to the cinema. However, where others focused on the elements that distinguished live theatre, Piscator went head-to-head with film, competing on its own ground.

During the brief two years before his theatre fell victim to the Great Depression, Piscator developed what has come to be known as documentary drama. The most complete example was his 1927 production of *Rasputin, the Romanoffs, the War, and the People who Rose up against Them*: a title which indicates the breadth of subject, as well as the four contrasting poles of the performance. Piscator's staging was the outline of an era, focusing—as with the *Storming of the Winter Palace*—on the 'driving forces of the Russian Revolution'. But, instead of magnified spectacle, his aim was to 'condense reality', giving concrete shape to such sociological abstractions as 'capitalism' or 'the class struggle'. This was technological theatre at its extreme, with a metallic half-globe that almost filled the stage. Divided into multiple acting spaces behind hinged flaps, this rotated or opened

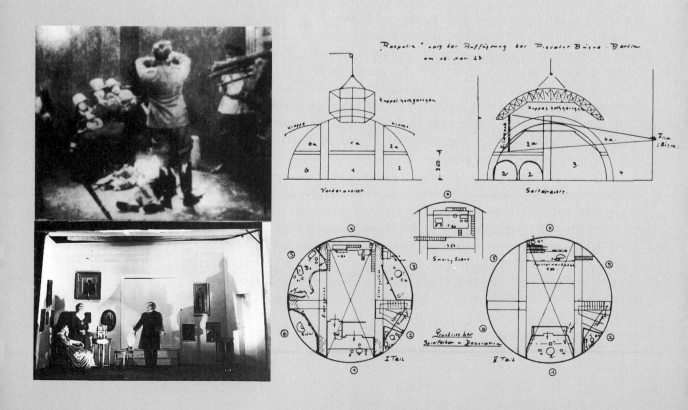

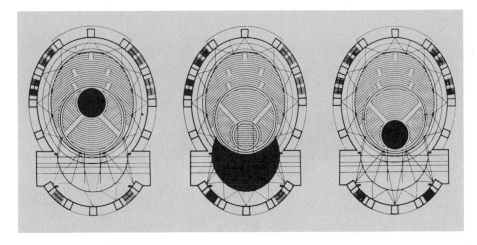

Plan for Piscator's 'total theatre', 1927. Complete mechanization of the building in Walter Gropius's design allowed the acting area to be transformed from proscenium to thrust stage, arena, or multiple stages during a performance. Audience seating could be swivelled round, and film projected over the whole wall and ceiling area of the auditorium to envelop spectators completely in the action.

up into segments, and film could be projected on to its curving surface as well as on to surrounding screens, juxtaposing different levels of reality. Thus under the title of 'Headquarters of the Tsar' Nicholas II divided his attentions between issuing inspiring directives to his troops and flirting with a nurse, while newsreel footage of battlefield devastation, as well as statistics of Russian casualties on a mobile 'calendar' screen at one side of the stage, showed the murderous effects of his grandiose orders. And his criminal egoism was underlined by one of his letters to the Tsarina, flashed on to the upper screen: 'The life I lead at the head of my army is healthy and has an invigorating effect.'

This devastating ironic counterpoint also 'alienated' the action. The artifice of the acted scenes was exposed by the authenticity of documentary film, while making up the actors as exact counterparts of the historical personages who appeared on the screen emphasized that it was not the characters but their context in class society that was artificial. The fluidity and movement of the 'globe' made the epic sweep of events and panoramic treatment possible, and unified the different types of material. Swivelling like a tank-turret from scene to scene, the construct served as a graphic image of the modern technology that gave the conflict its world-wide scope and intensity, as well as symbolizing the entrapment by events of those deluded into believing they controlled them. The factual nature and topicality of this type of drama is underlined by the response to the performance. The ex-Kaiser won an injunction for defamation of character—leading to one of the script-writers appearing in place of his figure and reading extracts from the court judgment (which reinforced Piscator's use of recent history as a perspective on the present)—and performances ended with spectators spontaneously joining the actors in singing 'The International'.

By comparison, the treatment of the First World War on the British or American stages tended to be limited to individual suffering on the battlefield, as with Maxwell Anderson's *What Price Glory?* and R. C. Sherriff's *Journey's End*, or a lyrical depiction of a doomed society on the eve of the war, like Shaw's *Heartbreak House*. Completed in the year of Verdun, one of the bloodiest battles in history, this 'fantasia in the Russian manner' was only performed after the armistice, Shaw having made himself so unpopular by publishing a pamphlet ridiculing the English war propaganda that in 1916 no theatre would stage his plays. Perhaps because of the vilification he had received, the political statement of *Heartbreak House* is oblique. The moral abdication of the establishment, responsible for the suicidal conflict, is projected into a poetic frame where coincidence reigns and bizarre behaviour flourishes. The self-alienated and romantically 'heartbroken imbeciles', who 'haunt' rather than inhabit this 'madhouse', are a microcosm of 'cultured, leisured Europe', literally fiddling while their civilization burns. The overriding metaphor, embodied in the quasi-nautical architecture of the setting, is the Ship of State which the emblematically named Captain Shotover—a failed Shavian superman, now senile, whose mental powers are nothing but a drunken delusion—is allowing to drift into shipwreck 'on the rocks' (the title of Shaw's 1933 'Political Comedy' dealing with the Depression and the General Strike). However, on this allegorical level, the play is paradoxically hopeful. The bombs that fall in the final act blow up the immoral financier who controls the system and the burglar who exploits it, along with the clergyman's house nearby representing the religion that sanctifies it, while the family look forward to the destruction of their 'home' as the only possibility of liberation from 'this soul's prison'. Thus for Shaw it must have been deeply disappointing to find that the war had left the structure of English society essentially unchanged, and this is reflected in his later plays.

His two major post-war works retreat from the present into mystical science fiction (in the 'metabiological pentateuch' of *Back to Methuselah*, which forecasts the eventual triumph of mind over matter 30,000 years in the future) and historical costume play (*Saint Joan*). However, Shaw remained a dominant presence in British theatre right up to his death in 1950. The Malvern Festival was established to showcase his drama in 1929—making Shaw the only dramatist apart from Shakespeare to win such recognition—and he continued to turn out new plays right up to 1949. He influenced both Noel Coward and Sean O'Casey; and this dominance, together with the survival of the social establishment, helps to explain the conventionalism of the English theatre between the wars.

Avant-garde forms of theatre already entrenched on the Continent were restricted to theatrical outposts, such as Peter Godfrey's tiny Gate Theatre, or the Mercury Theatre—where Ashley Dukes provided a venue for German Expressionism and established a 'Poets' Theatre', staging the expressionistic parables of Auden and Christopher Isherwood in the late 1930s. Terence Gray established a

flamboyantly symbolist style at the Festival
Theatre in Cambridge, using 'isometric scenic
design' for everything from Pirandello, Yeats,
and Elmer Rice to Shakespeare or Aeschylus
(*The Suppliants* being his farewell production in
1933). Gray was perhaps the most influential of
British experimenters, since he worked with
Norman Marshall (who took over the Gate
Theatre from Godfrey in 1934, eventually becoming the Head of Drama for Associated-
Rediffusion TV in the 1950s) and with Tyrone
Guthrie, who had a lasting influence on
classical productions in North America. But
such theatrical experiments were marginalized.
When Sean O'Casey abandoned the realistic
comedy of his 'Dublin Trilogy' in 1928 with the
only play in English to attempt an expressionistic depiction of the horrors of the First World
War, his *Silver Tassie* was rejected by the
Abbey. Similarly, a decade later, Priestley's expressionist *Johnson over Jordan* took eight years
to reach the stage. As for the effect of Pirandello's philosophical 'theatre-in-the-theatre',

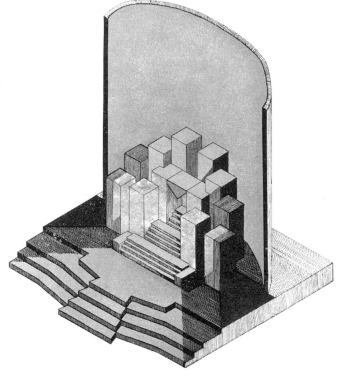

1930 Design for
Oedipus. Combining
Gordon Craig's
concepts in movable
neutral-toned or
internally lit luminous
columns with steps
and multi-level rostra
reminiscent of Jessner,
Terence Gray created
a uniquely English
equivalent of
Continental
expressionism.

that was only discernible in the popularized form of Priestley's 'Time Plays',
Dangerous Corner in 1932 and *An Inspector Calls* in 1946. Indeed, Priestley—along
with Shaw—became a mainstay of the repertory theatres that had sprung up all
across England in the wake of Annie Horniman's venture in Manchester and
Barry Jackson's outstanding company in Birmingham. Although this network
developed regional dramatists and gained reputations for high-quality ensemble
playing, forming the basis for the new civic theatres that appeared as part of the
urban renewal that followed the Second World War, their need to build local
audiences discouraged any but mainstream work.

In England, where the director had always been a less significant figure than on
the Continent, the theatrical tone between the wars was set by Noel Coward, who
leapt to prominence with a scandalously daring revelation of the hothouse relationship between a drug-addict son (identified by Coward with himself in a typically self-promoting gesture) and an adulterous mother in *The Vortex*. This
created much the same effect of shock and instant recognition in 1924 as *Look
Back in Anger* was to do just over a quarter-century later. Coward's characteristic
voice was that of light comedy—with himself, as both matinée idol and author,

and Gertrude Lawrence in the lead roles—but the sparkling glamour carries a sting. Brittle deadpan wit and artificial manners both epitomize and excoriate a trivial society; careless infidelities expose a moral and emotional vacuum while titillating the audience. Posing as a socially disengaged amoralist, whose only talent was to amuse, Coward extends the satiric line of Congreve and Oscar Wilde into a typically twentieth-century mode. All his comedies from *Fallen Angels* in 1925 to *Present Laughter* in 1942 are variations on the same theme: the irreconcilable tension between individual experience and the deadening mores of society, which is summed up in the title of *Private Lives*, his best-known comedy. Coward's plays and hectically dazzling musical revues encapsulate the Jazz Age of the 1920s and 1930s, yet he rejects its values even while voicing them.

No play of Coward's shows the dynamic between traditionalism and theatrical experiment more clearly, or has a closer umbilical connection to its time, than *Cavalcade*, which is one of the few English works to parallel the panoramic effect of Piscator's documentary drama. Its title and structure, running in an unbroken flow of contrasting scenes from the last day of the nineteenth century to New Year's Eve in 1929, are taken from the type of horseback procession that had paraded imperial glory at Queen Victoria's Jubilee. But however colourful the historical pageant—which was hailed as 'Drury Lane's Greatest Drama: Vast Crowds in Amazing Stage Pictures'—the progress from Victoria's funeral to the '*Night Club*—CHAOS' finale is presented as an accelerating decline. The theme is spelt out in ironic contrasts, as with the mother who has just received news of her son's death in the last days of the war, caught up in the Trafalgar Square celebration of victory over Germany '*like a sleepwalker through dense crowds of cheering, yelling people . . . She stands there cheering wildly, with the tears rolling down her face.*' At the same time, the emotional weight of the lives sacrificed as the price of Empire lend it an idealistic splendour; and the 1931 production was not only an immense popular success, leading to outbursts of patriotic fervour from audiences who rose to sing 'God Save the King' along with the actors in the final scene, but also directly influenced the outcome of the General Election that year. As a result the play is generally dismissed as chauvinistic flag-waving. Yet *Cavalcade* has become part of popular British culture in a way few plays achieve: the long-running 1970s *Upstairs, Downstairs* TV serial borrowed some of the characters' names as well as Coward's central idea, while key lines were even adopted (unacknowledged) by Mrs Thatcher as an electoral rallying call in 1983.

The sensational scale of scenic effects in *Cavalcade* and the mammoth cast—the original production required a company of over 250 actors and stage-hands—is representative of a trend to monumentalism that reached its apogee between the wars. Prefigured in some of Henry Irving's late nineteenth-century Lyceum productions, and Reinhardt's 1911 symbolist *Miracle*, it was promoted by the same advances in lighting and stage machinery that had made Piscator's documentary theatre possible. But here they were deployed for the most traditionalist purposes. Indeed *The Miracle*, revived by Max Reinhardt in 1924 at the Century Theatre in New York, which was gutted to provide a 'cathedral' 200 feet long and 120 wide with a height of 110 feet, literally became the emblem of such grandiose spectaculars. This production toured the United States until 1930; and Reinhardt remounted the pageant for the Salzburg Festival in 1925, at Budapest, Prague, and Vienna in 1927, and again in London (this time at the Lyceum) in 1932. The archaism inherent in this style of theatre is indicated by the American designer Norman Bel Geddes, who became obsessed with creating towering visionary settings for (unproduced) adaptations of classical literature after working with Reinhardt on the New York *Miracle*.

Bel Geddes also applied the same visual spectacle to naturalistic scenes, as with the myopically detailed three-dimensional streets and apartment-houses for Sidney Kingsley's *Dead End* in 1935—appropriately at the Belasco Theatre, since he was effectively extending the hyper-realism of David Belasco. Reinhardt, however, looked beyond the conventional stage to realize his monumental concepts. Paralleling the Russian Proletcult pageants of revolution, his open-air productions in the Cathedral Square at Salzburg dispensed with all the technology of the modern stage to recreate an archetypal theatre of ritual by integrating choreography, acting, and music with the natural grandeur of Renaissance buildings and the steeply wooded hills crowned by the medieval castle high above. His 1920 production of *Everyman* was repeated for the next eighteen years, and in revised form is still played each summer today.

The Salzburg Festival, which Reinhardt founded, was the model for the theatre festivals in Malvern, and at Canterbury, where T. S. Eliot's *Murder in the Cathedral* was the direct equivalent of the Salzburg *Everyman*. And Reinhardt was indirectly the progenitor of all festival theatres that have subsequently sprung up around Europe and North America. In this, Tyrone Guthrie was Reinhardt's

The prototype for festival theatre. Reinhardt's 1920 Salzburg production of Hoffmannsthal's *Everyman*, with the town turned into a stage: the banquet scene in front of the cathedral, and, high above, the medieval castle from which the voice of Death sounded eerily.

The apogee of monumentalism. Norman Bel Geddes's visionary project for staging Dante's *Divine Comedy*, combining religious symbolism, massed choirs, and towering flights of steps derived from Craig's concept for Bach's *St Matthew Passion*—unrealizable in any theatre, then or now.

natural heir. Founding Shakespeare festivals at Stratford, Ontario, in 1953 and at Minneapolis ten years later, he championed open staging as a modern equivalent to the Elizabethan relationship between actors and audience. His development of the thrust stage as an alternative to the proscenium arch revolutionized the way new theatres were built in the 1970s: more modest than Reinhardt, Guthrie's concept had more lasting influence. But the use of historic architecture for the setting (as at Canterbury) and the choice of a venue outside of the big-city theatre centres, the focus on classics and in particular Shakespeare, together with a return to open staging—all of which characterize contemporary festivals—were first developed by Reinhardt.

By any standards, Reinhardt's productions at the London Olympia, his 'Theatre of the 5,000' in Berlin, the Hollywood Bowl in Los Angeles, or turning the whole centre of Salzburg into a theatre-site qualify as epic in scale. But already, when his career was at its peak in the 1920s, a very different and still more influential concept of 'epic' theatre was being developed, which was the opposite of this monumentalism in almost every particular. Its originator was Bertolt Brecht, who moved rapidly from his early expressionist parody, through involvement in communist agitprop and collaboration with Piscator, to establish his own aggressively modernist style of staging. If Max Reinhardt epitomized the director as autocrat and impresario, Brecht represents the dramatist as director. Even though the sheer scale of the shows that Reinhardt directed gave a higher and more public profile to theatre as an art-form than possibly it had enjoyed since the theatre of ancient Greece, Brecht's career marks the re-emergence of the playwright as the creative centre of theatre.

As Peter Brook has asserted, Brecht is the key figure in twentieth-century theatre—although even in Germany, as well as the rest of Europe and America, his influence was delayed by dictatorship and exile until the 1950s. By that time he had achieved a comparable eminence to Reinhardt, as a dramatist who directed his own work at his own theatre in East Berlin with his own acting company, the Berliner Ensemble, which continued to spread his work even after his death in 1956. Since then, Brecht's theories (even more than his example) have influenced almost all dramatists, directors, and designers over the last three decades, from Arthur Adamov and Edward Bond to Giorgio Strehler or Peter Weiss, and from modern Shakespeare productions to Stephen Sondheim's version of the American musical. But Brecht was very much a child of the 1920s. Extreme in ideological stance as well as in opposition to the established theatrical conventions of the time, he dismissed both 'Aristotelian' forms and 'culinary' commercial entertainment. From the beginning his plays had attracted considerable public attention, due as much to the scandals he fostered around their production and his flamboyantly 'proletarian' life-style as to the literary prizes awarded his work or its iconoclastic qualities. In 1929 Brecht posed a question that remains the fundamental challenge for contemporary theatre: 'Can we speak of money in iambics? . . . Petroleum resists the five-act form; today's catastrophes do not progress in a straight line but in cyclical crises . . . Even to dramatize a simple newspaper report one needs something much more than the dramatic technique of a Hebbel or an Ibsen.' Over twenty-five years later, the same question—'Can the present-day world still be reproduced by means of theatre?'—was raised again by Brecht's Swiss follower Friedrich Dürrenmatt, and responses solicited from the leading German dramatists of the 1950s by the journal *Theater Heute*. Brecht replied, in what turned out to be his last public statement, asserting that the solution was political: representing the world 'as being capable of transformation'.

That had been the thesis of his first play to qualify as 'Epic Theatre', *Man is Man* in 1926, where personality was seen as being the product of social conditioning and thus completely changeable. This emphasis on environmental determinism is in fact contradicted in many of Brecht's protagonists, who have a core of natural goodness, which survives being perverted by capitalism in *The Good Person of Setzuan*, or which emerges even in a domineering landowner when the dictates of class and wealth are drowned by alcohol in *Puntila*. Yet the effect is still to create a radically different type of dramatic character from the coherent individual of naturalism; and this was accompanied by extreme simplification in staging, together with overt theatricality. Settings are stripped down and placards indicate the scene or give information to remove suspense. Machinery and lights are exposed to prevent illusion, stage-hands work in full view, and instrumental-ists playing the music for the songs that punctuate his plays are visible to the audience. A 'half-curtain' replaces the solid drapes that customarily close the proscenium arch to emphasize that the stage is not a special or magical space, but part of the everyday world. At the same time, his actors are frequently masked, and were required to demonstrate the act of acting instead of pretending to 'be' the characters.

His plays of this period are full of grotesques, with the soldiers of *Man is Man* being performed on stilts to embody their monstrosity, or a giant wooden clown being sawn apart limb by limb in the *Baden-Baden Cantata of Acquiescence*; and graphic theatrical display always remained central to his work. Its structure was just as much a departure from the naturalistic norm. Standard 'dramatic' theatre relies on linear plot and fully individualized characters, designed to evoke empathy. In deliberate contrast, Brecht used techniques to present events as narrative—hence his label of Epic Theatre—creating discontinuous action from a montage of scenes linked by their illustrative relationship to a central political theme, while his figures were intended to be representative. In addition, the various elements of theatre were to be separated: speech from gesture, voice from music.

Such techniques were inherently pedagogical. According to theory, all this was supposed to distance spectators from the drama, turning the audience into object-ive observers; and during the 1920s much of his output was 'teaching plays': short pieces performed at communist rallies or at the Baden-Baden musical festival. Brecht's key term for his theatre was 'rational'; and in order to inculcate such an attitude (as well as emphasizing the collective nature of the audience) he insisted that the house-lights remain on at the start of a performance, so that the spectators should remain aware of themselves as a group. In deliberate contrast to the standard cultural assumptions about going to a theatre, his ideal was the sort of sporting-audience round an arena, smoking and drinking, vociferous in support of their side, and expert in evaluating tactics; and indeed the stage area for *Man is Man* was roped off in a square as a boxing ring. However, in practice his plays have an exceptionally strong emotional charge, which is intensified by the 'factual' presentation. Spectators fainted in 1929 when the clown was dismembered. Later plays like *Mother Courage* or *The Caucasian Chalk Circle* contain moments of tear-jerking melodrama or nail-biting suspense. A dumb girl sacrifices herself to save the children of a besieged town, beating a drum despite the rifles of the assaulting troops levelled at her. A desperate flight, babe in arms and brutal pursuers at heels, over a rickety bridge, rivals such spectacularly melodramatic sequences as the escape over the ice in the nineteenth-century classic of melodrama *Uncle Tom's Cabin*. Indeed, at a 1989 Berliner Ensemble performance of *Chalk Circle* in North America many of the spectators actually did weep at the small boy's cry as he is torn from the arms of the loving peasant-girl to be given to the biological but unnatural mother who abandoned him as a baby. In much the same way the 1928 Berlin audiences of *The Threepenny Opera* were swept into singing the army recruiting song, even though enthusiasm for the militarism it expressed was the last thing Brecht wanted.

This revised version of the *The Beggar's Opera* with its brilliant musical score by Kurt Weill, which has become by far Brecht's most famous piece, is perhaps the clearest example of his Epic Theatre. The geographical and historical distancing, with a modern critique of post-war German society presented in a late nineteenth-century English setting, and referring back to John Gay's 1728 political satire, is intended to bring out the essential nature of social structures that might be lost in a contemporary close-up (Capitalism = Crime). It also demonstrates that, despite any superficial appearance of progress, nothing has changed over the centuries (provoking the reaction of 'Things can't go on like that any longer'). To rule out illusion and empathy, the songs are separated out from the action as 'set pieces'; and—picking up on the way Gay had subverted eighteenth-century popular songs—words contradict the tone of the music or contrast with the situation, exposing the artifice of performance and creating a theatrical dialectic. Thus, when Macheath's down-at-heels gang ask his upper-crust bride to entertain them at the wedding feast, she announces a song in which she impersonates a down-trodden barmaid dreaming of bloodthirsty revenge on her wealthy customers. Given the *Threepenny* (*Drei Groschen*) title to indicate the cheapness of the deliberately crude and basic staging, because this was all that beggars could afford—although in fact designed as the opposite to Wagnerian grandeur—this ballad-opera reveals the popular basis of Brecht's work.

The organ-grinder's 'Moritat' or 'dirty deed' song of Mack the Knife, which frames the action, is derived from a traditional type of fairground performance. Standing on a bare bench, the singer would illustrate his violent and exotic ballads by pointing to a sequence of garish horror pictures, while accompanying himself on a barrel-organ: all of which is picked up in the central elements of Brecht's theory. The parable-like 'histories' are narrated, with no attempt by the singer to impersonate their simplified folk-tale characters, who are sharply divided into moral categories. The performance as a whole is a montage of distinct 'scenes' (or tales), presented in daylight without any illusory pretence to a strolling audience, smoking or munching on frankfurters while they watched. Another popular source that Brecht drew on was the highly politicized German tradition of cabaret. An accomplished guitarist himself, Brecht followed in the footsteps of the expressionist playwright Frank Wedekind, who had played a banjo to the nasal rendition of his own satirical verses and appeared with the famous masked troupe of 'The Eleven Executioners'. Brecht's distinctively harsh voice was echoed in the singing style of his actors, while the discordance between the light tunes he composed and the cynicism or political extremism of his lyrics (several of which

Clowns. The oldest surviving form of popular theatre. Each of the Fratellini Brothers, warming up in their dressing-room, has his own character—ranging from traditional white-face to a prototype of the figure developed by Charlie Chaplin—while the elaborate array of props indicates the range of mini-dramas on which their circus-act was based.

appeared in *The Domestic Breviary*, a blasphemously scandalous collection of poems he published in 1925) carried over into Kurt Weill's music.

The clown figures in some of Brecht's didactic pieces and his use of masks, particularly the half-masks in later parable plays such as *The Good Person of Setzuan* or *The Caucasian Chalk Circle*, point to another facet of theatre between the wars. Copeau, who had already been experimenting with a stage stripped to a bare performing area at the Vieux-Colombier, explored a modern version of the *commedia dell'arte*. From *Les Fourberies de Scapin* in 1920, it was a short step to creating an improvisational drama, based on stock figures and using masks, which characterized Les Copiaus, the troupe he worked with from 1924. Drawing inspiration from circus clowns, in particular the acrobatic Fratellini brothers, whom he invited to teach at the Vieux-Colombier Acting School, Les Copiaus dramatized themes of rural life from the Burgundian village to which Copeau retreated, in an attempt to return theatre to the people. These ranged from *La Danse de la ville et des champs* (in which a leading role was played by Michel Saint-Denis, who went on to found the London Theatre Studio and eventually to revive the Old Vic) to *Arlequin magicien*.

Copeau's aim of creating a truly 'Popular Theatre' (the title of his last theoret-

Facing, above: from fairground to stage. Brecht (in characteristic working-class cap) with the cabaret performer Karl Valentin in Munich, 1924; and (*below*) Ernst Busch as a typical 'Bench-Singer' in the 1931 film version of *The Threepenny Opera*.

407

Above: blurring artistic boundaries. A distinguishing mark of modern theatre is the expropriation of other types of entertainment. Circus became Broadway spectacle in the 1928 *Jumbo*: subsequent more sophisticated borrowings have included Magritte's paintings and a whole symphony orchestra in one of Stoppard's plays.

Right: politicization of the archetypal clown. A pre-social Everyman figure in Peter Handke's *Kaspar*, 1968, demonstrating how language brainwashes the individual into a schizophrenic conformity.

ical book) led him to revive the medieval mystery play in the mid-1930s. These open-air productions—of a fifteenth-century Tuscan saint play in a monastery cloister, a modern text commemorating Savonarola's burning at the stake in the Piazza della Signoria at Florence, and Copeau's own version of a fourteenth-century French miracle play in the courtyard of the Hospice de Beaune—with their spectacle, crowd scenes, and multiple stages in evocative historical locations, were reminiscent of Reinhardt's monumentalism. Yet Copeau's purpose was closer to Brecht in that these performances were designed as catalysts for social regeneration: but with a religious instead of a Marxist basis. And, looking back on his career, Copeau concluded that the small-scale theatre he had earlier promoted, as a way of exploring dramatic art free of commercial pressures, blocked out the mass public. 'Those little theatres were only technical laboratories, conservatoires where the noblest traditions of the theatre could be resurrected, but which could not be called true theatres, because they lacked a true audience.' Even so, it is from Copeau's work that the 'laboratory theatres' of the 1960s and 1970s ultimately sprang—Jerzy Grotowski's Polish Laboratory Theatre, Peter Brook's Centre for International Theatre Research in Paris, La Mama, or Richard Schechner's Performance Group in New York—all of which were following Copeau's line in returning theatre to its roots, stripping the stage, and reviving ritual.

However, Copeau's more immediate influence was on the development of mime, which has been a strong element in French theatre since the 1930s. Jean-Louis Barrault, who shared his belief in a 'sacred theatre' and became the main proponent of Claudel's neo-symbolist plays of Christian redemption, first established his reputation as a mime. His concept of a 'total theatre' was a purely imaginative world 'more expressive than reality', created on the stage through a synthesis of stylized conventions from circus and *commedia dell'arte* to Japanese Noh drama. This was embodied in Claudel (with Barrault mounting multiple different productions of his five major plays between 1943 and 1972) and in his own adaptation of Faulkner's *As I Lay Dying* in 1935. Evoking a primordial existence in which man and animal were linked, *As I Lay Dying* relied solely on choreographed mime backed by choral chanting to create both the characters and their physical environment. The 'centaur-horse' that Barrault created by simultaneously miming a prancing stallion and its rider (as well as birds of prey hovering overhead), the high point of this performance, became the emblem for his total theatre: the naked actor expressing the whole of existence through gesture and physical movement. The strikingly imaginative and emotionally affecting quality of his art is preserved in Carne's film *Les Enfants du paradis*, where Barrault plays the role of Deburau, the nineteenth-century originator of French mime. And Barrault's lyrical rendering of Deburau's pale-featured and elongated Pierrot, farcical and macabre, became the basis for Bip: the philosophical clown in striped jersey and melancholy white-face, with a red rose decorating his battered top hat,

Giorgio Strehler's 1947 production of Goldoni. From Copeau onwards, reviving traditional forms has proved a vital part of theatrical renewal: *commedia dell'arte* figures and a booth stage on the naturalistic set of *The Servant of Two Masters* at the Teatro Piccolo, Milan.

which was the trade-mark and *alter ego* of Marcel Marceau. The most famous of all modern mimes, Marceau attracted disciples world-wide when he toured his highly sophisticated mimodramas to sixty-six countries after triumphing at the Berlin Festival in 1951.

Another type of popular theatre, related to Copeau's work in its revival of 'folk' forms and in taking performances out of the urban centres to country people, was García Lorca's La Barraca (The Barn). This touring group, formed after the establishment of the Spanish Republic in 1931 to promote its populist ideals, worked on a tiny portable stage with only the most basic of costumes and props. It mainly performed classical Spanish drama. But Lorca's major poetic peasant tragedies came out of that experience of an unsophisticated audience's enthusiastic response to the archetypal dramatic situations of Cervantes and Calderón's *Life is a Dream*. (This went into the repertoire solely because Lorca wanted to play the Shadow, which he interpreted as a surrealist ghost pinpointed by a cold blue spot: a memorable performance preserved in a brief film clip). A close friend of Salvador Dali and Luis Buñuel, Lorca combined a surrealist approach with traditional subjects—as with one of his early plays, where a folk ballad became the basis for a distinctively modernist physical poetry, combining lyrical language with carefully orchestrated light, colour, and music. The same qualities characterize *Blood Wedding*, which immediately established his reputation as the leading twentieth-century Spanish playwright when it was performed at Madrid in 1933. Using a traditional subject from his collection of poems *Romancero gitano* (*Gypsy Ballads*), it juxtaposes realistically conceived peasant characters with emblematic figures such as the Moon and Death—which Lorca (in startling contrast to convention) insisted be presented as an irresistibly attractive woman, the ultimate lover—to raise the themes of passion and revenge to a mythic level. His two subsequent tragedies play on the same themes, although progressively discarding surrealistic devices to achieve what Lorca called a 'photographic' representation of repressed and violent rural life. Perhaps because of this closeness to recognizable society, the opening of *Yerma* provoked a riot by outraged anti-Republicans. But beneath the naturalistic surface the characters embody elemental forces, and the action has an almost Greek simplicity.

One of the people who had most admired Barrault's 'centaur-horse' was his mentor Antonin Artaud, possibly the most influential of all the modern theatre-theorists. At the time, his tiny handful of productions—a hyper-realist staging of Strindberg's *Dream Play* and Vitrac's surrealist *Victor* in 1928, his own mythically inflated version of Shelley's *The Cenci* in 1935—seemed to have little significance outside of the explosive scandals provoked by each of their performances, which were deliberately designed to outrage his patrons and the bourgeois public. An extraordinary non-naturalistic actor, externalizing psychological states in exaggerated gesture and facial intensity that outdid the German expressionists, Artaud literally incorporated his 'theatre of cruelty' in the lead roles of his own productions. Their settings presented a fragmented, fractured world in which perspectives were hallucinatorily distorted, and normal time no longer applied, with frenetically speeded-up action juxtaposed to the 'statue-like pace' of figures in exaggeratedly slow motion. Many of these elements were derived from the techniques of silent film, and could only be fully realized on a cinema screen. But his ideal of a performance that would totally envelop and involve the audience physically, affecting the psychological being of spectators through the skin, like acupuncture, required above all the immediacy of live theatre. In practice, due partly to a paucity of resources, only the sound effects in his productions achieved this, projected through loudspeakers at all four corners of the auditorium to create an overwhelming 'net of sonorous vibrations'.

Despite technical limitations, the outline of Artaud's concept of theatre can be faintly discerned in a production like *The Cenci*. Strongly marked rhythms of movement, as well as sound, formed the basis of the action. A dynamic choreography, orchestrated on geometrical patterns, symbolically represents the 'circular and closed world' of a hierarchical society, while contrasting violently with the story, which Artaud had selected because it contained 'corporal and moral excess . . . to the extremity of instinct'. However, arguably nothing could measure up to the visionary ideal set out in *The Theatre and its Double*, his collection of essays published in 1938, which became the inspirational handbook of experimental groups more than thirty years later. This manifesto called for a theatre of physical action: the metaphysical 'double' of not only commercial entertainment, but also serious (naturalistic) or literary drama. It was uncompromisingly opposed to the whole theatrical output of Western culture. 'No More Masterpieces!' The 'chains' of logic and reason, which are the basis of a debilitating civilization founded on a religion (Christianity) that denies the body, are to be broken by 'images of energy in the unconscious and gratuitous crime on the surface'. These release a primal spontaneity and 'communicative delirium' in the audience. The 'reservoir of energies made up by myths that man no longer incarnates is incarnated in the theatre', which acts like a 'plague', causing 'all social forms to disintegrate' and returning people to their natural state. Thus theatre rediscovers the 'necessary cruelty' that is a law of nature, creative through destruction.

Like W. B. Yeats before him, Artaud anticipated later trends in taking oriental theatre as his model. But, instead of the aristocratic Noh tradition, he turned to the Balinese trance dances that he had observed at a colonial exhibition in 1931.

RALLYING-CALL FOR THE AVANT-GARDE

An extract from Artaud's trend-setting manifesto *The Theatre and its Double*. Aristotle's *Poetics* might have been influential—but in no period has theatre been so dominated by theory as in the twentieth century.

We need above all a theatre that wakes us up: nerves and heart . . . Everything that acts is a cruelty. It is upon this idea of extreme action, pushed beyond all limits, that theatre must be rebuilt . . . so we believe that the images of thought can be identified with a dream which will be efficacious to the degree that it can be projected with the necessary violence . . .

Hence this appeal to cruelty and terror, though on a vast scale, whose range probes our entire vitality, confronts us with all our possibilities.

. . . we want to resuscitate an idea of total spectacle by which the theatre would recover from the cinema, the music hall, the circus, and from life itself what has always belonged to it . . . a spectacle addressed to the entire organism . . . an intensive mobilization of objects, gestures, and signs, used in a new spirit. The reduced role given to the understanding leads to an energetic compression of text; the active role given to obscure poetic emotion necessitates concrete signs. Words say little to the mind; extent and objects speak; new images speak, even new images made with words. But space thundering with images and crammed with sounds speaks too.

For Artaud the high degree of formalization in conjunction with the trance state achieved by the Balinese performers expressed 'the automatism of the liberated unconscious' that he envisaged for his actors. Instead of representing individualized characters they were to be 'hieroglyphs . . . signalling through the flames' in a 'concrete language, intended for the senses and independent of speech'. In this way the actors—like the figures of his own plays—would become 'incarnations of great forces, outside of good and evil'.

The Cenci was the first—and last—production of Artaud's Théâtre de Cruauté, which had been intended to be the catalyst for a radical transformation of Western theatre. After its failure he abandoned European civilization to search for authentic primitive experience in Mexico, and was eventually committed to an insane asylum at Rodez, where he remained until two years before his death in 1949. Yet, even though it would be a further twenty years before his significance was realized, Artaud represents one pole of the post-First World War theatre. His theories are the opposite in every way to those of Brecht, although both started from an equivalent rejection of the conventional naturalistic stage. To take titles from their essays as examples, the pre-verbal irrationalism of Artaud's 'affective athleticism', which aims at liberating the inner nature of man repressed by civilization, is at one end of the theatrical spectrum. Brecht's 'rational theatre', intended to revolutionize society, is at the other. Equally they represent the two extremes of theatrical creation. Brecht achieves the apotheosis of the writer: the dramatist as director. In Artaud the director becomes a new kind of dramatist in his own right, creating images in space through the figures of depersonalized actors transformed into hieroglyphs, in conjunction with sound, light, and symbolically weighted objects.

The extremism of such contrasting approaches is largely what distinguishes European theatre from that of England and America between the wars. The relative security of their social orders—plus perhaps the degree to which dramatists held star billing, rather than the director—meant that stylistic experimentation had little impact on their theatrical scene. Even in Ireland, where the independence wrested from a weakened empire by civil war was celebrated ironically by O'Casey in strongly socialist comedies such as *Juno and the Paycock* or *The Plough and the Stars*, naturalism remained the dominant mode. However, in the USA—due at least partly to the fresh national confidence engendered by victory on the European battlefield and their first major intervention in world affairs—a new and specifically American voice emerged. This was Eugene O'Neill.

Although strongly autobiographical, his work was specifically intended to embody a panoramic dissection of the nation from its beginnings in the aftermath of the American Revolution to the materialistic present. Only the first in his projected seven-play cycle of 'The Possessors Self-Dispossessed', *A Touch of the*

Facing: the continuing expressionist mode. Distorted perspectives and destabilizing visions, a style revived to evoke the era of the 1920s: social repression in Peter Stein's 1986 production of O'Neill's *The Hairy Ape*.

Poet, was completed. But other plays from the 1920s fill in the general theme. The life-denying repression of the puritan ethic is documented in the post-Civil War trilogy of *Mourning becomes Electra*; the moral corruption of greed in *Desire under the Elms*; the progressive egoism and loss of all spiritual values in *Strange Interlude*, which traces social developments from the effects of war in 1919 to a future projected a whole generation after the play was written. O'Neill's breakthrough came after an apprenticeship in a small provincial drama group, the Provincetown Players, with *The Emperor Jones* in 1920. American theatre was still under the influence of Belasco, who continued to stage solidly realistic productions until just before his death in 1931. And naturalism remained the norm, with his mantle descending on the Group Theatre headed by Lee Strasberg, who promoted Stanislavski's approach. As director of the Actors' Studio from 1951 right up to 1982, Strasberg established the 'Method School' (exemplified by Marlon Brando and the films of Elia Kazan) as a distinctively American style of acting. Yet new European ideas were being disseminated by Sheldon Cheney's *Theatre Arts Magazine*, while Arthur Hopkins introduced the techniques of Craig and the German expressionists for his Broadway productions of Shakespeare; and almost every one of O'Neill's plays has an experimental aspect.

The Emperor Jones was not only one of the first American plays to employ a Black actor, Paul Robeson, in a lead role. It also used symbolic dream sequences and a compelling drumbeat, keyed to the pulse of the audience in an accelerating rhythm that gradually built throughout the performance, in order to bring home the superiority of primitive instinct over the materialistic rationalism of White society. In *The Hairy Ape* O'Neill borrowed expressionistic distortion (brilliantly realized in a 1986 production by the German director, Peter Stein) to attack the effeteness and inhumanity of American industrialism. And in subsequent plays during the 1920s he experimented with pantomime, masks, inventive scenic devices, and interior monologue. He also (more literally) extended the range of the theatre by the sheer length of plays like the monumental nine-act *Strange Interlude*—an attempt to combine drama with the novel that takes some seven hours to perform, yet broke box-office records in 1927 and still proved remarkably successful when revived with Glenda Jackson in the 1980s.

Although winning four Pulitzer prizes, followed by the Nobel prize in 1936, O'Neill withdrew his plays from the stage after critical assault on *Days without End* in 1934. The United States was already in the grip of the Great Depression; and the grittily realistic, sharply politicized tone of the Group Theatre was more in tune with the times. America's first true ensemble (with its socialist overtones), they were a key factor in creating a theatre of political consciousness, which found its expression in the plays of Clifford Odets; and it was the production of his first play, *Waiting for Lefty*, in 1935 that gave the Group Theatre its leading position. Although his subsequent work was strongly naturalistic, and has been compared

AN IMPROMPTU DADAIST PERFORMANCE-PIECE

Theatre as the tool of dictatorship. The communal rituals developed by the Nazi state found their fullest expression in the carefully stage-managed annual Nuremberg Rally, which Goebbels hailed as 'the Thing-concept made blood and spirit'.

Politics is sometimes literally theatre, as at a 1935 state banquet with Marinetti as guest of honour, documented by Moholy-Nagy's wife. Moholy had a train ticket to exile in his pocket, while Kurt Schwitters's 'Merz' art was publicly burnt by the Nazis.

Short of Hitler, all the Nazis were present . . . Schwitters and I were sandwiched between the head of the Organization for Folk Culture and the leader of the 'Strength through Joy' Movement . . . The more Schwitters drank, the more fondly he regarded his neighbour.

'You think I'm not worthy of sharing your art chamber for strength and folk, ha? I'm an idiot too, and I can prove it . . . I'm the great Aryan MERZ. I can think Aryan, paint Aryan, spit Aryan.' But before he could incite anyone to action, Marinetti had risen from his chair.

'My friends,' he said in French. 'After the many excellent speeches tonight'—the silent officials winced—'I feel the urge to thank the great, high-spirited people of Berlin. I shall recite my poem "The Raid on Adrianople".' There was polite applause . . . 'Adrianople est cerné de toutes parts SSSSrrrr zitzitzi . . .' roared Marinetti . . . He grabbed a wineglass and smashed it to the floor. 'Zitzitzit toum toum Patrouille tapie'—Marinetti threw himself over the table—'Vaniteeee, viande congeleeeeee . . .' Slowly he slid to the floor, his clenched fingers pulling the table-cloth downward, wine, food plates, and silverware pouring into the laps of the notables.

Schwitters had jumped up at the first sound of the poem. Like a horse at a familiar signal, the Dadaist in him responded. Breaking out into a roar that drowned the din of protesting voices and scraping chair legs, he thundered: 'Oh, Anna Blume / Du bist von hintern wie von vorn / A-N-N-A.'

to O'Casey, *Lefty* came out of the agitprop movement, and Odets later called it 'a machine-gun' that could be deployed at any strike meeting or picket-line. On a bare platform representing a union hall, its sequence of short scenes portrays the injustice and poverty that have fuelled the militancy of the workers who rise to speak at the strike meeting. The urban folk-idiom of the dialogue and the passionate emotions of the episodes carried the New York audience away, and they rose to join in the culminating call of 'Strike! Strike!', which became, as Lee Strasberg remarked, 'the birthcry of the Thirties'.

The most vibrant sign of this new politicization was the Federal Theatre Project—part of a programme to provide jobs for the unemployed—which financed over 800 different productions between 1935 and 1939. Many of these were 'living newspaper' documentaries, dealing with the plight of American farmers (*Triple-A Plowed Under*) or the homeless (*One Third of a Nation*) through illustrative scenes of 'little' people affected by the problem, intercut with film clips and quotations from news reports or government studies, leading to a call for specific action. The Federal Theatre also included Orson Welles's all-Black version of *Macbeth*; but the limitations of a government-directed political theatre became clear in 1937 when Welles tried to mount *The Cradle Will Rock*, a musical by Marc Blitzstein that lambasted the union-busting activities of steel companies. It was blocked as too explosive, and the level of social criticism in the 'living newspapers' eventually led to the programme being closed down. However, by that time politics had even infiltrated Broadway, which was now open to experimental drama like Thornton Wilder's Pirandellian *Our Town*. This was followed in 1942 by *The Skin of our Teeth*, Wilder's anachronistic and overtly theatrical parable of human resilience in the face of disaster (the ice age and Noah's flood, as well as war).

By contrast, in Europe theatre had already fallen victim to ideological repression and violence. Artistic experiment was crushed along with political dissent. Performances were banned or broken up, as well as books being publicly burnt; and, even more damaging, the avant-garde was co-opted by dictatorship.

Futurism, with its underlying glorification of power through the synthesis of man and machine, became the official art of Italian fascism; and, as Mussolini's cultural ambassador, Marinetti—whose work had been included in early Dada exhibitions—found himself guest of honour at a Nazi banquet in Berlin in 1935, to which he got his old comrades in artistic arms invited. The absurd performance-piece into which the evening developed was the death-rattle of post-war experimentation. Impromptu sound poems and satiric gestures were empty charades in the face of totalitarian force. Lorca was murdered by Franco's Falangists at the beginning of the Spanish Civil War in 1936. In Stalin's Russia the short era of theatrical liberty had already come to an end with Mayakovsky's suicide, to be shortly followed by the shootings of Meyerhold and Tretyakov. In Germany and Austria the avant-garde were being silenced by censorship, exile, or—as for Toller—suicide. It was a break as complete as when English theatres had been closed under Cromwell in the seventeenth century. A whole artistic élite was silenced: some, like the 36-year-old Lorca, when their careers had barely begun. And the next generation were blocked from building on their achievements. The consequences of this tragedy (later dramatized by David Pownall's *Master Class* with Timothy West's memorable and menacingly reasonable Stalin) were conservatism and the almost complete disappearance of the avant-garde until the late 1960s.

It is worth noting that some of the fascist leadership were as closely involved with theatre as early Bolshevik figures like Lunacharsky or Trotsky. In his first speech to the Reichstag on taking power, Hitler had declared 'Heroism is passionately asserting its claim to shape and guide the People's destiny . . . It is Art's task to be an expression of this decisive spirit of the time' (listing theatre first among all the media). The Nazi Propaganda Minister Goebbels had himself written an expressionist play, and lavishly funded his 'Strength through Joy' Movement to expand the (National Socialistically correct, and stylistically conservative) theatre beyond its position in the Weimar years. He also fostered the 'Thingspiel' movement, a form of quasi-religious community theatre that celebrated the mystical essence of the Nazi state. Typically representing the banishment of a traitor to the cause, these choral and choreographed performances in 'hallowed' spaces confirmed the Aryan masses in the 'Blood and Soil' cult. Mussolini too was a playwright; and his bombastically inflated Napoleon drama *The Hundred Days*,

Entertaining troops on the Russian front. Theatre's contribution to the war effort: stand-up comedy, popular songs, or dancing chorus-lines, on makeshift and mobile stages. Sustaining morale was easily confused with distraction from ugly reality helping to form the post-war taste for light entertainment.

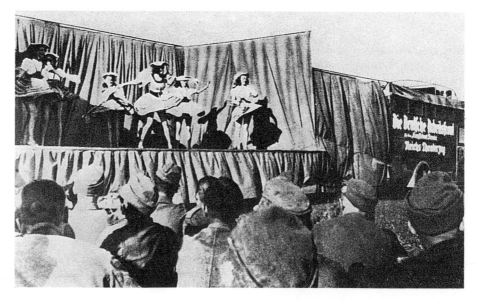

performed before Hitler and the Nazi leadership in 1932, had been a foretaste of things to come.

Indeed, both ideological extremes were convinced of the theatre's social significance and accorded it a special place in their state. But this degree of political integration, so desired by many of the artists, in the event proved disastrous. German theatre was reduced to mediocre sterility for the whole decade of Nazi rule; and the same creative darkness lasted in Russia until well after Stalin's death.

When hostilities opened in 1939, theatre everywhere was drafted for the war effort. ENSA and its German or Italian counterparts formed touring theatres to bolster their troops' morale. In England Noel Coward (now a naval officer) turned to the emotional patriotism of plays like *For Whom We Serve*; and in France, under Nazi occupation, politics could only be obliquely hinted through adaptations of classical myths, as with Anouilh's *Antigone* in 1944 or Sartre's modern version of the Electra story in *Les Mouches* a year earlier.

Even after the final defeat of the fascist powers in 1945 there was no immediate return to the earlier experimentation. The sense of artistic excitement or political significance that earlier made theatre the focus of public battles had gone. The European combatants were exhausted; and in Germany the occupying Allied forces imposed censorship. In addition, even though the fighting had stopped, the ideological confrontation of the 'Cold War'—already building in tension while the victory celebrations were still going on—was almost equally inhibiting.

In England the poetic comedies of Christopher Fry offered verbal fireworks, colourful frivolity, and sentimental escapism from post-war austerity, while serious drama was represented by Terence Rattigan, whose well-made problem plays harked back to the nineteenth century. And the change from the openness to experiment in the 1930s is summed up by T. S. Eliot. *Sweeney Agonistes*, performed by the Group Theatre in 1934, had been an excitingly eclectic mix of jazz, Noh conventions, and seedy modern characters that impressed Brecht. *Murder in the Cathedral* a year later was a conscious return to the ritual roots of drama, combining church liturgy with verbal patterns from the morality play of *Everyman*, involving the audience in Thomas à Becket's martyrdom as the congregation of the cathedral where he had been killed. Even *Family Reunion*, the first of his plays written for the commercial theatre in 1939, had experimented with expressionistic shifts of consciousness, incantation, and the juxtaposition of mythic figures and contemporary life. But when he returned to the theatre with *The Cocktail Party* after the end of the war, and after a break of ten years, his plays looked exactly like standard drawing-room comedies. Even the verse mimicked everyday dialogue to such an extent that Eliot himself questioned 'whether there is any poetry in the play at all'.

In France Barrault founded his own theatre company and set the tone, affirming traditional cultural values with brilliant productions of both Claudel and

French classics, while Anouilh—representing much the same escapism as Christopher Fry—returned to the bitter-sweet romantic comedies with which he had made his pre-war reputation. Sartre was almost alone among French dramatists of the time in tackling the issues of political expediency and moral responsibility raised by the German occupation in *Dirty Hands* (1948) or *The Condemned of Altona* (1959). Yet his most successful play was an adaptation of Dumas. The Italian stage was dominated until the 1950s by Visconti, who combined film work with (like Barrault in Paris) flamboyant revaluations of Goldoni and other classical drama. In Spain the theatre remained heavily censored right up to Franco's death in 1975. Of all Europe, only Switzerland, secure in its neutrality and still prospering, continued the line of development that had flourished in the 1930s.

In both World Wars the country had served as an artistic refuge. Dada had come into existence at Zurich during the first conflict; and it was there that the plays Brecht had written in exile were produced, like *Mother Courage* in 1941, Oskar Wälterlin at the Zurich Schauspielhaus having gathered actors and directors fleeing Hitler's persecution from 1938 on. The Austrian historical dramatist Hochwalder settled there; and when Brecht was forced out of America in 1947 by the McCarthy hearings he initially came to Zurich, where he staged his version of *Antigone* as well as *Puntila*. But there were (as yet) no Swiss playwrights to take advantage of such models, theatre in Switzerland having been largely limited to festival pageants commemorating historical events or the biennial staging of Schiller's *Wilhelm Tell*.

In contrast to the 1920s, and perhaps as a continuation of the close links that had developed between state and stage during the war, government involvement became one of the major elements in theatrical renewal after 1945. Particularly in England, where dramatists abdicated their trend-setting role, the state stepped in. The new Labour government designated a fraction of local taxes for entertainment, laying the foundation for a wave of civic theatre-building in the 1950s, while subsidy by the Arts Council (founded in 1946) freed regional companies from the exigencies of 'weekly rep', allowing increased rehearsal time and making it financially possible for them to risk new plays. This development of provincial theatre is a striking feature of the twentieth-century cultural scene in Britain: one which was echoed in other countries, notably Germany and America. Expressing civic pride, these new theatres were architecturally adventurous—the most striking being the Manchester Royal Exchange theatre, a glass module suspended from four pillars inside the former cotton market—and by the mid-1960s distinctive regional companies had evolved around them. At Stoke-on-Trent particularly strong links were created with surrounding communities by musical documentaries on local history and issues, such as the threatened closure of a neighbouring steelworks. Scarborough provided a base for Alan Ayckbourn, one of the founders of the Victoria Theatre at Stoke, where he was able to develop his unique brand of popular comedy through working with his own company. The Citizen's Theatre of Glasgow, set in what were in 1945 the worst slums of Europe, encouraged new Scottish playwrights, and after the arrival of Philip Prowse in 1970 gained an international reputation for eye-opening theatricality.

In America similar effects were created by the escalating costs of Broadway production, rather than subsidy, which led both to the emergence of new theatres 'Off-Broadway' and to the development of theatres outside New York. One of the earliest was the Alley Theatre in Houston, which began with amateur productions at a small dance studio in 1947, moved into a converted fan-manufacturing plant two years later, becoming a fully professional company in 1954, and raised private funding for a purpose-built theatre complex in 1968. Such development was typical: small semi-professional groups (some, like the Arena stage in Washington, coming from local university drama departments) performing in *ad hoc* settings, who established a reputation over the 1950s that brought funding for their own stage from local benefactors or the Ford foundation. Most remained commercial operations. But regional non-profit theatres emerged during the 1960s, some being based in universities (such as the Yale Repertory Theatre), or funded by federal grants, like the Trinity Square Repertory Company in Rhode Island— though not all were closely tied to a particular locality, with the American Conservatory Theatre moving from Pittsburgh to San Francisco.

Both in the relative lack of state involvement, and in its unbroken theatrical continuity, American theatre stands in sharp contrast to Europe and Britain. Fresh developments emerged even during the war. In particular, the musical achieved its mature form with the Rogers and Hammerstein *Oklahoma!* in 1943. Its unprecedented popularity came from the nostalgic idealizing of an earlier America. Yet the integration of musical numbers into a coherent story-line with fully developed characters, and the inclusion of a ballet dream sequence, points forward to Leonard Bernstein's *West Side Story* in 1957 and Stephen Sondheim's brilliantly inventive *Cabaret* in 1966, where the frame of a decadent Berlin nightclub exposes the seductively sinister psychology of fascism.

The leading pre-war dramatists overlapped with the new generation. O'Neill's most famous play, *Long Day's Journey*, was completed in the 1940s—a family portrait so searingly honest it was not released for performance until 1956, after his death—while Tennessee Williams's first play had already been staged in 1940 and Arthur Miller's *All my Sons* appeared in 1947. Their work built on the wide social relevance and stylistic innovation of American theatre in the 1930s. In each case the plays that made their reputations were directed by Elia Kazan, who had been a member of the Group Theatre; and the fusion of external realism with psychological symbols in their drama extends some of O'Neill's earlier experiments into a distinctive American style.

American naturalism: The psychological universe. Inner and outer worlds in Kazan's staging of *Streetcar Named Desire*: Stanley (Marlon Brando) watches his pregnant wife being comforted on the exterior balcony, his poker-playing buddies slouch at the parlour table, while Blanche is led away to an asylum from the interior bedroom.

Williams's 1947 *Streetcar Named Desire* encompasses all his major themes: the ambiguous nature of sexuality, the betrayal of faith, the corruption of modern America, the overarching battle of artistic sensitivity against physical materialism. These are given archetypal shape in the protagonists: Blanche Dubois, symbol of the decaying South, neurotic and corrupted, hiding from herself behind artificial illusions, yet the voice of civilized values as well as imagination and emotional truth; Stanley Kowalski, acted with brutal virility by Marlon Brando in his most famous stage part, representing the norms of a soulless society, crude and ruthlessly competitive as well as uncultured, yet sexually vital and dynamic. Two years earlier, Williams had been experimenting with overlapping time-frames, a narrator, and symbolist screen projection in the autobiographical memory play *Glass Menagerie*. With *Streetcar*—as increasingly in his later work—the surface is naturalistic, yet the cries of street-vendors and the streetcar routes (Desire, Cemeteries, Elysian Fields) give an archetypal depth to the story of rape and mental breakdown, while music echoes from within Blanche's tortured mind, transforming New Orleans into a psychological world. And this was reflected in the multiple set for Kazan's production, which showed simultaneously the inside and outside of the tenement.

Similarly in Miller's *Death of a Salesman* two years later, originally titled 'The

Inside of his Head', the setting Jo Mielziner designed for Kazan showed both the high-rise blocks that overshadow the small Loman home and its interior, the walls of which could be made transparent for characters to move through them in the memory sequences. And in both these productions the sense of subjective emotional vision interpenetrating the realistic outlines of material existence was reinforced by the acting style developed through the Actors' Studio, which emphasized the purely physical expression of intense psychological truth. Miller's 'low man' on the economic ladder of capitalism, the ageing commercial traveller discarded by the system to which he has mistakenly devoted his life, became the trade-mark figure of post-war American theatre. *Death of a Salesman* was performed world-wide, even (during the 1980s) in communist China; and Miller's devastating attack on the 'American dream' defined a characteristic theme for subsequent dramatists that was still being echoed in David Mamet's work almost forty years on. But no American dramatist has matched the moral integrity of Miller's work, with its passionate political involvement based on the irreconcilable conflict of the individual's imperative to preserve his personal truth against the demands of a materialistic or repressive society. This was most forcibly expressed in his one historical drama, *The Crucible*, where the Salem witchcraft trials are used as a parallel to the hysteria and betrayals of Senator Joseph McCarthy's anti-communist investigation (a subject Miller returned to a decade later in his highly autobiographical *After the Fall*).

After the mid-1950s Miller withdrew from the American stage for long periods, disgusted by the increasing commercialism of Broadway and the autocratic reviewers who were beginning to dominate it. Although Williams retained his position on Broadway for a little longer, after *The Night of the Iguana* in 1961 he too had difficulty in getting his plays performed. Several of his later pieces deal with despairing artists who have lost control of their work, and the final period of his life was a descent into alcohol and drugs. But in the immediate post-war period Williams and Miller served as models of a new stylistic freedom and social relevance that helped to liberate theatre elsewhere. Their work was staged in all the European centres, which had been cut off from foreign developments during the war (or in Germany's case since 1933, when Hitler seized power); and the challenge it represented is indicated by the British establishment's response. Tennessee Williams was limited to private club performances in London because of the supposed immorality of his plays, while the homosexual kiss in Miller's *A View from the Bridge* was seen as so subversive that it was banned altogether. In the 1920s German and Russian experiments had been imported into America. Now—confirming the dominance of dramatists over the directors—Miller and Williams, along with Thornton Wilder, were the general catalyst for a revival in Europe.

The first signs of this appeared in the early 1950s. Brecht, who had been offered his own theatre by the East German regime in 1949, began staging all his major works. In Berlin the bleakness of the bare stage in his *Mother Courage*, the distancing effects, clinical lighting, and 'alienated' acting, were designed as 'a withdrawal course for emotional addicts' from the narcotic excesses of German theatre under the Third Reich. Elsewhere the Berliner Ensemble performances were hailed as strikingly imaginative and theatrically inventive. When they appeared on the London stage in 1956, not only were individual playwrights like John Arden strongly influenced, but a radical change in Shakespeare production followed that was to condition perceptions of the classics for a whole generation. In Italy Brecht's epic principles were adopted by Giorgio Strehler, and came to form the basis of his unique directorial approach. In France Roger Planchon, together with the dramatist Arthur Adamov, evolved a Marxist theatre style from Brecht's work that combined anti-illusionistic theatricality and politically simplified characterization with meticulously realistic social detail. But, as might be expected from the Zurich theatre's involvement with Brecht since the early 1940s, it was in Switzerland that his influence first produced results and was most obvious.

Max Frisch and Friedrich Dürrenmatt were the first Swiss playwrights ever to win international recognition. Frisch began writing for the stage while still practising as an architect, and his early plays were highly philosophical commentaries on individual guilt and the problems of power in relation to totalitarianism, war, and Hiroshima. His breakthrough came with *The Fireraisers* in 1958 (on which Dürrenmatt collaborated). This 'openly unhistorical' microcosm of complicity in evil reflects both Hitler's take-over of Germany with the burning of the Reichstag and the more recent communist seizure of Czechoslovakia in 1948 through the scaled-down activities of a bourgois everyman. The strength of this 'morality play without a moral' is in its universal applicability; and exactly the same type of generalization distinguishes Dürrenmatt's satiric questioning of Swiss materialism and bourgeois complacency, which from the opening of his first play in 1947 repeatedly generated controversy and outrage in the Swiss audience.

Dürrenmatt's best-known play, *The Visit*, which was almost immediately adopted as a modern classic when performed in 1955, is a study of communal guilt and the corrupting power of money. The central figure is baroquely allegorical, 'the richest woman in the world' (whose name is abstracted from a compound of millionaires—Zacharoff, Onassis, and Gulbenkian): a fashion-plate Whore of Babylon, whose elegant limbs turn out to be artificial prostheses. Highly theatrical artifice transforms potential tragedy into mock-heroic farce, flattening individual characters into grotesques. Responding to the ethical dislocation of global war in much the same way as the Italian 'Theatre of the Grotesque' had to the previous conflict, Dürrenmatt asserted that 'comedy alone is appropriate to us. Our world has led to the grotesque as it has to the atomic bomb . . . the grotesque is our mode of perception, a physical expression of paradox.'

Beckett's overt theatricality. The tramps meet Pozzo and Lucky in Roger Blin's original 1953 production of *Waiting for Godot*: metaphysical clowns performing on the bare boards of the stage, occupied only by an obviously fake tree.

The other main line in this theatrical revival had its source in French existentialism, and was given its theoretical basis (the distinctive mark of almost all modernist art movements) by Albert Camus's philosophical essays on 'the absurd'. Camus had himself written plays: *Caligula*, produced in 1945, or *The Just Assassins* in 1949, focusing on the legitimacy of violent revolt as shown by Russian anarchists in 1905. But the kind of drama that gave life to his ideas was the opposite of such historical discussion plays.

In 1950, two radically new voices emerged with Eugene Ionesco's *The Bald Soprano* and Samuel Beckett's *Waiting for Godot*. For Ionesco, language itself was problematic; and his early one-act plays discard rational dialogue, along with logical plot and consistent characters. People are reduced to objects, speech to empty cliché, and all ideals to meaningless illusion, in a surrealistic and hallucinatory farce. Thus in *The Chairs*—parodying all 'message-drama' and the public that valued such work—seats are set out for an (invisible) audience, gathering to hear about the meaning of life, until the chairs fill the whole space of the stage. But the orator hired to deliver the message is a deaf mute. As Ionesco commented, the image itself is the reality: 'the chairs, that is to say the absence of people . . . the absence of God . . . the unreality of the world, the metaphysical void; the theme of life is nothingness.' *Waiting for Godot* can be seen as a more coherent and extended adaptation of these elements. Two clown-like tramps in a non-place (the theatre itself) trying to find a 'meaning' for their pointless existence are trapped in a circular action which always returns them to the same point; and at the end their attempted suicide turns into a stock farce routine. As with Ionesco, *Godot*'s meaning is in the image created by its performance. The only progression is entropy, the stripping away of hope, material objects, even extended speech—and a typical reaction (to the first English production) was 'nothing happens—twice!'

Both were ahead of their time. Beckett took three years to find a director prepared to stage *Godot*; and it was also only in 1953 that Ionesco began attracting attention with *Victims of Duty*, which generated a glowing write-up from Anouilh. But Beckett's reductive drama, his theatrical minimalism, and his undefined symbols left open for spectators to interpret became an instant sensation. Within five years of its opening at Roger Blin's tiny Théâtre de Babylone in Paris, *Waiting for Godot* had been seen by over a million people. As Tom Stoppard later commented, after it, nothing in the theatre could remain the same.

Beckett's following plays define other aspects of his vision. In *Endgame*, where there is no possibility of making an exit from the claustrophobic interior of what might be a nuclear shelter or the inside of a skull, since 'the earth is extinguished' outside, the central figure presents himself as a ham-actor. The function of the dialogue is simply to enable the characters to exist by prolonging the performance in displays of histrionic (dis)ability; and when Beckett directed the play himself he stressed the chess-metaphor of the title. The paralysed and red-faced Hamm was presented as the King, who plays out senseless moves 'conscious of their pointlessness', while his antagonist was made to move as a Knight in jerky zigzags. As Beckett explained to the actors, the significance of the action was 'plain play. Nothing less. No thought of puzzles and solutions.' For him, 'the value of theatre' was that 'one can set up a small world with its own rules, order the game as if on a chess-board'. By extension, all human activities are routines, rituals serving only to cover over the painful awareness that our lives have no purpose; and this is particularly clear in *Happy Days*. Buried to her waist in a featureless desert by a mound of sand that eventually rises to her neck, unable to make contact with her monosyllabic husband, the heroine prattles platitudes and goes through trivial social rituals. Shown up as comically incongruous, these keep her cheerfully unaware of her situation.

In pursuing his existential vision with unexampled rigour, Beckett continued to set new standards with each of the plays, unlike Ionesco, whose later work is

more conventional and related to Jungian psychology. Following the principle that 'less is more', full-length dramas were whittled down to one-act pieces, and in 1965 to a 'dramaticule'. A decreasing number of characters are capable of ever fewer physical activities, while dialogue gives way to monologue. Continually redefining accepted ideas of what constitutes 'drama', Beckett's figures are at an ever more extreme edge of existence, until it might have seemed that with *Breath* in 1969 (see p. 436) his reductive approach had reached its limits: no figures, no words, no action.

By comparison—although from an insular perspective *Look Back in Anger* might have seemed 'a landmine' that exploded what Peter Brook called the 'deadly theatre' of the post-war decade—the revival came late to England, and was surprisingly conventional. Revolutionary new styles were already well established throughout Europe by the time John Osborne's play was performed in 1956. Indeed, *Waiting for Godot* had already been staged in London and the Berliner

Look Back in Anger, 1956. A shocking contrast to the standard elegance in British stage-pictures of the time, the ironing-board became emblematic of a new social realism—the context of despair against which the younger generation of John Osborne articulated their anger.

Ensemble appeared there in the same year. Beside these, *Look Back in Anger* was a dated slice-of-life drama. However, in the context of Coward's stylish artifice, Fry's poetic fantasy, or even Rattigan's bourgeois gentility, the grubby working-class environment won immediate recognition. The theme of social conflict, projected into the sexual battleground of Jimmy Porter's marriage to a daughter of the establishment, had direct relevance in a situation where the social barriers were beginning to crumble. But what made Jimmy—and indeed Osborne, whose voice was closely associated with his frustrated hero—the spokesman for the younger generation was the passionate invective of his long diatribes attacking every aspect of post-war England that socialism had failed to reform.

This play set the tone of George Devine's English Stage Company, founded with the specific aim of fostering new British drama at the Royal Court Theatre (where Shaw and Granville-Barker had accomplished the same task in 1911). 'Kitchen-sink' realism became the mode: almost literally in Arnold Wesker's *The Kitchen*. Wesker's semi-autobiographical trilogy, pleading for a renewal of socialist commitment after the disillusion generated by the 1956 Russian invasion of Hungary, and John Arden's early plays dealing with corruption in local government or violence on council estates, were followed by Edward Bond's graphic depiction of working-class back streets as a brick jungle in *Saved* in 1965. Yet, despite its immediacy and direct relevance, this realism harked back to the beginning of the century; and the links with D. H. Lawrence's proletarian drama were implicitly acknowledged when Lawrence's plays (written between 1909 and 1912) were revived at the Royal Court in the late 1960s. But already by that time almost all the Royal Court dramatists had expanded their stylistic range.

Even Osborne abandoned the kitchen sink, acknowledging that *Look Back in Anger* was a 'rather old-fashioned play'. Its title continued to sum up the overriding theme of both his drama and his vehement newspaper pieces, lashing the flaccid conformism and hypocrisy of the British public, in outrage that the national ideals peddled by the jingoistic Tory press—for which so many had sacrificed themselves during the war—were an empty sham. But this becomes internalized in the psychological state of his increasingly alienated characters. Thus his treatment of Luther, the original non-conformist, focuses on the visionary phantasmagoria of an agonized mind. Osborne's attack on the psychologically crippling myths of an outdated social establishment was given its most memorable form in his second play, *The Entertainer*, where the family life of a broken-down comedian—a 'star' role that attracted Laurence Olivier—is framed in his sleazy on-stage appearance. The death of a 'war hero' son in the Franco-British invasion of Suez (the last gasp of colonialism) is paralleled by the dying music hall; and the decrepit theatre building in which the bankrupt entertainer performs is a graphic image of Britain at the terminal point in imperial decline.

By presenting his social criticism through the music-hall prism, Osborne had effectively created a home-grown equivalent to Brechtian theatricality, which was picked up by Joan Littlewood. Taking over the abandoned Theatre Royal in Stratford East, an impoverished industrial district of London, she had founded an explicitly socialist stage in opposition to the West End three years before *Look Back in Anger* hit the scene, winning wide recognition for the dynamic ensemble acting of her Theatre Workshop. This was best embodied by a 1963 show that again used Edwardian music hall as the example of authentic popular culture— an image developed in a 1956 radio series by Charles Chilton, who collaborated with the Theatre Workshop cast on the script. *Oh, What a Lovely War* returned to the horrors of the First World War, focusing on the criminal responsibility of the English ruling class, in a series of vaudeville 'turns' acted out as a pierrot-show. A British equivalent of black-face minstrels, this once popular seaside entertainment fulfilled the ideals of ensemble creation and improvisation which have become hallmarks of socialist and anti-establishment theatre; and it was combined with yet another form of specifically working-class culture, the songs of the troops in the trenches. 'Pack up your Troubles in your Old Kit Bag' or 'The Ragtime Infantry' expressed the common people's spirit and indomitable humour, while the white pierrot smocks with red pompoms and ruffs automatically brought out the wider significance of historical scenes and presented a satiric clown's-eye perspective on the machinations and self-serving rhetoric of the generals and politicians. Following Piscator, the pierrot scenes were accompanied by documentary photographs and a 'news panel' listing dates and statistics. Photos of

The last scene censored in Britain. The aggressiveness of the new political drama in Britain is epitomized by the brutal stoning of a baby that Bond uses to symbolize the moralized violence of a repressive society in *Saved*, 1965.

429

Evening Standard, 14 Feb 1966. As the only mass medium up to the advent of film and radio, theatre has always been recognized as a potential source of propaganda—and censored. The fight against censorship, which Shaw had led at the beginning of the century, came to a head when the Royal Court was prosecuted for staging Bond's *Saved*. As a direct result of the furore that followed, censorship of the English stage was abolished in 1968, clearing the way for new developments.

Evening Standard

44,058　　MONDAY, FEBRUARY 14, 1966　　●● 4d.

WEST END FIN—
CLOSING PRICES

Don't be round
the bend fit
Britax
safety belts
Double Send

'Take this play off' court plea rejected

THEATRE DEFIES THE CENSOR

Liz as Helen FIRST PICTURE

Elizabeth Taylor, seen for the first time in her Helen of Troy costume, rehearses for tonight's Dr. Faustus at Oxford. Her husband, Richard Burton, plays the name role in Christopher Marlowe's play. John Howard took this picture and the one in Pages 14 and 15, after flying from Rome for the occasion.

WEATHER—Cold.—See Page FOURTEEN

Magistrate warns of 'further penalties'

Evening Standard Reporters

The English Stage Company refused a request today by the Director of Public Prosecutions to suspend performances of the controversial play Saved at the Royal Court Theatre.

The play, by Edward Bond, contains a scene in which a baby is stoned to death

The request was made by counsel for the Director of Public Prosecutions in a court case at Marlborough Street when the English-Stage Company was summoned for putting on the play without a licence from the Lord Chamberlain.

The company was alleged to have broken the law six times last year during the performance of the play at the Royal Court Theatre.

Act of 1843

The defendants were the English Stage Company, which was incorporated in 1954; Mr. William Gaskill, 34, of North Side, Clapham Common, its artistic director; and Mr. Alfred Esdaile, of Connaught Place, Bayswater, licensee of the Royal Court Theatre.

The summonses, brought under Section 15 of the Theatres Act 1843, allege that the defendants "presented for hire at the Royal Court Theatre a new stage play entitled Saved before the same had been allowed by the Lord Chamberlain."

The wording is laid down in the 123-year-old Act which has rarely been tested in the courts in recent times.

The English Stage Company's management council has wealthy Mr. Neville Blond as chairman and Mr. Greville Poke as secretary.

The play was put on a club presentation last November.

A plea of not guilty was entered on behalf of the three defendants, who were represented by Mr. John Gower.

'Irrelevant'

Mr. Oliver Nugent, for the Director of Public Prosecutions, told the magistrate, Mr. Leo Gradwell, the background history involved after referring to the law on the subject.

He said after referring to cuts which had been made by the Lord Chamberlain in the original script:

"I would point out that whether the Lord Chamberlain is right or wrong about those cuts and whether you or anybody else think they should or should not have been allowed is totally irrelevant to this case."

He suggested that the words in the summons which might cause

▲ **Back Page, Col. Two**

HIGHER LONDON FARES THREAT

London's train travellers are faced with the threat of higher fares.

British Rail have applied for increases on journeys from 11 miles and upwards—in the London area only; removal of special fare concessions to working juveniles; and freedom to alter the charges on excess luggage.

The bid to increase fares will also apply to certain London Transport Underground lines. Fuller report—Page ELEVEN

Chelsea have Cup luck

AT HOME TO CARLISLE OR SHREWSBURY

Of London's three remaining clubs in the FA Cup only Chelsea got a home tie in today's fifth round draw—against Shrewsbury or Carlisle.

Tottenham travel to meet the winners of the Bolton v. Preston replay. West Ham, if they win their replay against Blackburn on Wednesday, visit Norwich in the next round.

Full draw and Bernard Joy's comments. — Page TWENTY-SEVEN

Wider ban on cigarette ads?
PAGE FOURTEEN

Russian writers jailed for 7 and 5 years
PAGE FIFTEEN

ANOTHER DOUBLE FOR THE EVENING STANDARD

CAMERON
Writer of the Year

FOOT
Critic of the Year

Once again the Evening Standard has scored an impressive double in the Hannen Swaffer Awards for Journalists (details Page Four).

JAMES CAMERON, whose despatches from Hanoi for the Evening Standard were so outstanding is voted Descriptive Writer of the Year.

MICHAEL FOOT, Evening Standard book critic, is voted Critic of the Year.

Thus the Evening Standard repeats its success of last year when GEORGE WHITING was voted Sports Writer of the Year, and ANNE SHARPLEY was voted Descriptive Writer of the Year.

The Evening Standard is proud to have a team of writers without equal in British journalism.

corpses in the mud, and such headlines as 'APRIL 22 . . . BATTLE OF YPRES . . . GERMANS USE POISON GAS . . . BRITISH LOSS 59,275 MEN . . . GAIN NIL', formed a starkly powerful contrast to the ironic theme song of 'Oh it's a Lovely War'.

Littlewood followed up with a more direct evocation of Edwardian music hall in *The Marie Lloyd Story*; and one of the recurring features of recent British drama has been the same radical use of traditional forms for social criticism. Bond's *Restoration* pastiche, Caryl Churchill's revival of eighteenth-century 'city comedy' in *Serious Money*, Peter Nichols's adaptation of pantomime to illustrate that the source of imperial Britain's wealth came from the heroin trade in *Poppy*: all were to exemplify this trend in the 1980s. But perhaps the most immediate inheritor of the Theatre Workshop's polemic comedy was Joe Orton, whose anarchic farce in many ways typifies the freewheeling sexual liberation of the mid-1960s.

Joan Littlewood's *Oh, What a Lovely War*. Traditional working-class entertainments adapted for a socialist critique of Britain in the 1960s: polemic pierrots replaying the recruiting shows of 1914–15.

Building on Ben Travers's titillating but morally conformist farces, which had filled the Aldwych Theatre uninterruptedly from 1925 to 1933 (and indeed continued to appear right up to 1975), and echoing Oscar Wilde in their epigrammatic style, *Loot* and *What the Butler Saw* turned all moral standards upside down. As his characters declare, for Orton 'sexual licence is the only way to smash the wretched civilization'. His plays were deliberately designed to outrage the bourgeois public. In *Loot* a funeral is treated as knockabout slapstick, with a woman's corpse dumped from the coffin by her son to hide stolen money, stripped naked, and paraded as a dressmaker's dummy as well as being subjected to incestuous necrophiliac jokes. In *What the Butler Saw*—a title referring to salacious seaside peep-shows—all the permutations of transvestism are milked for every possible sexual relationship, while the characters' convoluted situation is resolved by a joyful revelation of 'double incest'. Their first productions were greeted with vociferous hostility from spectators—a furore abetted by rabidly critical letters to newspapers from Orton himself under a pen-name. Society is presented as a madhouse controlled by lunatics; and his primary targets are the police and the Church. As one of his hypocritical authority-figures hilariously declares when faced with popular insurrection, 'We'll have a couple of verses of "Love Divine All Love Excelling", Padre. It's fire-hoses, tear-gas and the boot from then on.' This celebration of anarchic disorder projects the destruction of repressive social norms that Orton hoped to achieve through his plays. However, their subversive shock effect was rapidly defused by acceptance; and barely five years after Orton's gory murder by his homosexual lover they were being treated as modern classics.

If Orton's farce had little discernible effect on English power-structures, Rolf Hochhuth's accusation that the Pope had been a knowing accomplice of the Jewish holocaust under the Nazis in *The Representative* caused a global uproar when it reached the German stage in 1963. Although an anachronistically conventional five-act heroic tragedy in rhyming couplets, the painstakingly documented scenes and carefully researched portraits of immediately recognizable public figures gave it extraordinary impact. This massive play generated parliamentary debates, heated discussion among historians and theologians, as well as so many leading articles and editorials in journals and newspapers that they were published as a separate book—and in his next work Hochhuth widened his range to attack Sir Winston Churchill and the Allied leaders as war criminals for the fire-bombing of Dresden. It was no accident that Hochhuth's plays were developed and directed by Piscator, who (like Brecht) had returned from the United States in the early 1950s after being targeted by the communist witch-hunt of the McCarthy hearings. The political significance and scope of *The Representative* and *The Soldiers* fulfilled the criteria for the documentary drama he had hoped to foster in the 1930s, and which he was now—as director of the Freie Volksbühne in

Berlin—at last in a position to promote. Indeed under his influence many of the leading German writers produced highly topical documentary plays, the most stylistically interesting being *The Investigation* by Peter Weiss. Stripped of all imaginary elements—scenery, plot, and even individualized characters—it presented the machinery of Auschwitz in bleak and horrific clinical detail, using unadorned extracts from the Frankfurt war-crimes trial as dialogue to demonstrate that the Nazi death camps were not inhuman aberrations, but the 'logical and ultimate consequence' of economic exploitation. The focus of *The Investigation* is not the historical events but the need for destroying the present system, Weiss's aim being explicitly to 'brand capitalism' as 'the customer of the gaschambers'; and, in a deliberate attempt to restore the status of a 'moral tribunal' to the German stage, when first performed in 1965, the play was simultaneously staged by seventeen different theatres across the Bundesrepublik.

However, the politically involved or subversive drama of social criticism, which had been the dominant mode in Germany and Britain as well as America, was already being increasingly challenged. In Germany in 1966 Günter Grass, who had written a number of quasi-absurdist playlets, attacked Brecht as a character in *The Plebeians Rehearse the Uprising*, where the setting reproduces the stage of the Berliner Ensemble. Brecht stands condemned for a betrayal that revealed the hollowness of the political principles of his own drama in refusing to support the 1956 workers' revolt against (communist) dictatorship, which is brutally crushed by Russian tanks on the Berlin streets while he rehearses his version of *Coriolanus* in his rococo Schiffbauerdamm Theatre. In England and America too there were signs of a move away from left-wing politics in main-stage drama. True, Bond and Arden were becoming more strident and explicit in their attacks on the social system, while Wesker was involved with Centre 42 (named after a resolution passed by the Trades Union Congress in 1960, committing unions to developing socialist art specifically for the working classes). Similarly a radical Black drama emerged in the United States with James Baldwin and LeRoi Jones, while Sondheim politicized the musical. Even so, the major American dramatist of the decade was Edward Albee, who can be seen as the counterpart to Harold Pinter in England. In addition, Neil Simon's immensely popular comedies dominated Broadway throughout the 1960s, while Tom Stoppard's iconoclastic wit and Alan Ayckbourn's astringent but explicitly undidactic farces gained an immense following in England. In contrast, by the mid-1970s many of the English radicals had either retreated from increasingly unsympathetic audiences (like Wesker, or even Osborne) or become marginalized by their extremism (like Arden, or Peter Barnes).

Though not as obviously as Stoppard's *Rosencrantz and Guildenstern are Dead*, Pinter's early drama was influenced by Beckett. Rejecting all 'didactic or moralis-

The stage on stage. Life literally 'behind the scenes' in Thornton Wilder's 1938 *Our Town*; and Brecht's Schiffbauerdamm-theater reproduced in Günter Grass's 1966 *The Plebeians Rehearse the Uprising*. Self-reflexive metatheatre is a distinguishing characteristic of modern theatre.

tic theatre' as 'sentimental and unconvincing', Pinter presented deliberately ambiguous images of almost anonymous people victimized by nameless forces or threatened by apparently motiveless games of dominance and subservience. In the 1950s his plays were generally dismissed as incomprehensible and obscure. Yet the surrealistic dismantling of the protagonist's personality in *The Birthday Party* is not only conveyed in meticulously realistic dialogue, but structured on familiar, even clichéd patterns—riddles, children's games, music-hall cross-talk routines—and in his own 1964 production each line was greeted with gales of laughter. Pinter's breakthrough came with *The Caretaker* in 1960, where the audience could identify with the generalized symbolism of the struggle between a pretentious but decrepit old tramp and two brothers, each of whom stakes a claim to possession of a room piled with junk: a home of one's own, the competition for material possessions. The claustrophobically enclosed setting offers a reductionist image of existence, as in Beckett. But in place of featureless Beckettian wastelands, the context of Pinter's characters is naturalistic to an exaggerated, even myopic degree.

Self-referential theatre. Tancred Dorst's 1968 dramatization of the playwright Toller's life also parodied the documentary style, multiplying placards to expose the slogan-eering aspect of this characteristic technique, and dwarfing the actors with simplistic agitprop puppets. Other scenes set Toller within the text of his own plays.

****

THE MOST MINIMAL DRAMA POSSIBLE

Reducing the whole of life—from birth to the grave—to a bare thirty-five seconds, this self-negating piece is also a witty theatrical riposte. Invited by Ken Tynan to contribute to the erotic revue *Oh! Calcutta!* Beckett's derisive response exemplified his description of his work as 'a matter of fundamental sounds. No joke intended.' But in performance the tables were turned (and Beckett's bleak vision travestied) by Tynan's replacement of the 'unidentified rubbish' with nude figures rolling lasciviously across the stage.

Samuel Beckett—*Breath*

1. Faint light on stage littered with miscellaneous rubbish. Hold about five seconds.
2. Faint brief cry and immediately inspiration and slow increase of light together reaching maximum together in about ten seconds. Silence and hold about five seconds.
3. Expiration and slow decrease of light together reaching minimum together (light as in 1) in about ten seconds and immediately cry as before. Silence and hold about five seconds.

****

During the 1960s his 'Comedy of Menace' acquired an increasingly identifiable social background. But this always remained disturbingly off-key, as in *The Homecoming*, which generated considerable controversy (with spectators walking out of performances in disgust) when it was produced by Peter Hall and the Royal Shakespeare Company in the West End: a venue that marked Pinter's arrival in the mainstream of British theatre. A parody of the standard meet-the-family type of drama, it escalates from the conventional comedy of a new wife being introduced to her husband's brothers and father into an attack on the family (and by extension society) as prostitution, with all personal relationships being transactions between pimps and whores. Peter Hall's staging of this ironic reflection on such clichés as 'woman's place is in the home' was centred around the image of a 'human jungle'—building on the image of the male characters as, in their own words, 'a lot of dogs' fighting for dominance. But in this biological competition it is the woman who has control through dispensing sexual favours; and the question of who wins is highly ambiguous. In London the husband was portrayed as an intellectual wimp, who is forced out, neutered by his wife's rejection, while she liberates herself from a stifling marriage. By contrast, in New York he was seen as the victor, who has brought his wife back in order to discard her by exposing her worthlessness and confirming his family's hatred of women.

After this play, at the end of the decade, Pinter moved into a phase of more abstract, poetic drama—and a remarkably similar trajectory can be seen in the career of Edward Albee. Almost exact contemporaries, though coming from different theatrical traditions, Pinter and Albee represent the middle ground in the post-war revival, between political populism (derived from Brecht) and the poetic existentialism epitomized by Beckett. Their development also illustrates the competing pull of experimental versus mainstream theatre.

At the beginning of the century the aim of the 'Little Theatre' movement had been to reform the commercial stage; and the new styles or tendentious subjects it promoted were adopted by the establishment. Even Artaud, though totally antagonistic to Western drama as well as the dominant style of the time, still formed part of the standard theatrical scene, with a glittering array of high-society and artistic celebrities attending the opening night of *The Cenci*. And this continuum remained throughout the 1950s, with dramatists who had begun Off-Broadway or at the Royal Court transferring (like Tennessee Williams or Osborne) to Broadway, to the West End, or to subsidized state theatres, and with actors and directors moving interchangeably between the two. But by the middle of the following decade an unbridgeable gap was beginning to appear with the emergence of Off-Off-Broadway and the various 'theatre laboratories', exploring forms of experimentation that were radically incompatible with conventional stage performance: Environmental Theatre, Guerrilla Theatre, Theatre of Images.

Albee forms the bridge between these and the earlier generation of Miller and Tennessee Williams. His first pieces were staged Off-Broadway, and like Pinter's early work they were short, mixing absurdist elements with a recognizable social context. They also continued the attack on the materialistic ethos of the USA that had been the major theme of O'Neill and Miller, one title being *The American Dream*. Like Pinter too, Albee moved into the mainstream and achieved international recognition with a quasi-naturalistic play excoriating the family. Very much a counterpart to *The Homecoming*, which it preceded by three years, *Who's Afraid of Virginia Woolf?* also centres on the dominating and promiscuous wife of an emasculated college professor. But this laceratingly witty Strindbergian battle of the sexes is openly allegorical, with the protagonists named after George and Martha Washington, the college they are attached to being 'New Carthage', and their imaginary son the deluding American dream. These broad resonances presume the central significance of theatre to be a moral commentary on society (deriving from the American tradition of the 1930s) in a way that Pinter could not. Yet the solutions offered are anything but political. The play is a ritual of purgation, drawing spectators inexorably into the characters' world through 'Fun and Games' that build to the sado-masochistic revelations of 'Walpurgis Night' and end in a literal 'Exorcism', designed to dispel the audience's (and society's) illusions. However, Albee's subsequent plays became increasingly rarefied and evasively oblique, with surrealist elements undermining the naturalistic surface. By the end of the 1960s his work had moved beyond the scope of commercial theatre into brief, abstractly musical chamber pieces like *Box* and *Quotations from Mao-Tse Tung*: a progression followed by Pinter in his 1969 interior monologues *Landscape* and *Silence*.

Pinter and Albee were part of a wider group that included Günter Grass and Tankred Dorst in Germany, and Vaclav Havel or Slawomir Mrożek in Eastern Europe. But its most extreme and earliest member came from France. Jean Genet's short early plays, *The Maids* and *Deathwatch*, had anticipated Albee's ritualistic exorcism by almost fifteen years; but, rather than attacking the illusions of society, Genet's role-playing drama of ceremony seeks to negate reality itself. Perhaps the ultimate outsider, a self-proclaimed thief, murderer, and homosexual (at a time when homosexuality was still a crime in many countries), he started writing in prison and was elevated to a 'martyr' by Sartre. But *The Balcony*, initially banned, was an instant success when it hit the stage in 1960. Theatre is identified with a brothel, both being a 'house of illusion'; and the erotic fantasies played out by the actor/clients are transformed into symbolic reality when revolution wipes out all the state authorities. Image is all. Even the revolution is only 'someone dreaming'. Baroquely colourful symbols and poetic speech combine with violence and squalor. Everything cancels out in the interplay between different levels of illusion and the interchange of identity—except performance, which becomes a 'reflection of reflections'. This is equally true of Genet's more overtly political plays that followed: *The Blacks* and *The Screens*, which centred around Afro-American revolution and the very recent liberation of Algeria from French colonialism. In each case the revolutionaries are trapped in the same nihilistic cycle as the society they oppose, their actions inflated to fake gestures. Even so, *The Blacks* was greeted with shocked outrage and accusations of racism, while *The Screens* was considered so subversive that its production was delayed by Barrault for five years. When it finally reached the stage of the Odéon (the French equivalent of a national theatre) in 1966, it caused riots, with cries of 'traitor!' from the audience and protest marches by the Ex-servicemen's League that blockaded the theatre doors.

These two plays were topical in more ways than their inverted challenge to the political order. The ritual quality, the celebration of perversity and cruelty that made Genet's drama a 'black mass', also corresponded with a reawakening interest in Artaud, signalled by Peter Brook's 1964 experimental 'theatre of cruelty' season in London. This had been intended to develop an acting style capable of presenting *The Screens*. But although that was restricted by the censor to a private performance of selected scenes, out of it came one of the most brilliant productions of the period, *The Marat/Sade*. In this depiction of the world as a madhouse—where the Marquis de Sade turns a play-within-the-play about the French Revolution into an inmates' revolt against Napoleon, representing all authoritarian society—Peter Weiss had explicitly combined Brecht (represented by the rationalist Marat) and Artaud (in Sade, the advocate of instinct and anarchy). Brook placed the patients in the foreground; and the actors' powerfully convincing, grotesque portrayal of insanity formed a visual and emotional shock treatment for the audience. In addition, Genet's interest in both negro culture and Africa was very much a sign of the times. Emerging from the civil rights movement, American Black theatre became a potent force during the 1960s with the bleak and violent depictions of racism in the plays of James Baldwin and Amiri Baraka (the name taken by LeRoi Jones). Although the earlier avant-garde, in particular the expressionists and Artaud, had taken non-Western performance as their models, these had been valued as archaic survivals of 'primitive' art. Now, following the breakup of the European empires, a modern indigenous theatre was emerging in Africa.

Outside Europe and the United States, during the middle decades of the century theatre continued to be employed as a method of cultural colonization. As in the Victorian era, throughout the British Empire touring companies spread the gospel of Shakespeare, whether performing to the expatriate community or (particularly in India) to native audiences; and, where there were home-grown acting companies, they generally performed popular hits from the West End. From a Eurocentric perspective, nothing else counted as theatre. In countries where the colonized populations remained in the majority, a parallel drama continued to exist. Tamil or Bengali companies performed stories from their own cultures in buildings constructed on the lines of European theatres, which thus corresponded to standard European expectations. But in Africa the native performance art went unrecognized.

During the inter-war years, with the whole continent under colonial rule after the Italian invasion of Ethiopia in 1935, the only native African drama was a surviving tradition of tribal dance. Accompanied by rhythmic chanting and orchestrated around performers in symbolic masks (hence the name 'masquerade') that might represent ancestors, gods, natural forces, or even icons of the colonizers, these mass rituals of celebration or initiation reinforced community, commemorated historical events, propitiated spirits, or promoted a sense of harmony with nature. Such dances were not spontaneous—as early European anthropologists suggested—but had evolved over centuries, with each performance being rehearsed over a period of months. Not only communal expression, but also a teaching tool, these rituals increasingly served to create political identity. It was therefore logical that they should become a basis for indigenous drama once European theatrical models were adopted.

Generally recognized as Africa's greatest living playwright, Wole Soyinka can be seen as representative, even though his dramatic material is specifically Nigerian. At university in England during the 1950s, he became involved with the Royal Court Theatre, and combined the kitchen-sink techniques of the new wave British dramatists with the Yoruba masquerade. On returning to Nigeria, he founded the first native theatre company, which opened in 1960 with a play celebrating national independence, characteristically titled *A Dance of the Forests*. As with most of his work from this period, it deals with three coexisting and intertwined levels of reality: the past inhabited by the dead, the present of the living, and the future of unborn spirits. Contemporary reality (or in history plays the actions of living people) is generally represented realistically, with dances performed on stage as a metaphor for transcendence and the masquerade being a rite of passage to the spiritual levels. In a play such as *The Road*, dealing with the emergence of an impoverished urban proletariat, a Wesker-like naturalism is used to present false materialistic values adopted from the West, which are rejected for the native culture embodied in the Masks. His subjects are highly political: the abuse of power, corrupt politicians, the destructiveness of inter-tribal conflict. But, set in the context of a European style of drama, the traditional performance art stands for a revitalized, uniquely African moral consciousness, which will transcend both the tribal past and modern materialism. Indeed, for Soyinka the act of performance itself has the potential to transform time and space, making theatre a spiritual experience. This is most clearly expressed in his play *Death and the King's Horseman*, written in exile after the Nigerian civil war, where the action is centred around a dance-induced trance that has to be authentic for the play's theme to be realized.

The play can also be seen as a conscious attempt to establish a national dramatic canon in dealing with the same historical incident from shortly before Nigerian independence—a British District Officer's prevention of ritual suicide in 1946—that had already formed the subject of an indigenous opera by Duro Ladipo; and indeed Soyinka's work served as a catalyst for African dramatists in other newly independent countries during the 1960s and 1970s. Since resettling in Nigeria in 1976, Soyinka has moved towards more experimental dramatic models (political satires using not only Brecht but Genet's *The Blacks* as models), bringing the reciprocal relationship between Africa and the avant-garde full circle.

The winds of political change that blew through Africa in the late 1950s and early 1960s were also stirring up the cultural landscape in South Africa, Australia, and Canada. Arguably the commercial dominance of West End models, or the continuing Anglo-Saxon propensity to see England as a spiritual home, prevented the development of any identifiably national drama until the progressive disillusionment of being sacrificed as cannon-fodder in two World Wars. In any case, it was not until the late 1950s that authentically post-colonial voices from the White

Commonwealth emerged in the theatre. Even then, over the next decade these were limited to isolated plays—*Fortune in Men's Eyes* from Canada, *The Summer of the Seventeenth Doll* from Australia—and the only playwright from that period to win international status came from South Africa: a direct response to the uniquely repressive racial politics of apartheid.

The success of an 'African Jazz Opera' based on the life of a well-known Black boxer, *King Kong*, in 1959 gave the Union of South African Artists funding for an African Music and Drama Association, which opened its Rehearsal Room performances with Athol Fugard's *The Blood Knot*. Among the productions Fugard directed for students (mainly factory workers and servants) at Union Artists were O'Neill's *Emperor Jones*, as well as Pinter and Beckett. All are clear influences on his short, two- or three-character pieces that focus on a situation rather than developing a plot, reproducing a style of speech that reflects the history of violence, dispossession, and poverty in crude slang and ready-made phrases. Taken to an extreme in *Boesman and Lena*, the last of his 'family plays' in 1969, the stripped-down dialogue and lack of action on an empty stage graphically portray the limit of human endurance. Evicted from their bulldozed home and alienated from both communities by their mixed race, the couple repeat the ways society oppresses them in their own relationship, as they set up camp on the bare mud of a river bank with the few remaining possessions they carry on their backs.

Performed at a Bantu Men's Social Centre, in the minimally equipped Rehearsal Room that was part of an abandoned Johannesburg factory building, or on tour in the townships and in church halls, the bleakness of these plays was reflected in the almost totally non-existent scenic or lighting possibilities. Even the scripts were rough drafts rather than carefully worked texts; and in his first pieces Fugard worked with the actors, changing scenes and dialogue as they rehearsed, just as later in the 1970s he would build his 'workshop plays' out of improvisation. This 'poor theatre', along with ensemble creation, anticipated avant-garde developments in Europe and America. Whereas these elements were a natural result of Fugard's circumstances, they became a sign of ideological commitment from Grotowski and the Living Theatre.

Founded in the provincial town of Opole in the early 1960s, Jerzy Grotowski's Polish Laboratory Theatre came as a revelation when it appeared at the Paris Théâtre des Nations festival under Barrault's aegis in 1966. The texts were arcane: Wyspiański's symbolist *Akropolis*, and particularly Calderón's highly poetic drama of religious martyrdom *The Constant Prince*. But the way performances integrated the audience with the action, and the intense emotional authenticity achieved in the acting, gave a new meaning to 'total theatre'. Grotowski had evolved a training method to liberate actors from psychological blocks and penetrate their inner selves, 'surpassing limits' and creating moments of spiritual transcendence or 'translumination' through self-revelation. Linked with physical stances intended to evoke archetypal 'images deeply rooted in the collective imagination', this produced extraordinarily powerful effects. On a completely bare and tightly enclosed acting area, with the barriers between stage and spectator broken by close proximity, the stylized torture of the 'constant prince' and Ryszard Cieslak's physically extreme expression of inner states—sweat pouring from his brow like blood from Christ's crown of thorns, a red flush spreading all over his skin, tears flooding from closed eyes—gave an absolutely convincing portrayal of spirit triumphing over physical existence. This, together with highly mesmeric and ritualistic elements of the presentation, and the reduction of Calderón's rhetorically mannered verse to chants and pure vocal sounds, seemed the true fulfilment of Artaud's ideals. 'Poor Theatre' (the title of Grotowski's book of theoretical essays) became the model for Brook, and almost all of the 1960s' experimental groups.

With all historical reference removed, Grotowski's version of *The Constant Prince* portrayed the victory of the individual, purified by suffering, over a sadistic and spiritually castrating society—and this melded with the beliefs of the political radicals who were beginning to emerge in the underground theatre, particularly in America in reaction to the Vietnam War: Chaikin, Schechner, Beck, and Judith Malina. The earliest and most politically committed of these was the Living Theatre, who were also the most open exponents of the 1960s drug

The model for American 'alternative' theatre. A thrust toward transcendence motivated most radical theatre groups in the 1960s: its model was the mortification of the body to reach mystical ecstasy in the 1966 performance of *The Constant Prince*. The bull-ring stage and matador-pose of the 'torturer' display the symbolic basis of Grotowski's minimalism.

culture. Inspired by Artaud's theories, which provided material for a show like *Mysteries* in 1964, their scenario for *Paradise Now* four years later was based on Grotowski's acting exercises and Cieslak's 'transluminations'.

Merging politics and mysticism, *Paradise Now* was intended to revolutionize society through the psychological liberation of the public. Opening with vocal protests about laws restricting freedom (nudity, drugs, censorship) as the company moved among the audience, performances were 'spiritual voyages' with each step on the way establishing a more intense contact between actors and spectators. Graphic tableaux, like multiple pairs of actors posed as the Saigon Police Chief executing a Vietcong prisoner in Edward Adam's award-winning photo, alternated with improvised sections that made up over two-thirds of the whole. The audience were encouraged to form small groups as 'revolutionary cells', incited to smoke marijuana with the aim of entering a 'trance', invited to experience a symbolic vision of human potential and communal trust in 'flying'—launching themselves off a high platform to be caught in the interlaced arms of others below—and swept into a 'ritual of love', communal copulation, after which the actors were supposed to lead everyone in a naked procession out of the theatre to 'Free the streets'. At each stage they were exhorted to 'Free the theatre' as a paradigm of revolutionary action. In the Becks' view, if this happened then the performance would be broken off, 'paradise' having been 'consummated'. Theatre itself would become redundant: indeed they soon issued a call to 'smash the art barrier' and abandon the stage altogether. *Paradise Now* drew huge numbers of

Paradise Now: epitome of 1960s avant-garde. This extreme of improvisational performance and audience involvement, with its strong visionary thrust combining sex, drugs, and anarchistic politics, had immense appeal to the youth of the time and was echoed by dozens of small 'alternative theatre' groups.

443

enthusiastic participants in its tours of both Europe and the USA. It fed directly into the Student Revolution of May 1968; and the Living Theatre led the month-long student occupation of the Paris Odéon that brought Barrault's career to an ignominious end.

Other precursors of the Student Revolution, which formed the most radical and widespread challenge to the post-war order, also rejected conventional theatre: notably Peter Handke in Germany. Handke's first work, *Offending the Audience*, set the tone of the 1966 Experimenta festival in its verbal abuse of spectators as representatives of a repressive state. No plot, dramatic action, characterization, dialogue, or scenery—only a diatribe against theatre-going as an example of social conditioning. Although his subsequent plays were more conventionally dramatic, Handke also called for 'Street Theatre'. The opposite of 'Theatre Theatre' or performance in any sense, this was an extension of the 'Happenings' organized by Allan Kaprow and others during the 1960s—the most famous being an incident when dollar bills were thrown down into the trading-pit of the Wall Street Stock Exchange, exposing the greed of capitalism by the dealers' frantic scrabble for the money. But Handke envisaged something even more real, as in the orchestration of a protest march. People on crutches and pretty girls would be placed in the front ranks of the marchers, and a confrontation with the police engineered to generate public support against the system by providing the most extreme images of physical brutality over nation-wide television.

This political ferment, born of the 1960s counter-culture, focused by the anti-Vietnam War movement, and culminating in the Student Revolution, set the scene for the next phase of theatre. Out of it came a new generation of radicalized playwrights in England, directors like Ariane Mnouchkine in France, Sam Shepard and the avant-garde groups in America. Theatrical expression in Russia and Eastern Europe continued to be strangled by Stalin's inheritors, though in the politically neutral area of set design strikingly inventive artists like Svoboda had begun to emerge. But in Britain the final abolition of censorship in 1968 freed the stage for non-traditional work. All this coincided with the development of native drama in Africa, and with a new sense of national identity in Australia and Canada (where theatre was fostered in recognition of the 1967 centennial of confederation). Even if some elements carried through, the explosion of experimental activity set off by the First World War had come to a close. Just as Barrault, one of the leading radicals from the 1930s, had become a symbol of the authoritarian establishment in the eyes of the Living Theatre, so, in the aftermath of world-wide challenge to the cultural status quo, a new theatrical era was about to begin.

PART FOUR

World Theatre

12

ORIENTAL THEATRES

SOUTH ASIAN THEATRES

FARLEY RICHMOND

THE story of theatre and drama in south Asia begins and primarily centres in India even though lively traditions of performance are to be found in Pakistan, Bangladesh, Nepal, and Sri Lanka, not to mention the tiny kingdoms of Bhutan and Sikkim. Yet it is in India, with its massive population of nearly a billion people and diverse cultural variety, that the full flowering of the classical temple and court traditions, the broad spectrum of rural genres, modern, and contemporary explorations, is fully realized.

The mythological origin of theatre and drama is related in the first chapter of the *Natyasastra*, the world's oldest text of dramaturgy, attributed to the sage Bharata. According to the story, when the world was full of sensual pleasures, desire and greed, and jealousy and anger, Indra, the warrior-god, asked Brahma, the creator, to invent a form of amusement fit to be witnessed by all the Hindu caste hierarchy—priests and warriors, as well as tradesmen and peasants. Brahma fell into a state of deep meditation and drew together recitation, song, acting, and aesthetics from the four *Vedas*, the most sacred books of the Hindus, and composed a fifth sacred text, the *Nalysounatyasastra*. At first it was unclear who should preserve and maintain the new creation. Brahma charged Indra to utilize it but he and the other gods considered it below their dignity to do so and advised that men and women were properly suited to the task. And so Brahma taught the secrets of the art to Bharata, who along with his 100 sons and heirs was charged with preserving and maintaining it on earth.

Bharata set about rehearsing the first play, which concerned the victory of the gods over the demons. Obviously, the theme was not popular with the demons, who considered themselves much maligned. Through numerous disruptions they

hindered Bharata's actors from rehearsing. So Brahma charged Visvakarma, the heavenly architect, to design a performance space which could be sanctified through rituals and so protect the actors from harm. Eventually, in mediating a solution to the unresolved conflict between gods and demons, Brahma articulated the chief objectives of performance—to instruct and to entertain.

Whereas the mythological origins are compact and complete, the historical sequence of events is not so easy to follow. There are no ancient ruins of the physical theatre to turn to. The *Natyasastra* is the earliest important text concerning the performing arts in India but the date of its composition is conjectural at best. It probably came into existence sometime between 200 BC and AD 200. Important writings on other subjects—the Vedic hymns, the *Mahabharata* and *Ramayana* epics, and the Puranic tales—yield only vague references to the existence of drama in ancient India. Archaeologists trace dance, and by association music, to the earliest sites of civilization on the Indian subcontinent—the ruins of the cities of Mohenjodaro and Harappa, along the Indus River basin about *c.*2500 BC. Two small figurines, one of copper and the other of grey stone, have

Above: Indus Valley relic. The so-called dancing girl (*c.*3000–1500 BC) excavated at the ancient city of Mohenjodaro, one of the chief centres of Indus Valley civilization.

Right: palm leaf manuscript in Sanskrit. *An ankiya nat* play-text composed in Sanskrit incised on palm leaf.

been called female dancing figures although there is no evidence to support this claim.

The earliest surviving plays come from the first and second centuries AD and are the work of Asvaghosa, a north Indian playwright. They are written in Sanskrit, the classical language, and reveal that dramatic writing was fully developed by this time, conforming to rules already laid down in the *Natyasastra*, suggesting that the origin of theatre and drama occurred somewhat earlier in history.

We know from comments in Patanjali's *Mahabhasya*, a grammatical text of 140 BC, that all the elements for the formation of drama were present by this date: pantomime, graphic illustrations of stories, and the art of story-telling. This places the possible origin of theatre in India between 200 and 100 BC, several centuries prior to the plays of Asvaghosa. It is during this period that Indian theatre probably emerged.

According to the *Natyasastra*, ancient theatre companies had a troupe leader (*sutradhara*), regarded as the chief actor, who presided over a group of singers, dancers, musicians, assistants, and playwrights. Although men seem to have dominated the groups, women figured prominently in the female roles and as singers and dancers.

Acting in ancient India was an integrated art in which body, voice, costume, make-up, and emotions blended seamlessly together. Five chapters of the *Natyasastra* are devoted to the body, suggesting that its every aspect, from the feet to the head and especially the hands, eyes, and facial expressions, was considered important. Rigorous training was required of actors beginning with purgation of the bowels, dietary restrictions, and a regimen of exercises to build strength, flexibility, and endurance. Many of the same procedures are preserved today by artists in widely differing rural theatre genres, as well as by contemporary urban performers.

Character depiction was conventional and role categories included heroes, heroines, ministers, jesters, courtesans, and so forth. Physical movements, vocal patterns, properties, make-up, and costumes were all convention-

Kathakali actors' training procedure. Well before sunrise during the monsoon season, a foot massage is given daily to youthful actors by their teachers after a regimen of strenuous exercises. Massage is an essential part of the *kathakali* training discipline.

alized and considered extensions of role categories. Even emotional depiction and response was dictated by convention.

Except perhaps for Bharata, performers were considered outcastes and their names have long since been forgotten, but drama enjoyed a significant place in the literary tradition and playwrights are still remembered for their accomplishments.

The most prolific playwright of ancient India was Bhasa, who probably lived in the early fourth century AD in north central India. Among his thirteen extant plays, *The Vision of Vasavadatta* (*Svapnavasavadatta*) is regarded as his best. Fragments of this and his other plays are still performed in the *kutiyattam* of Kerala, to be discussed later, and by contemporary urban theatre companies.

Sudraka, who probably lived in the mid-fourth century AD, was a playwright-king whose reputation rests on a single work—*The Little Clay Cart* (*Mrcchakatika*). Sudraka comes closer than any other Sanskrit playwright to depicting the life of the common people. The plot of his monumentally long play concerns the intense love of Carudatta, a heroic brahman merchant, for Vasantasena, a ravishingly beautiful and intelligent courtesan, and their struggles to outwit the machinations of a jealous and powerful suitor. The play contains an array of interesting, humorous characters from all walks of life.

India's greatest playwright is undoubtedly Kalidasa, who is thought to have lived in the early to mid-fifth century AD. Little is known about this great master. He wrote a lyric monologue, two lyric narratives, and three plays. His most famous work is *Sakuntala* (*Abhijnanasakuntala*), regarded by many as the best Sanskrit play ever composed. The plot concerns the love of King Dushanta for Sakuntala, daughter of a heavenly nymph and a sage, raised and educated in the humble surroundings of a forest hermitage but destined to become queen and mother of Bharata, progenitor of the Indian race. The couple is prevented from achieving happy union due to a curse placed on Sakuntala by a recalcitrant sage. In the final act, their difficulties end and Dushanta is united with his beloved and his young son.

Although Sanskrit plays continued to be written after 1000 AD, none of the works achieved the stature accorded to many of those before that date. Among the important later playwrights are Bhavabhuti (*c.* AD 700), who wrote *The Latter History of Rama* (*Uttararamacarita*), and the playwrights Harsa, Visakhadatta, and Mahendra Vikrama Pallava.

Play composition fell into ten different types depending on subject-matter, length, character, and sentiment. Despite the broad range of choices afforded the writers the *nataka* was the favourite, with many excellent surviving examples. A *nataka* is a play in which the subject-matter concerns the exploits of a royal sage or king in a story that is well known and which has no fewer than five and no more than seven acts, and in which the dominant sentiment is either love or heroism. *The Vision of Vasavadatta, Sakuntala,* and *The Latter History of Rama* are all *natakas.* Sudraka's *The Little Clay Cart* is a *prakarana,* the second most popular play type, of which this is the most important surviving example.

The basic unit of a Sanskrit play was the act (*anka*), which is regarded as the 'body' of the drama. Acts were to be performed only by major characters and concern material directly related to the development of the plot. Plots were to have five stages: the beginning stage or planting of the seed or germ regarding the desire of the hero to obtain his objective; determined effort to achieve the object of desire; hope of success; assurance of success, if only some difficulty is overcome; and, finally, achievement of the objective. The *Natyasastra* lists a host of other elements related to the steady progress of the plot, suggesting that the structure of composition was to be rigidly followed. Preceding many published plays is a poetic benediction and sometimes a short introductory scene. These are but part of an elaborate set of eighteen preliminaries (*purvaranga,* literally 'before the stage') set down to bridge the pre-show events to the body of the play. A brief verse concludes many works, bringing the action to a ritualistic conclusion.

Performances of plays celebrated religious festivals and special court occasions, such as the coronation of kings, marriages, the birth of a son, the return of a traveller, and the confiscation of a town or state. Performances were judged by experts in various fields, scores were kept, and prizes awarded to the actors with the fewest faults.

The *Natyasastra* tells us that the appropriate performance place was a medium-sized rectangular building, 96 feet by 48 feet. Half of the space was allotted to the spectators and the other half to the performance space, which was raised and further divided in half (24' × 48'). A wall separated the acting area from the dressing-room and two doors provided entrances and exits. It seems the acting area was again divided in half (12' × 48'). So that they could clearly observe the action musicians occupied the raised area between the two doors at the back of the space. The floor of the acting area was smooth and slightly raised at the centre to allow water to drain off after washing. A roof supported by decorated pillars stood above the performance area. The ceiling of the entire edifice seems to have been designed to maximize the acoustical properties of the human voice and the musical instruments, particularly the drums.

Given the size of the audience area (48' × 48'), no more than 500 patrons at most could witness a production, a considerably smaller number than that which witnessed the Greek, Roman, and Elizabethan theatres at the height of their development.

Since theatre architecture was a branch of temple architecture rituals accompanied the selection of a proper site, cleaning the land for construction, and laying out the space. Rituals accompanied virtually every step of the construction process, including the reconsecration of the acting area before every show in order to protect the space and the performers from malevolent forces.

Among India's original contributions to world theatre is its aesthetic theory, which was developed for drama and applied to all branches of literature and performing arts. The theory of *rasa*, or sentiment, attempts to explain how a spectator perceives the performance experience. *Rasa* means 'taste' and, like the connoisseurs of *haute cuisine*, the spectators experience a performance as though they were dining on a fine meal. All the functionaries—actors, musicians, dancers, playwrights, and so forth—are expected to prepare and train themselves according to the rules and conventions for the moment when they will present their lavish feast for the eyes and ears of the spectators. One of eight basic sentiments is expected to dominate the experience. The eight sentiments are erotic, comic, pathetic, furious, heroic, terrible, odious, and marvellous. To give variety to the experience the sentiments act in permutations and combinations. But one sentiment is always expected to dominate the whole. The most common sentiment of the *nataka* and *prakarana* plays is the erotic.

Sanskrit theatre became inactive after the tenth century AD owing to the political and social unrest in north India, with the invasions of Mohmmed of Gazzni and successive waves of peoples from the Middle East. However, far from the scene of its declining fortunes in north India, Sanskrit drama and theatre had taken root in a remote region of the subcontinent. In what is now the state of Kerala, a green stretch of land running along the Arabian sea coast in southwestern India, in the tenth century AD, King Kulashekara Varman, with the help of his trusted minister Tolan, is reported to have recommended needed reforms in *kutiyattam*, the Sanskrit drama of the region. *Kutiyattam* survives today because of the tenacity of a small community of actors and musicians who have doggedly preserved it for more than 1,000 years.

The traditional actors are members of the *cakyar* community, a subcaste of temple servants whose ritual duty is to perform Sanskrit plays as a visual sacrifice for the presiding deity of a few Hindu temples, mostly in central and north central Kerala. They are accompanied in their effort by *nambiars*, another subcaste of temple servants devoted to playing the *mizhavu*, a large pot-shaped drum. The women of the *nambiar* community, known as *nanyars*, preserve the hereditary right to perform the female roles in *kutiyattam* and in *nanyar koothu*, another form of ritual performance. Owing to changing social conditions after Indian Independence in 1947, members of other castes and communities learnt the art of *kutiyattam* and *nanyar koothu* and now engage in public performances, except for those in the temple compounds which remain the exclusive prerogative of the hereditary castes.

The entire *kutiyattam* repertory includes a narrow range of plays, most of which were written by relatively minor playwrights such as Kulashekara and Shaktibadhra, although acts of a few works by Bhasa, Harsa, and Mahendra Vikrama Pallava are still to be seen. It is said that even Kalidasa's *Sakuntala* was once staged.

Kutiyattam theatre structure. Haripad temple theatre (*kuttampalam*) exterior in south central Kerala. The structure is characteristically located in the walled compound of the temple to the right of the central shrine and next to the golden flagstaff, seen here.

Kutiyattam is unique in that generally only single acts of plays are staged at any one time. And although the text of the act is important, it is not nearly as essential to the actors as the manuals (*kramadipika* and *attaprakara*) that describe how the act is to be staged. The manuals are absolutely necessary to performers before they set about producing any work.

The manuals prescribe rituals, properties, costumes, make-up, correct chronology of events to be acted, length of performance in terms of days, specific items of movement and rhythmic accompaniment, names of melodies to be chanted, where and when to move on the stage, and even the differences in prevailing performance practice. Most essential are the textual passages which are to be inserted into the scripts through symbolic gesture. These passages are not found in the Sanskrit plays.

Performance of one act of one play normally requires several days to complete. *The Farce of the Pious Courtesan* (*Bhagavad-Ajjuka*), King Mahendra Vikrama Pallava's one-act play about the mistaken exchange of souls between a priest and a courtesan, requires at least thirty-five days to present and the staging manual is many times larger than the play itself. Typically, the first day of performance begins with special rituals on a stage which is festively decorated with coconut fronds and stalks of freshly cut banana trees. Then each character is introduced separately over a period of days, expounding on past and related events to the main story. In this way, audience suspense is aroused. Most of the temple performances begin at 9 p.m., shortly after the evening rituals, and conclude around midnight, except on the final day of the show, when the entire act of the play is performed from beginning to end, finishing around 3 or 4 a.m., just prior to the morning rituals in the sanctum sanctorum.

Kutiyattam's unique contributions to Indian and world theatre architecture are the permanent theatre structures (*kuttampalam*) located in the temple compounds. At least nine rectangular theatre buildings have survived since the sixteenth century. They range in size from the tiny Guruvayur Temple Theatre to the impressive Vatukumnathan Temple Theatre of Trichur. Similar in many ways to the medium-sized rectangular theatres described in the *Natyasastra*, the *kuttampalam* are all rectangular in shape and have similar structural features. Although differing in detail, the stages are square raised spaces, with pillars supporting a roof. Often the surface beneath the roof is elaborately decorated with sculptural figures depicting scenes from the Hindu epics and deities appropriate to the various directions symbolically honoured by the actors during the opening rituals. Two doors separate the stage from the dressing-room. The door upstage left is normally used for entrances and that upstage right is normally reserved for exits. Between the doors, two large *mizhavu* drums are held in heavy wooden cases. Most of the acting takes place downstage centre in the direction of a large brass oil-lamp, which serves as the chief and often only source of illumination. In order clearly to see the faces and the eyes of the actors, enthusiastic patrons sit close to the front of the stage, women on house right and men on house left.

Kutiyattam performances are rare today owing to the declining fortunes of the temples that support them. However, the Vatukumnatha Temple and the Irinjalagauda Temple in central Kerala still support annual performances. The Margi theatre school in Trivandrum, south Kerala, manages to give short performances every Friday evening in its modest training centre. And the Kerala Kalamandalam, the state-run performing arts academy, sponsors a few annual shows.

Between the tenth and fifteenth centuries the old social orders began to crumble, making way for the consolidation of power in the Delhi Sultanate. The emergence of Islam as the state religion in much of India meant that theatre was no longer encouraged or condoned at court. With royal patronage waning, performers took to the road catering to the less sophisticated tastes of rural audiences. Performers survived as story-tellers, jugglers, acrobats, and singers, much as their medieval European counterparts did during the Dark Ages.

Then Vaisnavism emerged in the fifteenth century. Vaisnavism is a religious movement centring on devotion, with Krishna as the central figure of worship. Poetry, song, dance, and performance of drama became popular as means of proselytizing the faith. Because of the euphoric devotion of poet-saints engendering a tremendous outpouring of love among the common people a different genre of theatre emerged—*ankiya nat* in Assam, *jatra* in Bengal, *bhagavata mela* in Tamil Nadu, *kuchipudi* and *bhamakalapam* in Andhra Pradesh, *raslila* and *ramlila* in

Above: a live *kutiyattam* performance. The bloody, disfigured demoness Surpanakha enters through the audience. Blazing torches light the way. This entrance is a special feature of *kutiyattam.*

Left: ritual performance. *Teyyam* performers blessing devotees as they dance and swirl through expectant crowds of spectators on a ritual occasion.

various states of north India, and *krishnanattam* in Kerala, to name the more important.

The period between the fifteenth and nineteenth centuries also saw the emergence of interest in historical and semi-historical tales centred on popular regional cultural heroes and heroines. This was paralleled by a vigorous interest in social satire. Among the surviving genres in which historical or secular concerns abound are *bhavai* in Gujarat and Rajasthan, *burrakatha* in Andhra Pradesh, *khyal* in Uttar Pradesh and Rajasthan, *maach* in Madhya Pradesh, *naqul* in Punjab, *nautanki* in Uttar Pradesh, Punjab, Rajasthan, Hariyana, and Bihar, *svanga* in Harayana, Uttar Pradesh, and Punjab, *tamasha* in Maharashtra, and *veedhi natakam* in Andhra Pradesh.

In some regions dance plays a significant, if not dominant role; there are *kathakali* and *tullal* in Kerala, *yakshagana* in Karnataka, Andhra Pradesh, and Tamil Nadu, *therukoothu* in Tamil Nadu, *seraikella chhau* in Bihar, *mayurbhanj chhau* in Orissa, and *purulia chhau* in Bengal.

Although the genres mentioned above were created by and generally for Hindus, the *bhagat* of Agra and *bhand jashna* of Kashmir were developed by and for Muslims and the *cavittu natakam* of Kerala caters to Christians. This period also saw the continuation and development of solo and group dances, story-telling, singing, and a myriad of shadow, glove, doll, marionette, and rod puppet traditions, some of which still survive and thrill audiences even today.

Although each particular genre is unique and has its own peculiar characteristics, there are some broad generalizations that apply to many of them. Like the Sanskrit plays before them the tales that are told in the rural genres derive their inspiration from the following sources. The vast majority of them come from epic literature and focus on the exploits of Rama and his struggles to regain his abducted wife Sita, and tales of Krishna's youthful pranks and his support as an adult of the Pandava brothers, who have been dispossessed of their property and kingdom by their Kaurava cousins. Some stories centre on the rich historical and semi-historical legends, such as the tales of Amar Singh Rathor, the Rajput prince, or Shivaji, the Marathi warrior, for example. Then there is an array of material gleaned from rural life with thinly veiled allusions to real events and people, stories caricaturing the extremes of money-grabbing merchants, slow-witted farmers, contentious washermen, and wily, gluttonous priests, as well as nimble, seductive wives—all classes and castes of humble folk. And, finally, only recently mined for their dramatic potentials are urban dramas, some of which have their roots in the plots of foreign plays and the cinema, particularly the popular and fantastical Hindi cinema and its television counterpart.

Characteristically, music, song, and dance are integrated into the performance dynamic although each genre stresses one or the other of these aspects to a greater or lesser degree. Rhythm is the single most important aspect, setting the pace of

the action and adding extra emphasis when necessary. Each region and genre has its own peculiar drums and cymbals which characterize its rhythmic outpouring. Melody is provided by the human voice as well as a wide range of stringed and wind instruments. Numerous subsidiary instruments, such as conch shells, ankle and knee bells, rattles, and gongs, produce special effects. Also counted among the musical instruments are Western instruments, such as the harmonium, which is said to have been introduced by Jesuit missionaries. Even clarinets, saxophones, and violins assume an Indian flavour in the hands of Indian artistes.

Musicians normally have a strong visible presence in performance. Often they flank the performers and in a few instances they even take part in the dramatic action, moving freely toward and away from the performers to heighten the dramatic effect. This calls attention to the highly theatrical quality of the performance event.

Normally, music is improvised on the spot and not written down. Genres of north India use Hindustani classical music as a base and those of the south generally rely on the Karnatak system. Folk-songs and rhythms often find their place in both. Even popular film tunes have been freely adopted.

Dramatic moment in a rural *yakshagana* performance. The god Shiva, left, humbles the god Indra, right, at a high point of a *yakshagana* performance. The musicians flanking the performers become a lively part of the dramatic action.

Dance, too, contributes to the overall success of most performance events. Like music the degree and detail vary greatly among the various genres. Generally speaking, south Indians seem to emphasize dance more extensively than their north Indian counterparts. The *kathakali* of Kerala, for example, has developed a highly sophisticated system of dance in which physical movements combine with symbolic hand gestures and eye expressions to convey character and emotion. Years of training are required to master the techniques of such a system. On the other hand, the simple movements and patterns of the dances of north Indian genres such as *nautanki*, *tamasha*, and *bhavai* require no formal training and may be learnt by observation and a little practice.

Techniques of classical dance-forms, such as the *bharata natyam* of Tamil Nadu have been appropriated in *bhagavada mela* and elements of the *kathak* of north India are found in *tamasha* and *nautanki*. Folk movements may be detected in the less sophisticated dances of *bhavai*, *veethi natakam*, and *raslila*.

Dance is often used to begin a performance, whether it be the sophisticated and formal opening dances of *yakshagana* and *kathakali* performed by student actors or the seductive twists and turns of the adult female artists of *tamasha*. The 'curtain looks' (*tiranokku*) of animals and demons in *kathakali* are particularly relished by audiences in Kerala; during these the actor teases the audience for more than an hour with special patterns of movement behind, over, and around a colourful curtain. There is nothing more startling than the bursts of flame shooting from the mouth of a wildly twirling actor in *therukoothu* or the impertinent toss of the neck scarf of an enormously crowned demon character in *yakshagana*.

Many forms have devised special entrances for important characters. The prepubescent leading actors of *ramlila* from north India are ceremoniously borne to the playing area on the backs of local priests, where they are welcomed as though they were the incarnations of the gods they portray. And when Krishna arrives in an *ankiya nat* performance in Assam crowds of spectators fall before him and reverently touch his feet, believing him to be an incarnation of the god himself. In *therukoothu* the actor playing Bhima at one point in the performance makes a spectacular entrance running and wielding his mace after making a round of the village streets urged on by the shouts of exuberant enthusiasts. And in this same form the actor playing Kali, the bloodthirsty goddess, is carried to the playing area on the shoulders of devotees as she puffs fire from her mouth. Masked actors of *purulia chhau* of Bengal stamp defiantly to part tribals bearing spears as they enter the corridor leading to the playing area.

Performance spaces are normally defined by bamboo or wooden poles connected at the top by simple but striking decorations of fronds or paper, and sometimes crowned by a colourful canopy. The neutrality of the playing space, in which no scenery is needed, allows for the actors and singers to define the space

Facing: entrance dance of an *ankiya nat* actor. A youthful actor portrays Krishna in an *ankiya nat* performance in rural Assam. The actor mimes the gesture for flute, which is associated with his youth among humble villagers.

Rural performance space. The morning after an all-night *yakshagana* performance in south Kanara. The four bamboo posts linked by bamboo supports inscribe the performance area in a rice paddy field on the outskirts of a small village.

in words and to suggest a change of place quickly through their songs, speeches, or conventional movement patterns.

Simple, undecorated wooden stools, chairs, and tables are employed over and over again to represent a throne, a mountain peak, an aerial cart, a tree, or any other elevated place required. Because these items are too heavy and cumbersome to be taken on tour they are often obtained from the homes of villagers or townspeople. Properties are made of light wood and painted bright colours. Maces, swords, spears, tridents, and bows and arrows are staple props used by the actors in performance.

Acting companies are as different as the theatre genres they represent. There are amateur companies which come together only once a year to perform as a ritual sacrifice at a local temple or on the occasion of great national events, such as the *dussehra* celebrating the victory of Rama over Ravana in which hundreds of thousands of spectators annually engage. Itinerant bands of professionals may travel the length and breadth of a linguistic region all year round, halting their odyssey only during the monsoon season.

To travel with a company of itinerant players is not easy. Just after the company arrives in a village they take their evening meals, usually offered gratis by the village headman. After a quick nap the players unpack their heavy straw bags or wooden trunks and begin their preparations. Very often they are told less than an hour before a show what stories the villagers want to see, so there is barely any time to think about the plot of the story, much less to rehearse. Throughout the long hours of the night, normally from 9 p.m. to 6 a.m., the performance continues uninterrupted. Only after the final devotional song does it officially end. Then, in the first light of dawn, they repack their belongings, bathe, offer prayers at the local temple, have a bite to eat, and catch a bus or train to their next destination, whiling away the rough hours of travel playing cards or sleeping. Such is the hard life of professional players of the many rural theatre genres of India.

Modern theatre came into existence with the growth of cities. The process was gradual and shaped by the British presence in India. It was born in Calcutta, Bombay, and Madras, all cities founded by English settlers, who consolidated their commercial and political power by the early eighteenth century.

Playhouses had been constructed in all three English settlements by the mid- to late eighteenth century. The early buildings were simplified versions of sophisticated models current in London and bore colourful names such as the Chowringhee, the Dum Dum, the Boitaconnah, and the Sans Souci.

The birth of a modern theatre movement owes much to the spread of English education in the nineteenth century. At the insistence of Lord Macaulay, English education was propagated in colonial India so the British might better govern the country through a class of Indian people educated to have British ideas, tastes, values, and morals. A class of Indian intellectuals was nurtured on English classics, especially the plays of Shakespeare. It was during this time that Indians developed an abiding love and knowledge of British drama and theatre.

From this background drama in Indian languages acted by Indian performers emerged. The movement began in Calcutta in the homes of wealthy Bengali families. The Hindu Theatre was founded in 1831 by Prasana Kumar Tagore; it was among the

Early English theatre company in India. Backstage during a performance in the early nineteenth century, English actors prepare to enter. Indian servants are shown waiting on them. Eager Indian audience members sit in the front row of the theatre.

first *private* theatres to produce plays in the Bengali language even though the first plays were translations of selected acts of *Julius Caesar* and the Sanskrit classic *The Latter History of Rama*. Subsequently, more private theatres sprang up in the homes of other wealthy Calcutta families. Among almost a dozen private theatres which opened between 1835 and 1871 were the Belgatchia Theatre (1858–61), which introduced the work of the Bengali poet Michael Madhusudan Dutt to the stage, and the Jorasanko Theatre (1865–7), which provided the aesthetic environment out of which the great poet-playwright Rabindranath Tagore would emerge. The first private theatre in Bombay was constructed in 1842, to which Vishudas Bhave, considered the father of Marathi drama, brought his first play in 1853.

Modern Indian theatre started in earnest with the inauguration of Calcutta's first *public* theatre building in 1872—the National Theatre, which opened its doors with Dinabandhu Mitra's controversial drama *The Mirror of Indigo Planters* (*Nildarpana*). The play depicts the inhuman treatment of Bengali labourers at the hands of ruthless British indigo planters. While on tour to Lucknow in 1875, the production caused a riot among the European settlers, who unceremoniously sent the actors back to Calcutta before they had time to finish their show. Because of this and other signs of political unrest the British government imposed a harsh censorship law in 1876 called the Dramatic Performances Act. The Act gave the government the right to prohibit public performances of works it considered scandalous, defamatory, seditious, or obscene. Subsequently, all works that were to be presented in public had to be submitted for scrutiny before they could be performed. As a result political and social protest was forced underground and Indian writers and actors veiled their criticisms of the colonial government in the guise of mythological and historical incidents.

The late nineteenth and early twentieth centuries saw the flowering of Indian drama and theatre with the construction of numerous theatre structures and the development of theatre companies throughout the subcontinent. In Calcutta, the great actor-playwright Girish Chandra Ghose came to prominence, as well as the popular actress Binodini and in Maharashtra Annasaheb Kirloskar pioneered the Marathi musical and K. P. Khadilkar, the playwright, assumed a position of authority in the budding urban Marathi theatre.

Among the most famous and unique personalities of the age was Rabindranath Tagore. Tagore was born into a wealthy and influential Calcutta family of artists and poets. He wrote many plays, poems, and songs, some of which were originally performed by members of his own talented family in private recitals before invited guests. In 1913 he was awarded the Nobel prize for literature for his poems *Gitanjali* and *Gitali* and he has the distinction of writing India's national anthem. In 1935 he founded Santiniketan, a unique university dedicated to education in the fine and performing arts. Among Tagore's better-known plays are *Raja* (1910), *The Post Office* (*Dakghar*, 1913), and *Red Oleanders* (*Raktakarabi*, 1924). Tagore symbolized the ideal toward which many contemporary Indian theatre artists might strive.

In 1877, while Tagore was working in Calcutta, the Parsi theatre came into being in Delhi and Bombay. It grew out of a genuine love of theatre and drama, when a small but powerful community of businessmen practising the Zoroastrian faith established companies of players performing in Hindustani, the language of the common people of much of north India. They launched the Oriental Theatrical Company of Bombay and the Victoria Theatrical Company of Delhi. Numerous spin-off groups soon followed. The plays and musicals of the Parsis had a lasting influence on the acting style and songs of theatre artists throughout the subcontinent because of their extensive tours.

A scene from *The Post Office* a one-act play by Rabindranath Tagore with Chaiti Ghosal and Sombhu Mitra. The play was performed by the distinguished contemporary theatre group Bahorupee.

With the development of film in the early part of the twentieth century and the coming of television by mid-century, modern theatre had to readjust to greater competition and revised expectations. Nevertheless, significant theatre personalities marked the period from the late nineteenth century up to approximately the 1940s. Among the many renowned individuals and companies are Sisir Kumar Bhaduri in Bengal, Mama Warerkar and Balgandharva in Maharashtra, Jayashankar Sundari and Chandravadan Mehta in Gujarat, the TKS Brothers and Rajamanickam troupe in Madras, Bellary T. Raghavachari in Andhra Pradesh, the Gubbi Company in Mysore, and Jai Shankar Prasad in the Hindi-speaking regions of the north, to name but a few.

The 1930s and 1940s were a time of great unrest around the world and India had its share of war, famine, and political instability. During this period the country gained its Independence (1947). With Independence, theatre spread and grew in all the major Indian languages and in virtually every seat of political power in the country, in small towns, and even in some villages. About this time the Indian People's Theatre Association (IPTA) was born out of the Indian communist movement. Before long theatre companies came into being in most major metropolitan centres, attracting committed and talented people who expressed sympathy with Communism even where they did not actively promulgate its messages. The legacy of communist influence is felt in Bengal and in certain parts of Kerala even today.

Commercial theatre was hard pressed to sustain itself because of competition with the film industry, the high cost of doing business in the cities, and the fact that no company owned its own building or rehearsal space. And so amateur theatre, much of it as professional in performance as any professional theatre in the West, flourished particularly in Bombay and Calcutta, both cities which had established a tradition of leadership in modern theatre by this time. Besides the basic repertoire of conventional works, new plays were being written which challenged the formulas established in previous years and which were sensitive to the changing currents in play construction abroad. Experimental works required experimental companies that pushed outwards the envelope of what constituted good playwriting, acting, and staging.

Training in modern theatre principles was firmly established when the central government inaugurated the National School of Drama in New Delhi in 1954 under the leadership of Ebrahim Alkazi, the Bombay director, who with a talented staff of teachers began the job of training whole new generations of young theatre artistes from all over the country. Many of today's major artistes and executives in theatre, television, and film received their formative training at the National School of Drama.

Up to the 1970s, several characteristics distinguished the modern Indian theatre tradition from the theatre that prevailed in the rural areas of the country. Modern urban theatre relied on plays which were written in dialogue and which were not dependent on music or dance. For the most part the plays concerned a broad spectrum of topics which appealed to the tastes of urban people and not those of the villages. In many ways urban theatre rests in the hands of strong directors who hold groups of performers together by the sheer force of their artistic vision. The style of acting has been realistic since the early part of the century, even if the actors slip into melodramatic extremes at times. It has been the characteristic of the modern age that performers choose to enter the profession of theatre out of love or interest rather than because it is their hereditary right or obligation to do so. Training in physical movement and speech follows similar procedures to those found in the Western world. Only recently have methods from rural theatre traditions been adopted for use. Productions have been mounted in much the same way as they are in the West, with attention paid to scenery and properties, stage lighting, and sound. Box-office revenues are considered essential for the survival of most groups and so the extent of the emphasis on spectacle depends on how successfully the companies have fared financially. Besides the income from box-office receipts many groups depend on subsidies from government sources to survive. Proscenium arch theatres dominate the modern theatre, although a few thrust and arena-style stages have been constructed recently. Unique methods of lighting plays have been developed because of the high cost of lighting instruments and the primitive state of the facilities. The work of Tapas Sen of Calcutta is unique in that it achieves a simple yet effective compromise with few instruments and limited power sources. Lucrative tours of successful shows to smaller towns and cities have become a necessity for many groups struggling to survive. Newspaper criticism exists, but is not very sophisticated nor respected by the performers or the spectators. Although urban theatre exists in all sixteen major languages, there is little exchange of plays and companies among the various regions.

EAST ASIAN THEATRES

COLIN MACKERRAS

ALL three countries of east Asia—China, Japan, and Korea—have cultures heavily influenced by Confucianism, a puritanical and hierarchical political ideology emphasizing filial piety and the family. Although the cultures thus have many commonalities, they are also very different in texture.

Traditional Theatres

The traditional theatres of China, Japan, and Korea are in general notable first and foremost for their performance, not as literature. In all three civilizations, popular balladry and puppetry have formed an extremely important basis for more complex drama. The balladeers and puppeteers transmitted numerous stories, acting techniques, and melodies which frequently evolved into dramas. In China story-tellers date back to ancient times and their art developed into numerous different regional styles over the centuries. In Korea the *p'ansori* is a folk-ballad form performed by a single singer accompanied by a double-headed drum, his stories being largely drawn from popular novels.

China and Japan both have old, varied and distinguished theatre traditions, characterized by highly stylized movement, gesture, and costume, and by a tight integration between the various theatre arts such as singing and performance skills. In China, the first great age of drama was in the thirteenth century, with the *zaju* (literally 'variety drama') as the dominant form. In the sixteenth century Wei Liangfu and others forged from older southern theatrical styles a form of drama called *Kunqu*, meaning 'the tunes of Kunshan', that being the place, just west of present-day Shanghai, where he lived and worked. Meanwhile a large variety of popular regional styles developed during the Ming (1368–1644) and Qing (1644–1911) periods, of which the most important was the Peking Opera. In Japan, two theatre forms to gain special prominence were the Noh and the kabuki. Noh was the favoured form of the warrior *samurai* class, whereas kabuki came to be attached to the urban entrepreneurial classes of the period of the Tokugawa shogunate (1600–1868), which had its seat in Edo (present-day Tokyo). The two main types of Korean theatre were folk mask dance-dramas and puppets.

China

Although performance is extremely ancient in China, fully developed drama dates only from the Song dynasty (960–1279), probably early in the twelfth century, when a form called *nanxi* or 'southern drama' was found in the province of Zhejiang. According to the sixteenth-century writer Xu Wei, the 'southern drama' was based on 'the lyrics of the men of the Song with the addition of ballads of the alleys and folksongs'. Already there is evidence of the four categories of actors which dominated virtually all the main forms of theatre of later times. These were the *sheng* (male character), *dan* (female character), *jing* (painted face), and *chou* (clown).

Many regard the *zaju* of the Mongol Yuan dynasty (1279–1368) as representing the golden age of Chinese drama. In structure the *zaju* were considerably stricter than the *nanxi*. They had either four or, occasionally, five acts sometimes with short prologues or interludes. Each act had a particular function with regard to the plot, the last act resolving conflicts and restoring harmony. A major musical feature was the use of song-suites, meaning that the music of each act was a suite or set of songs, all of which used the same mode. There was only one singer in each drama, who was the main performer of the troupe.

The texts of nearly 170 Yuan *zaju* survive, although none of the music. The themes are love-stories, with the women being courtesans or high-class ladies, while some are about rebel-heroes, tyrants, venal bureaucrats, or gods. Others

Thirteen Great Chinese Actors. Shen Rongpu's painting shows Peking Opera actors flourishing from 1860 to 1908. Third from the left is Mei Qiaoling (1842–82), a *dan* actor and Mei Lanfang's grandfather. Sixth from the left is Cheng Changgeng, called 'the father of Peking Opera'. The white patch on the face of Yang Mingyu, fifth from right, is the distinguishing mark of clown roles.

concern historical episodes, especially wars of the past, or reversal of court judgments. The most important *zaju* playwright was Guan Hanqing, whose career, though his precise dates are unclear, spanned the second half of the thirteenth century. A representative item by him is *Injustice to Dou E* (*Dou E yuan*), which is a courtroom drama in which a young widow Dou E is executed for a crime she did not commit, but is cleared posthumously, both the real criminal and the corrupt first judge being punished.

In stark contrast to most other Chinese drama forms until the 1930s, women were major participants on the Yuan *zaju* stage. Most of the known stars were female courtesans, the association of the actress with the prostitute already being very clear. It is highly likely that women were able to perform not only female but also male roles.

The *zaju* drama spread to the south and the beginning of locally distinct styles dates from the earliest history of Chinese drama, but major proliferation dates principally from the sixteenth century. The stories, categories of actors, and stage techniques were generally similar from region to region, but the music and the dialect used differed more substantially. Wandering companies took their art from place to place, bringing about adaptations and amalgamations of artistic elements, which had given birth to some 300 regional drama styles by the time the Manchu Qing dynasty fell in 1911. Almost all of these were popular theatre and hence despised by the educated. Actors of these folk styles, virtually all of them male, were very low in social status and suffered serious discrimination before the law.

One style, the *Kunqu*, was appropriated by the aristocracy and educated élite.

It is noted for its regular rhythm, dominated by common time, and elegant timbre deriving from the wistful sound of the bamboo side-blown flute or *dizi*. In contrast to the popular dramas, which were anonymous and usually handed down from master to disciple, authors put their names to *Kunqu* items. These were very long and if played *in toto* without interruption took several days. As a result, the practice of performing just a few isolated scenes became prevalent. An educated or wealthy man might train his own slaves to perform, or could hire a local company for a banquet or other special occasion in his own mansion.

The pre-eminent regional style is that of China's capital, Beijing, which is normally known as Peking Opera, Peking being the formerly accepted romanized name equivalent to Beijing. A year of great significance in its evolution was 1779, when the *dan* actor (female impersonator) Wei Changsheng brought his troupe to the capital, immediately arousing enormous enthusiasm. Wei Changsheng and his followers came from Sichuan in the south-west, where a tradition of regional theatre had been developing for at least a century and probably more. It appears from the records that one of the sources of Wei's popularity was his *risqué* acting and humour. This brought down the ire of the authorities on his head and he was banned from the stage in 1782. However, a homosexual love affair with a Manchu powerholder, Heshen, kept him performing until 1785, when a further ban forced him to leave Beijing. Wei Changsheng made a brief return to Beijing in 1800, still commanding a considerable following, but died there a pauper in 1802.

In 1790, in order to take part in the celebrations of the great Qianlong Emperor's eightieth birthday, drama companies came from Anhui and other southern provinces to Beijing. Known in history as the Anhui companies (*Huiban*), these troupes survived in Beijing for about a century and were the main wellspring producing the Peking Opera. Whether it is proper to take 1790 as its birth-year is a matter of definition. However, it is more or less certain that the combination of two sets of melodies named *Erhuang* and *Xipi*, which to this day remains the musical base of the Peking Opera, was heard for the first time in Beijing in that year.

The best-known actors of the early period were all *dan*. The reason for this is that the names of these actors survive through love poetry dedicated to them. They entered the acting profession as virtual slave boys in the south, but were then taken to Beijing. If they were lucky, they might achieve a certain glory there and attract patrons of good family. The latter often described the boys or young men as 'flowers', a polite word for a prostitute, suggesting that actors were frequently involved in homosexual love affairs.

In the early 1830s, a new wave of actors from the south found its way to Beijing. While the *dan* continued performing through the nineteenth century, the records speak of more mature actors who excelled in the portrayal of ministers, bureaucrat-scholars, and military heroes or villains. Among them by far the most famous

Chinese theatre bill (1912). For the Sanqing Theatre, the handbill states the company which will perform, as well as the actors and, at the bottom, the list of dramas. The theatre is on the southern side of the Dashala Road just outside the Front Gate, the main theatre area of Beijing.

was Cheng Changgeng (1812–80), a player of *laosheng* roles, such as generals and righteous men, whose impact on the art and content of the Peking Opera was such that some have even called him 'the father of the Peking Opera'.

All items were set in China's past, being based on old novels, stories, and dramas. Their content featured heroic courage, stratagems, honesty or treachery in warfare, love or family relationships, righted injustice, or legend. The costumes were stylized, intricate, and very colourful, and the make-up very expressive as well, with the 'painted face' characters in particular showing an enormously wide range of designs and colours which indicated to the audience their personality: sly, brave, honest, loyal, or whatever. The gestures were likewise highly stylized, with miming common and expressive. Many of these performance features had applied also to the *Kunqu*, although they were more highly developed in the Peking Opera. One major feature of the Peking Opera differentiating it from the *Kunqu* was the complex and thrilling acrobatics, with somersaults, tumbling, and throwing, catching, and returning of spears being performed to split-second timing.

All the Anhui companies were entirely male. In the late Qing period there were two all-female companies in Beijing, but both were disbanded in the first decade of the twentieth century. Some tentative attempts at forming mixed companies were made from the second decade of the century. In 1930 a mixed training school was set up and men and women acted together on the Beijing stage, setting a

Painting of a Chinese theatre by George Chinnery (*c.*1846). The stage is covered and has a curtain at the back, but not the front. Musicians are seated on the stage itself. The audience shown is mainly from the lower classes; they sit in the open air. The figure on the far right, bottom, is clearly of a higher class; his seat and location are appropriately better and less crowded. All performers, musicians, and members of the audience are male, although one of the two main actors (centre) is dressed as a woman.

precedent which has proved permanent. Not until after the People's Republic of China (PRC) was set up in 1949 did mixed troupes become the norm all over the country.

A strong theatre culture developed in nineteenth-century Beijing which centred around public teahouse-theatres in one area in the south of the city. Price for entry was graded according to where one sat. Tea was readily available, but not full meals. Officials professed to despise the Peking Opera but often went anyway to keep an eye open for any seditious or disruptive behaviour from the audience or content from the stage. The richer classes could patronize another form of theatre which was actually more like a restaurant, where the drama formed the background to a banquet or celebration. This second type was much smaller and more intimate, and guests could state a preference for which item they would like to see. With rare and few exceptions, only men were allowed access to either kind of public theatre. When women were first permitted entry in the second decade of the twentieth century, they at first had to sit in a separate part of the theatre from men.

Japan

In Japan Noh developed from older theatre forms, and scholars have demonstrated strong links with shamanistic practices and ceremonies. Yet the origin of the style of theatre which currently bears that title can be attributed to a particular time and person. Kan'ami Kiyotsugu (1333–84) was a troupe leader in Nara, Japan's former capital. He forged the new style by crafting the climax scene from a narrative song-dance based on a combination of a popular narrative song and rhythmic dance. Kan'ami's style was to focus the story on the crucial event in the life of the main character, termed *shite*, which means performer or doer. Because his emphasis was on effective physical and vocal characterization, as opposed to dancing and singing, he is often regarded as the father of acting in Japanese theatre.

Kan'ami's son, Zeami Motokiyo (1363–1443), was also a magnificent actor who focused attention on the doer roles. In 1374 Kan'ami's troupe, including the 11-year-old Zeami, performed for the young shogun Ashikaga Yoshimitsu, who was vastly impressed. As a result the shogun extended his patronage to the art of Noh and invited Zeami into the palace. Over the next half-century or so the actor performed at court and wrote numerous treatises about Noh and about fifty plays out of a total of some 240 which have found their way into the standard repertory. Under the patronage of the Ashikaga court and Zeami's brilliance, Noh attained its classically defined form and highest level of literary excellence late in the fourteenth century. At the same time, what had been a form of popular theatre became very much the property and entertainment of the *samurai* class. Noh actors, including Kan'ami and Zeami, had previously been very low in social status, but now rose enormously within society, even becoming accepted as members of the *samurai* class.

The official religion of the Ashikaga court was Zen Buddhism, the artistic principles of which Kan'ami and Zeami absorbed into their art. These included restraint, austerity, and economy of performance. It is said that a Noh play 'aims to catch the mood of a fleeting moment by complete rejection of realism and a subtle fusion of technical resources'. The principles of Zen Buddhist art are reflected also in the understatement and quietism of the expression, however intense the emotions of the characters, the stark stage, simplicity of properties, the refined and suggestive masks, and, with the significant exception of the lavish costumes of the doer roles, the simple dress. Kan'ami and Zeami were both Buddhist names assumed on entering court service.

Noh dramas fall into five categories according to content: gods, warriors, women, demons, or realistic items. Zeami was keen to dramatize the lives of personalities in Japanese history, especially the twelfth-century civil war between the Heike and Genji clans, as well as Chinese legends, but many of the plays are not located in a particular period. The Noh drama is totally unconstrained by unity of time and place, with the climax scenes often taking place years after the earlier ones.

A representative example is Zeami's *Takasago*, which falls into the first of the five categories, and is noted more for its mood than its story. Its theme is the

Facing, above: folk Korean drama. Note the use of different styled masks and the open-air performance, both features typical of this style.

Facing, below: Tibetan drama performance, Kumbum Monastery. Tibetan drama is strongly characteristic of the Tibetans and totally different from Chinese drama styles. It is a highly integrated form in which singing, dialogue, dance, mime, mask, and other arts play a part. The costumes are very colourful and elaborate. Masks are extremely varied and feature spirits and animals. The great monasteries are among the sites of Tibetan drama, the audience standing or sitting in a circle around the performers. The Kumbum Monastery is in Qinghai Province, China.

Japanese Noh drama performance (1848). The ordinary people sit below the stage and are shown in the front of the picture. The higher classes are shown in the covered section opposite the stage. The stage itself is covered and there are four square pillars. A formalized painting of a pine tree is always found on the rear wall of the stage.

Facing: audience at a Japanese kabuki performance. The kabuki was the entertainment of pleasure, and the audiences included people of virtually all classes. Food could be had in a kabuki theatre, as can be seen from the fish and rice-bowls at the bottom centre.

legend of the twin pines of Sumiyoshi and Takasago, which are personified by a devoted couple, symbolizing longevity and conjugal fidelity. The second, final, scene is set in a completely different place from the first.

The Tokugawa shogunate was a period of flourishing urban culture spawned by a confident and assertive mercantile society, with the country as a whole being closed off by deliberate policy from the outside world. The most important forms of theatre to take shape in this environment were the kabuki and the *bunraku* puppet theatre, both in the seventeenth century.

Japan's most famous dramatist is Chikamatsu Monzaemon (1653–1724), who initially wrote for the kabuki but later turned his attention to the puppet theatre. The topics of Chikamatsu's plays are of basically two kinds, historical dramas and domestic pieces, the former extolling the actions and values of the *samurai*, the latter focusing on sexual love and frequently ending in double suicide.

Theatres, brothels, teahouses, and bathhouses made up the thriving amusement quarters. This world of gaiety and pleasure stood in sharp contrast to the prevailing and highly disciplined Confucianized *samurai* society of the Tokugawa period. Kabuki was actually wider in its appeal than Noh. Despite government restrictions, the *samurai* and their ladies were not above going to kabuki performances, with the audience including also merchants, workers, monks, and servants.

The earliest kabuki performers were women, but later all roles, including female, were played by men. This was because the government banned women

from the stage in 1629, their policy being that nobody should follow more than one profession; this prevented women from being both prostitutes and actresses. Following the ban, boy actors also functioned as catamites, but a further ban against such behaviour in 1652 meant that it was adult males who entered the performing profession.

Ichikawa Danjūrō I (1660–1704). The great kabuki stars became popular idols, even though the authorities professed to despise them. Great dynasties of actors arose, at the peak of which was that of the Ichikawa Danjūrō. The current dynast is Ichikawa Danjūrō XII, born in 1946.

The authorities regarded kabuki performers with distaste, according them no civil rights, but ordinary urban folk saw them very differently, as idols. The acting profession became hereditary, with great dynasties establishing themselves. Ichikawa Danjūrō I (1660–1704) developed a flamboyant acting style marked by bold red and black make-up and exaggerated costumes. Even in the 1980s and 1990s Danjūrō XII (born in 1946) continued his distinguished family's tradition of acting.

The heyday of the kabuki was the eighteenth century, with the main centre being Edo (present-day Tokyo). The puppets had initially had a strong influence on the style of kabuki acting, but in the course of time there was a gradual shift in taste, no doubt reflecting the maturing urban society, and one result was that in the eighteenth century kabuki replaced puppetry as the main theatrical attraction. The most popular dramas were those on erotic, grotesque, or mysterious subjects, or on the exploits of outlawed heroes. There was also a move from stylization to realism.

Chinese culture exercised an enormous influence over Japanese before the modern age. Yet what is striking in their various forms of theatre is how different they are, how little sign there is of any Chinese impact, let alone derivation, in Japanese theatre. In the case of the Noh, Zeami makes clear that Chinese antecedents played but a minor role. Even though many items are based on Chinese myths, they were already heavily Japanized before being adapted as Noh. As for kabuki, the basis was Japanese dances with virtually no relation to Chinese, while the urbanized bourgeois Edo culture was quite unique to Japan, China producing nothing even remotely similar.

Korea

In Korea, the origins of theatre probably go back to religious ceremonies, including shamanistic rites, and folk observances. There were various regional forms of folk mask-dance dramas, termed *sandaegŭk*, which consisted of dance, singing, musical accompaniment, pantomime and dialogue. They were handed down orally and only very recently have any been written down. None of the authors has been identified. The stories were very simple, with different acts sometimes containing quite separate characters and plot. The *sandaegŭk* plays satirized four types of people: the corrupt local official who stole somebody else's woman, the renegade and lascivious Buddhist monk, the aristocrat who oppressed the people, and the tyrannical husband. The costumes were colourful, with those worn by servants or loose women being especially gaudy, and the masks grotesque. Sites of performance included the village square or a temporary stage. Actors were male and of a very low social status.

Modern Theatres

During the last two centuries the cultures, including theatres, have changed enormously under the impact of several forces. The first is European colonialism and imperialism, which introduced new cultural forms and transformed the socio-economic and political contexts within which east Asian cultures operated. Secondly, Asian nationalism changed profoundly and irrevocably what Chinese, Japanese, and Koreans themselves expected of culture. In China and North Korea, ruling Marxist-Leninist parties have imposed very special demands on artists and the kind of work they produce. Marxist thinking has even to some extent influenced the theatre of Japan, which never underwent a leftist revolution.

While Japan was subject to major influences from the West, it also imposed its own brand of imperialism on nearby countries. Japan colonized Korea from 1910 to 1945, suppressing its culture in general and in particular its traditional forms of theatre. Because Japan modernized itself far more successfully than China or Korea, its adoption of Western trends in its arts was much more from a position of strength than that of either of the other two countries. On the other hand, it was different from them in having to sustain the humiliation of defeat in the Second World War, a fact which was reflected in its cultural life during the post-1945 period.

In China, the New Culture Movement from 1915 on involved attempts, especially by the youth, to change or uproot traditional Chinese culture, which was seen as a relic of a Confucian and reactionary past. Although firmly anti-imperialist, it was also heavily influenced by Western cultural and literary patterns. There was simply no way around the dilemma that, if China was going to modernize and reject Confucianism, it would have to adapt itself to Western influence.

Marxist-Leninist parties have had no monopoly of censorship of the arts in east Asia. But they have been much more direct in their demands that the arts in general and theatre in particular should serve the state and socialism as a vehicle of propaganda. Marxist-Leninist parties in China and North Korea have attempted to build up a cultural cadre which can create a 'proletarian' mass culture. However, the cultures' attachment to a class does not mean they are not nationalist. All Marxist-Leninist parties have demanded that the arts remain 'patriotic' and build creatively on the traditions of their particular nation. In theoretical terms an important landmark in Asian Marxist-Leninist thinking on the arts was Mao Zedong's account put forward at a major forum held in the Chinese Communist Party (CCP) headquarters in Yan'an in 1942. Mao's focus was on the propaganda and revolutionary functions of the arts and the creation of works which would appeal to the masses, not merely an educated élite. In October 1949 Mao and the CCP took power in China, founding the PRC. Although he immediately implemented his views on the arts, the period when they were interpreted most rigidly was during the Cultural Revolution (1966–76). Despite emphasis on patriotism, the effect was to bring all but total destruction to China's arts, especially the traditional, and to make boring stereotypes of those that survived.

At the end of the Second World War, Korea was divided into a socialist north and capitalist south. It also experienced an extremely bitter war from 1950 to 1953, during which theatre was used by both sides, and especially by the left, for propaganda and nationalist causes.

China

In China the most important form of non-traditional theatre is termed 'spoken drama' (*huaju*) because dialogue is carried on more or less entirely through speech, with hardly any song or music. The earliest spoken drama in Chinese which was also performed and produced entirely by Chinese was staged not in China but in Japan. On 1 June 1907 Chinese actors put on in Tokyo a Chinese-language dramatization of Harriet Beecher Stowe's novel *Uncle Tom's Cabin*. Called *The Black Slave's Cry to Heaven* (*Heinu yutian lu*), it was strongly hostile to slavery and to all forms of oppression.

The spoken drama has played a highly political role in China throughout the twentieth century. One authority aptly calls it 'the most assertive form of innovative literature in modern Chinese society', its writers mostly regarding it as 'an iconoclastic statement, designed to convey broader views for social reform and revolution'. China's most famous dramatist of the twentieth century, Cao Yu, wrote several very important plays during the 1930s and 1940s on progressive and nationalist themes. His most popular items are probably the tragedies *Thunderstorm* (*Leiyu*) and *Sunrise* (*Richu*), the former attacking the traditional family system in very strong terms, the latter urban corruption and misery in the 1930s.

At first, traditional drama was only marginally affected by the revolutionary tide sweeping China in the first half of this century. Many young people may have opposed such reactionary culture during the May Fourth Movement, but it nevertheless retained a substantial following. The most famous actor China has ever produced was Mei Lanfang (1894–1961), an exponent of *dan* roles in the Peking

Opera. He was not only wildly popular in his own country but travelled abroad several times to Japan, in 1919, 1924, and 1956, the United States, in 1930, and Europe, in 1935. It was during this tour to Moscow, London, Berlin, and Paris that Bertolt Brecht saw him act, and was enormously impressed.

Except during the Cultural Revolution from 1966 to 1976, the PRC has generally patronized the performance of theatre, both traditional and modern, and encouraged the survival of folk troupes. However, it has also sought to control and reform the content of dramas, with a view to ensuring that no traditional items performed are 'against the people' or anti-socialist, and that all recently or newly written items are progressive or revolutionary.

Mei Lanfang, China's greatest actor (1894–1961). Here he plays the role of Yang Yuhuan in *Drunken Beauty* (*Guifei zuijiu*), one of his best parts. The story concerns Yang Yuhuan, favourite concubine of the Emperor Minghuang (reigned 712–56). The lovers have a falling-out, and she drowns her sorrow in drink. Mei Lanfang arranged the drama to suit his particular skills.

One of the mechanisms in the CCP's style of government is to establish models. In the theatre this was done through major festivals showing representatives of specially approved dramas. In October and November 1952 the CCP's Ministry of Culture sponsored a Festival of Traditional Music-Drama, the first event of its kind ever held in China. There were about 1,800 participants, including Mei Lanfang, with eighty-two dramas being performed belonging to over twenty regional styles. About three-quarters of the items were traditional, but there were also eleven newly written historical dramas, and a few with modern themes. Two items given special praise by cultural commissar Zhou Yang in his speech at the end of the festival were the rearranged traditional love-stories *Liang Shanbo and Zhu Yingtai* and *The White Snake* (*Baishe zhuan*), his reason being that both dramatize strong opposition to the old custom of arranged marriages and show the women as strong and determined to stand up for their own and their lovers' rights. The Ministry of Culture also sponsored a Spoken Drama Festival in Beijing in March and April 1956, in which the themes of solidly modern plays promoted the revolutionary struggle against imperialism and functioned as propaganda for contemporary life in the factories and countryside under socialism. Cao Yu's *Thunderstorm* was among the few plays given by famous dramatists.

'Newly arranged historical drama' is a form featured in the 1952 Festival and one much favoured by the CCP. The story is set in the dynastic past, with costumes and music being generally classical in style. However, in contrast to pure

'Model' drama *Red Lantern*. Most famous drama of China's Cultural Revolution (1966–76), it is set in China's north-east during the war against Japan and concerns three generations of revolutionaries, namely grandmother (right), son, and granddaughter (left). The revolutionaries suffer but win victory. Note the stereotyped poses and facial expressions.

traditional theatre, there is complex décor. The plot expresses conflict and tension, and rises to a denouement which should ideally show the wretched of the earth victorious over the rich and powerful among the feudal ruling classes, or Chinese patriotism winning out over treachery. Favoured themes include peasant rebellions of the past, lovers who overcome the feudal system of arranged marriages, women who stand up for their rights or make a public contribution to society, or patriotic wars against foreign aggressors.

Another drama festival of a kind very different from those of 1952 or 1956 took place in June and July 1964: the Festival of Peking Operas on Contemporary Themes. About thirty troupes from most parts of China participated, giving some thirty-seven pieces, all of them dealing with the revolutionary struggle. One of particular note was *The Red Lantern* (*Hongdeng jî*), which tells the story of three generations of revolutionaries, two of them women, and their struggle against the Japanese. The wife of Mao Zedong, Jiang Qing, who had once been a minor film star and held strong views on the theatre, gave a speech called 'On the Revolution of the Peking Opera', which advocated the indefinite suspension of all traditional items, no matter whether newly written, reformed, or purely classical. The speech was suppressed at the time, due to the very strong resistance to such a radical view among other leaders, and was not published until 1967.

By that time, the Cultural Revolution was in full swing and she had succeeded in gaining support for her experiments. She rearranged some of the pieces performed at the 1964 Festival into 'model' dramas, in which characterization must reflect the class struggle, with the proletarian characters being very good indeed, selfless and strong supporters of the CCP, but the bourgeois the epitome of evil. In Jiang Qing's 'models', Western musical instruments were added to the orchestra and the music made more staccato, on the grounds that it thus became more revolutionary and heroic; but some of the traditional melodies were retained. From early 1966 until the middle of 1977, no dramas on traditional themes were publicly performed, the stage being totally dominated by Jiang Qing's 'models'. Even an item such as *The Red Lantern* was not revolutionary enough for Jiang Qing, who had it revised and its characters reshaped to fit her desired paradigm more closely. In its 'model' form it was not published until May 1970. It was this writer's impression from seeing *The Red Lantern* several times in China that the item was initially quite popular, but it was repeated so often that, like the other 'models', it soon lost its appeal and became regarded as a monumental bore, especially since the number of 'models' was very small and the available dramatic diet at the time was thus narrow in the extreme.

Japan

In Japan in the early years after the Meiji Restoration of 1868 there was a move to reform the traditional kabuki theatre by modernizing it in line with Western practices. In 1872, the new Japanese Meiji government placed actors under the control of the Ministry of Religious Instruction in an attempt to use the theatre to improve public morals and in 1887 it even set up a committee to reform kabuki to make it acceptable to Western audiences by 'cleaning it up' to remove the eroticism and violence which had characterized it, and to reform it to suit nineteenth-century European realism. On the whole this was not a great success, the appurtenances of the modern world proving incompatible with the non-realistic movement and elocution of the kabuki. One feature of this new-style kabuki which did remain permanent was that literary figures outside the theatrical world, men such as the right-wing nationalist Mishima Yukio (1925–70), wrote new scripts for the kabuki, as well as for Noh. But even this has rarely inspired great enthusiasm among theatre-goers, and these newly written kabuki rarely run for very long.

Plays with political themes began to be seen in Japan in the late years of the nineteenth century. Termed *shinpa* (literally 'new school'), these plays won some popular favour because of their novelty and patriotic appeal. Actresses, who had been banned from the stage by the Tokugawa regime, reappeared on the 'new school' stage and were tolerated by the new government. In 1906 Tsubouchi Shōyō (1859–1935) and others established the Literary Arts Society, launching a new form of Japanese literary drama known as *shingeki*, which means literally 'new drama'. The point of the 'new drama' was to adapt Western drama for Japan, to translate many of its items into Japanese but more importantly to use its structure, emphasis on conflict, characterization, and literary and performance conventions for a new Japanese drama. Shōyō's principal model was Shakespeare, whose works he translated into Japanese, expressing his belief that 'research into Shakespeare might be the most useful means of improving the Japanese drama, not only for me personally, but for the Japanese as a whole'.

The 'new drama' has never had an enormous audience, but it has exercised substantial influence in the twentieth-century Japanese theatre. Alone among the arts, it never succumbed fully to the dictates of the militarist government. Indeed, Kubo Sakae (1900–57), one of the most famous of Japan's playwrights, was a Marxist whose aim was to create a proletarian theatre. His *Region of Volcanic Ash* (*Kazambaichi*), from the late 1930s, is an extremely long play which describes the society of an impoverished area of Japan in great detail. Kubo himself described it as 'anti-capitalist realism', but in 1940 the government suppressed the left-wing theatre and Kubo was imprisoned.

The period since 1945 has seen something of a reaction against excessive dependence on Western forms in the 'new drama'. The 1960s were ushered in by demonstrations in 1960 against the renewal of the USA–Japan Security Treaty. They were a vigorous period in Japanese intellectual life and one dominated by Marxism. The left wing continued its pre-eminence in the 'new drama' move-

Scene from a Mishima Yukio play. *The Hall of the Crying Deer* (*Rokumei kan*), written in 1956, was Mishima's most frequently performed play. Called after a dance pavilion erected in the 1880s to show the Japanese élite's enthusiasm for Western-style dance parties, the play illustrates the aristocratic pretentiousness of the 1880s, material suited to Mishima's talent for extravagant romance and intrigue.

ment, one method it adopted to keep itself afloat commercially being to form the Workers' Theatre Council, Rōen, in co-operation with the Japanese Communist Party.

The Japanese drama of the 1960s has been described as 'post-*shingeki*', and a 'return to the gods': dramatists were not only moving away from a form of drama which was derived from the West but also attempting to 'return to the Japanese popular imagination . . . molded by . . . folk religion'. The main characteristic of post-*shingeki* drama is the metamorphosis of a character or characters into an archetypal, transhistorical figure. Thus, Akimoto Matsuyo's *Kaison the Priest of Hitachi* (*Hitachibō Kaison*) of 1965 shows the metamorphosis of a young man of the 1960s into a twelfth-century warrior called Kaison, himself an immortal reputed to wander around Japan doing penance for sins committed centuries before.

A major specific example illustrating the reaction against Western influence in the 'new drama' and return to Japanese models is the work of Suzuki Tadashi. In 1961 he and others founded the Free Stage (Jiyū butai), which was reorganized and renamed in 1966 as the Waseda Little Theatre (Waseda shō gekijō). The first great success came in 1970, when *On the Dramatic Passions II* (*Gekiteki naru mono o megutte II*) was produced. It is a collage of dramatic scenes ranging from kabuki to Beckett's *Waiting for Godot*. Shortly afterwards, he began exploring and experimenting with notions and techniques of traditional Japanese theatre, such as Noh and kabuki, with the aim of welding Western drama concepts into a Japanese paradigm.

While neither kabuki nor Noh has done particularly well since the Meiji Restoration, including the period since the Second World War, neither shows any sign at all of dying out and both will remain part of Japan's theatrical diet indefinitely. What exists on the Japanese stage is a mixture of traditional and modern forms, foreign and Japanese, which one authority believes 'points to a pluralism of Japanese culture which . . . is not sufficiently recognized'.

Korea

In Korea, the early years of the twentieth century saw the introduction of a form of opera based on *p'ansori* but with more than two singers, and Western-influenced forms of drama, in particular 'new school' (*shinp'a*) and, somewhat later, 'new drama' (*shingŭk*), which followed their Japanese counterparts quite closely. Among the 'new school' items were military plays in which patriotic soldiers were glorified as national heroes. However, these forms did very badly under Japanese rule, which suppressed any play with a nationalistic theme. In 1941 all existing professional theatre companies were organized into a single one, which was allowed to perform only those dramas pleasing to the Japanese rulers, especially those glorifying their military policy.

North Korean drama
A True Daughter of the Party. Set during the Korean war, this picture shows Kang Yon Ok, the true daughter of the title (centre) taking a mission from the head of a field hospital to escort wounded soldiers to hospital. Note the elaborate scenery and realistic costumes.

In the Republic of Korea (South Korea) the years since 1945 have seen a major revival of traditional forms as well as the new beginnings of a modern theatre. In 1950 the National Theatre was established, but less than two months later the Korean War broke out, severely hampering its work. The Shinhyŏp Troupe, a breakaway from the National Theatre, was very influential in the 1950s and into the 1960s. However, the number of dramas actually written by Koreans and concerning Korean problems was disappointingly small, most of the main productions being foreign, especially Western, plays. In 1962 Yu Ch'i-jin set up the Drama Centre, with the aim of reviving the fortunes of the Korean theatre. However, five out of six of its first productions were of foreign plays and audience response remained unenthusiastic. For these and other reasons, the Drama Centre failed within a year. Initiatives in developing a specifically Korean theatre have increased greatly since the 1970s, meeting with considerably more success.

In the Democratic People's Republic of Korea (North Korea) the main theatre is revolutionary opera set in Korea and using Korean dance and melody but mostly Western musical instruments and heavily influenced by Soviet propagandistic style. The content of all items features the struggle against the Japanese during the colonial period or the wonders of life under Kim Il-sung, who dominated the North Korean state from 1945 until his death in July 1994. The Kim personality cult looms very large in the revolutionary operas. On the other hand, there are still items set in the distant past, although the old mask-dance drama has never been revived. A typical revolutionary opera is *Sea of Blood*, which is set in the 1930s and concerns the struggle against the Japanese. The main character is a middle-aged woman who becomes revolutionized and assumes leadership of her fellow villagers, the climax being the seizure of a walled town by the revolutionaries from the Japanese. In North Korea the state controls not only the arts but the training of artists, including performers and directors. Professional artists enjoy good social services and a high social status, while ordinary people, especially women, are encouraged to take part in amateur performing arts and theatre.

SOUTH-EAST ASIAN THEATRES

LEON RUBIN

THE region known as South-East Asia is saturated with numerous and varied forms of theatre throughout the countries of Burma, Cambodia, Laos, Indonesia, Malaysia, the Philippines, Thailand, and Vietnam. These countries have long traditions of performance that can be traced back, in several cases, over many centuries. In recent years, research has made good progress in clarifying the tangled story, especially that by Faubion Bowers, James R. Brandon, and Jukka O. Miettinen. We now know that most styles of performance originated in India and, in some cases, China. Between the different countries of South-East Asia there are many similar and inherited traditions; there are sometimes styles that have travelled thousands of miles over a period of time and have been transformed and adapted to each culture *en route*. In every country of the region there is a surviving influence of the Indian epic *Ramayana* and often the *Mahabharata*, whose stories appear in forms of theatre presentation of dance-drama, puppet theatre, and classical dance. Unlike the text-based theatre traditions of the Western world, South-East Asian theatre is almost always a complex blend of dance, song, movement, and recited text. In many of these countries the languages do not possess a word that describes a purely spoken dramatic form. The earliest origins of theatre in every country of this region seem to have parallel roots in religious ritual and ceremony, related to the world of spirits in animistic practices. It is as though the birth of performance in the East was to fulfil the need of man to communicate with the spirits; performance was the bridge between the human world and the spirit world. The performers, often in trance, allowed themselves to be possessed by spirits, in order that the other members of the tribe, the audience, could ask the spirits for help in their human predicaments. The spirits who possess the performers, or shamans, eventually became characters, later portrayed by actors as the various genres developed. Indeed, perhaps most acting is, in fact, a form of possession by another spirit or character. The ritual movements, emphasized and encouraged by percussive music, common to trance rituals in many parts of the world, later became studied, controlled dance-forms.

These rituals also linked the earliest rulers of kingdoms to the spirit world, thus reinforcing their earthly authority. Court traditions of performance probably grew up from these animist rituals, carefully nurtured by the rulers to remind the people of their own divine link. This magical link seems to have survived in performance despite the arrival of Hinduism, Buddhism, and Islam. The healing rituals, always involving trance and possession, are directly related to the same origins and may have given birth to many more popular performance forms outside the court, as the healing was directly beneficial to anyone in need; it was not only for humans themselves, but for the rice crops on which they depended. These pre-Sanskrit rituals were probably incorporated into Sanskrit drama, which grew up from the first century AD. Masks and puppets, which are manifested in numerous forms throughout the region, were also always associated with magic and ritual, and many still are today. In many of these countries there is a pattern of human performers mimicking puppet performances and developing dance and dance-dramas in a similar style. Court or classical dance and dance-drama, masked dance-drama, folk-opera, and puppet theatre are the main genres that emerged later in the region, all coming originally from the pre-Sanskrit rituals.

Burma (Myanmar)

The origin of the present-day Burmese people dates back to the ninth century (the earlier inhabitants were the Pyus and Mons). The city of Pagan was the centre of the Pagan dynasty (1044–1287) and the focus for a flourishing period of performance. The work was probably a mixture of Mon, Pyu, and Burmese traditions. These traditions were effectively destroyed by the Mongol invasion of 1287, and Burmese culture only revived in the fifteenth century. In the sixteenth century the Burmese captured the Thai capital of Ayutthaya for the first time; their power increased and expanded over the next 150 years until Ayutthaya was once again taken in 1767. This time artists, dancers, and musicians were captured and taken back to Burma and, consequently, many traditions of performance borrowed and adapted. In particular, the *Ramakien*, the Thai adaptation of the *Ramayana*, and stories from the Javanese *Panji* plays entered the Burmese performance repertory. Thai classical dance movements were also adopted at this time. This next major change was when the British defeated and began to rule Burma in 1886 and badly damaged traditional performance culture. After Japanese occupation and the end of the Second World War, Burma became independent. In recent times it has been mainly closed to outsiders and, consequently, outside influences.

This history has left a legacy of diverse influences on Burmese theatre. The style known as *nat pwe* originates in animism and relates to the spirit world. A *nat pwe* performance is often still part of a religious ritual, originally involving a medium, usually a *nat kadau*, female medium, who is possessed by a spirit whilst in trance; the ritual involves dance, music, and recitation. There are thirty-seven major *nats* or spirits that are honoured in such performances. A typical performance would begin with the orchestra striking up to indicate the commencement of performance; then food offerings are placed by the *nat* statues. The shaman, dressed in a shawl indicating the *nat* to be communicated with, begins to dance, and dances him- or herself into a trance state, when the *nat* takes possession. Whilst dancing and writhing the shaman, in the voice of the *nat*, communicates thoughts to the audience. There is in modern times a more refined version of the dance that does not use the trance state. Another style is known as *Anyein pwe* and centres on a female performer and two clowns. A third surviving style, *nibhatkin*, has not animist but Buddhist origins. These plays depicted stories concerning the life of Buddha and the performances moved from one village or part of one town to another on carts and have parallels in Western mystery plays. Humorous interludes punctuated these performances and were known as *zat pwe*; the interludes depicted stories from the *Jataka*. Later, this developed into the classical dance-drama which has been the most popular form of theatre from the beginning of the nineteenth century until the present day, and being particularly common up to the middle of the eighteenth century. The clown character, *lubyet*, became especially loved and survived in his own right to the 1970s and beyond.

Court dance-dramas developed rapidly after the capture of Ayutthaya in 1767. Like Thai performances these divided into two basic types of male masked and female unmasked, *yein*, performances; the *Ramayana* remained the main source.

In addition to *zat pwe* and *yein*, a new style termed *pya zat* developed in the twentieth century; Western influences in lighting and scenic styles became intermingled with traditional approaches. The plots were based on new scripts by playwrights, often on very contemporary themes, and in the 1970s many star performers emerged.

The other form of performance important in Burmese culture is puppet theatre. *Yokhtre pwe*, or marionette

Performance of *Yarma zat* from Burma. The *zat*, or play, *Yarma* is an ancient one from the *Ramayana* telling the story of Prince Rama and his brother in search of a golden deer in the forest. As in many such stories there is a princess, left at home and kidnapped by the evil ogre *Dhatagiri*. After many battles, also involving *Hunan* the monkey, the princess is saved.

theatre, developed in the late eighteenth century soon after the flourishing of the court dance-dramas; it was devised to re-enact the sacred *Jataka* stories. In a typical performance the play could last all night. The opening sequences describe the birth and dramatic destruction of the universe twice, once by flood and once by fire. Then the world of the spirits takes over with the arrival of the *nats*; these various creatures and demons then dance. Next the court puppets dance exquisitely, on the bamboo stage, with complex hand, head, feet, and arm movements as the strings themselves dance up and down. Then the plot itself begins, loosely based on a *Jataka* or *Ramayana* story. The puppets are manipulated from high above with sometimes up to sixty strings for one marionette, in order to allow subtle dance-type movements of the whole body, whilst the performances are accompanied by an orchestra. Some of this tradition still survives in rare performances in temples. The movements of the puppets and their fixed smile have influenced Burmese classical dance style and still have magical and spirit associations.

Cambodia

The first references to performance in Cambodia go back to AD 243, with reference to musicians from Funan, an early kingdom in the country. It is thought that the origins of dance and theatre pre-date even this period. There are various references in the sixth century to temple dancers, suggesting influences from India. Cambodian or, more accurately, Khymer dance flourished during the Angkor period (AD 802–1431). The founder of the Khymer kingdom, Jayavarman II, came from Java in AD 802, bringing with him Javanese dance, *bedaya*, and masked dance, *topeng*, which themselves had been formed, probably, with Indian influence. The dance traditions of the Khymer court were related to court life and ceremony, performed by hundreds, and at one point in the late twelfth century over 3,000, dancers. As in other countries in the region dance-drama also evolved as an important part of performance tradition, again based on the *Ramayana*, here known as *Ramker*.

In 1431 Angkor fell to the Thais and the court-dancers, and orchestra were taken to Ayutthaya, where, according to some scholars, Thai classical dance was developed, based on Khymer style and traditions. From that point in history, Cambodian traditions were weakened and revitalized from time to time, in the nineteenth century and again in the 1980s, ironically by the Thai classical performers. Therefore, it is difficult to describe accurately which traditions came from each of the two key sources. The Cambodian performance of *lakhon khol* for example is very similar to the Thai *khon*, masked dance-drama, based on the *Ramayana*. This confusion is added to by the lack of documentary evidence between the fifteenth and nineteenth centuries.

The other main performance style, based on the *Ramayana*, is *lakhon krabach boran*, the female court dance cycle. In these dance-dramas action is mimed whilst the chorus sings, and music accompanies. The orchestra has strong percussive sounds using various drums, *samphor* and *skor thom*, cymbals, *ching*, and gong-type instruments, *kong thom* and *kong touch*.

The poverty of the court during the late nineteenth century caused a diminishing of the court arts, and the first troupe to be formed outside the court was in the 1930s, during the French protectorate. In the 1950s and 1960s dance and dance-drama briefly flourished again with encouragement from the court; however, following the Khymer Rouge take-over in 1975, most performers were murdered or disappeared abroad. In the 1980s retraining with Thai help has taken place.

There are two other performance traditions in Cambodia, folk or village theatre and puppet theatre. *Ayay* is a form of story-telling by two groups of male and female singers accompanied by a musician. The puppet tradition, as in Thailand, is for shadow puppets known as *nang kalun, ayang* or *nang shek thom*. The Thai tradition itself probably derives from the Indonesian *wayang kulit*.

Indonesia

Throughout the numerous islands of Indonesia are found a diverse and rich collection of performances of dance, dance-dramas, and puppets, unrivalled anywhere else in the region. The main focus of these performances is, and was historically, on Java and Bali, although several forms evolved on other islands.

At the root of most styles of performance are the Indian classical dance traditions and, prior to that, spirit-trance rituals. Puppet theatre is thought to be the earliest performance form and is closely related to communication with the spirit world. The *Ramayana* is the main source for stories for puppets and dance, and other tales come from the *Mahabharata* and the Javanese *Panji* stories.

Trance is particularly important in Indonesia and manifests itself in varying degrees throughout the performance repertory. In Bali *sanghyang dedari* is performed specifically when help is needed by the spirits; in Java a doll is used in a trance dancing ritual to help crop growth. Masks extend the concept of possession and performers believe themselves possessed as the mask takes over their identity.

Recorded information of the history of theatre there is scarce and oral passing-on of information suspect; however, certain historical changes have left clear markers of the journey of performance from the indigenous early culture to the present day. From the seventeenth century onwards it was the Indian, Hindu, and the Malaysian Buddhist kingdoms that dominated the islands. Influence from these two powerful sources competed and blended with indigenous culture. By the fourteenth century Majapahit rule was dominant, until the fifteenth century, when Islam began to penetrate this region. Java was Islamized, but Bali kept to the old pre-Islamic Javanese, largely Hindu culture. It is claimed that the Islamic invaders used shadow puppets to communicate the ideas of Islam, even though the stories were based on the *Mahabharata*. Around the beginning of the seventeenth century European influence came with the Dutch and, later, the British, and by 1830 the Dutch had taken power throughout Java and had included Bali by 1906. By the 1930s Western cultural influences had become strong in Bali, in the form of settlers and tourism; then, after the Japanese occupation of the Second World War, came the war of independence against the Dutch. The diverse invaders and their cultures have all left deep impressions on performance traditions.

The oldest structured form in Indonesia is *wayang kulit*, shadow puppets. Its origins are, probably, pre-Sanskrit although the *Ramayana* and *Mahabharata* are important story sources. It is thought to have spread from Java to Bali in the tenth century, alongside other Indian court cultural influences. In Java it is believed that the shape and style of the puppets were transformed into their present almost abstract forms due to early Islamic influence that forbade the creation of human images. In Bali, however, the older style survived and still exhibits more specific realistic elements, in particular in relation to facial characteristics. In Java other performance forms have distinct derivation from *wayang kulit*, whereas in Bali the puppets are, on the whole, seen as a completely separate genre. The puppeteer, *dalang*, manipulates all the characters, often forty to fifty, aided by assistants, and he also functions as narrator. The story is, as in many performance forms in Indonesia, accompanied by a gamelan orchestra and there are still many ritual elements before and during the performance.

Although usually performed at night, daytime presentations are also possible without the shadow screen. A typical *wayang kulit* performance will begin late at night and continue to the early hours of the morning, at sunrise. The performance commences with a ritualistic narration and the 'tree of life' puppet dance symbolizing the creation of the world. Then the action begins with a scene in the courts of rulers in conflict. The atmosphere among the audience is quiet, at the outset, as the plot is established, seriously, in a magical atmosphere created by the moving shadows and lights. After various scenes of coming and going of various characters, some scenes serious and some verging on slapstick, the hero enters into a heroic battle. The villain Cakil is defeated and the great battle is fought splendidly, brilliantly manipulated by the now tired *dalang*. All ends happily and the tree puppet goes back to his resting place as the audience, weary and happy, after perhaps ten hours of watching, go home to sleep.

There are also other puppet theatre styles: *wayang klitik*, which uses two-dimensional wooden puppets, and *wayang golek*, which uses three-dimensional wooden puppets and dates from the sixteenth century.

Closely related to *wayang kulit* is Javanese *wayang wong*, human puppet; it is an early dance-drama form in which dancers perform the same stories. The stylized gestures reflected shadow puppet movements; sometimes masks were worn and a gamelan orchestra accompanied. This form evolved in Bali in the seventeenth century at the court. There are a large group of dancers, who play numerous human and animal characters, in particular monkeys. There was also an active *wayang wong* tradition in Java associated with important court ceremonies, the last of which was performed in 1939. In Java it is perhaps the most important classical dance-drama form and has spawned many related forms.

Topeng is another important mask-drama developed at the same time in Bali. *Topeng wali* or *topeng pajegan* is the oldest form and is enacted by a single performer using several masks in a complex ritual performed for special events. The dancer places his basket of masks before the gamelan and enacts each character in turn as each of the five masks is worn. The final mask used is particularly sacred and known as *Sidha Karya*, and with this grotesquely smiling mask on the Sanskrit prayers are recited. A more recent variation on this is the *topeng panca*, which utilizes five performers and was developed in the nineteenth century. In Java *wayang topeng* goes back to the eleventh century and evolved in an elaborate court style as well as in a popular version; it flourished until the eighteenth century.

Another early court dance-drama from Bali is *gambuh*, which probably dates from the sixteenth century and was performed for special events. The basic movement style has influenced all the other dance-forms on Bali and is part of the *bebali* dance tradition; the main story source was the *Panji* stories. Full-scale performances ended in 1906, although shorter adaptations still survive. *Legong* is another form developed at the beginning of the nineteenth century and is also usually connected to the *Panji* plays; it contains elements drawn from *gambuh* and *sanghyang dedari* and early sacred trance dance style. *Legong* is performed by three young girls, who perform until puberty. The gamelan orchestra accompanies the dance and also provides singers of the narration.

The other important use of mask in Bali occurs in the performing of the *Barong*, a mythical lion-like creature with ritual, spiritual associations. In the drama *Calonarang* he appears alongside Rangda the witch. The origins of the *Barong* are thought to be connected to the Chinese lion dance. The word *Barong* refers to the huge body mask itself, in which

Barong performance in Bali. The *Barong* is a mythical creature who is the ruler of the forest. In this performance of the drama *Calonarang* he fights with Rangda the witch. During the action the *kris* dance occurs, in which performers stab themselves with knives whilst in trance under the power of Rangda.

two men perform with carefully co-ordinated and complex movements. The front dancer manipulates the head, including moving jaws, while the dancer at the rear moves the rapidly swinging tail. The overall appearance is of an anthropomorphic creature fierce in its protection of good against evil. Its elaborate movements contrast with those of the evil Rangda, who moves little and stares through her horrifying bulbous eyes, as she casts her black spells to the delight and discomfort of the noisy audience. The *Barong* also appears in other celebrations and rituals in village life.

Laos

It is difficult to discuss the main performance traditions of Laos without linking them closely to those of Thailand. The court tradition is based on the *Ramayana* and *Jataka* stories and the styles of presentation are very similar to those of Thailand. It is generally believed that these were imported to Laos, partly from the Khymer court in the fourteenth century, but mainly from Thailand itself in following years. The main court dances from Thailand, *lakhon fai nai, khon,* and *nang yai* were adopted, with some minor variations, by Laos.

The most popular village drama is centred on a style known as *mor lam*, which is found in Laos and in north-east Thailand, whose peoples are related. The earliest form of *mor lam* probably had origins in China and involved a single story-

Mor lam performance in Laos. The costumes show signs of influence from the bright, colourful designs of *likay*, which originated in the south of Thailand. During the performance, the actors sing the text to the accompaniment of a small orchestra, featuring the *khaen*, usually upstage of the action.

teller reciting tales from the *Jataka* or other myths and legends. The male singer/reciter was accompanied by a wind instrument made of bamboo known as a *khaen*; the story was improvised around a basic rhythm, and the verses followed a structured pattern; a simple scarf was worn in different ways to indicate changes of character. Some performers of this basic style still continue today, but variations of *mor lam* are now frequently seen in Laos and north-east Thailand. A male/female love duet version developed later known as *lum glorn*. More recently, since the Second World War, *mor lam plun*, *mor lam luong* and *mor lam moo* have developed, versions which involve a full company of performers acting out, in mainly improvised song and dance, *Jataka* stories, Thai myths, or new plays on contemporary themes. The *khaen* is now supported by other instruments, in particular modern drums, and since the 1980s, particularly in Thailand, there has been a strong blending of traditional *mor lam* music and dance with modern pop music. One of the greatest qualities of *mor lam* has been its enduring ability to adapt and change its form, reflecting social changes, whilst simultaneously preserving the older ways of presentation. Since the 1920s the style of performance, in particular through the costumes, has been influenced by the popular opera performances from Thailand known as *likay*. There is also another variation that is performed in order to cure the sick. Old women in contact with the spirit world chant the verses as the curing ceremony is performed, and this type of *mor lam* has roots going back to early, animist times.

Mor lam singing and *khaen* playing are also part of a Lao shadow puppet performance style known as *nang daloong*. Adapted from Thai forms, this popular theatre depicts tales from Laos and Thailand; its origins date back to the 1920s.

Malaysia

The Malaysian performance traditions exhibit influences from all the surrounding countries of the region. The complex ethnic mix of the population adds to the direct geographic influences of nearby states: Thai, Arab, Indian, Chinese, and Indonesian influences are found throughout the culture of Malaysia, in addition to the indigenous Malay culture itself. Islamic, Buddhist, and Hindu performance cultures are all present in the dance-drama. One of the earliest forms of performance in Malaysia is the recitation of epic stories. There are several variations of style, some involving the performer accompanying the verses with instrumental playing, and the stories are often derived from the *Ramayana* and Malay stories, *hikayat*. There is also a form of courting performance, which has parallels to those of Laos and Thailand and involves a sung dialogue between male and female, and there is a sung performance style to cure the sick. The best known of these trance performances is *main puteri*. As with the Laotian version of curative performance the singers are possessed, aided by the musical accompaniment, in this case the instrument known as *sebat* or lute.

The relationship between performers and the spirits affects much of the theatre in Malaysia, including puppet theatre and various types of dance. In *ulek bandul* the dancer contacts the spirit of rice; in the better-known dance *kuda kepang*, horse dance, extraordinary feats are undertaken whilst the performers are in trance, and in the *silat* dance flesh is pierced with sharp objects with no apparent pain to the performers.

The main dance and drama traditions derive from pre-Islamic sources, particularly those from Java and India. Probably the most important influence was the *Sriwijaya* kingdom of Indonesia, which ruled over the region for many centuries from the seventh century onwards. The Islamic traditions added to these when the Sultanate of Malaysia became a dominant influence in the fifteenth century; then, after 1511, there were European influences following the Portuguese victory in Malacca. These mixed origins are clearly seen in puppet theatre.

In shadow puppet theatre the Javanese origins are strong in *wayang malayu*. In *wayang jawa* the puppets are stylized after the traditional Javanese manner, and the stories are normally based, as in Java, on the *Ramayana* and the *Panji* plays, although Malay is the language used for performance. The third, popular shadow puppet tradition is *wayang siam*, and is linked to Thai culture. This form comes from the Patani region, now within Thai borders, probably dating from as far back as the twelfth century, and the style of performance is still similar to that of *nang talung*, the Thai equivalent. Only one *dalang*, puppet operator, controls all the puppets; a gamelan orchestra accompanies. The opening rituals contain text that draws on Javanese, Malay, and Thai words and has spirit associations.

The two main dance or dance-drama forms in Malaysia are *Manora* and *mak yong*. The extraordinary and beautiful form Manora also originates in Thailand, where it is known as *lakhon nora* and is part of the *Jataka*. The *mak yong* female dance-drama has its origins in Kelantan, although the date of origin is not known: it probably began as an ancient folk-drama and gradually became a court performance; in recent times it has again acquired a folk form. Male and female principal roles are played by women in a collection of plays concerning princes and princesses; the dialogue is improvised, accompanied by an orchestra. The performances are still thought to have magical associations, the opening rituals, *buka pangung*, emphasising these connections. These dance-dramas are closely connected to female dance-dramas and dance in the Thai and Khymer traditions.

There is a popular theatre form known as *bangsawan* that has its origins in the 1870s, deriving from an Indian popular theatre that toured to Penang in 1875. The Parsi touring company played performances from many sources including Indian, Arabian, and Western plays. The first *bangsawan* company was formed in 1885 in Penang: the performers were male and female and the theatres usually proscenium. The settings were colourful and spectacular and the style a mixture of improvised spoken and sung text with eclectic musical accompaniment. The companies toured widely, but popularity faded during the Second World War and never revived strongly thereafter.

Since the 1950s a new popular theatre form has developed known as *Sandiwara*. The style was rooted in Western traditions of fully written, rather than improvised scripts. New plays have been written by Malaysian dramatists, largely on contemporary themes.

Philippines

The dominant theatre forms in the Philippines are drawn largely from Christian sources, rather than the Hindu, Buddhist, and Islamic traditions. These influences date back to the Spanish colonization of the Philippines from the seventeenth to the twentieth centuries. The early plays concern battles and conflicts between the Christians and the Muslims and are known as *komedya*. The style was distinctly Spanish, but some academics in the region detect pre-Christian influences in the accounts of productions.

There is also evidence of performance forms prior to Spanish colonization and

Scene from *senakulo* Holy Week performance in the Philippines. The photo is from the Moriones Festival of Marinduque during a Holy Week presentation depicting the apocryphal story of Longinus, the Roman centurion who pierced the left side of Jesus Christ, crucified on Calvary. The blood which spurted from Christ's wound restored the sight of his impaired eye. A huge audience watches the open-air performance.

some elements of those still survive, mainly in dance performances involving trance rituals and possession. Epic stories are also known to have been sung before Christian colonization and the *tutol* stories of Muslim peoples still survive. Muslim dances are accompanied by the *kulintang* and tell stories of princesses and queens.

The Christian-dominated theatre forms also included extravagant presentations at key festival times: *senakulo* for Lent and *panunuluyan* for Christmas. They were probably used by the early Spanish Church to promote Christian teachings and help in conversions. The best-known style of performance was the *senakulo* that was performed in Holy Week and dealt with the theme of Christ on the cross. The staging was spectacular, involving large-scale processions and tableaux, and the style of performance was grand, studied, and declamatory, interspersed with occasional comic interludes from minor characters. A dancing, seductive Mary Magdalene added a humorous element in an otherwise powerful, serious performance. The genre has parallels in Western mystery cycle plays, although the performance style included elements of pre-Christian indigenous performance.

From the beginning of the nineteenth century indoor theatres appeared and the Spanish *zarzuela* form became dominant; then, after the American take-over of the Philippines in 1898, other Western styles were introduced and text-based plays, usually involving songs, were the norm. By the 1920s vaudeville had arrived and been adapted locally. After the Second World War English became used more frequently in the theatre. Since the 1950s musicals have become particularly popular.

Thailand

It is perhaps invidious to compare qualitatively the theatre traditions of South-East Asia but Thailand, alongside Indonesia, probably offers the richest selection. Many modern Thai dances and ritual performances are closely related to the life of the temples, and consequently the spirit world. There is a mixture of animist, Buddhist, and Hindu influences on these performances and the overall effect suggests a strong link to an ancient history of ritual.

In the north-east, *mor lam* is still the most popular form (see Laos), and has connections back to ritual healing ceremonies. The earliest form in the south is *lakhon nora* or *Manora* (see Malaysia); this is connected to the early Malay kingdom, Ligor. (The word *lakhon* itself probably has a Javanese origin.) The style of performance has heavy emphasis on finger movements with long artificial fingernails and the poses and movements of the dancers suggest an ancient Indian influence. Originally, the performances involved three male dancers, but later female dancers were included. The story derives from the *Jataka* and focuses on the three characters of the prince, originally the Buddha, the princess, and the clown, who takes on other characters and creatures. The story-telling is accompanied by a small percussion and reed orchestra. The atmosphere conjured up during a performance is magical even though at times the mood changes to rough comedy. The dancers, decked out in colourful costumes of glass beads, a bird's tail, and golden crown, move elegantly with acrobatic poise, angular movements, and sliding feet. The story of the *Manora*, the princess who is half-bird and half-human, unfolds to the inevitable end when she flies away to freedom and is joined by her prince, who has suffered in her pursuit. There are strong spirit associations with *lakhon nora* and performances contain many ritual elements. Some scholars believe that this form is in fact that from which all other *lakhon* types evolved.

There are numerous other *lakhon* styles that developed over many centuries. *Lakhon nok* is an early popular example, probably related to *lakhon nora*; its style was bawdy and energetic with emphasis on humour. It, too, was performed only by males until the nineteenth century; it has faded in popularity during the twentieth century.

The court form of *lakhon*, which may also be related to l*akhon nora*, is *lakhon nai*. This is, perhaps, the best-known dance form in the classical Thai tradition. Some scholars believe that its main origins date back to 1431, when the Thais conquered Angkor. Others, however, believe that it greatly pre-dates this period and was only modified by the Khymer court style. Either way, Khymer and Thai classical court dance have origins in Indian classical dance, in spite of significant differences in music, rhythm, and gesture. In turn, Thai classical dance was imported to Burma after the capture of Ayutthaya by the Burmese in 1767 (see Burma). The style of performance of *lakhon nai* again emphasizes delicate hand and arm movements. The accompaniment is by an orchestra, *piphat*, dominated by percussive instruments, and dialogue is mainly sung by a chorus off-stage. The performers were traditionally all female, apart from the clowns.

The male court dance-drama tradition is known as *khon* and has probable origins in early Khymer *khol* dramas based on the *Ramayana*, although the first record of performance in the Thai court dates from 1515. This masked dance-drama has a performance style of acrobatic poses and sideways profiled movement that suggests origins in shadow puppet performances, *nang yai*. As with *lakhon nai* the performers are accompanied by a *piphat* orchestra and a narrator chants or sings to the mimetic dance movements.

The *nang yai* shadow puppet performances also have a long history dating back, at least, to the take-over of 1431. The Khymer shadow puppet tradition itself is connected back to the shadow puppets of Indonesia, although there are significant differences in the puppets themselves and the performance styles. *Nang yai* puppets are much larger than in parallel shadow puppet traditions, approximately 4 or 5 feet high. They are manipulated in silhouette against a high, wide screen 30 feet long and 12 feet high. The stories, based on the *Ramakien*, the Thai adapted version of the *Ramayana*, are related by two narrators, *khon pak*, to the accompaniment of a *piphat* orchestra. The puppet-eers dance with their puppets, emerging to the view of the audience. The popularity of *nang yai* continued until the 1940s but it has rarely been seen since.

In addition to the court traditions, long sponsored and nurtured by the royal family of Thailand, there is also a tradition of popular theatre. As well as *mor lam* from the north-east there is *likay* from the south of Thailand. The earliest performances of this popular operatic form were probably influenced by Islamic singing known as *dikay* and it displays elements of Indian and Malay culture. The brightly coloured pantomime style of costume and basic acting style make it easily accessible to all types of popular audience. It has stayed popular, especially in Bangkok, and the painted scenery and glittery costumes have influenced modern performance styles of *mor lam* in the north-east.

Likay performance in Thailand. The performance takes place on a small stage. The scenery is usually painted on to screens across the upstage performance area. The lurid colours of the costume are similar to those on the painted screens. The performance includes mainly singing with some moments of action.

Vietnam

Vietnam borders on China, Laos, and Cambodia, but throughout its history has mainly been influenced by China and Chinese culture. Its predominant performance forms reflect that influence, unlike many countries of the region, which owe a heavy cultural debt to India. The Hindu kingdom of Champa became part of Vietnam in the fourteenth century, thus producing one of the few Indian roots of Vietnamese performing traditions, mainly through music. It is Confucianism, Taoism, and Mahayarian Buddhism that have left the strongest marks on Vietnamese culture and society; but from the nineteenth century onwards Western influence has been strong, particularly under French rule 1862–1945.

Chinese opera is the most important source for Vietnamese court performance. It is generally thought that this began in 1285 when Chinese opera performers were captured and their craft taught to locals. The Vietnamese version is called *hat boi,* which developed mainly for the court until the seventeenth century and gradually became a more popular form. *Hat boi* has many similarities to Chinese opera, with the performers singing or speaking the text in a highly stylized manner with mimetic movement and character-typing in the Chinese tradition. However, the music is distinctive to Vietnam, combining Chinese traditions with indigenous music that has some origins in the music from Champa. The costumes and elaborate make-up of *hat boi* are generally similar to those of Chinese opera, but have local characteristics and variations; the stories also deal with Vietnamese tales as well as those of Chinese origin. The performances were popular and continued with the court company, *phuong nha tro*, until 1946. Modern commercial performances of *hat boi* have relatively small audiences.

Within the court performance traditions were the dances *van vu* (civilian dance) and *vo vu* (military dance), which originated alongside *hat boi* from Chinese ceremonial tradition. These large-scale dance performances still took place well into the twentieth century and faded in popularity at the same time as *hat boi*.

The most popular theatre form in Vietnam emerged during the 1920s; it is called *cai luong* and owed some of its development to *hat boi*. It is a type of musical theatre that centres around a number of songs, between which dialogue is improvised, following a basic story-line. The songs are mainly melodic songs from South Vietnam, well known to the audiences. *Tu dai* ballads are at the centre of a *cai luong* performance: one song in particular of the 1920s, 'Vong Co', was, and still is, featured in most performances. Productions mixed elements of *hat boi*, traditional Chinese opera costumes, and modern dress; scenic devices were used reflecting Western influences. In the 1930s particularly extravagant techniques were imported from various cultures, including use of foreign popular music. *Cai luong* was very popular until the Vietnam War in the 1960s, but is still found in Saigon and Hanoi today.

Almost alone in South-East Asia, Vietnam also developed a spoken theatre form, *kich noi*. It dates back to the influence of the staging of Molière's *The Miser* in translation in 1907. Many European plays were performed in translation and in 1921 the first Vietnamese play was written and performed (*A Cup of Poison* by Vu Dinh Long); many theatre companies for spoken drama have developed since then.

There is also a tradition of puppet theatre in Vietnam. The most interesting form is that of water puppets, *muá rôi nuoc*, for which the first reference dates from AD 1121; but it is believed to have much earlier origins. There are early references to water puppets in China, but Vietnam may have

Hat boi performance from Vietnam. This operatic performance style has origins in Chinese opera, through the costumes and make-up; in performance, this is evident in the character-typing and vocal delivery. However, the stories depicted are specifically related to Vietnamese history. An orchestra accompanies the action throughout.

been the point of first origin. The puppets are controlled underwater by long rods operated by puppeteers hidden behind a screen; an orchestra at the side of the stage accompanies the narrators of stories taken from Vietnamese history and contemporary domestic village life. The puppets include human and animal characters often performing with fireworks and fire coming from their mouths; creatures shoot up out of the water as battles are enacted. These extraordinary, unique, and delightful performances take place in village ponds throughout North Vietnam. There has been a recent revival of interest in these performances, largely due to successful foreign touring, particularly in Paris.

The other form of village theatre performance is *hat cheo*, which dates back to the first century AD; its origins are connected to humorous celebrations at harvest time. The performances involve music, dance, song, and gesture, and originally took place next to temples. The clown figure was dominant. *Hat cheo* remained popular until the 1940s but has significantly declined since, although, since the 1980s it has made a gradual come-back with the arrival of video culture in Vietnam.

13

THEATRE SINCE 1970

JOHN RUSSELL BROWN

FOR most people working in established and publicly subsidized theatres, the years after 1970 were a time for cut-backs, expedience, and survival. Many old ways of producing plays were proving too expensive. Set and costumes, together with suitable cast and adequate rehearsals, cost far more money than could be recouped during a run of three or four weeks—always supposing sufficient numbers of the public could get themselves from their homes to congested city-centres, pay baby-sitters, obtain something to eat, and be bright enough for an increasingly pressurized job the next day. In British regional theatres productions became fewer, casts smaller, and sets less ambitious. The National Theatre in London, trying to live up to expectations in its new buildings on the South Bank, reorganized its company three times during Sir Peter Hall's directorship (1973–88) and closed one of its three theatres for a season. The Royal Shakespeare Company shut down its entire London operation for months in 1990–1. There was talk of amalgamating the two companies.

Commercial theatres felt the same pressures. In the West End of London and other metropolitan entertainment districts, solvent impresarios, plays with more than a handful of characters, and dramatists commanding a wide public following became ever more scarce. By casting leading actors from popular films or television series a producer might reduce the risk of loss, but could not ensure the extended runs with which personal fortunes had previously been built and audiences assured of a good evening out.

In January 1986, years after the crisis had first been noticed, the Arts Council of Great Britain commissioned an 'Enquiry into Professional Theatre in England', ordering it to report with all possible speed by September of the same year. Huge

A film star on Broadway. David Mamet's *American Buffalo* (1975–7) ran on Broadway when it returned to New York with Al Pacino in the lead as 'Teach'; the production had originated at Long Wharf Theatre, New Haven, Connecticut (October 1980).

problems became evident. The slice of the subsidy 'cake' given to the two national companies in London had increased from 30 per cent in 1970–1 to 47 per cent in 1985–6. This enabled them to maintain programmes good enough to play regularly to 80 per cent capacity or more, but theatres in the regions, with subsidies reduced proportionately, seldom achieved more than 61 or 62 per cent, and often very much less. Statistics like these led to the demand for another £13,000,000 a year to be injected into subsidized theatres, or for one of the national companies to be closed. Neither solution was adopted and the crisis continued.

Elsewhere in Europe, subsidies were often far more generous, but in the 1980s a similar malaise began to be apparent in shorter seasons, fewer productions, and fewer performances. In the United States of America generous sponsorship from large corporations and wealthy individuals cushioned regional theatres against discomfort, and successive governments were content to endorse this leadership by making such gifts tax-deductible. But an underlying anxiety became very noticeable. In 1985 the Trustees of the Theatre Communications Group called for an inquiry. In their view, non-profit theatre was struggling 'to ensure not only its institutional survival, but its future as a vital art form'. Something had to be done to reduce the 'artistic deficit'—a term invented to explain how economic priorities had begun 'to take precedence over aesthetic concerns. . . . The gap between what had been achieved and the potential for future achievement had never been greater.' After discussions conducted across the country, *The Artistic Home* was published in 1988 and, predictably enough, it called for more money and fatter organizations: theatre directors should not be overworked; programming should be more flexible; and, in the bureaucratic language of such organizations, 'theatres and artists alike need to find creative ways to address the chronic undercompensation in the field'. By the 1990s subsidized theatres throughout the world, and the smaller commercial theatres, had followed those in Britain and were in a 'hold position', clinging with more or less desperation to production processes and organizations that had worked well enough in earlier and braver days.

Meanwhile another sector of theatre was enjoying a boom, as spectacular musicals settled into the larger commercial houses and played nightly to thousands for years on end. Productions, such as *Jesus Christ Superstar* (1972), *Cats* (1981), *Les Misérables* (1985), *The Phantom of the Opera* (1986), or *Miss Saigon* (1989), started life in London and were then reproduced in major cities around the world, making huge fortunes for producers, authors, composers, directors, designers. Actors proved not so lucky, since producers found that they could recast major roles without appreciable effect on the box office. This new kind of entertainment took advantage of the latest developments in technology and in publicity and sales management. Earlier forms of musical were largely displaced; they had usually been American in origin and depended on star performers, dance-routines, and songs that would subsequently be sung, recorded, and played everywhere. *Chorus Line* (1975–90) was the last big hit of that kind, but lacked a great star, like Ethel Merman in *Gypsy* (1959) or Topol in *Fiddler on the Roof* (1964), and supplied few ear-catching songs. From the 1960s onwards, popular singers, song-writers, and composers no longer worked for theatre; television, videos, and live concerts in huge stadiums were where money and fame could most readily be won.

So the 'mega-musical' was born. Sensational spectacle, using mobile scenery, computerized lighting, 'sound-enhancement', and well-drilled actors, overwhelmed audiences with crashing chandeliers, helicopters flying on to the stage, whole cities going up in flames. *Blitz*, designed by Sean Kenny in London in 1962, was perhaps the beginning of this mechanized vogue; but it was micro-technology that delivered its full power, so that by the 1980s and 1990s it dominated commercial theatre.

The New York staging of *The Phantom of the Opera* by Andrew Lloyd Webber meant increasing the height of the proscenium frame of the Majestic Theatre by

6 feet to 32 feet. New steel beams were installed for the theatre's grid and roof, the stage was extended backwards, and the basement was excavated a further 7 feet to accommodate machinery. Then audiences could be shown the gas-lit auditorium of the Paris Opera House in the 1880s; and the scene could move to its rooftops and grand staircase, to a graveyard and the Phantom's subterranean lair. The show-deck had ninety-six trapdoors, through which scenery could emerge and the Phantom make his mysterious entrances and exits. More scenery tracked in from both sides, and a 'travelator'—a 36-foot flexible ramp, suspended between two 25-foot towers—tracked up and down the entire depth of the stage. A gondola floating on an underground lake was radio controlled. A 43- by 29-foot portcullis was constructed of aluminium, but still weighed more than a ton. For moving all this hardware accurately, automation was essential, and for lighting its dizzying vistas instruments of computerized finesse. Once constructed, of course, and skilfully manned, the whole complex machine could be operated tirelessly and over 1,600 people thrill to the show's melodramatic excitements night after night. In some ways this was a return to Victorian scenic spectacle, but now managed with amazing fluidity, visually and aurally. As intended, it proved irresistible.

These changes came during a recession in England and America. Between 1970 and mid-1992, the dollar and pound had both fallen by over 65 per cent against the D-mark and yen, and wholesale prices had risen twice as fast as in Japan and Germany. At a time when unemployment was threatening new sections of society, powerfully effective scenic fantasy, bizarre characters and plots, and sustained musical and other sound input offered sure returns on the audience's money by transporting them out of everyday concerns. *Les Misérables* was advertised in the 1990s as 'The Show of All Shows' and 'The Musical Sensation'. 'Fight to get a ticket,' the public was told. Older musicals were now revived in newly sumptuous and irresistible productions, and prices rose fast: in 1970 the best seats for musicals in New York City were around $10, in 1980 $22.50, and in 1990 $55—considerably more if obtained at short notice through a ticket agency or other third party.

The mega-musical's success was not lost on directors working in other kinds of theatre. Resident companies stretched their resources to copy its sensationalism. The Royal Shakespeare Company filled the stages at Stratford and London with mobile scenery, lighting it in fantastic ways—sweet, sinister, or blatantly impressive at choice. By the 1980s a 'sound designer', as well as a composer, was named among the production credits. Possibly the influence was the other way around, for Trevor Nunn, Artistic Director at the RSC in the 1970s, went on to have a notable share in the development of the musical; he and John Napier, his designer, may have learnt the necessary skills in the classical repertoire, having to devise ways to augment the eye-catching resources of a dwindling cast and boost its vocal delivery. By the late 1970s, extended foreign tours of Stratford produc-

tions were accompanied by vast pantechnicons filled with custom-built apparatus that could deliver a shattering visual impact. Many hours would be spent at each location adjusting the complicated, computerized lighting plot; in Paris in 1979 eight different lighting boards had to be used simultaneously for Terry Hands's production of *Coriolanus*. More time was needed to ensure correct operation of moving gantries and three-dimensional scenic units.

Other companies also found the means to profit from staging devices that could overwhelm audiences regardless of whether they could understand what was spoken on stage. 1991 saw a British tour of a Japanese *King Lear* played on a brilliant red stage, an example of 'total theatre' that included explosive percussion and sensational lighting. In Germany, where bigger subsidies were provided, hugely impressive stagings of Shakespeare had been developed, often in the same theatre buildings as lavish opera productions. Indeed, throughout Europe, without the subtleties of the original dialogue to hold attention, expressionist and social realist directors had already used Shakespeare's plays as opportunities for making large-scale statements about their own worlds. The effect of these productions was now boosted by the full resources of new technology. Yury Lyubimov's *Hamlet*

Spectacle in musical theatre. On metal scaffolding in strong cross-lighting and stage-smoke, revolutionaries sing at the barricades in the sensational *Les Misérables*, which played to capacity audiences in London from 1985 into the 1990s.

(1971) had a star central performance by Vladimir Vysotsky, in his own right a popular protest-poet and singer, but the production swept the play's narrative along by using a huge curtain that was driven across the stage on a multi-directional tracking system. In France, in the 1980s, alliances between strong directors and their designers (or scenographers, as they were called by now) brought about 'étonnantes' and 'formidables' productions of Molière, Marivaux, Racine, and (again) Shakespeare: Roger Planchon with Ezio Frigerio, Patrice Chéreau with Richard Peduzzi, Antoine Vitez with Yannis Kokkos. With such support the classics dominated each season and comparatively few new plays were produced.

Technical advance was also seized upon by small companies which could operate cheaply and so spend months or years perfecting the staging of a show. Sensing that everyone's perception and thought-processes were changing under the influence of the electronic media and in response to the pace and pressures of everyday life, independent theatre directors sought to free themselves from fixed theatrical habits and so create small-scale productions expressing the new sensibility. In New York City in the late 1960s and early 1970s computerized control over light and sound came to be used with an almost messianic sense of mission. In number of performances or size of audiences, the effect of these experiments was small; but for artists of that time, the future of theatre rode with them.

Robert Wilson trained as a painter, but became impatient with what could be expressed on canvas; he needed three-dimensional space, solid objects, people, and the possibility of manipulating sequence and simultaneity. By 1970 he had gathered a small company with whom he created theatre shows presenting carefully chosen and sustained images, sometimes accompanied by a spoken text, music, or sounds, and sometimes performed in silence. Actors or objects might move very, very slowly. Light might be severely limited or be directed from an unlikely direction; or it might suffuse the entire stage space, as if from nowhere. Every detail of what was shown on stage was deliberate and self-sufficient. No attempt was made to interact with the audience, whose members were left free to pay attention or not, to day-dream or make their own

Facing: charming, spacious, and studied presentation of Shakespeare. Barry Kyle's production of *Love's Labour's Lost* for the Royal Shakespeare Company played at their London theatre in the Barbican during 1985; the set was by Bob Crowley.

Hamlet in Soviet Russia. Backed by a vast curtain, constantly changing the emphasis of space on stage, Vladimir Vysotsky's Hamlet stands at the margins of power in Yury Lyubimov's 1971 production at Moscow's Taganka Theatre.

more intellectual responses. For Peter Brook, Wilson's work had an 'indefinably potent fascination . . . [allowing us] to see that a continuous flow of images is a language'. Control of light was crucial to these shows: 'for me,' said Wilson, 'light is the essential element in the theatre.' Complex and precise, his lighting-plots were geared meticulously to the movements of the actors, and their performances were also controlled with split-second timing: perhaps only a single hand would be lit, or a whole body merely etched in all the colours of a rainbow. Lighting rehearsals would take weeks, adjusting the composition moment by moment, every change computer-programmed. The extraordinary and expensive stage world of *Great Day in the Morning* (1982) or *Death and Destruction in Detroit* (1979) would have been impossible only a decade earlier. For Wilson's actors, the challenge was to fit in without sign of difficulty and to be totally present so that everything seemed to be necessarily what it was. The audience was not led towards moments of intensity and absorption, but watched a whole process that might seem like a journey or a pursuit of rebirth.

In the 1970s and 1980s a large pool of theatre people lived in and around New York City. When out of work, as in difficult times they frequently were, they kept their art alive in various studios, paying for classes or working on group projects without pay. From time to time this activity would surface in the form of a small and independent producing group, or it would move into an existing organization. At the Public Theatre, directed by Joe Papp and funded by royalties from *Chorus Line* that had originated under his aegis, several different scale auditoriums were available. By using both avenues to production, Richard Foreman was able to pursue a deeply personal objective: 'I want to make a theater that closes your eyes to the specific reality of where it's supposed to take place, so that all other possible locations can bleed through it, one which models itself on the strategies of poetic syntax and structure, the better to handle the materials out of which the life-force builds subjective human experience.' His own company was called the Ontological-Hysteric Theater; his plays included *The Cure* (1986) and *What Did He See?* (1988). Julie Taymor and Anne Bogart were among directors in the 1980s and 1990s who mixed independent projects with work for established companies, the former specializing in interplay between puppets and actors, the latter in group shows with inventive and highly energized actors whom she choreographed within a carefully constructed 'soundscape'.

The Wooster Group was founded in 1975 as an association of more or less independent theatre-artists of differing skills. Together they staged an average of one show a year, each a step forward in a series of experiments which sought a new theatre for the new technological age. In 1991, the group premiered *Brace Up!*, a version of Chekhov's *Three Sisters* 'composed and conceived . . . under the direction of Elizabeth LeCompte'. Work on this show continued throughout the following two years in the course of an international tour. Huge labour was

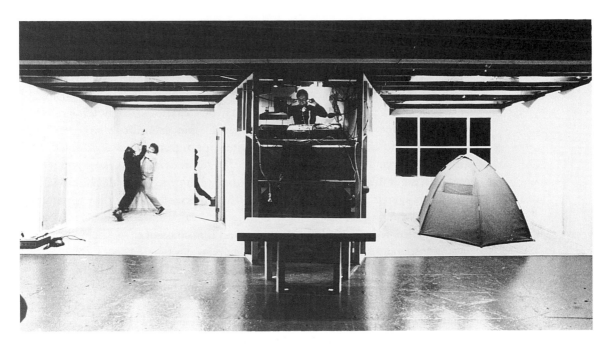

involved, which was financed by the National Endowment for the Arts, the New York State Council on the Arts, the New York City Department of Cultural Affairs, the Ford Foundation, the AT&T Foundation, the Greenwall Foundation, the NPN Creation Fund, and the Inter-Arts Program of the NEA—and more besides. The group used this support to think through Chekhov's text afresh, applying their own sensibilities, modes of performance, and technical expertise. The result was highly disciplined and idiosyncratic: 'I was trying to do a production of Chekhov as true to the author as I could conceive in my heart of hearts,' said LeCompte. The world of the original play became an extravagantly yet coolly obsessed theatrical mechanism, geared to carry an audience forward by its persistent inventiveness. Within the stage area was a platform, with chairs, miscellaneous properties, TV monitors, and microphones, and with batteries of lights above. An ancient, slow, and non-theatrical Anfisa, the family nurse, appeared only on the monitors, as did the infant Bobik. A narrator was by turns character, supernumerary, technician, and stage manager. Prerecorded music was used variously, to control mood and tempo; but there was a build-up of chanting, calling-out, thumping, and electronic sound for the fire in Act III. Some of the text was spoken with brilliant clarity, often in a totally 'unreal' tempo, or at a strange pitch or volume, or in a contrived silence. Some was not delivered to the characters who are meant to be addressed; some was spoken into a microphone and to no one on stage—Vershinin, the colonel, often addressed the audience in this way. Irina, the youngest sister, was played by a white-haired older actress, so that her words

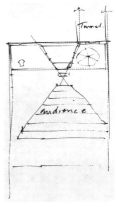

Treble-vision drama by the Wooster Group. Under Elizabeth LeCompte's direction, the setting for *Rumstick Road* (1977) ensures that the drama can remain simultaneously in focus at three different locales.

seemed wilful and contrived, believable only as an attitude of mind. Much of the setting and some necessary properties were not present on stage, but were suggested in mime by the actors as occasion demanded. However, actual exterior scenes, showing falling snow or crowds, were to be seen on the TV monitors, and so were silent characters, evocative images of flowers or soldiers, and analogous happenings from present-day reality. Much of Chekhov's Act IV was presented only in résumé, but there were farewells and a dance; Andrey's last speech was given in sound only, while Natasha his wife appeared at a window on video, with Bobik, her son. The whole show presented a group of highly refined and intelligent artists doing what they, and only they, could do with Chekhov's text: a theatrical exercise of modern sensibilities that managed to be both strenuous and self-indulgent.

In France, Marguerite Duras, novelist and film-maker (*Hiroshima mon amour*, 1959), wrote plays during the 1970s in which past and present, and thoughts and speech, dissolve into each other. *Eden Cinema* (1977) is about a French settler, who lives alone with her son and daughter and struggles for survival and justice; but the story is not told clearly. Technical resources are used in ways that bring a filmic sensitivity to the theatre and enable Duras to weave fragmentary perceptions into a subtle net of understanding: recorded voices and music are heard, while actors move slowly on a mysteriously illuminated stage; location and time shift easily; a silent figure holds attention while others talk. The audience gains a heightened and sensuous consciousness which turns a particular tale into an experience which is like a discovery each time it is played; and so it gains a general and timeless validity.

In Germany, some of the larger and more established theatres participated in such innovations. In East Berlin at the Deutches Theater, Heiner Müller's production of *Hamlet/Maschine: Shakespeare/Müller* (1990) used batteries of technical equipment: loudspeakers moving in a bulky contraption above the stage broadcast passages from Müller's own *Hamlet-Machine* into a performance of Shakespeare's play; television monitors were positioned at the side of the stage; the sound of water trickling into a large puddle signalled a thaw of political power; Ophelia re-entered towards the end of the seven-and-a-half-hour performance to move Hamlet's body centre-stage. Here technical innovation was at the service of one particular revision of Shakespeare's play and much of the experiment centred on the actors' imaginative and expressive powers, as guided by scenographer and director-author.

As directors experimented with sophisticated control of stage images and sound, another line of what was by now called 'theatre research' concerned itself with the nature of performance. In November 1970, growing ill at ease within the established theatre of England and with the restrictions imposed by its production schedules and need for long-running successes, Peter Brook moved to Paris and founded an International Centre for Theatre Research. In the spacious exhibition hall of a government-owned workshop in the Gobelins district, he was joined by experienced actors from a dozen countries, four other directors, and a designer, and together they set about trying to 'uncover the roots of theatre through a long series of experiments'. The Centre's first work shown in public was the mythic *Orghast* (1971), created for open-air performance at the Shiraz-Persepolis Festival of the Arts in Persia. It was lit by fire, and most of its text was in Orghast, a new language devised for the company by the poet Ted Hughes. According to the *Observer*'s reviewer, most of the audience that had travelled into the desert were 'bewildered'; but 'what we saw and, above all, what we heard was riveting, beautiful and disturbing'. At that time Brook said the 'result we are working towards is not a form, not an image, but a set of conditions in which a certain quality of performance can arise'. Over subsequent years he was to define more closely what he required from actors: they had to achieve 'crystal clarity' and 'openness', and to be 'concentrated . . . but accessible'; and so they might touch 'at certain short moments a quite genuine but fleeting experience of what could be a higher level of evolution'.

Brook continued to work in many forms. He travelled with his actors to Africa to perform for people who had not been influenced by earlier theatre experience. He moved headquarters to Les Bouffes du Nord, an old theatre stripped of its gilding and fitments, and closer to the centre of Paris. His productions toured the world, while he made a film in Afghanistan and London. In 1978 he fulfilled a promise and returned to the Royal Shakespeare Company to stage an *Antony and Cleopatra* which seemed out of place in the repertoire and ill at ease with itself. He lectured, conferred, conducted experiments, led workshops, wrote about theatre and his own work, and sought in all this to refine and deepen his understanding of what theatre could achieve in his own time. Shakespeare's plays he staged repeatedly with his own company and on his own terms: *Timon of Athens, Coriolanus, Measure for Measure, The Tempest*. Between 1981 and 1988, he directed Chekhov's *Cherry Orchard*, in French and then in English. He filmed and staged versions of Bizet's *Carmen*. He produced a nine-hour version of *The Mahabharata*; work had started on the Sanskrit original in 1982, and parts of the show were premiered in a quarry near Avignon during the 1985 festival, the full work being shown in Paris later that year. In 1987 an English-language version started a world tour and the production was subsequently filmed. But a tally of Brook's work does not indicate the full measure of his achievement; at every stage of his journey since setting up the Centre for Research, the demands that he has made upon theatre and its performers have inspired many others—directors, actors, designers, and writers. He became known around the world through his writings, especially *The Empty Space* (1968) and *The Shifting Point* (1987), and an autobiography is likely to become equally sought-after among theatre people.

PETER BROOK'S PRODUCTION OF *THE MAHABHARATA*

Irving Wardle reviews the first production at the Avignon Festival for
***The Times* of London, 13 July 1985.**

As with *Orghast* at the Shiraz Festival in the early seventies, Brook's spectators undergo a pilgrimage—a long walk up a dusty hillside—before being rewarded with a location hardly less magnificent than Persepolis: an amphitheatre facing a blasted cliff-side, with a lofty upper pathway for the actors, and a desert floor divided by a canal. Surrounded by the most sophisticated of lighting rigs, the show is rooted in the basic elements of the universe—water, stone, and fire.

What Brook and his adapter, Jean-Claude Carrière, have drawn from this immense narrative is the story of the struggle between the five Pandava brothers and their cousins, the Kauravas, for power over the country. As that country constitutes the entire known world, and as the title embraces both sides within the family of man, there is no missing its application to the modern world.

The seeds of discord are sown in the first play, *The Game of Dice*, and ripen in the second (*Exile in the Forest*) where the world is split into two opposing camps. As the work advances towards a conflict of total devastation (*The War*), it jumps the centuries from swords and clubs to a blinding magnesium flare that erupts in the rock face as the ultimate weapon.

Brook, however, is not an artist who ever allows himself to be driven into a corner: nor is he the man to regurgitate an Asian masterpiece as a Western editorial. His purpose is more modest and vastly more ambitious than that. In line with all his public work since he set up the International Centre of Theatre Research, he seeks out the blackest evidence in human history as a means of exalting the human race. . . .

Played in clear, simple French, the performance is a melting-pot of national differences. Among the musicians, percussion combines with the barbaric blare of the nagaswaram (a harshly funereal quarter-tone trumpet) and the homely sound of trombones. Boots and trousers mingle with robes and skirted battle costume. And the accents of the cast vary between German, Greek, Polish, Japanese, and declamatory Parisian French. . . .

Brook the magician surpasses himself in this production. For his ceremonies, trails of fire race across the sand and ignite the waters. For his battles, you can trace the flight of the mimed arrows, and visualize the horses from the sight of chariot drivers whipping a single heavy wheel over the desert. As before in his work, the immediate impact produces amazement and pleasure: the long-term effect is the imprint of images of tenderness, triumph and death that lodge in the memory for ever.

But the experimenter, analyser, and inspirer was also the consummate show-man. No matter how adventurous or fragmented the work in hand, Brook was able to draw it together and create a theatrical event which held attention by his authoritative use of space, time, and performance. When the public is due to see a show, the explorer and enquirer becomes, as he has said, an 'interventionist': in short time, orders are given and a show is made that works for its audience. Most of the cast for *The Cherry Orchard* at the Majestic Theatre in 1988 were new to the production and to Brook's ways of working; the result made it appear that Chekhov's play was peopled with individuals who were actors half-lost in studied seriousness, each one intent on his or her own performance. Brook had used the *ad hoc* conditions of rehearsal to reinterpret the text. The refined expertise and textual consistency of this visually grand and austerely crafted production was to be sharply contrasted a few years later by the Wooster Group's handling of Chekhov's *Three Sisters*: these were two almost totally opposed ways of using theatre research to make a classic text viable for new audiences.

Other directors in the new high-tech. world have won world-wide attention by concentrating on innovative work with actors. Grotowski's 'laboratory' theatre of the 1960s had little direct impact after the early 1970s, when he began to operate in social rather than theatrical contexts, but his strictly controlled ensemble had many successors, especially in Poland. Among these were Tadeusz Kantor's Cricot II, which had also been founded in the 1950s (*The Dead Class*, 1975, and *Let the Artists Die*, 1985) and the Gardzienice Theatre directed by Wlodzimeirz Stan-iewski, a former member of Grotowski's company. The first few minutes of a Staniewski production seem to offer a mere copy of the master's work, with nearly uniform dress, strong rhythms of corporate performance, heavy boots used for percussive effect; but soon the vigorous individuality of the actors establishes that the show is about passionate search, rather than suffering and endurance; and their entire bearing, as if inured to hard labour, reveals an attempt to free theatre and its audience from the limitations of intellectualism. Each production is devel-oped to be performed for the rural people living and working in the remote part of Poland where the company has its base.

After three years of study with Grotowski, Eugenio Barba founded the Odin Theatret, first in Oslo and then at Holstebrö, in Denmark. He was seeking a mode of performance dominated by neither commercial considerations nor the singular vision of a director. As well as visiting major theatre festivals, his company worked in remote areas of Sardinia and South America, and subsequently adopted popu-lar and traditional modes of performance that they had found there. In 1979 Barba established an International School of Theatre Anthropology, publishing the results of his research in articles and books, of which the most substantial is the lavishly illustrated *Dictionary of Theatre Anthropology: The Secret Art of the Performer* (published in English translation 1991). Here theatre work is related to performance in combat, religious ritual, and other highly charged public activities.

Performance research and actor-training were fundamental to a number of companies founded in the 1970s and 1980s. Ariane Mnouchkine had started Le Théâtre du Soleil in 1964, but *1789*, its distinctive 'création collective' about the French Revolution, was first staged in 1970. Performances continued to be given until 1973, by which time the company had moved to the Cartoucherie at Vincennes in the Paris suburbs, a former cartridge factory which was to be their capacious home, accommodating many different audience/stage relationships. Before all this, Mnouchkine had travelled extensively in Asia, where she discovered a physicality and musicality in traditional performances which have influenced much of her subsequent work. She may be said to have founded a neo-folk-theatre: her actors were members of society at large who staged plays for fellow members. They were encouraged to bring their own life-experiences and individual ideas to the making of a show. Performance-texts were developed from the actors' improvisations and individual historical researches. As director Mnouchkine shaped whatever the actors found, claiming to have no knowledge before rehearsals of what the play would be. From 1968 onwards, everyone in the company received the same wages—unless their long rehearsals had left no money to pay anyone. Lines of demarcation between various production and administrative tasks, an accepted feature of unionized theatres, were swept aside. There were specialists, of course, but an actor might work front of house, meeting and serving the audience before a show, or might sweep the stage, or set out costumes and properties, or rig the lights; and so might Mnouchkine herself. The audience was not isolated from backstage activity, but allowed to view the actors preparing and making up before a show. Performance could take place on a number of different stages around the sides of the hall, so that the audience in the centre might more readily occupy the same imaginary 'world' as the performers. When the company tackled political issues, as it consistently did—notably in *1789* (1970), *1793* (1972), and *L'Age d'Or* (*The Golden Age*, 1975)—the performers had both a collective and an individual commitment to what was presented. For the fall of the Bastille, actors were scattered among the audience so that each one might engage with those within reach, giving an account of the event as understood by the character he or she had by this time established; and they did so as if to fellow citizens of Paris at that crucial moment in the past.

These actors were also highly skilled performers: speakers, mimes, dancers, singers, puppeteers. They had studied clown techniques (both before and after *Les Clowns* of 1969), and also the use of half-masks and *lazzi* in the traditions of the *commedia dell'arte*. For repertoire they returned time and again to those classic texts (especially Shakespeare's) which drew on traditional expertise. The major project of the early 1990s was *Les Atrides*, a ten-hour, four-part show to be seen over two or three days and consisting of Euripides' *Iphigenia in Aulis* and Aeschylus' *Oresteia* trilogy. The variety of style used for the Revolution plays, where each scene took whatever form arose from actors' explorations, now gave way to a strongly marked group-style, an exotic and expertly performed blend of Asiatic and European dance and declamation, accompanied by specially composed music played on a great variety of unfamiliar instruments. The sense of individual involvement which occurs in Aeschylus' text when members of the chorus engage with the protagonists, and the contrasting power of speaking in unison, were both neglected in favour of a Chorus Leader to speak on behalf of

Asian influence on Western theatre. In Ariane Mnouchkine's Théâtre du Soleil production of *Les Atrides*, Iphygénie is about to be sacrificed by her father, Agamemnon: Asian dance, drama, costumes, and make-up are adapted to the demands of Euripides' text.

them all. In this production the effectiveness of the chorus depended on music and dance, and on a single voice—perhaps because these were found to be the most accessible elements of the complex style of the ancient plays. In the main roles, the decisiveness and sensitivity of individual performers, moving on a huge bare stage some 22 metres wide and 19 deep, and the consequent power of both silence and stillness, drew upon the company's corporate and by now deeply engrained strengths. Their style could also encompass, in a non-Asiatic way, moments in which intense feeling was directly expressed.

Over almost three decades this company's folk-theatre ideal had waned. Named after the sun, as if that were the source of its natural power and health, it could now be known more fittingly by the name of its director. At the time of *Les Atrides*, Mnouchkine spoke of her 'inner knowledge' of Asian theatre: 'we in Europe have authors, but they have acting. They know what an actor should be. And they have known that for two or three thousand years now. We don't. We still think an actor should be like someone in ordinary life.' Her vision of what actors should be had dominated this show, rather than the variety of their skills and individuality of their engagement.

A concern for what acting has been and can be was at the root of many developments towards the end of the twentieth century, as if theatre had to be rediscovered or created all over again. The company of Anatoli Vasiliev in Moscow was organized like a school, entrance to which required specific qualities of which musicality had special priority. His productions toured beyond Russia (e.g. *Six Characters in Search of an Author* in 1987 and the mysterious *Cerceau*, by Victor

Slaukin, in 1989), but the exploratory nature of his work was perhaps more evident in a show based on a non-dramatic text by Oscar Wilde (1989). At 11 p.m., after the main theatres of the city had shut their doors, fifty or sixty invited guests assembled in a cellar-studio with a deep, narrow stage of polished wood and the very same words were repeated time and again. Young actors, in carefully chosen street-clothes, entered a few at a time, but in widely different relationships and moods, and in a sequence of scenes shared the text in various ways, testing its structure and rhythms, altering physical circumstances, finding hesitations, confusions, and momentary clarity. This was not an improvisation, but a finished piece of theatre that had been polished until it shone. The audience could sense individual tensions and excitements living within the discourse and was drawn into a maze of intentions and contrary feelings, held in the calm of considered artistry.

In an old farmhouse at Toga, a small village in the Japanese Alps several hours by air and road from Tokyo (and at least a day's journey by any other means), Suzuki Tadashi has built a theatre, in the form of an old kabuki stage. Seating a little over 200 on the floor of its auditorium, it is maintained in meticulous cleanliness and order. Here Suzuki has trained the Suzuki Company of Toga (or SCOT), and prepared productions which have then travelled the world. On stage the actors of SCOT seem a special breed of mankind with great reserves of power, absorbed in a timeless drama and emanating unshakable purpose. In a series of essays, published in English as *The Way of Acting* (1986), Suzuki has told of his quest for a theatre that could speak beyond the bounds of any verbal language: a physicality in performance that is firmly grounded and yet flexible; vocal resources that can sound, and can be controlled, like music. His actors function like priests answering to some power above or beyond the stage, so that, facing their audience, they do not address them. They are trained to be aware always of unseen presences which circle and threaten them; and to concentrate on outfacing this opposition. Suzuki has said that his inspiration came from seeing a master of Noh theatre performing on a small stage in Paris, at Barrault's Théâtre des Nations; without the usual refined and complicated context of a Noh theatre, communication had been complete and the performance greeted with thunderous appreciation. Here was an internationally viable theatre and so he set out to develop a way of acting which could learn from ancient techniques and respond to his own imaginative awareness. By privileging performance and seeking to escape from the isolation caused by the limited geographical reach of the Japanese language, Suzuki, like Mnouchkine, Brook, and Grotowski, was creating an 'intercultural theatre'.

The result of Suzuki's research and training method is an impressive style which does not suit a complicated or quick-moving narrative, and in keeping with this he has broken up the classic texts which attract him for both their poetry and their themes. His *Tale of Lear* (1990 and 1993) was the dream of Shakespeare's Lear as experienced by a patient in a contemporary hospital or asylum. *Waiting for Juliet* (1993 and 1994) was a self-absorbed reverie on past desire. Six Greek tragedies yielded a *Clytemnestra* (1983) about isolation and the disintegration of society. Suzuki's earliest distinctive productions, *On the Dramatic Passions, I, II,* and *III* (1966–75), were conceptual and accumulative, not concerned to present any single story or situation. These productions offered haunting evocations of an entire world, at once contemporary and traditional. Even in the heath scenes of *The Tale of Lear,* the king-figure is never free from the presence of Goneril and Regan, whose strength, as played by male actors, equalled his own: in ninety minutes, a slow nightmare of self-assertion is realized on stage, and its self-inflicted pain.

The expressive power of actors is at the heart of many directorial innovations in the last decades of the twentieth century, even as technical advance has given access to still more sensational spectacle. In his seventies, Giorgio Strehler cast himself as Goethe's Faust, although he had scarcely acted at all since his earliest years. While the great new Piccolo Theatre in Milan remained unfinished because of rising costs and political wrangling, Strehler was glad to work in the Teatro Studio, newly formed out of a second-rate cinema. Here a central stage is encircled on two sides with four tiers of wooden seats which accommodate, without much comfort, a little over 400 persons, all having an intimate view of the stage. With a newly trained, young company and some older associates, Strehler worked on 'The Faust Project' from 1982 to 1992. He had envisioned a production playing over six evenings, but by 1991 only *Fragments* had been shown in two successive years. These were collections of scenes and episodes in differing directorial styles and differing interpretations. The work culminated in 1992 with the *Faust Festival,* spread over two evenings. The art of the director was on display almost continuously in the varying ways that the text had been used, but the director was also on display physically, as an actor at the centre of that changing and demonstrative world. The maker stood revealed, or nearly so; he was ensuring that his instinctive exploration of Goethe's text was subjected to the test of performance. The production was based on the earth of his own life in theatre.

By the 1980s one idea had gained a new currency: if a theatre-show is well acted—truthfully, clearly, physically, co-operatively, passionately, skilfully: whatever makes good acting—then an audience will be entertained and enlightened. Performance had become both test and vindication, the utmost reach of imagination and its definition. Giorgio Strehler wrote: 'My profession is to tell stories to others . . . I don't care about being understood. It's enough that people listen to me.' In 1981, near the end of a long rehearsal period, Peter Brook was reported as saying:

We are a small group of human beings. If our way of living and working is infused with a certain quality, this quality will be perceived by the audience, who will leave the theatre

subliminally coloured by the working experience we have lived together. Perhaps that is the small contribution we can make, the only thing we have to convey to other human beings.

According to Suzuki, an actor can create a 'stillness' within a performance, and a 'vision beyond sight'. In different ways, all three claimed that theatre, through its actors, creates its distinctive mode of consciousness and offers an experience found nowhere else.

Directors known in the 1960s and early 1970s for their conceptual grasp of a text and for staging the classics in new and often contemporary settings, became increasingly concerned with the actors' relation to the public. Ingmar Bergman developed a form of staging that concentrated action on a small platform right downstage, close to the audience. There could hardly be a greater contrast with his work in film or his own earlier theatre productions. He had come to realize: 'the theatre calls for nothing. TV includes everything, film includes everything, there everything is shown. Theatre ought to be the encounter of human beings with human beings, and nothing more. All else is distracting.' With Gorki's *Summerfolk* of 1974, Peter Stein gave actors at the Schaubühne in Berlin more individual life on stage than audiences could readily take in at one viewing. In a collective creation, *Shakespeare's Memory*, two years later, he required them to present their own research and encounter members of the audience as they walked around to see whatever most interested them. These experiments led on to an *As You Like It* of 1977 which started in one large studio but, when the play's action moved to the forest of Arden, the audience had to follow the actors through a long passageway to reach another large space where they found a great beech tree, a pond, and, spread out on all sides, the various locations required for the forest scenes. Having shared a journey with the actors the audience had become privileged eavesdroppers in close and immediate contact with the characters.

How an audience sees the actors was again central to the work of the French-Canadian director Robert Lepage, but to quite different effect. His *Midsummer Night's Dream* (1993) at the National Theatre, London, was played as a dream within a dream. Theseus and Hippolyta entered on a bed that then took its place centre-stage under a single light bulb, surrounded by a wide pool of watery mud: and the whole play became their fantasy. This was a conceptual production, in the style of the 1950s and 1960s, but it made new and large demands upon the actors. The actress playing Puck had to be a contortionist and acrobat, while everyone playing the fairies and lovers had to perform in the mud and with the mud. For *Coriolanus* (1993), Lepage set the action behind a solid curtain-wall, so that it was viewed through an opening above floor level and little more than a metre high. The stage behind the curtain could be raised and lowered, but the actors' whole bodies could seldom be seen. By manipulating the audience's perception and by

Facing: the forest of Arden in Berlin. Peter Stein's production of *As You Like It* (1977) at the Schaubühne in Berlin placed the audience as eavesdroppers around the broad expanse of its setting.

ordering the actors' strange activities, the director had awoken images from these plays never seen before.

Sometimes actors were used to create a specific style and draw a play into unfamiliar guises. In Joachim Burkhardt's *Potsdam,* premiered in Stuttgart in 1991, the director Jurgen Bosse used performative devices from high farce, berserk activity, caricature, and weird dance to convert a play about past times into a loud and inescapable metaphor for present anxiety. From the old East Germany, Leander Haussmann staged *Romeo and Juliet* (1991) so that the balcony scene became a 'magnificent slapstick routine involving ladders, makeshift ropes and much bare flesh'; quality of performance was again the essence of the reinterpretation.

Not all directors have come to use performance so aggressively. In the small studio-theatres which many established theatres opened during the 1970s and 1980s a strong focus upon the actors could be sustained without wilful intervention. Intended as laboratories for testing new or unfamiliar scripts, these secondary facilities have been used increasingly to stage major classics normally seen in large theatres. In 1974 at The Other Place, Stratford-upon-Avon, Buzz Goodbody directed a modern-dress *Hamlet* in conditions so intimate that many watchers could have touched the actors, if they had bent forward in their seats. The immediacy gained by this staging led to a series of small-scale productions of Shakespeare by the RSC. Its Artistic Director, Trevor Nunn, was himself responsible for *Macbeth* (1976) and *Othello* (1989), both cast with actors of sufficient power to fill a theatre seating thousands. In this new proximity, subtle playing could reveal contradictions within the text and communicate the inward consciousness of its characters by physical concomitants—change of breathing, tension, posture, pulse-rate. But at The Other Place, after the *Macbeth,* a more familiar brand of broad theatricality began to reassert itself in performances and designers moved in with ingenious sets, negating those different possibilities. Most established theatres would need radical reorganization, as well as new finance, to capitalize on such experiment.

Below: Ben Kingsley's intense and anxious Hamlet in Buzz Goodbody's modern-dress 1974 production at the Royal Shakespeare Theatre's Other Place crouches in close encounter.

Elsewhere innovation is easier. Increasing interest in the unique qualities of live performance, as opposed to the easily promoted, profitable, and artfully edited shows of film and television, has sustained a rising tide of entirely new companies, most of them very small in scale and insecure in finance; almost their only resource has been their actors. The numbers of such companies became as astonishing as the range of their work. They have remade old plays and commissioned new ones geared to their particular abilities and interests; and they have created their own texts in rehearsal, or done without words altogether. They have worked with masks or puppets, trained as clowns, comics, mimes, dancers, acrobats, singers, musicians. They might perform in the open air with motorbikes, or on stages so minute that the actors can scarcely move. During the three weeks of the Edinburgh Festival, so many small companies come together for its Fringe that no one person could see a tenth of the shows, and many are so poor in quality that no one would want to. With very little money to spend, the actor has to be the centre and shoulder almost all the burden; but that is no deterrent, since in most cases a discovery of the actor's powers has provided the drive and obsession behind the enterprise.

In 1990, when the post-war Edinburgh Fringe was in its forty-fourth year, 9,000 performers came to the city for the three-week festival, from the British Isles and North America, and from Venice and Tokyo, Mombasa and South Africa. With little official backing, the Fringe had grown to over 500 shows in some 140 venues. The same surge of output is found almost everywhere. By June 1993 *The London Fringe Directory* listed seventy-seven different venues. During each Avignon Arts Festival the city's central square becomes an arena in which performers recommend their shows among café tables and idling crowds. Pavements become littered with discarded handbills. Outside the Palace of the Popes, where a grand production will draw thousands of spectators, an improvised stage, home-built props, and very basic costumes are used to stage a melodrama, while a little way off a solo performer in loincloth 'acts' a series of classical statues and a death's head on stilts tries to draw spectators away to another new creation. Some fringe companies are active for most of the year, but all become more visible during summer festivals; some of this work may hold the seeds of a new theatre relying chiefly on the resources of its performers.

By the 1990s the welter of new activity was most evident in Tokyo, where theatre traditions are not entrenched or ossified, and union regulations do not rigorously divide professional and amateur theatre, or paid professionals from unpaid. On arriving in the city and consulting *PIA* (the equivalent to *What's On* in London, New York, or Sydney), a visitor finds advertisements for some 250 different shows, not including vaudeville or 'traditional' dramas. Tokyo must be ranked as the 'New Theatre Capital of the World', for most of these shows are premières or revivals from a year or so before. Theatre buildings tend to be small,

Sold-out performances in Tokyo. Noda Hediki's *Comet-Messenger Siegfried* played to capacity audiences three times in one day when revived in 1988 in Yoyozi Gymnasium; it then travelled to the New York International Arts Festival.

but the younger companies almost always draw crowded audiences. (Fire regulations seem lax enough to allow standing or sitting in aisles.) A few have the aura of pop musicians, their fans buying videos of earlier shows, and sweatshirts, buttons, photos, posters, scripts, magazines. When Noda Hideki restaged his *Suisei no Siegfried* (*Comet-Messenger Siegfried*, 1985) before taking it to New York in 1988, its set was erected in an open-air stadium and so many applied for tickets that three sold-out performances had to be given on the last day available.

In 1976, aged 21 and still a law student at Tokyo University, Noda had started his own company. Sixteen plays later, in 1981, they left the campus and by 1983 were fully professional. Noda wrote, directed, and starred in at least one new play a year, as well as undertaking guest productions of Shakespeare as one of the most sought-after directors in Japan. With a company of around eighteen young and energetic actors, his plays explore a theme, rather than tell a clear narrative or present commanding characters. With plenty of verbal wit (for this he has been compared to Tom Stoppard), with circus tricks, music, song, dance, and acrobatics, with quotations from Noh and kabuki, many role-changes, and many distinct episodes in varying times and locations, the audience is kept alert and busy, sharing an adventure which seems spontaneous and close to their own wildest dreams. Lighting is fluid, and the supporting music partly pop and partly classical. Tempos shift abruptly, sometimes into the very quick speech and action which is a Noda speciality and used for reading instructions, engaging in an argument, fighting, or expressing exhilaration. Critics in the West and older theatre people in Japan have sometimes condemned Noda's productions as frenetic, sensational, fashionable, and childish, preferring the superbly crafted productions of Shakespeare or Euripides of Ninagawa Yukio which draw together elements of Western and Japanese theatricality. Sitting among the avid Tokyo audience, however, one can believe that Noda is being faulted for the very qualities he most values: he has made performative sense out of what he finds irrational within and around himself. He enjoys active visual images—his *Twelfth Night* (1986) had Malvolio running endlessly inside a huge wheel like a mouse in a cage—and he likes to superimpose one image on another, transforming them as if by switch of television channel. The fun of a game seems to be at work, and also a sense of celebration or ceremony—but if a god is being worshipped, it is one offering a return to open-minded freedom, without denying the complexity of twentieth-century living.

Each year this company (in 1993 renamed Noda Map) comes together for about four months to create a new show. At first rehearsal there is almost no script, and nothing will be fixed until some ten days before opening. In *Zendajo no toriko* (*Prisoner of Zenda*, 1981 and 1992) much of the action is very childish. Boys and girls of the Children's Crusade get mixed up with events from the adventure-novel. As troops they are easily enthusiastic and quarrelsome, so making a mockery of warfare as a means of achieving anything. They experiment by rushing around on all fours or moving like automata; they fight, engage in chases, dress up in toy-like costumes. Lucifer arrives and changes into Alice in Wonderland. Trite lessons are given about sexuality. Physical exhilaration is sustained by restless and bright invention, by make-believe, caricature, verbal dexterity, transformation, and a sense of accomplishment. The intelligence at work is both iconoclastic and playful, ambitious to stay afloat among a mass of impressions and anxious to be the winner in every game. The audience is infected by the good spirits and shares in the intellectual and mischievous adventure.

These shows do not travel well and their appeal may not last; but, in comparison with the expediency and uncertainties of theatre elsewhere, Noda's creation of a series of innovative and hugely successful plays based on vital live performance is highly significant. Except to finance foreign tours, this company has never needed public subsidy, and in Tokyo it is only the most notorious among other young and iconoclastic theatres. Noda's chief rival has been Kokami Shoji, whose Third Stage company is also continuously sold out. Although not a performer like

Noda, he also is a public figure, forging a mutually supportive link between theatre and his own film, radio, and journalistic work. For these highly productive, popular artists, theatre is at the centre of a thriving new world and a means of becoming alert to its possibilities and dangers.

In the West, some theatres have also pursued a popularity that could entertain and challenge their audiences. Most frequently, however, they have sought inspiration in old-fashioned theatre rather than in the entertainment industry of the late twentieth century. In the United States, Charles Ludham's Ridiculous Theatre has specialized in outrageous farce and obvious melodrama. He himself performed in drag and that was entirely at home in the company's style: 'rather than *hiding* anything,' he has said, 'I *show* everything.' Productions were blatantly theatrical and unashamedly entertaining—such as *The Mystery of Irma Vep: 'A Penny Dreadful'* (1984) or *The Artificial Jungle: 'A Suspense Thriller'* (1986)—but they were also teasing hieroglyphs, sustaining such themes as hypocrisy, abuse of authority, individualism. In France during the 1970s Le Grand Magic Circus, directed by Jerome Savary, avoided 'high culture' and sought to rediscover a rough, brash and offensive theatre. In Australia, the Pram Factory was home in Melbourne for productions based on vaudeville and pantomime, and for the Soapbox Circus (later called Circus Oz). In Brazil, in the late 1980s and 1990s, Gabriel Villela's Melodramatica Brasileira performed with exaggeration and excess. In his *A Guerra Santa* (*The Holy War*, 1993) a convicted murderer, called Dante, kills Virgil, whose spirit then becomes a narrator-clown and attempts, with Beatrice, to investigate his own death. In Villela's words, the aim was to create a 'lyrical and grotesque' piece of theatre-trash which was 'an absurd and fragmented picture of our country'. Again a carnival theatre of violence, laughter, and sensation was being resurrected for moral and reforming purpose.

Another route toward popular theatre has been through exploitation of personal success as comic entertainer and a lively sense of political reality. Dario Fo and his wife Franca Rame were known throughout Italy when, in 1970, they formed a new company. The popular and urgent farces that followed—such as *Morte accidentale di un anarchico* (*Accidental Death of an Anarchist*, 1970) and *Non si paga! Non si paga!* (*Can't Pay, Won't Pay*, 1974)—were written and performed to express outrage at the ways of a capitalist society and to provoke strong reactions, but they were also robust and traditional enough to be adapted and produced in many countries.

In one respect a politically motivated theatre can be dangerously popular. At times of outright political repression, even a very oblique reflection of public concern can quicken an audience. In Eastern Europe under Soviet occupation, a character on stage had only to say, 'I want to get out of here,' for applause to erupt. In Prague during the 1970s, an entire audience would rise to its feet in Act IV of Chekhov's *Three Sisters* to acclaim Masha's line 'The army is leaving.' Children's plays were written as disguised comments on the regime. Hidden in the bravado and fantasy of broad comedies, which a censor could scarcely take seriously, far-reaching criticism was voiced. An 'Aesop's Theatre' developed, especially in Moscow, where what was seen and heard on stage lacked any overt connection to everyday politics, and yet could be received as a defiant challenge to political oppression. Theatres prepared to take these risks were full to overflowing and theatre people became heroes. Vaclav Havel was imprisoned and under surveillance for much of the 1970s and 1980s, but he continued to write plays which

THEATRE'S ROLE IN BERLIN, 1989

Robert Weimann, theatre scholar and consultant, reports on events immediately preceding the fall of the Berlin Wall.

Demonstrations were brutally suppressed in Berlin on 7th and 8th October. In Leipzig they continued, but there was a show of indecision, even ambiguity, about the balance of power until 4th November. On that day several Berlin (East) theatres had applied for the right to hold a demonstration so as to give 'cultural workers' a platform on which to speak out and make constructive proposals. This had been denied to them in the past, but when at last the demonstration was held, something like 500,000 (many say considerably more) turned up, from all sorts and classes of people. Although well-prepared, the theatre people were flabbergasted at the sheer size of the 'performance'. But the preparation was such that plenty of theatre workers (actors, directors, designers, dramaturgs, electricians, drivers, and so forth) were equipped with *Armbinden* which gave them authority to arrange for and conduct some kind of orderly proceedings. Among the theatres in charge of the organization was *Deutsches Theater* (formerly Max Reinhardt's), the one that had introduced Ibsen and Strindberg to the continent, and had produced Gorky, Hauptmann, and the complete Shakespearean history cycle long before any other theatre. This theatre took the lead, and the demonstration was absolutely disciplined and peaceful throughout.

After this successful 'production', the precarious balance of power shifted. Shortly afterwards, the Central Committee of the Socialist Unity Party dissolved itself: the days of General Secretary Krenz and the rest of the Politburo were numbered. They simply panicked: under pressure the gates in the wall were flung open. The whole massive, carefully guarded construction was crumbling on 9th November, five days after what was to remain the largest of all the demonstrations. Without the crucial initiative and participation of East German theatre people, all this could not have happened in the way it actually did.

MASK WORN ON
TOP OF HEAD WITH
HOOD. PUPPETEER NOT
SEEN.

JOINT

BLACK FABRIC
2 RODS THROUGH
POCKET AT TOP.
BRIGHTLY COLORED
STRIPS OF FABRIC RANDOMLY
ATTACHED TO BLACK WING.
WINGS ATTACHED TO EACHOTHER
BY BLACK BRAIDED ROPE.

BREASTS ARE WATER
SACKS THAT SQUIRT
STRONG STREAMS
WHEN SQUEEZED.

DRESS BODICE HAS
ELASTIC NECKLINE
THAT IS PUSHED
UNDER BREASTS.

FLEXIBLE HOOP IN
HEM OF SKIRT.
LARGE OVAL KEEPS
IT VERY WIDE.

SKIRT IS A
LONG NARROW
OVAL.

BONING OR
HOOP KEEPS
IT OUT. VERY
FLEXIBLE.
LIGHT FABRIC.

Street theatre in Colombia.
The Teatro Taller specializes
in performance on stilts, so
that actors are clearly visible
above surrounding crowds.
G. W. Mercier's sketches show
how puppet-like effects are
contrived.

could be performed only in the living-rooms of like-minded people or in theatres abroad (*Audience*, 1975, and *Largo desolata*, 1985). In 1989 he was elected President of Czechoslovakia and oversaw the withdrawal of Soviet troops from his country. Theatre's influence was far-reaching throughout Eastern Europe as a training ground for political change.

In countries with huge populations, theatre has long had an important role in furthering political debate, their dramatists often taking Ibsen, Shaw, and Brecht as models. Under Communist Party rule, however, most new plays in China during the 1970s and 1980s did no more than criticize the shortcomings of individual officers, stopping short of demanding radical change. After suppressing the protest centred in Beijing's Tiananmen Square in June 1989, the Party found theatre too unreliable to support further, dramatists having been among those arrested or seeking asylum overseas. Funding was switched to sports and competitive displays of skill in dance or music, activities with little opportunity for ideological engagement. Before 1989, Shanghai had boasted some two dozen theatre companies, but by the early 1990s only one 'Comedy' theatre was regularly able to mount new productions. Even traditional Operas, a source of tourist-money, were struggling for survival.

Western forms of drama continued to be adapted for Indian political purposes. In 1992 in Calcutta, the veteran dramatist, actor, and director Utpal Dutt staged his *Janathat Aphein* (*Opium of the People*), a Shavian debate play about Hindu and Muslim claims to the same temple site; this was at a time when such a conflict was leading to large-scale violence. Indigenous forms began to be used more frequently for the same purpose, despite the argument that revivals of any form of drama created originally to serve an authoritarian religion or autocracy could only obscure current issues. At Kerala a *kathakali* dance-drama *Manavavijaya* (*People's Victory*, 1987) showed World Conscience defeating Imperialism in his attempt to dominate the world by threat of nuclear war. Starting in 1975, at Imphal in the north-east, Ratan Thiyam staged a series of plays taken from old texts but newly interpreted. His Chorus Repertory Theatre used Manipuri dances, rituals, and martial arts, but handled by the director, and costumed and lit, so that the shows have the attraction and compulsion of Western musicals. In *Karnabhaaram* (*The Burden of Karna*) the hero is torn between being son of the Queen and the Sun-God, and being the son of his poor foster-mother. In large cities, small companies continued to perform in the streets. Since 1968 in Delhi, for example, Aloke Roy has run Jagran, a group specializing in mime so that it can speak to people of many languages and little education. Plays are noisy, funny, and highly topical, as they demonstrate such evils as drugs, crime, and extortion.

Throughout the developing countries of Asia, Africa, and Latin America many small companies were formed in the 1970s and 1980s to work as agents for political change, and they too depended increasingly on local traditions of puppetry,

Celebratory theatre in rural USA. At Glover, Vermont, Peter Shuman's Bread and Puppet Theatre draws an audience with a neo-folk carnival, using traditional circus devices. (© 1993 by Peter Moore.)

story-telling, music, and dance. A few samples must stand for a vast range of work that no one person could see or assess. In Colombia, Teatro Taller developed a style using stilts, so that performance could be seen above the heads of a standing crowd. In Lima and Chiclayo in 1973, Augusto Boal used a government Literacy Programme to develop a theatre operating as if in continual rehearsal:

The spectators feel that they can intervene in the action. The action ceases to be presented in a deterministic manner, as something inevitable, as Fate. . . . Everything is subject to criticism, to rectification. All can be changed, and at a moment's notice: the actors must always be ready to accept without protest any proposed action; they must simply act it out, to give a live view of its consequences and drawbacks.

Dictators and military police were often shown as monsters but, in General Pinochet's Chile, Juan Radrigán wrote a series of plays for bare stages in which the only characters are society's marginals; the oppression they suffer and the humane past they have lost are made apparent in the tensions experienced between themselves. In Africa in 1977 Ngugi wa Thiong'o was detained by the Kenyan government for scripting politically conscious plays with members of the Kamiriithu Community Centre at Limuru. By the early 1990s travelling theatre companies, based in Lesotho, Malawi, Zambia, and Nigeria, had built reputations for raising and explicating political issues. In South Africa in the early 1970s Barney Simon co-founded the Market Theatre in Johannesburg. He had worked backstage at Joan Littlewood's Theatre Workshop in London, and inherited her vision. He not only staged plays, but also provided a 'market-place' in which others could show their work and participate in cross-racial theatre. Productions, such as the all-black *Woza Albert* (1981), toured the world from this base.

In the 1970s and into the 1980s political theatre also thrived in the West. With a rapidly growing circuit of non-traditional venues, companies such as Welfare State, 7:84, Red Ladder, and Triple Action toured around Britain, but the movement was to suffer from withdrawal of Arts Council funding. In 1982 its Secretary-General had questioned why artists should 'expect public money to advocate the overthrow, not of the particular party in power, but of the whole system of parliamentary democracy'. Few companies survived, but interest remained, evidenced, for example, in translations of Boal's textbooks, *Theatre of the Oppressed* (1979) and *Games for Actors and Non-actors* (1992). Elsewhere sturdier political theatres included Sidetracks in Sydney, Australia, and the investigatory Het Werkteater in Amsterdam. In the United States, Peter Shuman's Bread and Puppet Theatre, founded in New York in 1969, had by 1974 moved to a farm at Glover, Vermont,

American blues on Broadway. Set in 1927, August Wilson's *Ma Rainey's Black Bottom* (1984) showed the power of the 'Mother of the Blues' and its sources in her life.

✳✳✳✳✳✳✳✳✳✳✳✳✳✳✳✳✳✳✳✳✳✳✳✳✳✳✳✳✳✳✳✳✳✳✳✳✳✳✳

FEMINIST THEATRE

At the Foot of the Mountain of Minneapolis, one of the first feminist theatre ensembles sets out its ideology and quest in a brochure of 1976.

At the Foot of the Mountain is a women's theatre—emergent, struggling, angry, joyous. . . . We are asking: What is a woman's space? What is a women's ritual? How does it differ from the theatre of the patriarchy? We struggle to relinquish traditions such as linear plays, proscenium theatre, non-participatory ritual and seek to reveal theatre that is circular, intuitive, personal, involving. We are a theatre of protest, witnesses to the destructiveness of a society which is alienated from itself, and a theatre of celebration, participants in the prophesy of a new world which is emerging through the rebirth of women's consciousness.

✳✳✳✳✳✳✳✳✳✳✳✳✳✳✳✳✳✳✳✳✳✳✳✳✳✳✳✳✳✳✳✳✳✳✳✳✳✳✳

where every year a communal carnival-festival is held. Home-baked bread is shared between actors and visitors, and monster puppets lead a *Domestic Resurrection Circus* and other shows about militarism, ecology, capitalism, and other issues of the day. Puppet theatre, wrote Shuman in 1990, is 'an anarchic art, subversive and untameable by nature . . . representing, more or less, the demons of [its] society and definitely not its institutions'.

One form of political theatre not only survived but continued to grow. The Black Theatre movement in the United States had started back in the late 1950s, creating plays about Afro-Americans and so raising public awareness of their inheritance, present achievements, and civil rights. By the early 1970s Gay Theatre, Feminist Theatre, Hispanic and Native American Theatres, Theatre for the Deaf, and for other special constituencies had all found their individual voices. By showing that theatre can illuminate the nature of society and provide opportunity for self-recognition, these movements were soon copied throughout the world. As Ntozake Shange's *For Colored Girls Who Have Considered Suicide/When the Rainbow is Enuf* (1974–5) and Tony Kushner's *Angels in America: A Gay Fantasia on National Themes* (1990–3) found their way to Broadway without compromising originality of form and manner, so companies like Gay Sweatshop and Split Britches thrived in Britain and the United States, Aboriginal and Maori plays were staged in Australia and New Zealand, and plays about immigrant workers in Germany and Switzerland.

Theatres wishing to align themselves with social and political issues turned instinctively to writers to make issues clear in words as well as stage image and performance, and by responding to this opportunity, towards the end of the century, numerous new writers developed into practised dramatists. Having co-founded the Black Horizons Theatre in Pittsburgh in 1968, August Wilson had by the late 1980s become a writer regularly produced on Broadway. After 1970 the number of women dramatists increased as never before, especially in America and Britain; by the 1990s, the plays of Beth Henley, Marsha Norman, Wendy Wasserstein, Caryl Churchill, and Timberlake Wertenbaker had become standard offerings in established theatres on both sides of the Atlantic and beyond. In Germany during the 1970s and 1980s, Franz Xaver Kroetz and Tankred Dorst wrote plays with brutal and unavoidable insistence on political purpose. In Britain, Howard Barker found shocking form and pungent words with which to confront audiences with social issues; without reaching a wide public, his career in theatre was both sustained and adventurous at a time when other dramatists would turn aside to work on films, television, novels, or journalism. In Australia, John Romeril has written plays that express his own vision of society, but their texts are not definitive: *The Floating World* (1974) was 'written to express and extend the humanity of the people working on it, not oppress and entrap them'.

Theatre explores traditional attitudes. This group from Act 1 of Caryl Churchill's *Cloud Nine* at the Royal Court Theatre, London (1979), shows Joseph, the Black servant, played by a White actor, Betty by a male actor, the young Edward by an actress, and the younger child by a doll: these 'miscastings' were intended to show what the persons wanted to be, or were wanted to be.

The complications of *A Small Family Business*. Alan Ayckbourn's comedy at the National Theatre, London (1987), used the same multi-levelled set to represent different houses, all belonging to the same family and furnished by the family's firm.

But in this period theatre has not been dominated by dramatists, as it had been as recently as the 1950s and 1960s. Writers who directed their own plays on a regular basis have made the strongest showing, among these Alan Ayckbourn at Scarborough, England, and Noda and Kokami in Tokyo. Those who have directed occasionally, in film as well as theatre, have grown in effectiveness; for example, David Hare, David Mamet, Sam Shepard, Len Jenkin, Tankred Dorst, Heiner Müller. Alternatively a dramatist at this time seems to have needed a close and sustained relationship with a particular director to ensure that his or her plays were produced regularly and promptly: so, in the United States, Lanford Wilson has been staged by Marshall Mason; in France, Bernard-Marie Koltes by Patrice Chéreau; in Japan, Inoue Hisashi by Kimura Koichi. Even writers who were established by 1970 or soon after have tended to stay loyal to one director, as Tom Stoppard to Peter Wood, or Peter Shaffer to Peter Hall. Some plays have remained unproduced until the dramatist found a new allegiance; an extreme example is Michel Vinaver's *Iphigénie Hôtel*, written in 1959, but waiting until 1977 for Antoine Vitez to undertake its production. The staging of a play has become such a complex project, technically and organizationally, that attention to the originality of its text may well take second or third place.

Staging and performance have become the dominant elements of theatre. Directors and theatre companies (and occasionally a star actor) are the attractions at festivals, and most often they use an old text, or one 'developed' during re-hearsals, or one contrived by the director working together with a writer of his choice on a project which is also of his choice. A script by a new writer will be subjected to readings and workshops, during which dramaturgs and directors, or individual actors, will suggest ways in which it could be changed and so yield a production in the theatrical terms with which they are most familiar. Even a play by an established writer is liable to be 'developed' before a final decision to produce; and then the published text will not be the dramatist's unaided work. A dramatist seldom works alone and then delivers a finished script to a producer. An original script was never all that went to the making of the première of a new play, but in earlier times—in ancient Greece, Elizabethan London, in Louis XIV's France—it was far more likely to be the crucial element, and the dramatist the centre of attention.

Perhaps writers have retained hold of one vital function: the quick response. On paper any number of corners can be turned, new territories annexed, and topical matters assimilated, and with requisite speed; whereas to alter the working of a company, change scenography or schedules, or develop new ways of actor-training takes months or years. In response to the hour, dramatists still keep the initiative. David Hare has worked in television, film, and theatre, and an alert sense of

Theatre shows realities from a film-world. David Hare's *Map of the World* at the National Theatre, London (1983), showed the making of a film about an international conference on World Poverty in a luxury hotel in Bombay; cameras, lights, crews, and director were on stage.

interplay between these media is evident in his theatre scripts; he has also gravitated towards issues of the hour, exposing the current working of established religion, law, and politics in Britain. Botho Strauss was editor of *Theater Heute* and dramaturg for Peter Stein at the Berlin Schaubühne before turning dramatist; in structure, his plays from 1971 onwards are uncommonly ambitious and intellectual in ways that could originate only from a single acute and experienced mind. Richard Nelson has written plays that are keenly sensitive to 'translated' customs and international tensions; his Americans abroad and his Europeans in America are shown with a clarity and an exploitation of contrasts which could only be the products of one person inhabiting both worlds. Living in Manhattan, Wallace Shawn has written plays full of familiar images of city life, but demanding subtle playing as sustained as their author's engagement in the writing. Spalding Gray, an original member of the Wooster Group, has developed a style of monologue for his own performance which appears improvised but has the structural ironies and emotional strengths of well-made literature. None of these authors was an acknowledged leader in the 1990s, like the directors of the time or the few dramatists who had 'arrived' earlier and were still produced commercially, but their works bring sensations to theatre which are special to the century's last decades and not readily accessible to directors working on old plays or building up collective creations. Although a playwright's innovation may not achieve full force in production, time may show the worth of its direct and concentrated response to lived experience, as opposed to the reliance on theatrical experiment which had become a major source of 'new theatre'.

But however strong the influence of dramatist, director, or performing company might be, the most pervasive influence at the end of the century probably belonged to theatre's administrators, the bureaucrats, development officers, subsidizers, and producers. These enablers had become the constrainers, and sometimes even the originators. The political or independent masters who give the subsidies have set up various committees and inquiries with an inordinate appetite for paperwork that must be satisfied. Every new venture has to provide budgets, audience projections, and artistic justifications in order to compete with other projects. Every success or failure has to be explained. Theatre itself has followed suit: by the 1990s every member of a company belonged to one or more professional organizations and had learnt to compete for benefits such as bursaries and project-funding. Grant-giving associations and councils provide injections of much-needed cash, but they also demand advance planning in great and punishing detail. Even more dangerously, they also favour 'excellence', which usually means that the money goes to projects which follow established principles, use established artistes, or strive for previously acceptable standards. Innovation is by definition not assessable in advance and therefore does not provide the right sort of answers for the forms which must be filled out and filed in triplicate. Individual artistes may, in fortunate circumstances, make a good deal of money from theatre and especially from subsequent promotions in television, print, or film, but a growing proportion of the theatre's basic income goes to sustain the huge organizations which private patronage and public surveillance both require. Around 1990, a dramatist discovered that if he had written all the plays produced by the Manhattan Theatre Club, a company with a deserved reputation for staging new work, and if he had also directed all of them himself, he would have received less money in that year than either the theatre's administrator or its development officer. The business of running a theatre in these times absorbed great energies, so that anxieties about money had become so pressing that they threatened to overtake the primary artistic concerns. Freedom of manœuvre and originality were found more readily among smaller, lighter, and less permanent operations than among the well-established and securely funded theatres.

Yet organizational stability has had its rewards. The fine and welcoming buildings of the National Theatre in London or the City Theatre in Helsinki, the large subsidies of many theatres in Germany and elsewhere in continental Europe, the growth of elaborate and expensive festivals in Chicago, New York, and Philadelphia, in Hong Kong, Adelaide, and Perth, Istanbul, Avignon, and Edinburgh—the line stretches around the world—are all the products of theatre's new bureaucracy, and they have immeasurably enriched theatre-going possibilities for theatre fans. At the end of 1993 a visitor to Berlin could have seen three premières among productions at the Volksbühne, the work of twenty different dramatists at the Gorki-Theater, and of a further six at the Schaubühne; at the Berliner Ensemble were three plays by Brecht, one directed and adapted by Heiner Müller; at the Deutsches Theater and Kammerspiele during November could be seen plays by Lessing, Hofmannsthal, Strindberg, Turgenev, Klaus Pohl, Ostrovsky, Bernard Shaw, Ionesco, Sartre, Goldoni, and more. Such an international galaxy is unique to the theatre of the end of the twentieth century, and perhaps its most accomplished or most visible achievement.

Its most naturally thriving event is not so easily identified—that achievement which does not depend on huge organization and investment, but is able to reach out to new audiences and change continually with the times. Many of the traditional or folk theatres that have accepted new techniques, new writers, and new themes might vie for that distinction. One such is Orissa Opera, a Jatra (or 'travelling') company in India to the west and south of Calcutta. Its traditions are centuries old, but this theatre has contemporary plays in its repertoire, and actresses among the male players who have hitherto played all female roles. It uses popular modern music and dance, together with traditional music and ceremonies. When the actors come to town with a repertoire of five or six plays, an estimated 3,000 or 4,000 people will gather nightly an hour or so before midnight and stay to watch the performance until after dawn the following day. The stage is like a garishly decorated boxing-ring. The play's story is long and complicated, but can be taken as a heightened version of everyday life. The company is self-supporting and works all year round except for harvest time. To quote Balwant Gargi, novelist and film-maker:

A Jatra actor, can be recognized by the way he stands—a tilted tower. He does not hold himself back but throws his weight forward. Passionate, charged with energy, he explodes into fiery dialogue. . . . He is superbly aware of the four-sided audience and is sturdily graceful from all angles. There is speed, action, flamboyance. Sharp turns in mood, abrupt flares and sudden drops in pathos are underlined by the orchestra. Drums clatter and thump and rumble.

After the first few nights in town, the audience will have learnt to recognize the actors in their new roles, and will greet them as victors after the six hours of performance. Away from modern publicity and all but the simplest technology, theatre has changed and thrived in Orissa, with a general acceptance that is very rare in the West.

To identify what was the most innovative theatre in the years since 1970 is not at all easy. Much of the 'new' has grown out of the old. Mega-musicals rely on the drawing power of spectacle which all theatres have recognized and used according to their means. Many famed productions have been reworkings of old plays. An

Westernized traditional theatre from Manipur. Based on ancient texts, Ratan Thiyam's productions for his Chorus Repertory Theatre in Imphal use traditional dances, rituals, and martial arts together with emphatic lighting, glamorous costumes, and sophisticated use of stage space; this production of *Chahravgnka* travelled to the Edinburgh Festival in 1987.

emphasis on the quality of performance has brought back a close focus on actors that has always been enjoyed by many simple theatres. 'Performance art', the development of shows out of the actors' own explorations, sometimes without any textual basis, sometimes borrowing from dance or acrobatics, and often using music, is a reclaiming of the earlier freedom of performers in cabaret or music hall, or in independent travelling troupes where the actors were in charge—a rediscovery too of personal initiative. Perhaps what has been most innovatory is not a single kind of theatre, but a general tendency to draw disparate elements together, to mix old and new, to use electronic reproduction, film, TV, dance, music, literature, painting, sculpture, sport, journalistic reportage, drawing any of these elements together with actors doing what they have always done, to travel with one show across continents to perform for audiences who understand only other languages, to work as an artist in theatre and also in film, TV, radio, journalism, sport, or education: the intercultural multi-competence of theatre at the end of the twentieth century is both new in itself and hopeful for the future. The ability to transcend difficulties of verbal language and customs, to use the apparently simplest of means—a person's hand or face, an inarticulate cry, an empty space—or the most complex—a decaying citadel alive with rumours and echoes from the past or electronic equipment costing millions and needing highly trained operators: this diversity is truly amazing. Yet theatre by the end of the twentieth century has learnt how to draw these means together for moments of performance that can seem an enhancement of an audience's own thoughts and desires.

❧ CHRONOLOGY ❧

Items shown in bold indicate political events

BC

*c.*3000–1500	**Indus Valley Civilization**
*c.*2600	Earliest inscriptional evidence for festival performances in Egypt
*c.*1918–1875	The Ramesseum Dramatic Papyrus
508	**Introduction of new 'democratic' constitution in Ancient Greece**
	Possibly first performances of tragedies at the City Dionysia in Ancient Greece (trad. date 534)
499	First production by Aeschylus
490	**Persian invasion of Greece repulsed by Athenians at Marathon**
480	**Naval victory, led by Athens, over second Persian invasion at Salamis (victory on land at Plataea in 479); victory of Greeks in Sicily over Carthaginians**
461	**Further democratizing reforms at Athens; Pericles' era of influence begins**
458	Aeschylus *Oresteia*
456	Death of Aeschylus; Euripides' first production
430s	Sophocles at maturity, elected a general; *Oedipus the King* probably produced; Comedy also becomes important at the City Dionysia (trad. date of introduction 486)
431	Euripides *Medea*
431–429	**Major war between Athens and Sparta begins; plague at Athens; death of Pericles**
425	Aristophanes (aged 20) *Acharnians*
415–13	**Athenian expedition to Sicily ends in disaster**
412–11	Euripides *Helen*; Aristophanes *Thesmophoriazousae*
406–5	Deaths of Euripides (*Bacchae* produced posthumously), and of Sophocles (*Oedipus at Colonus* produced posthumously)
404	**Athens finally loses the war against Sparta, but recovers quite quickly**
388	Aristophanes dies; Plato the philosopher at maturity
386	Regular performances of 'old' tragedies at Dionysia begin
338	**Philip, king of Macedon, defeats Thebes and Athens and takes over mainland Greece; Theatre at Epidaurus built about this time**
336	**Alexander succeeds Philip** Aristotle active at Athens (dies 322)
*c.*330	Official texts of the 'canon' of old tragedies made
323	Death of Alexander after conquests as far as India

319	First production by Menander
311	Division of Alexander's territories into the Hellenistic kingdoms
292	Death of Menander; Artists of Dionysus probably set up
240	First Greek-style play performed in Latin
218	**Hannibal invades Italy**
206	Guild of actors and writers established in Rome
*c.*200–AD 200	*Natyasastra* composed
*c.*200–100	Period of composition of first Sanskrit plays
191	Plautus *Pseudolus*
160	Death of Terence
*c.*110	The text of a traditional Egyptian performance, *The Triumph of Horus*, is engraved on the Ptolemaic temple of Edfu
55	Dedication of Pompey's stone theatre
30	**Rome ceases to be a republic and Octavian (Augustus) assumes monarchical power**

AD

65	Death of Seneca
67	Nero performing in Greece
*c.*100–200	Plays of Asvaghosa
*c.*300–400	Bhasa and Sudraka lived and wrote
330	**Constantine, the first Christian emperor, makes Byzantium capital of the Roman empire**
*c.*400–500	Kalidasa composed plays and poems
476	**Deposition of the last Roman emperor in the West**
526	Justinian I withdraws state support from Byzantine theatre
549	**Last recorded games in Rome**
679	Council of Rome orders English Church to ban plays
692	Justinian II legislates in Byzantium against theatre and worship of Dionysus
802	Jayavaraman II travelled from Java to Khymer kingdom, bringing performance traditions with him
*c.*960–70	Hroswitha writing in Germany
*c.*1000	Approximate end of classical Sanskrit drama and reformation of *kutiyattam*
*c.*1110	First record of a saint play in England
*c.*1150	Hildegard *Order of Virtues*; first surviving Passion play (Italy)
*c.*1160	*Play of Antichrist* (Germany)
1204	**Christian forces pillage Constantinople: end of the Byzantine empire**
1287	**Mongol invasion of Burma and consequent destruction of early Pagan culture**
*c.*1350–75	Biblical cycles first performed in Britain
1374	The young Japanese shogun Ashikaga Yoshimitsu invites Kan'ami's Noh troupe to perform at court and decides to extend shogunal patronage to this form of theatre
1431	**Fall of Angkor to the Thais** Cambodian tradition imported to Ayutthaya

1476 First evidence of Corpus Christi plays on pageant wagons in Spain

1480 Poliziano *Fabula di Orfeo*

1492 **Completion of the Reconquest; discovery of America**

1496 First publication of plays by Juan del Encina

1508 Ariosto *La cassaria: commedia*

1511 **Portuguese victory in Malacca and the beginning of European influences in Malaysia**

1513 Bibbiena *La Calandria: commedia;* **Giovanni de' Medici becomes Pope Leo X**

1515 Trissino *Sofonisba: tragedia;* Rucellai *Rosmunda: tragedia;* First recorded performance of *Khon* masked drama in the Thai court

1517 **Luther initiates the Reformation in Europe with his 95 theses**

1519 **Invasion of Mexico**

1527 **Sack of Rome by imperial forces of Charles V**

1531 Congrega dei Rozzi founded; Academy of the Intronati, Italy, produces *Gli inqannati*

1533 **Excommunication of Henry VIII of England**

1539 Last carnival at Nuremberg

1540 **Society of Jesus chartered by Pope Paul III**

1541 Giraldi *Orbecche: tragedia*

1542 Speroni *Canace: tragedia*

1545 First extant contract among professional actors

1545–63 **Council of Trent implements the Counter-Reformation**

1547 Passion play at Valenciennes

1548 The Confraternity of the Passion acquires the Hôtel de Bourgogne as a site for the first permanent Paris theatre; an edict of the Paris Parlement bans the performance of mystery plays

1558 Theatre at Valladolid, Spain proposed (apparently) by Lope de Rueda
Accession of Elizabeth I to the English throne

1559 Publication of Spain's first *Index of Prohibited Books*

1561 **Capital of Castile established at Madrid**

1567 John Brayne builds a playhouse on the Red Lion farm

1568 First description of a *commedia dell'arte* performance, improvised by Orlando di Lasso and others in Munich

1571 *Comici dell'arte* go to Paris; first Italian company performs at the French Court

1573 Tasso *Aminta: favola pastorale;*

1574 Elizabeth I of England issues a patent to Leicester's Men

1576 James Burbage builds the Theatre in Shoreditch, London

1579 Last performance of the Coventry cycle; Corral de la Cruz opened in Spain

1583 Elizabeth I appoints the company of Queen's Men; Corral del Príncipe opened in Spain

1585 Teatro Olimpico at Vicenza completed

1588 **Defeat of the Spanish Armada against England**

1646–55 Ottonelli *Della Christiana moderatione del theatro*

1649 **Execution of Charles I of England**

1652 First palace productions by Baccio del Bianco

1656 Performance of *The Siege of Rhodes* at Rutland House, London

1658 Molière's company begins to perform in Paris

1660 **Restoration of Charles II** Charles II issues warrants to Davenant and Killigrew; first appearance of actresses on the professional stage

1661 **Beginning of Louis XIV's personal rule in France** The Duke's Company opens the Lincoln's Inn Fields playhouse, London

1668 Ottavio Burnacini's Imperial Theatre at Vienna inaugurated with spectacular music drama *Il pomo d'oro*, culminating two years of festivities for wedding of Leopold I and Marguérite of Austria

1671 The Duke's Company opens the Dorset Garden playhouse, London

1672 Lulli takes over the privilege for the Paris Opera

1673 Molière's death leads to the first amalgamation of Paris theatre companies by royal decree

1674 The King's Company opens the Drury Lane playhouse, London

1677 Racine ceases to write for the public theatre after *Phèdre*

1680 The second amalgamation of theatre companies by French royal decree leads to the foundation of the Comédie-Française

1681 Uniting of the Duke's Company and the King's Company

1688 **Expulsion of James II and accession of William III and Mary II in England**

1692 Foundation of the Society for the Reformation of Manners

1693 Christopher Rich assumes effective control of the United Company, England

1695 The rebel company of players, headed by Betterton and Barry, open *Love for Love* at Lincoln's Inn Fields

1697 The Italian players are expelled from Paris

1699 Perrucci *Dell'arte rappresentativa, premeditata e all'improvviso*

1701–14 **War of Spanish Succession**

1715 **Death of Louis XIV of France** Term 'opéra-comique' first appears

1716 Riccoboni invited to reopen Comédie-Italienne in Paris

1720 Steele edits first theatre journal in England, *Theatre*

1722 Première of Ludvig Holberg's *The Pewterer Who Wanted to be a Politician* in Copenhagen heralds birth of modern Danish theatre

1728 Première of Gay's *Beggar's Opera* in Lincoln's Inn Fields

1731 Premières of Lillo's *The London Merchant* at Drury Lane Theatre and Voltaire's *Zaïre* at Comédie-Française

1737 'Licensing Act' reduces number of London theatres to two and sharpens censorship

1745 Première of Goldoni's *The Servant of Two Masters* in Venice

1747 Garrick becomes joint manager of Drury Lane Theatre; première of Sumarokov's *Khorev* in Petersburg, the first authentic Russian tragedy

1753 Maria Theresa bans improvisation and knockabout pieces in Vienna

1756–63	**Seven Years War between France and England**
1759	Banishment of spectators from stage at Comédie-Française
1761	Parisian première of Diderot's *The Head of the Family* (*Le Père de famille*) in Marseilles
1767	**Fall of Thai capital, Ayutthaya, to the Burmese; Thai traditions imported to Burma** Première of Lessing's *Minna von Barnhelm* in newly-founded National Theatre in Hamburg
1770	**Boston Tea Party**
1771	Accession of Gustav III to Swedish throne leads to vigorous promotion of theatre in Stockholm; première of Cumberland's *West Indian*
1773	Première of Goldsmith's *She Stoops to Conquer* in Drury Lane
1776	**US Declaration of Independence**
1777	Première of Sheridan's *School for Scandal;* Klinger's play *Sturm und Drang* gives name to Storm and Stress movement
1782	Première of Schiller's *The Robbers* in Mannheim
1783	**England recognizes independence of USA**
1784	First use of oil-lamps for stage lighting for public première of Beaumarchais's *The Marriage of Figaro* at Comédie-Française
1789	**French Revolution begins**
1790	Entry of the Anhui companies into Beijing in celebration of the Qianlong Emperor's eightieth birthday
1791	Goethe becomes director of Court Theatre at Weimar (until 1817)
1793	**Louix XVI of France executed in Paris**
1799	**Napoleon Bonaparte named First Consul in France**
1800-1900	Colonial period with origin and development of modern urban drama and theatre (India)
1806–7	Napoleonic decrees establishing the number of major and minor theatres in Paris: four in each category
1809	Drury Lane burns; enlarged Covent Garden reopens after fire previous year Old Price riots, London
1816	First introduction of gas into stage lighting, at the Chestnut Street Theatre, Philadelphia; début of Jean-Gaspard Deburau at the Théâtre des Funambules, Paris
1826	First Western account of an Apidan masquerade
1827	Victor Hugo publishes his romantic manifesto in the Preface to *Cromwell*
1831	First showboat for river touring launched at Pittsburgh by William Chapman
1833	Dramatic Copyright Act and foundation of Dramatic Authors' Society: first English copyright protection for plays performed
1837	**Queen Victoria ascends the English throne**
1840	Karl Immerman's open-stage, architectural production of *Twelfth Night* in Düsseldorf
1843	Theatre Regulation Act: abolition of Drury Lane and Covent Garden monopoly and extension of Lord Chamberlain's censorship powers to plays performed anywhere in Britain

1848 **Revolutions in Europe**

1849 Début of Edwin Booth at the Boston Museum

1850 Ibsen's first produced play, *The Warrior's Barrow*, at the Christiania Theatre; the text of *Rabinal-Achi*, a Mayan pre-conquest drama preserved through oral tradition, is recorded by Brasseur de Barbourg

1851 Dingelstedt's productions of Sophocles in Munich on an imitation Greek stage

1852 Canterbury Hall opens in Lambeth, London, the first purpose-built music hall **Proclamation of the Second Empire under Napoleon III**

1864 Abolition of the monopoly of the Comédie-Française over the classic French drama

1870 Strindberg's first produced play, *In Rome*, at the Dramaten, Stockholm

1872 Sarah Bernhardt's come-back at the Comédie-Française; The new Japanese Meiji government (established 1868) places actors under control of the Ministry of Religious Instruction in an attempt to use the theatre to improve public morals

1875 Tour of Indian popular theatre company to Penang leading to the creation of *bangsawan* performance style in 1885

1876 First production of Ibsen in Germany, *The Pretenders*, at Meiningen (later touring to Berlin)

1878 Henry Irving assumes management of the Lyceum Theatre, London

1879 Opening of the technologically advanced Madison Square Theatre, New York

1880 Squire Bancroft installs the first picture-frame stage at the Haymarket Theatre, London

1882 Legal monopoly of the imperial theatres abolished in Russia

1884 Buffalo Bill's first open-air Wild West show, in Nebraska

1887 Antoine starts the Théâtre Libre in Paris

1888 Opening of the new Burgtheater on the Ring in Vienna

1889 Brahm starts the Freie Bühne in Berlin

1891 Paul Fort starts the Théâtre d'Art in Paris; 31 March: Independent Theatre opens with Ibsen's *Ghosts* in London

1893 Lugné-Poë starts the Théâtre de l'Œuvre in Paris

1894 William Poël opens the Elizabethan Stage Society in London

1896 First night of *Ubu roi* at the Théâtre de l'Œuvre

1898 Moscow Art Theatre opened; **US take-over of the Philippines**

1904 The Abbey Theatre opens in Dublin

1904–7 The Vedrenne–Barker management at the Court Theatre

1905 Max Reinhardt takes over the Deutsches Theater, Berlin

1906 Final full-scale court performance of *gambuh* dance-drama in Bali

1907 First staging of a Western play in translation in Vietnam (Molière's *The Miser*); Strindberg opens his Intimate Theatre in Stockholm; the Chinese spoken drama form is born with the performance of *The Black Slave's Cry to Heaven* in Chinese in Tokyo

1909 First visit of the Russian Ballet to Paris

1912 The Republic of China replaces the overthrown Manchu Qing dynasty of China, established 1644

1913 Jacques Copeau opens the Théâtre du Vieux-Colombier, Paris

1914 **Outbreak of First World War** Lilian Baylis takes over at the Old Vic, London

1918 **End of First World War; revolution in Germany**

1919 Reinhardt opens the Grosse Schauspielhaus, Berlin; First Production in Reinhardt's 'Theatre of the 5,000' (*Oresteia*); Copeau remodels Vieux-Colombier stage

1920 **Irish civil war begins** Inauguration of the first Salzburg Festival *Every man* played in Salzburg for the first time; first O'Neill play (*Emperor Jones*) produced in New York

1921 First text-only original Vietnamese play written and performed, *A Cup of Poison* by Vu Dinh Long

1923 Meyerhold Theatre founded (Moscow); 'Blue Blouse' agitprop groups formed

1924 Stanislavski *My Life in Art* published; opening of the Tsukuji Little Theatre, significant in the development of modern Japanese theatre

1925 Teatro d'Arte founded by Pirandello in Rome

1927 Piscator takes over the Volksbühne in Berlin

1928 Brecht *Threepenny Opera* opens in Berlin

1929 **Wall Street crash**

1931 Group Theatre founded by Clurman and Strasberg in New York

1933 **Burning of the Reichstag; Hitler takes power in Germany** Brecht, Reinhardt, and Piscator leave Germany

1935 Federal Theatre Project inaugurated in USA; death of Japanese Tsubouchi Shōyō, a major leader of the 'new drama' movement (b. 1859)

1936 **Spanish Civil War begins**

1939 **Second World War begins** Last major court ceremonial performance of *wayang wong* in Java 1939

1940 Meyerhold shot

1942 Mao Zedong's 'Talks at the Yan'an Forum on Literature and Art'

1945 **Hiroshima; Second World War ends; start of 'Cold War'**

1947 Strehler's first productions at the Piccolo Theatre in Rome (incl. *Arlecchino*) **Indian Independence**

1949 Brecht's Berliner Ensemble opens in East Berlin **The Chinese Communist Party wins victory in China, setting up the People's Republic of China**

1950 Establishment of Korean National Theatre

1952 First All-China Festival of Traditional Music-Dramas held in Beijing; first 'Off-Broadway' success (Tennessee Williams *Summer and Smoke*)

1953 Blin directs Beckett's *Waiting for Godot* in Paris

1954 Joe Papp founds New York Shakespeare Festival; inauguration of the National School of Drama in India

1956 English Stage Company founded at the Royal Court, London, and performance of *Look Back in Anger;* first All-Chinese Spoken Drama Festival held in Beijing

1958 'Off-Off-Broadway' movement opened with Joe Cino's Caffe productions

1961 Death of famous Peking Opera actor Mei Lanfang (b. 1894)

1964 All-China Festival of Peking Operas on Contemporary Themes held in Beijing; Ariane Mnouchkine founds Le Théâtre du Soleil in Paris

1965 **Vietnam War begins**

1968 **Student Revolution in Europe** Abolition of stage censorship in Britain; Living Theatre tours Europe with *Paradise Now;* Dario Fo and Franca Rama found Nuova Scene theatre company in Milan

1969 **Man walks on the moon for the first time**

1970 Ritual suicide of Mishima Yukio, famous right-wing Japanese dramatist and literary figure (b. 1925); Peter Brook founds the International Centre for Theatre Research in Paris

1971 Yury Lyubimov's production *Hamlet* starts its ten-year run in the repertoire of the Taganka Theatre, Moscow

1972 *Jesus Christ Superstar* opens at the Palace Theatre, London

1973 Peter Hall succeeds Lord Olivier as director of the National Theatre in London and prepares the company to move into its new building on the South Bank

1975 **Khmer Rouge take-over in Cambodia** Performance traditions destroyed and performers murdered in Cambodia; the Wooster Group is founded and *The Chorus Line* opens in New York

1976 Robert Wilson's *Einstein on the Beach* has its first American production in New York City

1977 Peter Stein directs *As You Like It* at the Schaubühne, Berlin

1982 Suzuki Tadashi founds the Toga International Arts Festival in Japan

1986 Arts Council of Great Britain's *Enquiry into Professional Theatre in England*

1989 **Protests in Tiananmen Square, Beijing; the Berlin Wall is demolished** Dramatist Vaclav Havel elected President of Czechoslovakia

1992 Giorgio Strehler directs the *Faust Festival* at the Piccolo's Teatro Studio in Milan

1993 In Tokyo, after a year's absence, Noda Hideki re-forms his theatre company as Noda Map

✣ FURTHER READING ✣

CHAPTER ONE

Umberto Albini, *Nel nome di Dioniso: vita teatrale nell'Athene Classica* (Milan, 1991); a useful survey of the practicalities and texts.

Simon Goldhill, *Reading Greek Tragedy* (Cambridge, 1986); an application of post-structuralist methods.

Eric Walter Handley, 'Comedy', in *The Cambridge History of Classical Literature*, i: *Greek*, ed. P. Easterling and B. Knox (Cambridge, 1985).

C. J. Herington, *Poetry into Drama: Early Tragedy and Greek Poetic Tradition* (Berkeley, Calif., 1985); relating the emergence of theatre to pre-existing poetic and musical activities and contests.

Richard L. Hunter, *The New Comedy of Greece and Rome* (Cambridge, 1985).

Bernard M. W. Knox, *The Heroic Temper* (Berkeley, Calif., 1964); on the central intransigent figures of Sophocles.

—— *Word and Action* (Baltimore, 1979); collected articles and reviews of a great critic.

H.-J. Newiger, *Metapher und Allegorie* (Munich, 1957); on the centrality of the literalization of metaphors and abstracts in Aristophanes.

A. Pickard Cambridge, *The Dramatic Festivals of Athens*, 2nd edn., rev. J. Gould and D. Lewis (Oxford, 1968; addenda 1988); the basic collection of evidence.

Rush Rehm, *Greek Tragic Theatre* (London, 1992).

Erika Simon, *The Ancient Theatre* (London, 1982); a succinct account of the archaeology; translated from German.

Oliver Taplin, *Greek Tragedy in Action* (London, 1978; rev. 1985); performance criticism applied to nine tragedies.

Jean-Pierre Vernant and Pierre Vidal-Naquet, *Myth and Tragedy in Ancient Greece* (New York, 1988); collected articles by the two leaders of the approach through the structures of Athenian 'mentalities'; translated from French.

David Wiles, *The Masks of Menander* (Cambridge, 1991); an approach to New Comedy through performance, especially the significance of the masks.

John Winkler and Froma Zeitlin (eds.), *Nothing to Do with Dionysos?* (Princeton, NJ, 1990); a collection of the best essays on Athenian drama in its social and ideological context.

R. Winnington-Ingram, *Studies in Aeschylus* (Cambridge, 1983).

CHAPTER TWO

R. Axton, *European Drama of the Early Middle Ages* (London, 1974); the best study of early medieval theatre.

Richard C. Beacham, *The Roman Theatre and its Audience* (London, 1991); covers the entire Roman period.

Peter Burke, *Popular Culture in Early Modern Europe* (London, 1979); an analysis of carnivalesque phenomena.

A. S. Gratwick, 'Drama', in *The Cambridge History of Classical Literature*, ii: *Latin Literature* ed. E. J. Kenney (Cambridge, 1982), 77–137; an essay on Roman comedy.

Peter Dronke, *Women Writers of the Middle Ages* (Cambridge, 1984); a good essay on Hroswitha, though nothing on staging.

Erich S. Gruen, *Studies in Greek Culture and Roman Policy* (Leiden, 1990); the cultural and political background to Plautus.

John Wesley Harris, *Medieval Theatre in Context: An Introduction* (London, 1992); a readable introduction, good on church drama.

David Bevington (ed.), *Medieval Drama* (Boston, 1975); an anthology of texts, admirably annotated.

Allardyce Nicoll, *Masks, Mimes and Miracles* (London, 1931); explores the unscripted dramatic tradition which links the Roman period to the medieval.

Erich Segal, *Roman Laughter: The Comedy of Plautus* (2nd edn., New York, 1987); a lively analysis of the psychology of Plautine comedy.

Eckehar Simon (ed.), *The Theatre of Medieval Europe: New Research into Early Drama* (Cambridge, 1991).

Martin Stevens, *Four Middle English Mystery Cycles* (Princeton, NJ, 1987); the social context for English mystery cycles, and their artistic structures.

William Tydeman, *The Theatre in the Middle Ages* (Cambridge, 1978); survey of staging methods, with a European perspective.

Glynne Wickham, *Early English Stages* (London, 1959–); closely documented volumes exploring different modes of medieval performance.

CHAPTER THREE

Anne Chapman, *Drama and Power in a Hunting Society* (Cambridge, 1982); an account of the Selk'nam of Tierra del Fuego and the importance that the Hain ('comparable to the theatre in our society') had in their social life.

Michael Etherton, *The Development of African Drama* (London, 1982); an account of traditional and modern performance in Africa, with particular emphasis on drama as a process of social development.

H. W. Fairman, *The Triumph of Horus* (London, 1974); a translation of the Egyptian festival play engraved on the Ptolemaic temple of Edfu, with a discussion from a literary viewpoint as to whether there *was* drama in Ancient Egypt.

Theodor H. Gaster, *Thespis: Ritual, Myth, and Drama in the Ancient Near East* (2nd rev. edn., New York, 1961); an examination of the evidence for ritual theatre in the Ancient Near East, which includes an English translation of the Ramesseum Dramatic Papyrus.

Kacke Götrick, *Apidan Theatre and Modern Drama* (Stockholm, 1984); an account of Apidan (Alarinjo) theatre and its influence on contemporary Nigerian playwrights.

Anthony Graham-White, *The Drama of Black Africa* (New York, 1974); a seminal study of African plays and playwrights, which includes an important survey of traditional theatre forms.

V. Laski, *Seeking Life* (Philadelphia, 1958); a description of the Tewa Raingod Drama.

Miguel Leon-Portilla, *Pre-Columbian Literatures of Mexico* (Norman, Okla., 1969); includes an account of pre-Columbian theatrical traditions.

Miriam Lichtheim, *Ancient Egyptian Literature*, i (Berkeley, Calif., 1973).

Carroll E. Mace, *Two Spanish-Quiche Dance Dramas of Rabinal* (New Orleans, 1970); a transcription of two indigenous dance dramas, together with an account of the persistence of pre-Columbian elements in such performances.

Wole Soyinka, *Myth, Literature and the African World* (Cambridge, 1976).

CHAPTER FOUR

Richard Andrews, *Scripts and Scenarios: The Performance of Comedy in Renaissance Italy* (Cambridge, 1993).

Christopher Cairns (ed.), *The Commedia dell'Arte from the Renaissance to Dario Fó* (Lewiston, NY, 1989); essays on various aspects and phases of *commedia dell'arte* style.

Silvia Carandini, *Teatro e spettacolo nel Seicento* (Bari, 1990); a concise survey of 17th-century Italian theatre.

Louise George Clubb, *Italian Plays (1500–1700) in the Folger Library* (Florence, 1968); description and history of a representative collection of about 800 printed plays of the Italian Renaissance.

—— *Italian Drama in Shakespeare's Time* (New Haven, Conn., 1989); essays on the major genres, theories, and structural principles of post-medieval Italian theatre.

Pierre Louis Duchartre, *The Italian Comedy*, trans. Randolph T. Weaver (New York, 1966).

Thomas F. Heck, *Commedia dell'Arte: A Guide to the Primary and Secondary Literature* (New York, 1988); an exhaustive annotated bibliography of the *commedia dell'arte*.

Marvin T. Herrick, *Italian Comedy in the Renaissance* (Urbana, Ill., 1960); description and classification of selected comedies.

—— *Italian Tragedy in the Renaissance* (Urbana, Ill., 1965); description and classification of selected tragedies.

Kathleen M. Lea, *Italian Popular Comedy: A Study in the Commedia dell'Arte 1560–1620, with Special Reference to the English Stage*, 2 vols. (Oxford, 1934).

J. R. Mulryne and Margaret Shewring (eds.), *Theatre of the English and Italian Renaissance* (Basingstoke, 1991).

Marzia Pieri, *La nascita del teatro moderno in Italia tra XV e XVI secolo* (Turin, 1989); a concise survey of Italian theatre from the 15th to the 16th centuries.

Nino Pirrotta and Elena Povoledo, *Music and Theatre from Poliziano to Monteverdi*, trans. Karen Eales (Cambridge, 1982).

Kenneth Richards and Laura Richards, *The Commedia dell'Arte: A Documentary History* (Oxford, 1990).

Ellen Rosand, *Opera in Seventeenth-Century Venice: The Creation of a Genre* (Berkeley, Calif., 1991).

CHAPTER FIVE

John J. Allen, *The Reconstruction of a Spanish Golden Age Playhouse: El Corral del Príncipe 1583–1744* (Gainesville, Fla., 1983); a detailed, illustrated account of the history and structure of this theatre.

Walter Cohen, *Drama of a Nation: Public Theater in Renaissance England and Spain* (Ithaca, NY, 1985); a history of the development of drama in England and Spain, from a Marxist perspective.

Margaret Rich Greer, *The Play of Power: Mythological Court Dramas of Calderón de la Barca* (Princeton, NJ, 1991); a reappraisal of Calderón's court drama, including detailed discussion of several plays.

Barbara E. Kurtz, *The Play of Allegory in the Autos Sacramentales of Pedro Calderón de la Barca* (Washington, DC, 1991); a study of the kinds of allegory exploited by Calderón.

Donald R. Larson, *The Honor Plays of Lope de Vega* (Cambridge, Mass., 1977); a study of Lope de Vega's treatment of the theme of honour.

Melveena McKendrick, *Theatre in Spain 1490–1700* (Cambridge, 1989); a comprehensive study.

—— *Women and Society in the Spanish Drama of the Golden Age* (Cambridge, 1974); a study of different types of *comedia* heroines.

Alexander A. Parker, *The Mind and Art of Calderón: Essays on the Comedias* (Cambridge, 1988); a study of Calderón's *comedias*, both general and specific.

N. D. Shergold, *A History of the Spanish Stage from Medieval Times until the End of the Seventeenth Century* (Oxford, 1967); a detailed account of all aspects of Golden Age staging.

Ronald E. Surtz, *The Birth of a Theater: Dramatic Convention in the Spanish Theater from Juan del Encina to Lope de Vega* (Princeton, NJ, 1979); a study of sixteenth-century Spanish drama.

John E. Varey, *Cosmovisión y escenografía: el teatro español en el Siglo de Oro* (Madrid, 1987); a collection of twenty-two studies, in Spanish, devoted mainly to the performance of seventeenth-century drama.

Edward M. Wilson and Duncan Moir, *The Golden Age: Drama 1492–1700* (A Literary History of Spain, ed. R. O. Jones; London, 1971); a general survey of the period, including detailed reference to many playwrights and plays.

CHAPTER SIX

G. E. Bentley, *The Profession of Player in Shakespeare's Time* (Princeton, NJ, 1984); an appraisal of the social and professional life of actors, carefully based on available documentation.

Martin Butler, *Theatre and Crisis 1632–1642* (Cambridge 1984); a radical re-evaluation of Caroline drama in its social and political context.

Andrew Gurr, *The Shakespearean Stage, 1574–1642* (3rd edn., Cambridge, 1992); a judicious summary of the social and physical theatre of the age.

—— *Playgoing in Shakespeare's London* (Cambridge, 1987); a non-partisan assessment of the evidence on the composition and tastes of the various audiences at London's play-houses.

Peter Holland, *The Ornament of Action* (Cambridge, 1979); a stage-centred study of Restoration drama.

J. W. Krutch, *Comedy and Conscience after the Restoration* (New York, 1924; rev. edn. 1949); an investigation of the moral climate that released Collier's ferocious attack on late 17th-century drama.

David Lindley, *The Court Masque* (Manchester, 1984).

Stephen Orgel and Roy Strong, *Inigo Jones: The Theatre of the Stuart Court*, 2 vols. (Berkeley, Calif., 1973).

Jocelyn Powell, *Restoration Theatre Production* (London, 1984); a valuable combination of theatre history and stage practice.

Richard Southern, *Changeable Scenery: Its Origin and Development in the English Theatre* (London, 1952).

David Thomas (ed.), *Restoration and Georgian England: A Documentary History* (Cambridge, 1989); original documents, succinctly contextualized.

Peter Thomson, *Shakespeare's Theatre* (2nd edn., London, 1992); an analysis of the likely priorities of the Chamberlain's/King's Men at the First Globe.

Erica Veevers, *Images of Love and Religion: Queen Henrietta Maria and Court Entertainments* (Cambridge, 1989); an examination of the central role played by Charles I's queen in the complex theatrical manœuvres of the period.

Glynne Wickham, *Early English Stages*, ii. *1576–1660* (pt. 1, 1963; pt. 2, 1972); a detailed study of playhouse architecture in the first years of the professional theatre.

CHAPTER SEVEN

W. D. Howarth, *Molière: A Playwright and his Audience* (Cambridge, 1982); relates Molière's comedy to the spectators for whom he wrote.

Brian Jeffery, *French Renaissance Comedy, 1552–1630* (Oxford, 1969); a thorough analysis of the major comedies of the period.

H. C. Lancaster (ed.), *Le Mémoire de Mahelot, Laurent et d'autres décorateurs de l'Hôtel de Bourgogne et de la Comédie-Française au xviie siècle* (Paris, 1920); contains facsimile reproductions of the forty-seven sketches of the designer Mahelot.

Thomas E. Lawrenson, *The French Stage in the XVIIth Century* (Manchester, 1957); sub-titled 'A Study in the Advent of the Italian Order', this important work was republished in revised form after the author's death as *The French Stage and Playhouse in the XVIIth Century*, ed. D. Whitton (New York, 1986).

John Lough, *Paris Theatre Audiences in the 17th and 18th Centuries* (London, 1957); a pioneering study of the sociology of the Paris theatre in the neo-classical period.

—— *Seventeenth-Century French Drama: The Background* (Oxford, 1979); the best short study in English of the theatrical conditions in which the classical masterpieces were produced.

G. McAuley (ed.), *From Page to Stage: L'Illusion comique* (Sydney, 1987); a unique attempt at a truly theatrical reading of a Corneille play.

David Maskell, *Racine: A Theatrical Reading* (Oxford, 1991); breaks new ground in Racine studies.

Barbara G. Mittman, *Spectators on the Paris Stage in the Seventeenth and Eighteenth Centuries* (Ann Arbor, Mich., 1984); studies an important aspect of the theatre history of the period.

G. Mongrédien, *Daily Life in the French Theatre at the Time of Molière*, trans. C.-E. Engel (London, 1966); an excellent evocation of the atmosphere of the seventeenth-century theatre.

William Leon Wiley, *The Early Public Theatre in France* (Cambridge, Mass., 1960); perhaps the most readable survey of the way in which material conditions in the theatre developed through the period.

CHAPTER EIGHT

Emmett L. Avery *et al.*, *The London Stage 1660–1800* (Carbondale, Ill., 1960–8); a day-by-day calendar of all known performances at London theatres, in the period, with superb introductions.

Richard W. Bevis, *English Drama: Restoration and Eighteenth Century, 1660–1789* (London, 1988); a critical study of drama in the period.

George W. Brandt and Wiebe Hogendoorn (eds.), *German and Dutch Theatre, 1600–1848* (Cambridge, 1992); a useful collection of documents relating to German and Dutch theatre 1600–1848.

Marvin Carlson, *Goethe and the Weimar Theatre* (Ithaca, NY, 1978); the best study in English of the Court Theatre at Weimar under Goethe's directorship.

—— *The Italian Stage from Goldoni to Annunzio* (London, 1981); a well-written survey of 18th- and early 19th-century Italian drama.

N. Glendinning, *The Eighteenth Century (A Literary History of Spain)* (London, 1972); a general history of Spanish literature, including drama.

Philip H. Highfill *et al.*, *A Biographical Dictionary of Actors, Actresses . . .* (Carbondale, Ill. 1973); a multi-volume dictionary covering anyone who worked in the theatre.

Robert D. Hume (ed.), *The London Theatre World, 1660–1800* (Carbondale, Ill. 1980); a collection of essays on a wide range of aspects of London theatre.

Eleanor F. Jourdain, *Dramatic Theory and Practice in France 1690–1808* (London, 1921); an old but still valuable work.

Simon Karlinsky, *Russian Drama from its Beginnings to the Age of Pushkin* (Berkeley, Calif., 1985); a detailed study of the strange development of Russian drama.

Allardyce Nicol, *The Garrick Stage* (Manchester, 1980); a well-illustrated study of stages, staging, and audiences in the 18th century.

Michael Patterson, *The First German Theatre* (London, 1990); a study of major plays by Schiller, Goethe, Kleist, and Büchner in performance.

Cecil Price, *Theatre in the Age of Garrick* (London, 1973); a vivacious short study of drama and theatre in the mid-18th century.

Laurence Senelick (ed.), *National Theatre in Northern and Eastern Europe, 1746–1900* (Cambridge, 1991); a companion volume to Brandt and Hogendoorn; a compilation of theatre documents.

George Winchester Stone and G. M. Kahrl, *David Garrick: a Critical Biography* (Carbondale, Ill., 1979); a large, authoritative study organized to explore different facets of Garrick's career.

David Thomas and Arnold Hare (eds.), *Restoration and Georgian England, 1660–1788* (Cambridge, 1989); a superb collection of documents about the theatre.

Simon Williams, *German Actors of the Eighteenth and Nineteenth Centuries* (Westport, Conn., 1985); contains a study of Ludwig Devrient.

CHAPTER NINE

Robert Baldick, *The Life and Times of Frédérick Lemaître* (London, 1959); contains much information on the boulevard theatres of the 1820s and 1830s.

Michael R. Booth, *Theatre in the Victorian Age* (Cambridge, 1991).

—— *Victorian Spectacular Theatre 1850–1910* (London, 1981).

J. S. Bratton (ed.), *Music Hall: Performance and Style* (Milton Keynes, 1986); a collection of essays on the performance side of music hall.

Marvin Carlson, *The French Stage in the Nineteenth Century* (Metuchen, NJ, 1972).

—— *The German Stage in the Nineteenth Century* (Metuchen, NJ, 1972).

Joseph Donohue, *Theatre in the Age of Kean* (Oxford, 1975).

Douglas Gilbert, *American Vaudeville* (New York, 1940).

Russell Jackson (ed.), *Victorian Theatre* (London, 1989); contemporary documents relating to audiences, actors, production, management, and the work of dramatists, with commentary.

George Henry Lewes, *On Actors and the Art of Acting* (London, 1875); essays on English, French, and Italian actors, and on the state of the theatre in France, Germany, and Spain.

Bruce A. McConachie, *Melodramatic Formations: American Theatre and Society 1820–1870* (Iowa City, 1992); stresses the economic, political, and social context.

Frederick J. and Lise-Lone Marker, *The Scandinavian Theatre* (Oxford, 1975).

J. P. Moynet, *French Theatrical Production in the Nineteenth Century* (*L'Envers du théâtre*, 1873), trans. and augmented by Allan S. Jackson and M. Glen Wilson (Birmingham, NY, 1976); an explanation of the technology and production methods of the Parisian stage, well illustrated.

John Osborne, *The Meiningen Court Theatre 1866–1890* (Cambridge, 1988).

Marc Slonim, *Russian Theater from the Empire to the Soviets* (London, 1963); only 60 pages on the pre-Stanislavski nineteenth-century theatre.

Robert C. Toll, *Blacking Up: The Minstrel Show in Nineteenth Century America* (Oxford, 1974).

CHAPTER TEN

Richard C. Beacham, *Adolphe Appia, Theatre Artist* (Cambridge, 1987).

Jean Benedetti (ed.), *The Moscow Art Theatre Letters* (London, 1991); a collection, arranged to tell a narrative of the opening seasons of the Art Theatre, from correspondence by

Stanislavski, his colleague Nemirovich, some actors, Chekhov, Bulgakov, and others; the story is then briefly sketched up to 1938.

Jean Chothia, *Andrè Antoine* (Cambridge, 1991).

Edward Gordon Craig, *On the Art of the Theatre* (London, 1962); this brings together many papers published elsewhere and is the most basic account of his theories available.

Martin Esslin, 'Reinhardt, Creator of the Modern German Theatre', in *Mediations* (London, 1981).

Michael Holroyd, *Bernard Shaw*, 3 vols. (London, 1988–91); the most comprehensive biography available and a lively introduction to theatrical and cultural life of the period.

Dennis Kennedy, *Granville Barker and the Dream of Theatre* (Cambridge, 1985); a study which establishes Granville-Barker as leading dramatist as well as director, theoretician, scholar and actor.

David Magarshack, *Stanislavsky: A Life* (London, 1950).

—— *Chekhov, the Dramatist* (London, 1980); the first study of the structure of Chekhov's plays available in English; it has remained a classic account.

Hesketh Pearson, *The Last Actor-Managers* (London, 1950).

John Rudlin, *Jacques Copeau* (Cambridge, 1986).

Bernard Shaw, *Our Theatre in the Nineties*, 3 vols. (London, 1932); a collection which establishes Shaw both as a critic and a visionary of the theatre.

Peter Whitebrook, *William Archer: A Biography* (London, 1993).

CHAPTER ELEVEN

Antonin Artaud, *Theatre and its Double*, trans. Mary Caroline Richards (New York, 1958); essays and manifestoes for 'theatre of cruelty', calling for an anti-literary theatre and attacking the dualism of Western civilization: it formed the basis for avant-garde theatre in the 1960s.

Christopher Bigsby, *Twentieth Century American Drama*, 3 vols. (Cambridge, 1982–5).

David Bradby, *Modern French Drama 1940–1990* (Cambridge, 1991); an account of directors and theatre companies, as well as of dramatists.

Bertolt Brecht, *Brecht on Theatre*, trans. John Willett (London, 1978); a complete collection of Brecht's major writings on theatre, outlining the basis of his epic theory and its development into 'dialectical theatre'.

Ruby Cohn, *Just Play: Beckett's Theatre* (Princeton, NJ, 1980).

Martin Esslin, *The Theatre of the Absurd* (Harmondsworth, 1968); the first and most influential critical analysis of Existential drama, showing the links between Ionesco, Beckett, and dramatists in England, Germany, Poland, and Spain.

Christopher Innes, *Avant Garde Theatre: 1892–1992* (London, 1993); an analysis of the development of 'alternative' theatre from Jarry and the expressionists to Peter Brook and Robert Wilson, setting the movement in its political and cultural context.

—— *Modern British Drama: 1890–1990* (Cambridge, 1992).

George Rowell and Anthony Jackson, *The Repertory Theatre Movement* (Cambridge, 1984).

Raphael Samuel and Ewan MacColl, *Theatres of the Left: 1880–1935* (London, 1985); a history of British working-class and agitprop theatre between the wars, documenting productions and including scripts.

J. L. Styan, *The Dark Comedy* (Cambridge, 1961); overview of the cross-over between traditional genres that characterizes modern drama, covering developments from Chekhov to the 1950s.

—— *Modern Drama in Theory and Practice*, 3 vols. (Cambridge, 1981).

John Russell Taylor, *Anger and After* (London, 1962; rev. edn., 1969); an examination of the upsurge in British drama initiated by John Osborne's *Look Back in Anger* in the 1950s.

Kenneth Tynan, *A View of the English Stage* (London, 1975); the most significant critical voice in modern Britain, embodying the vision and evaluating the achievements of the post-1950s.

John Willett, *Theatre of the Weimar Republic* (New York, 1988).

CHAPTER TWELVE

SOUTH ASIAN THEATRES

Rachel Van M. Baumer and James R. Brandon (eds.), *Sanskrit Drama in Performance* (Honolulu, 1981). Numerous essays by various authorities collected from a major international conference.

Bharata, *Natyasastra*, trans. and ed. Mannohan Ghosh (Calcutta; vol. i, rev. edn., 1967; vol. ii, 1961); the most authoritative edition of this seminal work of ancient dramaturgy.

Rustom Bharucha, *Rehearsals of Revolution: The Political Theater of Bengal* (Honolulu, 1983); an account of selected writers, directors, and actors of the modern Bengali theatre, principally in Calcutta.

Balwant Gargi, *Folk Theater of India* (Seattle, 1956; Calcutta, 1993); an early illustrated study of selected genres of rural theatre.

John Stratton Hawley, *At Play with Krishna: Pilgrimage Dramas from Brindavan* (Princeton, NJ, 1981); an introduction and sample plays illustrating *ras lila*.

Goverdhan Panchal, *Kuttampalam and Kutiyattam* (New Delhi, 1984); a detailed and thoroughly illustrated study dealing principally with the architecture of *kutiyattam*.

Farley P. Richmond, Darius L. Swann, and Phillip B. Zarrilli (eds.), *Indian Theatre: Traditions of Performance* (Honolulu, 1990); survey from ancient to modern Indian theatre with selected examples of rural theatre genres. Well illustrated.

Kapila Vatsyayan, *Traditional Indian Theatre: Multiple Streams* (New Delhi, 1980); an overview of many rural theatre genres.

Phillip B. Zarrilli, *The Kathakali Complex: Actor, Performance, Structure* (New Delhi, 1984); most thorough study of *kathakali* in print, extensively illustrated.

EAST ASIAN THEATRES

Asian Theatre Journal is a twice-yearly scholarly periodical on Asian theatre, beginning in 1984, published by the University of Hawaii Press. It includes specialist articles, reports, book reviews, and occasionally translations of plays.

James R. Brandon, *The Cambridge Guide to Asian Theatre* (Cambridge, 1993); invaluable summary of the major performance forms throughout the region.

J. I. Crump and William P. Malm (eds.), *Chinese and Japanese Music-Dramas,* Michigan Papers in Chinese Studies (Ann Arbor, Mich., 1975).

David G. Goodman, *Japanese Drama and Culture in the 1960s: The Return of the Gods* (New York, 1988); translations of five plays written in the years 1964 to 1970, with essays of explanation and commentary.

Kawatake Toshio, *Japan on Stage: Japanese Concepts of Beauty as Shown in the Traditional Theatre,* trans. P. G. O'Neill (Tokyo, 1990); a translation of a work published in 1982 which focuses on Japanese attitudes to beauty.

Donald Keene, *Dawn to the West: Japanese Literature of the Modern Era* (New York, 1984). The first volume covers fiction, while the second covers poetry, drama, and criticism; part of a multi-volume history of the whole of Japanese literature since the Meiji Restoration.

Colin Mackerras, *Chinese Drama: A Historical Survey* (Beijing, 1990); this covers all main forms of Chinese drama from its beginnings in the twelfth century to the late 1980s.

—— (ed.), *Chinese Theater from its Origins to the Present Day* (Honolulu, 1983, 1988). This is a multi-authored work covering Chinese theatre from the predecessors of drama to the early 1980s.

—— *The Chinese Theatre in Modern Times: From 1840 to the Present Day* (London, 1975). The three parts of this study cover the theatre in Beijing before 1949, in various specified regions of China in the same period, and throughout the country from 1949 until the Cultural Revolution.

—— *The Rise of the Peking Opera, 1770–1870: Social Aspects of the Theatre in Manchu China* (Oxford, 1972).

Benito Ortolani, *The Japanese Theatre: From Shamanistic Ritual to Contemporary Pluralism* (Leiden, 1990). This 'outline of the state of knowledge in the West about the Japanese theatre' is comprehensive in its coverage of theatre forms and includes an account of Western research on Japanese theatre.

A. C. Scott, *The Classical Theatre of China* (London, 1957); still a very fine study of the performance aspects of the Peking Opera, covering the techniques of the actor, the categories of roles, as well as some material on the content of the plays.

—— *The Theatre in Asia* (New York, 1972); a survey of the theatres of India, China, Japan, and the Islamic world by one of the leading post-war authorities on Asian theatre in general and Chinese in particular.

SOUTH-EAST ASIAN THEATRES

Faubion Bowers, *The Drama in the East: A Survey of Asian Dance and Drama* (New York, 1956); a general reference book, especially good in relation to dance and dance-drama.

James R. Brandon, *Theatre in South East Asia* (Cambridge, Mass., 1974); comprehensive and detailed history.

—— *Brandon's Guide to Theatre in Asia* (Honolulu, 1976); an account of surviving performance forms throughout Asia.

Buurman, *Wayang Golek: The Entrancing World of Classical Javanese Puppet Theatre* (Singapore, 1988); containing an account of the ritual origins of puppet theatre.

Paul Cravath, 'The Ritual Origins of the Classical Dance Drama of Cambodia', *Asian Theatre Journal*, 3/2 (1986), 179–203.

Jukka O. Miettinen, *Classical Dance and Theatre in South-East Asia* (Oxford, 1992); well-illustrated survey of performance in the region with detailed historical information.

Urs Ramseyer, *The Art and Culture of Bali* (Oxford, 1977).

Mattini Rutnin (ed.), *The Siamese Theatre* (Bangkok, 1975).

T. Zarina, *Classical Dances of the Orient* (New York, 1967).

CHAPTER THIRTEEN

John Allen, *Theatre in Europe* (Eastbourne, 1981); an account of the organization, working conditions, and subsidy of theatres throughout Europe, based on a report commissioned by the Council of Europe.

C. W. E. Bigsby, *Modern American Drama, 1945–1990* (Cambridge, 1992).

Peter Brook, *The Shifting Point, 1946–1987* (New York, 1987); the director's account of his career in theatre, and his quest for fuller understanding.

David Edgar, *The Second Time as Farce: Reflections on the Drama of Mean Times* (London, 1988); the dramatist's account of the theatre in which he has worked.

Peter Hall's Diaries, ed. John Goodwin (London, 1983); a blow-by-blow account of the struggle to work adventurously in theatre at the same time as directing the National Theatre's move into its new buildings on the South Bank in London.

Andrew B. Harris, *Broadway Theatre* (London, 1993); a critical report on the progress of major plays, by O'Neill, Williams, Miller, Mamet, and others, on their way to Broadway openings.

John Harrop, *Acting* (London, 1992); a careful examination of what acting may be at the end of the twentieth century, in the United States and Britain.

Christopher Kamlongera, *Theatre for Development in Africa: With Case Studies from Malawi and Zambia* (Bonn, 1989).

Dennis Kennedy, *Looking at Shakespeare: A Visual History of Twentieth-Century Performance* (Cambridge, 1993); a copiously illustrated account of the staging of Shakespeare's plays in Europe and North America, and a valuable introduction to scenography at the end of the twentieth century.

Todd London, *The Artistic Home: Discussions with Artistic Directors of America's Institutional Theatres* (New York, 1988); the official report of the Theatre Communications Group investigation into the state of theatre in the United States.

Iain Mackintosh, *Architecture, Actor and Audience* (London, 1993); a scrutiny of theatre-building, especially in the last decades of the twentieth century, with an account of modern thinking about the relationship between stage and audience.

Tsuboike Eiko (ed.), *Theater Japan* (2nd edn., Tokyo, 1993); a *Who's Who* for theatre in Japan, with critical and biographical entries, issued by the Japan Foundation.

Eugene Van Erven, *The Playful Revolution: Theatre and Liberation in Asia* (Bloomington, Ind., 1992); this survey covers the Philippines, South Korea, India, Pakistan, Indonesia, and Thailand, accounting for a wide range of political theatre.

GENERAL SURVEYS

Martin Banham (ed.), *The Cambridge Guide to World Theatre*, (Cambridge, 1995).

Oscar G. Brockett, *History of the Theatre* (Boston and London, 1968); the standard textbook for the study of theatre, especially in the USA.

Enciclopedia dello spettacolo, 9 vols. (Rome, 1954–68); an international encyclopedia of all aspects of theatre.

Phyllis Hartnoll (ed.), *The Oxford Companion to the Theatre*, (4th edn., Oxford 1983; reprinted with corrections 1993).

Phyllis Hartnoll and Peter Found (eds.), *The Concise Oxford Companion to the Theatre* (2nd edn., Oxford, 1992).

Cesare Molinari, *Theatre Through the Ages*, trans. Colin Hamer (New York, 1975); a superbly illustrated history with emphasis on performance.

Allardyce Nicoll, *World Drama* (revised edn., London, 1976); still the most comprehensive account of dramatic literature.

Simon Trussler, *The Cambridge Illustrated History of British Theatre* (Cambridge, 1994).

* * * * * * * * * * * * * *

✥ ACKNOWLEDGEMENTS ✥

Preparation of this history has involved its editor in an adventure for which customary gratitude is more than usually inadequate. To catalogue all debts would lead back many years and across all continents. In a brief note only the most recent and closest to home can be acknowledged. First of all, by their generous support and ready discussion, the influence of the contributors has extended far beyond their own chapters of the book. Then for over five years the personal interest taken in this project by many individual colleagues and students at the University of Michigan has enlivened and informed progress. For consultation on scholarly problems, assistance with some of the inevitable chores, and wide-ranging discussion, some persons, besides contributors to the volume, cannot be left unnamed: these include Pratibha Agrawal, Imade Badra, Samik Bandyopadhyay, Martin Banham, Alexie Bartoshevich, Hilary Brown, Ruby Cohn, Stephen di Benedetto, Ding Luonan, Josh Gasster, Hiroyuki Kondo, Molly Mahood, Anthony Morris, Smita Nirula, Pornat Damruny, Talia Rodgers, Wayne Silka, Yasunari Takahashi, Biswaskesh and Biyot Tripathy, Robert Weimann, Stanley Wells, John Wright.

Without financial and administrative backing from the University of Michigan, the task could not have been attempted; the editor is indebted for special funding from the Office of the Vice Provost for Research, the Horace H. Rackham School of Graduate Studies, the School of Music, and the Department of Theatre and Drama.

At the Press, the project was proposed and guided by Michael Cox and its progress assisted by many others. I cannot imagine a volume of this complexity being served better and I am very grateful to all who had a hand in it. I especially wish to thank Angus Phillips and Alysoun Owen, for facilitating progress and coming to my assistance both as a matter of course and at times of unforeseen difficulties, and Sandra Assersohn, for highly skilled picture research.

J.R.B.

ILLUSTRATION SOURCES

The editors and publishers wish to thank the following who have kindly given permission to reproduce the illustrations on the following pages:

tion; **401** Salzburg Festival Archive; **402** Norman Bel Geddes Collection, Theatre Arts Collection, Harry Ransom Humanities Research Center, The University of Texas at Austin by permission of Edith Lutyens Bel Geddes, Executrix; **406** (top) Berliner Ensemble GmbH, (bottom) Märkisches Museum, Berlin; **407** Annie Fratellini; **408** (top) Billy Rose Theatre Collection, the New York Public Library for the Performing Arts, (bottom) Günter Englert; **410** Syndics of Cambridge University Library (René Hainaux 'Stage Design throughout the World since 1950' 1964); **415** Ruth Walz; **416** Ullstein Bilderdienst; **422** Eileen Darby; **425** Agence de Presse Bernand; **427** Photo by Houston Rogers from the collections of the Theatre Museum. By courtesy of the Board of Trustees of the Victoria & Albert Museum; **429** Snowdon; **430** Evening Standard; **431** Romano Cagnoni (Report); **434** (top) Billy Rose Theatre Collection, the New York Public Library for the Performing Arts, (bottom) Ilse Buhs; **435** Werner Schloske; **442** Agence de Presse Bernand; **443** Gianfranco Mantegna/ University of California, Davis; **448** (top) National Museum of India, New Delhi, (bottom) Farley Richmond; **449** Phillip B Zarrilli; **453** Farley Richmond; **455** (top) Madhavan Cakyar, (bottom) Phillip B Zarrilli; **457** Farley Richmond; **458** Farley Richmond; **460** Farley Richmond; **461** Farley Richmond; **463** Bahorupee; **466–7** Chinese Arts Research Institute's Traditional Music-Drama Research Section; **469** 'Pictorial History of Beijing Opera' Beijing Yanshan Publishing House 1990; **472** Nogami Memorial Noh Theatre Research Institute, Hosei University; **473** Anonymous Gift, courtesy Museum of Fine Arts, Boston; **474** Collection of the Tokyo National Museum; **477** Xinhua News Agency; **478** Greenhill-SACU; **480** Bungakuza; **482** 'Juche Art' Foreign Language Publishing House, Pyongyang 1976; **485** Erle Taylor; **489** ZEFA; **490** James Brandon/ Syndics of Cambridge University Library (Theatre in Southeast Asia' 1967); **493** Philippine Dept of Tourism; **496** James Brandon/Syndics of Cambridge University Library (Theatre in Southeast Asia' 1967); **497** Embassy of Vietnam; **500** William B Carter; **503** Michael Le Poer Trench; **504** Donald Cooper © Photostage; **505** Taganka Theatre; **507** (top) The Wooster Group, photo by Ken Kobland, (bottom) The Wooster Group/Syndics of Cambridge University Library (David Savran 'The Wooster Group; Breaking the Rules' 1988); **513** Magnum Photos/Martine Franck; **516** Ruth Walz; **518** Shakespeare Centre Library, Joe Cocks Studio Collection; **520** © Noda Map; **524** (centre) Jorge Luis Vargas Echeverry; **527** William B Carter; **529** photo by John Haynes from the collections of the Theatre Museum. By courtesy of the Trustees of the Victoria & Albert Museum; **530** Nobby Clark; **531** Nobby Clark; **534** William Burdett-Coutts.

In a few instances we have been unable to trace the copyright holder prior to publication. If notified, the publishers will be pleased to amend the acknowledgements in any future edition.

Picture research by Sandra Assersohn

The text quoted from *Collected Shorter Plays of Samuel Beckett* on p. 436 is by kind permission of Faber and Faber Ltd; text on p. 510 by Irvine Wardle is from his article 'Images of tenderness, triumph and death', © Times Newspapers/Supplements Limited, 1985.

✤ INDEX ✤

Note: sub-entries referring to periods of theatre are listed in chronological, not alphabetical order. Page numbers in italics are to captions to illustrations; textual references may also appear on these pages.